MATHEMATICS RESEARCH DEVELOPMENTS

WAVELETS

CLASSIFICATION, THEORY AND APPLICATIONS

MATHEMATICS RESEARCH DEVELOPMENTS

Additional books in this series can be found on Nova's website
under the Series tab.

Additional e-books in this series can be found on Nova's website
under the E-book tab.

MATHEMATICS RESEARCH DEVELOPMENTS

WAVELETS

CLASSIFICATION, THEORY AND APPLICATIONS

MANEL DEL VALLE
ROBERTO MUÑOZ GUERRERO
AND
JUAN MANUEL GUTIERREZ SALGADO
EDITORS

Nova Science Publishers, Inc.
New York

For permission to use material from this book please contact us:
Telephone 631-231-7269; Fax 631-231-8175
Web Site: http://www.novapublishers.com

NOTICE TO THE READER

Library of Congress Cataloging-in-Publication Data

Wavelets : classification, theory, and applications / [edited by] Manel del Valle, Roberto Muqoz Guerrero, and Juan Manuel Gutierrez Salgado.
p. cm.
Includes index.
ISBN 978-1-62100-252-9 (hardcover)
1. Wavelets (Mathematics) I. Valle, Manel del. II. Guerrero, Roberto Muqoz. III. Gutierrez Salgado, Juan Manuel.
QC20.7.W38W377 2011
515'.2433--dc23
2011032451

Published by Nova Science Publishers, Inc. † New York

CONTENTS

PREFACE

This book presents an integrated view of the current use of wavelets in science and engineering. This tool has been used substantially in many different application fields, especially for identification or classification purposes. Various authors from around the world have collaborated in this edited collection. The book opens with an introduction of the involved theoretical concepts, starting with the applications related to image processing. Applications in the fields of biology, food, agriculture and environment are included, as well as the medical and chemical uses of wavelets.

Chapter 1 – This chapter extensively deals with the applications of wavelets. Due to their compatibility with non stationary random process, wavelet transform (WT) has the ability to analyze rapidly changing transient signals. It is used to decompose random process into localized orthogonal basis function and provides a convenient format for modeling, analysis and simulation of non stationary process. As wavelet transform provides time and frequency information of transient signals more accurately, it has been applied with great success to a wide variety of problems. The areas of application differ from science, medicine, finance, music and the list is endless.

Chapter 2 – This chapter presents a novel quaternion wavelet transform using the atomic function. Atomic functions are compactly supported and infinitely differentiable solutions of differential equations. Consequently, they constitute a promising kernel for wavelet multi-resolution analysis. This work makes use of the phase concept for the detection of key features. For this goal, the authors steer the mother atomic function wavelet through a multi-resolution pyramid to detect quaternionic phase changes, which can be used for the feature detection of lines, edges, centroids, and orientation in geometric structures up to affine transformation. The experimental analysis gives more insight into the authors' techniques and stimulates further applications.

Chapter 3 – 3D meshes are a piecewise-linear surface approximation and are the most popular representation of real scanned and modeled object used in computer graphics, medicine, games, simulators, scientific visualisation, CAD and virtual reality systems. Such meshes can be set of millions of polygons, vertices and edges. Complex three-dimension objects must be processed, stored, transferred, animated and analysed which is very expensive. Multiresolution models arise a lot of interest because they make it possible to represent and process geometric data in different levels of detail (LOD) depending on application needs. This chapter presents a survey of the different wavelet-based multiresolution methods proposed for triangle surface meshes. First the preliminary background on triangle meshes and multiresolution analysis is presented. Then the different proposed methods for regular, semi-regular and irregular meshes are described. This survey consists also description of Author solution based on second generation wavelets for decomposition and reconstruction of irregular, triangle mesh. The generalisation from the first generation of wavelets to the second one gives an opportunity to construct wavelets on irregular meshes. This solution is based on lifting scheme as a wavelet construction tool. The conclusion is the different application of result multiresolution representation in algorithms of digital geometry processing.

Chapter 4 – Biogeographical studies are often based on a statistical analysis of data sampled in a spatial context. If data sampled at adjacent locations are more likely to be similar to each other than distant ones, this is called spatial autocorrelation. However, in many cases standard analyses such as regression models may yield incorrect results, because the presence of autocorrelation violates the assumption of independently and identically distributed errors. First, the authors show that the theory of wavelets provides a method to remove autocorrelation in generalized linear models. Autocorrelation can be described by smooth wavelet coefficients at small scales. Therefore, data can be decomposed into uncorrelated and correlated parts. Using an appropriate linear transformation, the authors are able to extend generalized linear models to autocorrelated data. They illustrate their new method, called the wavelet-revised model, by applying it to multiple regressions with response variables conforming to various distributions. They recommend wavelet-revised models as a method for lattice data and applicable without any prior knowledge of the real autocorrelation structure. Second, the authors deal with the problem that statistical models may suffer not only from autocorrelation, but also from collinearity, i.e., predictor variables in a multiple linear regression can be highly correlated with each other. In the presence of collinearity among predictors, methods modelling, e.g., with latent variables such as partial least squares regression are recommended in literature. They introduce a novel approach coping simultaneously with both problems: collinearity and autocorrelation. Therefore, they develop an extension to partial least squares based on wavelet-revised modelling. They illustrate their new method, called wavelet-revised model for partial least squares (WRMPLS), by applying it to multiple regressions with responses of both normal and binary distribution. Its predictive performance is evaluated for simulated data.

Chapter 5 – In this chapter the discrete wavelet transform (DWT) is shown to accurately enumerate and visually demonstrate synchronisation of a flowering time series. The DWT is applied to flowering intensity records (covering 1940 – 1970) of *Eucalyptus leucoxylon*, *E. microcarpa*, *E. polyanthemos* and *E. tricarpa*. The DWT is used as a correlational method, and proposed as a new metric for synchronisation and visualisation for a multidimensional time series. The wavelet based results are compared to Moran's traditional synchronisation test and to a self organising map time series clustering approach. The Maximal Overlap DWT analysis successfully identify four subcomponents in each flowering series characterised as a (i) non-flowering phase (d1), (ii) duration (d2), (iii) annual (d3) and (iv) intensity (d4) cycles. The wavelet correlation (WCORR) signatures are shown to exhibit a common pattern for synchronous species pairs, for example d2, d3 and d4 WCORR values being significant and positive for *E. leucoxylon* and *E. tricarpa* whereas for species pairs that are asynchronous the d2 and d3 specific WCORR values are significant and negative for *E. tricarpa* and *E. polyanthemos*. The profiles of wavelet cross correlation (WCCORR) enable us to further understand how synchrony may relate to the subcomponents of flowering (peak annual cycles, start, cessation, duration, overlap) and climate. The positive periods of the WCCORR profiles for d2 reflect flowering overlap between species pairs. The d3 WCCORR's peak at the lag which is the difference between the species pair's peak flowering months. Distinctly different patterns for d3 WCCORR are associated with asynchronous and synchronous time series. The sinusoidal and cycling of the d3 specific WCCORR profiles mirror the cyclic 6 monthly climatic influence on species flowering, where for 6 months if temperature is positively correlated with flowering, rainfall is negative and vice versa.

Chapter 6 – The theory of the wavelet transform using the complex Morlet basis in the context of discrete signal sampling is thoroughly explored. A sequence of steps is developed relating the Fourier transform in the continuum to the wavelet transform on a lattice, including remarks on the admissibility criteria. The wavelet domain is extended to include all coefficients touched by convolution with the data. Nearly perfect reconstruction is obtained for bandwidth limited test signals with the inclusion of a correction factor on the norm of the basis wavelets. The Gibbs effect is observed in the residual near the edges of a finite duration of data, and the conditions required for perfect reconstruction are evaluated. An analysis of a coastal sea level time series is performed to demonstrate the utility of the extended wavelet transform.

Chapter 7 – Recent developments in sensing and optical instrumentation have shown tremendous potential for characterization and classification of agri-food material using non-destructive techniques such as gas sensors, near-infrared spectroscopy, ultrasound, X-rays, color machine vision, and hyperspectral imaging. Statistical methods are generally used to extract the features from the recorded signals of these instruments that describe the signal property. Wavelet transformation provides an alternative powerful tool to compress the signal and extract the features describing the signal. These extracted features can be used as input in pattern recognition, prediction, and supervised classification model development. In this review, various commonly used wavelet transforms are described and their applications in characterization and classification of agri-food products are discussed.

Chapter 8 – In this chapter of the book, first the wavelet transform and its corresponding inverse are defined in the Banach spaces in the three separate cases (I) Unitary representation of group (II) Homogeneous space representation for G/H and (III) Projective representation space. Here the intimate connection between the Banach space wavelet reconstruction method for each of the above three representations and some of the well-known quantum tomographies is established. For a unitary representation of a given group using wavelet transform and its corresponding inverse some of quantum tomographies such as: tomography of rotation group, spinor tomography and tomography of unitary group are presented. On the homogeneous spaces with both singular and nonsingular vacuum vectors some of the well-known quantum tomographies such as: Moyal-representation for a spin, discrete phase space tomography, tomography of a free particle, Homodyne tomography, phase space tomography and SU(1, 1) tomography are presented. The connection between the wavelet formalism on Banach space and Q-function is discussed. Wavelet reconstruction based on the projective representation with both singular and nonsingular vacuum vectors, and some of the quantum tomographies as Discrete phase space tomography, SL(2, z) tomography, Permutation group tomography, Homodyne and SU(1, 1) tomography is revealed. However, both the atomic decomposition and the Banach frame nature of these quantum tomographic examples are also revealed in details.

Chapter 9 – Biologically inspired technologies for artificial sensing have been investigated in the last years using different kinds of sensors and applying novel computational paradigms, in order to emulate the behavior of their biological counterparts. In this sense, the use of advanced processing techniques is essential in interpreting the obtained information. Among the available processing tools, the combination of Wavelet Transform (WT) and Artificial Neural Networks (ANNs) has been adopted as innovate strategy for processing complex signals, especially when they are composed of a large number of measures and include non-linear content. Wavelet Neural Networks (WNNs) are a new class of networks that combine wavelets and neural networks in order to improve tasks such as classification, prediction and modeling, taking benefit of both paradigms. This chapter describes the fundamentals of WNNs and their advantages in the treatment of signals, especially when feature extraction stage and/or signal preprocessing are mandatory to build an appropriate calibration or classification model. This chapter will highlight the successful applications of these networks in areas such as electronic tongues and myoelectric analysis. To demonstrate the computing power of the WNNs, two case studies from the authors' laboratories are presented. The first relates to the quantification of analytes using a voltammetric electronic tongue and the second involves the classification of motor unit spikes at different levels of muscle contraction from myoelectric records obtained with a needle electrode.

Chapter 10 – Biologically inspired technologies for artificial sensing have been investigated in the last years using different kinds of sensors and applying novel computational paradigms, in order to emulate the behavior of their biological counterparts. In this sense, the use of advanced processing techniques is essential in interpreting the obtained information. Among the available processing tools, the combination of Wavelet Transform (WT) and Artificial Neural Networks (ANNs) has been adopted as innovate strategy for processing complex signals, especially when they are composed of a large number of measures and include non-linear content. Wavelet Neural Networks (WNNs) are a new class of networks that combine wavelets and neural networks in order to improve tasks such as classification, prediction and modeling, taking benefit of both paradigms. This chapter describes the fundamentals of WNNs and their advantages in the treatment of signals, especially when feature extraction stage and/or signal preprocessing are mandatory to build an appropriate calibration or classification model. This chapter will highlight the successful applications of these networks in areas such as electronic tongues and myoelectric analysis. To demonstrate the computing power of the WNNs, two case studies from the authors' laboratories are presented. The first relates to the quantification of analytes using a voltammetric electronic tongue and the second involves the classification of motor unit spikes at different levels of muscle contraction from myoelectric records obtained with a needle electrode.

Chapter 11 – It is well known that Walsh functions can be identified with characters of the Cantor dyadic group. The authors give a review of recent results on wavelets and frames for this group. Among the main subjects to be discussed are the following: algorithms to construct orthogonal and biorthogonal wavelets associated with the Walsh polynomials; estimates of the smoothness of dyadic orthogonal wavelets of Daubechies type; an algorithm for constructing Parseval dyadic frames. Orthogonal and biorthogonal wavelets for Vilenkin groups are also discussed.

Chapter 12 – A new stable numerical method, based on Chebyshev wavelets for numerical evaluation of Hankel transform is proposed in this paper. Chebyshev wavelets are used as a basis to expand a part of the integrand, $rf(r)$, appearing in the Hankel transform integral. This transforms the Hankel transform integral into a Fourier-Bessel series. Truncating the series, an efficient, stable algorithm is obtained for the numerical evaluations of the Hankel transforms of order $\nu > -1$. The method is quite accurate and stable, as illustrated by given numerical examples with varying degree of random noise terms $\varepsilon\theta_i$ added to the data function $f(r)$, where θ_i is a uniform random variable with values in [-1,1]. Finally, an application of the proposed method is given in solving the heat equation in an infinite cylinder with a radiation condition.

Chapter 13 – Constrained portfolio optimization determines the optimal proportions of capital that the investor should invest on each asset of the portfolio to meet the bi-criterion objective of maximizing return and minimizing risk, subject to constraints such as basic, bounding, cardinality, asset class and short sales. Initially, the application of wavelets as noise filters for the financial covariance matrix, which is one of the key input to the portfolio optimization problem is discussed. The reliability of the portfolios obtained by the wavelet filtered method is measured and statistical analyses of the performance results, when compared with its counterpart's viz., Random Matrix Theory based filters and k-means cluster analysis are undertaken. Later, the application of wavelets in finding a neural network based solution to a complex constrained portfolio optimization problem is discussed. Usually, the bi-objective problem is converted into a single objective problem by introducing risk aversion parameter. When the constraints reflecting investor preferences and/or market frictions are included, the problem turns into a complex mixed integer-quadratic programming problem, making it difficult for direct solving by analytical or numerical methods. Although there are many heuristic algorithms, Evolution based Wavelet Hopfield Neural Network (EWHNN) is proposed for its solution, because of the following advantages a) handles diversification in both large and small portfolios. b) Intrinsically reports faster convergence and yields reliable optimal portfolios when compared with other heuristic algorithms. c) Incorporates more constraints that are of practical use, in one lot. The major limitations of single objective formulation are 1) need to form minimum N different problems to find N different Pareto-optimal solutions that lie on the efficient frontier. 2) uniformly spaced set of risk aversion parameter values may not produce uniformly spaced Pareto-optimal solutions. 3) different parameter values may produce an identical solution. These drawbacks are rectified in a multi objective formulation. Hence, Pareto-archived Evolutionary Wavelet Network (PEWN) is introduced for the solution of complex constrained multi objective portfolio optimization problem. The results obtained using the aforementioned two wavelet network based algorithms are evaluated using different quantitative measures. All the experimental studies are undertaken on the Bombay Stock Exchange, India (BSE200 index, period: July 2001 – July 2006) and Tokyo Stock Exchange (Nikkei225 index, period: March 2002 – March 2007) data sets.

Chapter 14 – Three-dimensional (3-D) surface shape measurement and sensing by using a projected fringe pattern has been widely used considering the Transform Based profilometry. Several articles show that these methods require a single image with a sinusoidal fringe pattern projected on it. Usually, the projected pattern has a known spatial frequency and this information is used to avoid discontinuities in the fringes with high frequency. Among single projected fringe pattern, most of the methods use Fourier or Wavelet transforms to extract the phase information. However, they focused only on the transform method and not on the phase unwrapping algorithms. It is well known that wavelet transforms are used to analyze signals whose frequency contents varies over the duration of the signal (wrapped phase), also, wavelet techniques provide a flexible, elegant, and a computational efficient solution to these type of problems. In this work, a 1-D wavelet profilometry method is presented considering two different wavelet transforms. Later, different phase unwrapping algorithms are used to extract the depth information considering local and global analysis. Several computer simulations and experiments are carried out to validate the proposed method. The merits and limitations of each of these variations on the method are indicated and the error is estimated.

Chapter 15 – Acoustic emission (AE) signal has been used in several process monitoring applications in manufacturing. AE is generated when atoms are torn from the surface during the material removal process in manufacturing. Of particular interest is the AE signal captured from nanomanufacturing processes, which have been shown to contain critical information related to the health of the process. Understanding the properties of AE signal could yield a myriad of process information, however, the task to process AE signal is challenging due to its non-stationary, high frequency, and heavy noise characteristics. This chapter presents a novel diffusion wavelet based AE signal processing method that is capable of efficiently extracting the underlying nano-scale process characteristics better than the well-known classical wavelets such as Haar and Daubechies wavelets. Results from applying diffusion wavelet in analyzing AE signal from a Chemical Mechanical Planarization (CMP) process indicate that the wavelet and scaling signals generated by the diffusion wavelet method has higher magnitude and sharp features to clearly detect nanomanufacturing process anomalies, and therefore provides greater insight into the AE signal characteristics. In particular, the classical wavelets fail to detect the end-point of the CMP process using AE signal. However, the end-point is clearly detected by the diffusion wavelet methodology. The diffusion wavelet method is made suitable for online application by developing a moving block data processing strategy, which matches the rate of data acquisition.

In: Wavelets: Classification, Theory and Applications
Editors: Manel del Valle et al.
ISBN: 978-1-62100-252-9

Chapter 1

APPLICATIONS OF WAVELET TRANSFORM

Vipula Singh[*]
Electronics and Communications Department, RNSIT, Bangalore, India

ABSTRACT

This chapter extensively deals with the applications of wavelets. Due to their compatibility with non stationary random process, wavelet transform (WT) has the ability to analyze rapidly changing transient signals. It is used to decompose random process into localized orthogonal basis function and provides a convenient format for modeling, analysis and simulation of non stationary process. As wavelet transform provides time and frequency information of transient signals more accurately, it has been applied with great success to a wide variety of problems. The areas of application differ from science, medicine, finance, music and the list is endless.

1. INTRODUCTION

Wavelet transform is extensively used in signal processing applications which are signal filtering for noise, speckle reduction, edge detection, multi-resolution editing, data compression, storage, recognition, enhancement and synthesis of speech. In data mining, it is used for efficient data processing, similarity search and data retrieval. Wavelet transform is a powerful tool for analyzing biomedical signals like ECG (electrocardiography), EEG (electroencephalography), and phonocardiography for early detection of the disease symptoms. Recent developments in medical imaging include medical image compression, noise reduction, enhancement, reconstruction, acquisition system, registration and analysis of medical images.

Wavelet transform has emerged as a tool to detect, classify and analyze defects in pipes, railway tracks, integrated circuits, concrete beams etc. It also helps in tasks such as monitoring structural health of civil engineering infrastructure, in power engineering

[*] E-mail address: vipulasingh@yahoo.com. (Corresponding Author)

applications, where wavelet transform is used for detection, localization, identification, classification, compression, storage and analysis of power disturbance signals. It is also a powerful tool for numerical analysis and computationally very efficient. It is used to identify and characterize transient random process involving ocean engineering, wind and earthquakes. In each application, we provide the reader with some general background information and a brief explanation.

2. COMPRESSION

Compression forms the backbone for several applications such as storage of images in a data base, picture achieving, TV, facsimile transmission and video conferencing. Compression of images involves taking advantage of the redundancy in the data present within an image. At present, image compression technology consists of three important processing stages: image transform, quantization and entropy encoding.

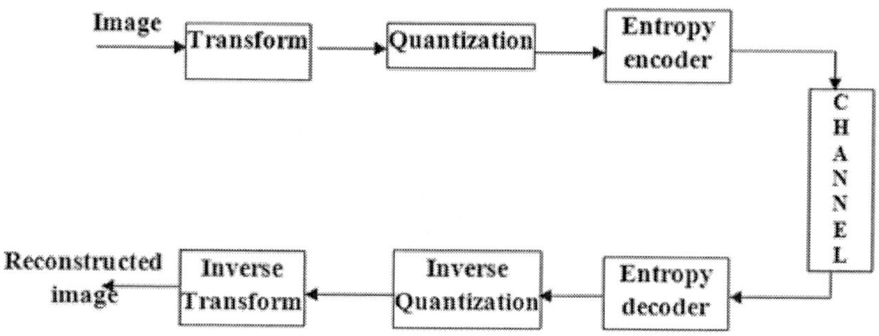

Figure 1. Typical transform based image coding system overview.

Typical transform based image coding system is shown in Figure 1. Input image is taken without dividing it into smaller sub images. Wavelet transform is applied on image pixels and the coefficients are arranged in a specified order. These ordered coefficients are quantized, entropy encoded resulting in a stream of bits. The reverse procedure is applied for decoding. The encoded bit stream is entropy decoded, inverse quantized and the resulting wavelet coefficients are reorganized for inverse transform.

2D wavelet analysis and synthesis filter banks are used for images. Low frequencies represent flat background and are extracted by high scale wavelet functions. Texture regions represented by high frequencies are analyzed by low scale wavelet functions. A p-level filter bank is used for signal compression. Approximation and detail information is extracted from original image using low pass and high pass sub-bands respectively. For signal compression, fewer bits are allocated to the sub images with low energy levels. Image compression is also achieved by rounding all wavelet coefficients below a certain threshold to zero. The choice of wavelet functions, the filter order, filter length and decomposition are chosen on the compression level, image quality and computational complexity. The wavelet decomposition is useful because most of the energy lies in the low-frequency sub-bands for natural images. Wavelet decomposition of images shows a unique property i.e. if the energy at a specified location at a low frequency band is low, then it is likely that there is low energy at higher frequency bands at the same spatial location.

Various techniques have been suggested in literature for quantization and coding of wavelet coefficients. One such technique is embedded zero tree (EZW) [1] and Set Positioning In zero trees (SPIHT) [2], which is an improvement over EZW and the latest is embedded block coding with optimized truncation (EBCOT) [3]. All these coding algorithms are iterative such that after initial transmission of bits, further refinement bits are transmitted. These refinement bits can continue till the desired compression rate has been achieved.

2.1. JPEG 2000

The existing JPEG standard needed some change because of the high quality and large size of modern digital images. Thus the JPEG2000 standard was introduced, so that it could be used in different types of still images such as bi-level, grey scale, color with different characteristics such as natural images, scientific, medical, remote sensing, text, graphics etc. and on different image models such as client/server , real time transmission, image archieval and limited bandwidth resource, all in a single frame work. New standard caters to all kinds of applications such as internet, color facsimile, printing, scanning, digital photography, medical imaging, mobile communications, remote sensing, digital libraries, archives and e-commerce. The salient features include: Improved low bit rate performance, both lossy as well as lossless compression, progressive transmission that allows images to be reconstructed with increasing pixel accuracy and resolution, region of interest coding which allocates more bits to certain regions of interest, random code stream access, better error resilience for wireless communication channels, support open architecture, content based description is used in medical imaging, forensic and digital libraries, image security is provided by watermarking, encryption etc., side channel support is also provided [4].

The key charaterstics of the new standard is that it has better image quality than JPEG at compression rate less than 0.25 bits per pixel (bpp). Figure 1 shows the block diagram of JEPG 2000 encoder. Before taking wavelet transform, the image is partitioned into non overlapping equal size tiles. Though tiling reduces peak signal to noise ratio (PSNR) for a given compression rate, memory and computational requirements are considerably reduced. This also allows selective encoding and decoding. Discrete Wavelet Transform (DWT) is applied to each tile using either biorthogonal 9/7 filter for lossy coding or integer coefficient filter for lossless coding. Wavelet coefficients are quantized and entropy encoded. JPEG 2000 was tested on standard natural images and compared with JPEG standard. The new standard outperforms the existing standard by 2 dB for compression ratios between 0.25 bpp and 2 bpp. JPEG 2000 software is freely downloadable from web [5].

2.2. Video Compression

Video compression is an extension of image compression by just adding one more dimension i.e. time. Wavelet analysis is expanded from 2D to 3D. Motion estimation is another method for video compression, where separate images with motion vectors are generated through a video which are compressed individually. In the synthesis filter bank, separate images are reconstructed and motion vector is used to form the subsequent frames [6].

2.3. Audio Compression

Audio compression is effectively achieved by DWT. Filter banks used in DWT for compression applications need to be close to regular in first few octaves of the sub band decomposition [7]. The limitations of HAS (Human Auditory System) is used for effective speech and audio compression algorithms. Human hearing is associated with critical bands because a masking effect is observed around frequency f_m. A frequency is masked by f_m and not audible, if its magnitude is below $T(f_m - f)$.

$$T(f_m, f) = \begin{cases} M(f_m)\left(\dfrac{f}{f_m}\right)^{28} & f \le f_m \\ M(f_m)\left(\dfrac{f}{f_m}\right)^{-10} & f > f_m \end{cases} \tag{1}$$

where, $M(f_m)$ is the masking threshold

For effective audio compression algorithm, the frequency chosen is near to the critical bands of the human ear. The frequency f_m is chosen which has greater power and $T(f_m, f)$ is computed for masking envelope. Masking curved is generated for determining which wavelet coefficients are retained and which are removed to achieve signal compression. This algorithm can be used in mobile communications to reduce transmission time. A four level dyadic filter band can be used to further improve the compression algorithm.

2.4. FBI Fingerprint Compression

Finger print images are digitized at a resolution of 500 dpi with 256 levels of grayscale / pixel. One image needs about 0.6 Mb for storage. FBI has around 114 million criminal finger print images requiring 65 Tb of image data. Transmission of such images without compression also needs a lot of time. The main steps in the algorithm are:

a) DWT decomposition of source finger print image.
b) Scalar quantization
c) Entropy encoding of quantized values.

In Wavelet/Scalar Quantization (WSQ) algorithm [8], DWT is implemented using a two channel perfect reconstruction linear phase filter bank. To apply filter near image boundaries, symmetric extension techniques are used. The analysis filter bank is a 2D composition filter. DWT is first applied on image rows and then to columns. The analysis filter bank is cascaded several times to generate the 64-sub band frequency decomposition. The 64 DWT sub bands are then partially quantized with uniform scalar quantizer. The output of the quantizer is entropy encoded using run lengh coding and Huffman coding. The encoder requires a table of wavelet transform specifications, table for scalar quantizer and Huffman coders. The decoding process also requires the same tables to generate a reconstructed image [9].

3. AUDIO APPLICATION

An audio is composed of three parts : a harmonics part (tonal), a transients part and noise.

$$s(t) = s_{tonal}(t) + s_{transient}(t) + s_{noise}(t) \quad (2)$$

The time frequency resolution property of wavelet transform is used to decompose the audio signal and then denoising, enhancement and recognition algorithms can be applied on the decomposed signal. Wavelet packet decomposition is applied on the audio signal.

3.1. Denoising

The denoising algorithm uses a hard thresholding within the various sub-bands according to the equation

$$T_{tk}(x) = \begin{cases} 0 & if\ |x| < t_k \\ x & if\ |x| \geq t_k \end{cases} \quad (3)$$

where, the thresholding t_k is chosen adaptively.

The wavelet coefficients below a threshold value are set to zero. The signal is reconstructed using the remaining coefficients. White noise which is present at all frequencies can be handled by denoising [10].

4. SPEECH PROCESSING

Various speech processing algorithms such as speech analysis, pitch detection and speech recognition can be successfully implemented using wavelet transform. Mother wavelet based on Hanning window is used for speech recognition as bandwidth is increased without affecting the time resolution [11].

Wavelet transform is used for splitting of input speech into sub-bands in two ways: equal bandwidths for sub-bands and dyadic bandwidths for sub-bands. Wavelet thresholding technique is used to denoise the signal. The signal in each sub-band is encoded into a stream of acoustic vectors which are passed to a Hidden Markov Model (HMM) based recognition system. The generated likelihoods by HMMs are given to a recombination module that delivers a unique answer to the recognition task.

4.1. Speech Enhancement

In a noisy car environment where the speech of the driver needs to be separated from the surrounding noise, e.g., speech of other passengers, car engine noise, music, etc. For a speech recognition algorithm, speech enhancement is used. Blind source separation (BSS) algorithm is used to separate the driver's speech from the background. The algorithm uses the time correlation of speech signals recorded by two micro phones and can be used for real time applications. This algorithm does not separate noise from the speech signal as noise is spread out in time. Thus denoising using Wavelet transform is used to remove noise. The filter choice depends on noise level, car speed and other parameters. A look up table consisting of pre-computed filters is used to determine the optimal filter and noise level is estimated by analyzing a signal during speaker silence [12].

5. Image Fusion

Image fusion refers to a process that integrates image data recorded at different resolutions or by different sensors to obtain more information from a single or more images. WT is used to improve the spatial resolution of a multi spectral (MS) image by performing fusion using Panchromatic (P) image. The main objective of fusion of high resolution panchromatic (P) and low resolution MS satellite images is to keep maximum spectral information from MS images which increase the spectral resolution. The fused image keeps maximum spatial and spectral information from the original image. Wavelet transform is well suited for image fusion.

Precise geometric correction and registration of two images is the most important prerequisite for accurate image fusion. Size of both images is made to be the same by re-sampling. Then WT is used to decompose the digital images into a set of multi spectral images. For each level of decomposition, the coefficients contain spatial differences between two successive resolutions levels. There are two ways in which the wavelet based image fusion can be performed.

1) Substitution method where some wavelet coefficients of low resolution image are replaced by corresponding coefficients of high resolution.
2) Additive method where high resolution coefficients are added to the corresponding low resolution coefficients.

Following steps are followed for wavelet based image fusion [13]:

1) Decompose a high resolution P image into a set of low resolution P images with wavelet coefficients for each level.
2) Replace a low resolution P image with a MS band at the same spatial resolution level.

3) Perform a reverse wavelet transform to convert the decomposed and replaced P set back to the original P resolution level.

6. Watermarking

Watermarking is the process that embeds data called a watermark into an image, audio or video. The watermark can be detected and extracted later from the carrier (cover). It can contain information such as copyright, license, authorship etc. A simple example of a digital watermark is a "seal" on the image to identify the ownership. Any watermarking algorithm consists of three parts:

a) The watermark, which is unique to the owner.
b) The encoder for embedding the watermark into the data.
c) The decoder for extraction and verification.

Ellinas [14] reported a robust watermarking algorithm using Wavelet transform and edge detection whose efficiency depends on preservation of visually significant information. This is carried out by embedding watermark in those sub-band coefficients that lie on the edges, where distortions are less noticeable. The technique is robust to common signal processing operations such as compression, filtering, enhancement, rotation, cropping and translation.

Figure 2 shows the overall process of watermark insertion. Initially, input image is decomposed to four levels by using Daubechies 8-tap filter. Then from each sub-band perceptually important wavelet coefficients are detected by sobel edge detector. These edges are classified into two groups with respect to a threshold value. Coefficients containing the region around the edges are separated using a morphological dilation operation. The watermark is inserted in detailed sub-bands that contain edge information or to the high frequency coefficients, thus making the watermark invisible to human eyes. Watermark detection is performed by correlating the watermarked coefficients of possibly watermarked image with the watermark to be tested for presence.

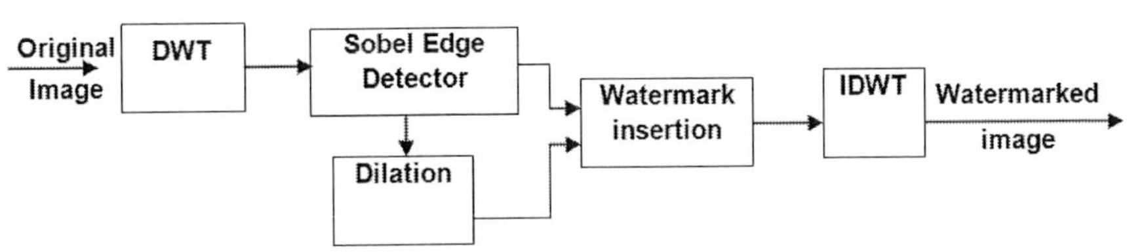

Figure 2. Block diagram of watermark insertion process [14].

Lee et al [15] proposed a high capacity reversible image watermarking scheme based on integer-to-integer wavelet transform. First X x Y input image is divided into M x N non overlapping blocks. A set of B_m message bits are embedded in this block using forward invertible integer to integer wavelet transform. The location map L is a binary matrix that indicates which blocks are watermarked. As a part of side information, this is sent to decoder to retrieve the message bits and to reconstruct the original image.

In the decoding process, the decoder has to retrieve the location map first. The watermarked image has to be divided into non overlapping blocks with the dimensions M x N. Each block is transformed using the same wavelet used in embedding process. Then the blocks for which least significant bit (LSB) can be changed (found in the location map) are searched in a predefined order. Based on the location map, the blocks into which the watermark is embedded is sequentially searched. Finally the entire payload is extracted which includes the original LSBs and the location map. Thus original image block can be reconstructed back exactly.

7. Texture Analysis

Texture analysis has emerged as a powerful visual primitive to describe the image content. Texture analysis and characterization has been widely sited for various applications e.g. medical imaging, remote sensing, image compression, industrial inspection and texture based retrieval. Traditional texture classification relies on extracting statistics from image (using methods like variance, histogram, autocorrelation and moments) in order to derive features for classification. G Fohmay *et. al.* [16] proposed a perceptually based joint compression and classification method for textures that exploits phase coherence within textural images. The scheme exploits phase coherence for image compression and then uses them to classify the compressed texture. The human visual system (HVS) relies heavily on texture analysis for interpretation and analysis of images. DWT is a useful tool for automated texture analysis and clarification. Wavelet filter based feature extraction algorithm is very less computationally demanding and covers the complete time frequency. Edges in the wavelet planes are detected by zero crossings. Wavelet detail coefficients can give zero crossing information horizontal, vertical and diagonal direction sub-bands are generated at each level using DWT. Feature vector is extracted from the transform to represent the texture. Average number of zero crossing is evaluated by counting the number of zero crossings in the various sub-bands. Texture is classified as homogenous if average number of zero crossings is more. Linear phase symmetric wavelets are used to minimize distortion effects leading to fewer decomposition levels and improved classification.

B S Manjunathan et al [17] used Gabor filters for texture analysis. It has been found that the response pattern of cells in visual cortex of brain can be modeled with Gabor function. Gabor filters are used for texture analysis. They provide multi scale and multi resolution processing of textural information. Wavelet filters also perform multi-scale texture analysis at a very low computational cost. Psycho-visual studies indicate that the Human Visual System (HVS) processes visual information in a multi scale manner, similar to what WT does. 2D DWT is used to decompose the image into its constituent bands. The authors discard phase data in LH and HL bands which continue low pass horizontal data and High pass vertical data and vertical direction phase data. The algorithm is only useful in LH and HL bands. Following are the steps being used:

1) A 5 level 9 x 7 wavelet is used to obtain a dyadic decomposition of the image.

2) The correlation across all bands is measured by
   a) Counting the number of zero crossings at each point across all 16 spatial frequency bands.
   b) Threshold the summation at a chosen value.

3) The phase information is discarded by replacing each pair of coefficient at 900 shift point with square root of the sum of the square of coefficients according to

$$mag(m) = \sqrt{w^2(m) + w^2(m-3)} \tag{4}$$

where, $w(m)$ = wavelet coefficient at 0^0 phase shift

$w(m-3)$ = wavelet coefficient at 90^0 phase shift

The basic idea is that only 2 coefficients having phase shift of 90^0 can be replaced by 1 and then phase can be assigned arbitrarily at the decoder side. This scheme successfully exploits the phase coherence between different spatial frequency bands of the image.

8. DENOISING

Denoising is a process to remove noise present in the signal. Wavelet transform is extremly useful for denoising applications such as wavelet filtering, real time Wavelet denoise, and adaptive wavelet denoise.

A noisy signal can be expressed as

$$f(t) = s(t) + \sigma\, e(t) \quad t=0, 1, 2........n-1 \tag{5}$$

Being s(t) = real signal, e(t) =noise, f(t) = noisy signal and σ = noise level coefficient

The goal of denoising algorithm is to obtain an estimate $\hat{f}$ of the original signal which minimizes a risk function, normally the mean square error $E\left[\left\|f-\hat{f}\right\|^2\right]$. Generally all practical and useful signals are low frequency signals but noise is a high frequency phenomenon. To eliminate or reduce noise, a suitable threshold is selected for Wavelet coefficients. The following steps are used for denoising :

1) Wavelet decomposition of the signal is done using FWT.
2) A suitable threshold is selected and the thresholding is done on high frequency coefficients of decomposition layer.
3) Reconstruction of wavelet coefficients is done using inverse DWT.

To perform thresholding, a suitable threshold is selected. Zhai Hong [18] suggested several techniques to select a suitable threshold. Threshold function is of 2 kinds. Hard threshold function and soft threshold function. Hard threshold is discontinues at the threshold point which will cause concussion during signal decomposition. Soft threshold is a relatively smooth function which causes distortion phenomenon such as fuzzy edges.

Hard threshold function is $$Y = \begin{cases} X & |X| \geq T \\ 0 & |X| < T \end{cases} \quad (6)$$

Soft threshold function is $$Y = \begin{cases} X - T & X > T \\ X + T & X < -T \\ 0 & -T \leq X \leq T \end{cases} \quad (7)$$

X = high frequency Wavelet coefficients T= threshold $= \sigma\sqrt{2\log_2 N}$

N= signal strength σ = noise standard deviation

Noisy signal is transformed into the multi-level approximation and details using forward DWT. Then hard and soft thresholding is performed on wavelet coefficients using equations 6 and 7. After hard or soft thresholding inverse DWT is used to generate the denoised signal.

9. Biosignal Analysis Using WT

Medical vibrations are generated by pumping of heart which generates pulses, which can be heard through stethoscope. A microphone is placed inside the heart via a catheter to record these sounds. These sounds are analyzed to get an insight of any cardiac activity. Heart sounds and murmurs are the two types of sounds generated by the heart. Heart sounds are short, impulse like waveforms whereas murmurs are caused by blood flow turbulence and can detect cardiac disease such as aortic stenosis or valve defects. Wavelet transform is a useful tool for characterization of these heart sounds. Wavelet transform can pick certain sound components (the aortic and pulmonary valve components of the second heart sound) that can not be detected by any other method.

ECG signals are the changes in electrical potential between the surface electrodes on the body during the cardiac cycle. Figure 3 shows the ECG signal where P wave is the initial port of the signal which is due to the internal excitation starting at sinus node and then spreads to the atria. QRS complex is generated when the cardiac excitation reaches ventricles. ST segment represents when ventricles have been completely stimulated and then they are repolarized to generate T wave of ECG. The amplitude and the timing of these waves are important for diagnostic purpose. QRS complex, P waves and T waves of ECG contains information of human heart. ECG signal has strong random property with background clutter. The signal is first decomposed into different frequency bands and then the decomposed signal is filtered by matched filter. This surpasses high frequency clutter in the output signal and enhances R waves. Spline wavelets can be used for QRS wave detection of ECG signals. Wavelets are also used for extraction and recognition of ECG signals.

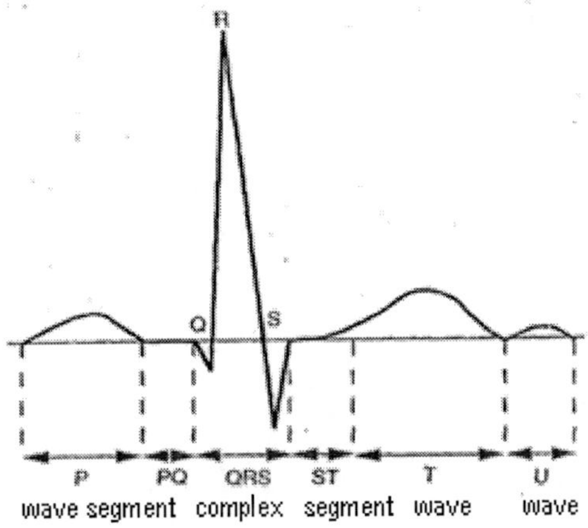

Figure 3. Schematic representation and labeling of ECG signal (one cycle).

9.1. Cardiac Signal Processing

ECG (Electro Cardiac Graphic) signals are well suited for analysis by wavelet transform. ECG signal has time varying characteristics identified as P-QRS-T complex having a) Low frequency P and T waves b) Moderate frequency, QRS-complex.

Elevation or depression of ST segment can be seen when heart muscle becomes infracted. To detect these changes, the signal bandwidth needs to be extended to 0.05 Hz, making it susceptible to artifact errors. Continuous wavelet transform is used to analyze signal ECG cycle where P-Q-R-S-T complex is plotted on one axis. A very common disorder is the occlusion of coronary vessels causing cardiac ischemia and later infraction. In the treatment, the vessels are opened using a laser based catheter and blood supply is restored. This is called percutanous transluminal coronary angioplasty (PTCA) [19].

"Late potentials" are caused by fractionation of the depolarization front after myocardial infraction [20]. They can predict life threatening reentrant arrhythmias. They occur in terminal portion of QRS complex and have small amplitude and higher frequencies than in the normal QRS complex. Wavelet transform provides a very accurate plot of localized time scale features indicating late potentials. Late potentials are characterized by spreading of high frequencies in the later part of QRS complex. Use of wavelet transform for detecting late potentials and predicting likelihood of life threatening arrhythmias in the heart is very effective.

P-QRS-T complex undergoes significant morphological changes when arrhythmias such as premature ventricular contractions (PVCS) and tachycardia occur. In PVC, abnormal beats have different time scale analysis and QRS complex may widen or invert. For widened QRS complex, the power spectrum shows less contribution at a higher frequency which can be localized easily on time features by wavelet analysis. Wavelet transform is used to separate dynamice activities embedded in a time series. Dynamical analysis of ECG before and during ventricular tochycardia is done using wavelet transform. Prior to the onset of life threatening arrhythmias, heart rhythm becomes rigidly constant. During disorganized rhythms like fast ventricular tachycardia, ventricular flutter and ventricular fibrillation, the correlation dimensions which measure randomness, increases to a value greater than one. This fact is used to identify risk in patients and arrhythmic episode can be predicted. ECG signal is corrupted by noise and artifacts which leads to inaccurate detection of QRS complex. While recording, ECG signal is prone to contamination by noise which can be caused by power line interference, electrode contact, muscle contaction, Electromyography (EMG) and electronic noise in the instruments. For correct analysis, it is important that ECG signal is noise free. Wavelet transform helps in removal of electrical disturbances from ECG. Wavelet analysis naturally decomposes the signal into different scales thus eliminating artifacts [21].

9.2. Neurological Signal Processing

EEG signals are measured from synchronous discharge of nerve cells creating rhythmic potential fluctuations on the surface of the head. In response to the external simulation, the signals are recorded from the brain which are called 'evoked potentials'. These signals are generally very small in amplitude whereas the background noise (EEG activity) is quite large. Signal-to-noise ratio(SNR) needs to be enhanced for the signal to be used. Stimulus artifacts and undulating patterns are seen in the signals which are characterized by one or more peaks at specific latencies beyond the stimulus. Analysis of a complex signal generated by brain injury or disease requires wavelet transform to identify features localized within the waveform indicating brain's response. The magnitude of the detail components is sensitive to the cerebral hypoxia during its early stages and the magnitude of the fourth level detail component drops rapidly when brain becomes hypoxic [22]. Thus, detail components of the response act as an indicator of early stages of brain injury. These are also useful in patient monitoring during surgery and neurological critical care.

9.3. EEG and Seizures

Slow moving events like a person going to sleep etc. can be diagnosed from EEG signal using spectrum analysis, but transient events such as epileptic seizures which are characterized by sharp spikes or a series of bursting events present in EEG are the early signs of seizure. As seizure progresses, transient activity develops into regular high amplitude quasi periodic oscillations reflecting abnormal discharge of a large group of neurons. Only wavelet analysis of such a signal can identify a seizure burst [23].

9.4. Wavelet Applications in Hearing Aids

Hearing loss at old age is a very common phenomenon. It becomes increasingly difficult to hear with age as cochlea becomes less responsive to sound analyzer and speech processor. This is mainly because of the structural damage of the sensors cells. Hearing impaired listeners have very poor speech recognition rate as they require louder speech to perform at par with normal hearing listeners. Wavelet based system to extract speech from back ground noise is very effective. Wavelet transforms provide both time and frequency information about the signal. Wavelet transform is very effective for noise removal, thus improving speech-to-noise ratio [24]. Wavelet packet algorithm for noise removal acts as a perfect preprocessor for hearing loss compensation in digital hearing aids. Local discrimination basis is used to discriminate between the desired signal and noise.

9.5. Early Detection of Breast Cancer

Breast cancer is one the main cause of death for women. Early detection and treatment can significantly reduce the mortality rate. Small tumors and micro calcifications are very difficult to detect in examination by a physician because both the malignant and the normal tissues look similar. Mammographic images have a significant amount of background noise which makes its diagnostic very limited. Wavelet transform helps in noise reduction and detection and classification of tumors in digital mammograms. Non-linear processing of Wavelet coefficients at each scale and wavelet filtering helps in noise reduction while preserving the edges in various images regions. Non-linear Wavelet shrinkage with adaptive thresholding is used for denoising. A threshold is decided and below that threshold all the details coefficients are rounded off to zero. The data set is reconstructed from the remaining coefficients with much less noise and prominent features intact. Wavelet transform enhances the mammographic images which increases early detection of breast cancer. One of the primary warning signs of breast cancer is the detection of clusters of fine granular micro calcification by computer-aided mammography. This can be achieved by thresholding in wavelet domain [25].

10. Applications of WT in Medical Images

Medical imaging plays an important role in clinical diagnosis, therapy, teaching and research.

10.1. Medical Image Compression

Despite the technical advances in data transmission rates and increase in the data storage capabilities, medical image compression is a very important aspect of medical imaging. Medical images are generally very high resolution images and image compression method has to be suitably chosen so that medically important information is not lost. Transmission of medical images is limited by band width and storage space. Day-by-day growing size of image data base necessitates the need for data compression methods. Transmission of such large images drastically slows down the hospital's communication network. Run length coding, dictionary coding, transform coding, entropy coding are some of the lossless techniques which do not provide much compression. As wavelet transform coefficients are localized both in space and frequency, wavelets are better suited for image compression method as compared to other methods. The compressed images retain the features necessary for medical diagnosis despite the presence of some degradation and artifacts. Use of lossy images compression schemes for medical images is very controversial. The use of these techniques also raises some delicate legal issues. Wavelet packet is a multi resolution decomposition of the signal. Wavelet packets analyze all subsystems of coefficients at all scales, thus generating a full binary tree, for orthogonal Wavelet transforms with an evenly spaced frequency resolution. Rich orthogonal structure that allows quick adaptation to a particular signal is provided by wavelet packet system. An optimal basis is found which efficiently represents the signal by minimizing some functions of the signal sequence thus achieving signal compression [26].

For medical images, a prior knowledge about image contents makes it suitable for object based compression algorithms. Medical images generally consist of a part of body under investigation with some background noise. A fully 3-dimentional object based coding scheme exploiting diagnostic relevance of different regions to allocate variable bit rate is discussed. 3D DWT is used to decorrelate the data and lifting steps scheme allowing integer–to-integer mapping enables lossless coding. The coding scheme assigns disjoint segments of the bit stream to the different objects which is independently accessed and reconstructed without any loss of quality. Each object is coded independently, which reduces computational cost to generate a segment of bit stream. The implementation of DWT via lifting scheme in the non-linear version and the embedding of the encoded information allow reconstruction of each object at a progressive up to lossless quantity.

A digital mammogram with a resolution of 4000 x 500 pixels with 50 μm spot size and 12 bits results in 40 Mb of data. Detection of isolated micro calcification clusters requires high resolution and transmission. Archiving such images is very expensive and slow. Border detection and segmentation of tissue area and radiological background of mammograms is suggested in [27]. The compression of mammography images is achieved by compressing only breast regions and discarding the radiological background which has no diagnostic interest.

10.2. Biomedical Image Enhancement

In low contrast or low brightness image, it is difficult to identify image features that are clinically relevant and very difficult to visualize under normal viewing conditions. e.g. in mammograms the difference in contrast between the tissues of the breast are very small and to identify any minor changes to detect the presence of malignant tumor cells is very difficult without image enhancement. Wavelet based enhancement methods have been proved very useful. Reversible wavelet decomposition is used for enhancement by selective modification of few wavelet coefficients. Then all the wavelet coefficients along with the modified ones are used to reconstruct an enhanced image.

The contrast in X-ray mammograms is limited by the strength of X-ray and the presence of X-ray scatter from the dense regions in the breast. Malignant tissues are hidden in the low contrast mass of mammogram and are undetectable unless contrast enhancement is done. The contrast in MR images is dominated by the relaxation times of the tissues imaged. Extra scanning time or reduced spatial resolution is needed to increase the chances of detecting low contrast structures. Contrast enhancement algorithm [28] using multistage edge gradient representation has the following steps:

1. The multi scale of a low-contrast image is computed using WT to detect the edges.
2. Local gradient maxima is detected at all scales which are collected in a representation set containing the position, magnitude and phase angle of local

gradient maxima and the low pass approximation of the image at the coarsest scale is also collected.

3. The magnitude of gradient maxima is stretched either by a constant or by a variable at different scales. Modified magnitude value is replaced in place of all gradient maxima.
4. The modified gradient maxima are retained and the rest of the WT coefficients including the coarse scale data are updated with iterative projections.
5. Contrast enhanced image is produced by taking inverse wavelet transform.
6. Rescaling is done, if the contrast of the enhanced image is out of range.

10.3. Registration of Medical Images

A preprocessing step in object identification and object classification is registration. Registration involves identifying some points in an image to be matched with similar points in the other image using translation and rotation parameters. Wavelet modulus maxima based medical image registration is very fast and accurate. It uses wavelet to obtain control points. The structures determined by wavelet high pass image provide necessary information for fast convergence. This method uses the correlation coefficients to find out how well the images are registered. This method works very well with very high registration accuracy.

Content based retrieval of medical images is very useful in finding similar images from a large database with the help of some key attributes associated with the image. The main goal is to provide diagnostic support by retrieving images of the past cases along with the pathology reports. This is also helpful to train medical students and researchers. Image registration is used in combining information from multiple image sources, monitoring changes; image guided surgery and compare different images. Medical image registration and retrieval require similar processing steps. Multi modal image registration is required for medical images retrieval as medical images are multi modal and heterogeneous with temporal properties. Medical images are generally of higher dimensionality (3D or 4D). However, these higher dimensional images are generally visualized as collection of 2D slices. The speed and computational requirements for the registrations is limited by the size of the image. Thus, fast, robust and fully automated registrations technique requires wavelet analysis [29]. Following are the steps for 2D image registration:

1) Identification of 2D rigid registration.
2) Calculation of mutual information (MI) in multi scale wavelet domain.
3) Sub-sampling of data at a higher scale to achieve higher speed of computation, variable bin size at multiple scales is used.

Multi scale edge detection and representation is done through dyadic wavelet transform which provides finer details at lower scales or coarse details at higher levels. Dyadic wavelet decomposition is invariant under translation rotation and scaling. Images from different modalities have to be fused or combined to improve the diagnosis in medical image processing. Matching of different structures present in the image produces a good registered image. Mutual information is computed from normalized joint histogram of the two images which are being registered. Search for maximum correlation is done on a coarse scale and refinement is done at finer scales.

11. WT Applications in Communications Systems

Wavelets are used in modulation schemes (e.g. pulse amplitude modulation PAM, orthogonal modulation, fractal modulation), in demodulation schemes (feedback equalizers and receivers in co-channel interference) and in multiple access systems (wavelet packet and perfect reconstruction quadrature mirror filter based (PR-QMF based).

Introduction of wavelets in PAM provides excellent spectral efficiency. Because of this either the system's bit rate can be increased or it can be reduced. Channel's distortions can be matched with much greater ease if Wavelet packet modulator is used which results in less distorted signal at the receiver and increased average bit rate if fractal modulation is used. Because of the use of wavelets in modulation schemes, system performance improves dramatically with no or minimal additional cost. Livingston et. al. [30] used orthogonal Wavelets in a traditional PAM scheme which increases the spectral efficiency of the system. As the Wavelet uses band-pass filters it can be sent using a single side band modulation. The extra bits can be sent using Wavelet system which increases the data rate while maintaining performance of the system.

Lindsey [31] discussed wavelet packets for orthogonal multiplexed communication systems. In some cases, wavelet packets show significant improvement over QAM MSM (multi scale modulation). Wavelet packet modulation (WPM) takes advantage of wider selection of time frequency tiling, thus providing better channel matching capability. WPM arranges the time-frequency tiling in a manner which minimizes channel disturbances. One major advantage is that it doesn't create ISI (inter symbol interference) or CCI (co channel interference) because wavelets are orthogonal across scales and time shifts.

When channel's bandwidth and duration parameters are either not known or unavailable to the transmitter, fractal modulation can be used. It uses margin as performance criterion to achieve higher average bit rate by allowing variable data rates with a constant bit error rate (BER). Fractal modulation is an alternative scheme when the system is required to operate efficiently over a broad range of rate-bandwidth combinations using a fixed transmitter. The information can be transmitted on multiple time scale, fractal modulation divides the available transmitted spectrum into multiple adjacent octave-spaced bands and source stream is then modulated into these bands at corresponding rates [32].

The inclusion of wavelets in feedback equalizers, blind equalizers and receivers for CCI improves system's BER by removing part of channel's distortion. The delays in radio transmission can be improved because of fast convergence time of the Wavelet based decision feedback equalizer. The improved delay allows the system to utilize more complex receiver or to increase the amount of interleaving present. Feng *et. al.* [33] proposed wavelet based decision feedback equalizers (WBDFE) having faster convergence time and suitable for time varying channels because of their time frequency localization. The authors used feed forward filter (FFF) which is based on a weighted sum of translates and dilates of mother wavelet, where feedback filter remains unchanged.

A blind equalizer does not know the transmitted sequence. Blind equalizers are non linear which makes them suitable for matching non linear channel distortions. A recurrent wavelet neural network (RWNN) is suitable for real time signal processing due to fast convergence time and its ability to dynamically configure to the channel distortions over time. RWNN is highly recommended to be used in high speed channel equalization because of its small size and high performance.

Wavelets are providing significant performance improvements in Multiple Access Communication Systems (MACS). Wavelet packet based MACS exhibits decreased multiple access interference (MAI) and computation complexity while showing promise for increasing user throughput. The goal is to design systems which maximize the channel throughput by allowing maximum number of users. Learned *et. al.* [34] in their research paper showed the increase in multiple access (MA) throughput. The received signal is jointly demodulated using wavelet packets and optimal matched filter provides higher performance, but the system becomes complex and impractical as the number of users is high.

A spread spectrum communication system using PR-QMF filter to perform spreading and de-spreading functions is suggested in [35]. This system reduces code complexity and increases code variation. A synthetic filter bank using IWT, implements spreading function. Number of filters divided by number of taps in each filter gives the spreading ratio. The location of non-zero element and basis factor is determined by pseudo random noise (PN) sequence. Spreading code is varied by hopping PN sequence. Analysis filter bank using WT is used as a de-spreading function. Quadrature mirror filters are used to have de spreader to be a mirror image of the spreader. The de spreader isolates the energy to a single bit using correct PN sequence and translates it into desired message bit.

12. Data Mining

Wavelets have been applied in computer vision, network management and data mining. Data mining is a process of automatic extraction of novel, useful and understandable patterns from a large collection of data. Vanishing moments, hierarchical, multi-resolution decomposition structure and de-correlation are some of the properties of Wavelets that make mining process more efficient and accurate. Wavelets can be incorporated into the kernel of many data mining algorithms. Non-trivial extraction of implicit, previously known and useful information from collection of data is data mining. Data mining major components are:

a) Data management is related to the mechanism and structure of accessing, storing and managing data. The implementation of data mining systems is data management.
b) Data preprocessing is an important step to improve the quality and efficiency of mining process. Preprocessing generally involves removal of noisy data and outliners, data integration to integrate the data from multiple sources and data reduction to reduce the dimensionality and complexity of data. Data transformation to convert the data into suitable form for data mining is also a part of preprocessing step.

WT provides a way to estimate the function from data. Few significant coefficients are selected by vanishing moment property of wavelets which are used for denoising and dimensionality reduction.

12.1. Denoising

Removing noise from data is a process of identifying outliners from the data and estimating the unknown data from available noisy data. Binning method, clustering [36], outline detection are some of the smoothing techniques for noise removal. Steps to follow are:

- Transform the data to wavelet domain
- Round off those wavelet coefficients which are lesser than a threshold to zero.
- The remaining coefficients are transformed back to spatial domain.

12.2. Dimensionality Reduction

Original data set is reduced to a smaller set of data with or without loss of information. Wavelet transform is used to transform the data in Wavelet domain. Following two methods are used for dimensionality reduction

- Keep largest k coefficients and approximate the rest to zero.
- Keep the first k coefficients and approximate the rest to zero.

First technique is more accurate and is generally used. Initially wavelet transform of data is taken to compute the wavelet coefficients. Then these coefficients are stored in order of decreasing magnitude. Best M coefficients are chosen from the N coefficients.

12.3. Data Mining Algorithms

They are applied to perform the data mining tasks. Tasks like visualization, classification, content retrieval, regression, clustering etc are done using algorithms based on neural network, Principal Component Analysis (PCA) etc. Each task solves a different kind of problem in data mining.

Clustering

It can be described as portioning of n data points into k classes such that the points within each class are similar to each other. Multi resolution property of wavelet transform identifies classes at different scales from a very large data base. N-dimensional feature space represents spatial data objects and feature vector represents numerical attributes of the object. If the data space is partitioned by a grid, it reduces the number of data points without inducing many errors. Rapid changes in distribution of objects (boundaries of clusters) are represented by high frequency part and high amplitude feature space of object concentration (clusters) is represented by low frequency part. Wavelet transform is applied to the signal to decompose it into different frequency bands and clusters are identified by connecting the components in the transformed space. Arbitrary shaped clusters are effectively identified at different scales with great accuracy because of the multi resolution property of wavelet transform [37].

Classification

Classification is to identify the characteristics that indicate the group to which it belongs. It can also be used to predict future instances. Multi-resolution property of wavelets can be effectively used for classification. A progressive classification based on wavelets is suggested where DWT is used to obtain a multi-resolution pyramid representation of data. Low resolution representation of data is classified using a generic classifier. At each level of multi-resolution pyramid representation of k x k pixel blocks, the algorithm decides if the coefficients correspond to a block or not. The data is re-examined at a higher resolution level if the coefficients do not belong to the block. This is an iterative process which repeats till all the coefficients are classified. As image pixels are highly correlated, this method produces accurate results as DWT produces a weighted average for the block.

Blume *et. al.* [38] classified an image classifier based on Learning Vector Quantization (LVQ) and Haar wavelet transform features. One feature vector per image pixel is generated using Haar wavelet transform which provides information about brightness, color and texture of surrounding area. Codebook is generated for vector quantization algorithm using optimal learning rate and hand labeled images. The classifier shows 99% results for small number of classes. Texture analysis based on Wavelet transform is suggested by where texture classification and image segmentation is done by orthogonal multi-resolution decomposition. Gray level and color texture features are extracted and rotation invariant features are used for classification. Textural features based image retrieval for geographical images is suggested. Wavelet transform is used to obtain multi resolution decomposition of images and features are extracted for classification and clustering. Efficient clustering and retrieval is possible using this technique.

Principal Component Analysis (PCA)

PCA at multi scales can be obtained from multi scale property of Wavelets without loss of key features. Multi Scale PCA (MSPCA) de-correlates the data to extract deterministic MSPCA features at each scale and then the results are combined at relevant scales. In MSPCA only those scales are combined where significant events are detected and the residual events are filtered out by adjusting the limits for easiest detection of deterministic changes in measurements.

12.4. Post-Processing

It is done in the end to refine and evaluate the knowledge derived from mining process. Extraction, evaluation and visualization of knowledge are done for the end user.

Data mining deals with huge amount of data hierarchical structure and multi dimensional data representation is required for managing such huge data which is provided naturally by Wavelet transform. To improve the efficiency of multi level trends and surprise queries on time sequence data, Shahabi *et. al.* [39] proposed a novel Wavelet basis tree structure. Rising and falling trends and abrupt changes at multiple level of abstraction are used to identify frequent queries on time series data. A large amount of data is retrieved and processed to support multi-level queries. Trends and Surprise Abstraction (TSA) tree expedites the query process. TSA tree is constructed by taking DWT of the data starting with the root as original data. Data is decomposed into a low frequency part (trend) and high frequency part (surprise) at first level of decomposition. Low frequency part of first level is further decomposed into low frequency part and high frequency part. This process is repeated as shown in Figure 4. As we go down the tree, level of abstraction on trends and surprises increase and size of the nodes are reduced by half. As wavelet transform is lossless, only leaf nodes (wavelet coefficients) need to be stored and the internal nodes and root can be computed from the leaf nodes.

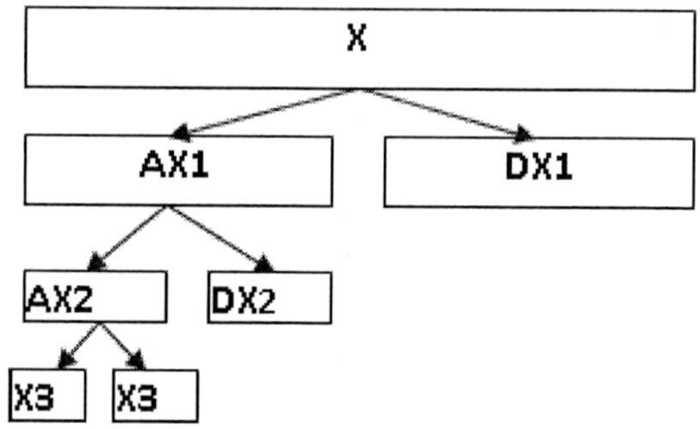

Figure 4. 1D TSA tree structure. X is the input sequence, AXi and DXi are the trend and surprise at level i.

13. Detection of Rail Defects

Maintenance of the railway tracks has been one of the most important aspects in rail safety which prevents the service failures. Defects in rail tracks can happen due to crystallization of cast-iron and due to strain under severe climatic changes or sudden shock loads. There are five kinds of defects found in rail racks

1) Transverse defects (TDs)
2) Longitudinal defects such as vertical split heads (VSHs)
3) Horizontal split heads (HSHs)
4) Burnt rail
5) Shelling defect

TD is a progressive fracture in transverse direction that occurs in the head of a rail which causes majority of rail failures.

VHS is a longitudinal defect and a progressive fracture in the head of the rail where as HSH is a progressive longitudinal fracture in the head of the rail where separation along a seam spreads horizontally parallel to the running surface.

Burnt rail is the rail which is scarred on the running surface because of intense friction heating which causes tread metal to be displaced on the running surface. Under repeated traffic gradual chip out and roughing of burned rail occurs.

Shelling defect, progressive longitudinal separation, that cracks out on the gauge side.

These internal defects are sometimes visible on the surface of the rail which is overlooked in the visual inspection. In recent times, ultrasonic detection of these defects is used. Gabor filter bank, wavelet transform and quadrature mirror filter are also used for defect detection. H. A. Toliyat *et. al.* [40] proposed wavelet transform based approach for detection of the faults in rail tracks. Due to orthogonality of wavelet functions, a set of decomposed signal at different frequency bands is obtained by multi-resolution decomposition of the signal. Fault diagnosis and condition monitoring is done on wavelet packets using auto progressive spectrum, energy monitoring. Magnetic coil and Hall sensors are used to acquire signals from rail tracks. Then wavelet transform is applied on these signals to obtain wavelet coefficients. Comparisons of these coefficients indicate the health of the rail tracks where healthy ones can be distinguished from the defective ones. The authors define an index Z which is obtained by adding all the coefficients at each level of decomposition. Z signifies the amount of energy in the signals at a specific decomposition level.

All the defects can be identified by the energy coefficient Z. The value of Z is higher in case of defected (burnt, shelled, VSHs, HSHs) than the same coefficients in a healthy rail. Deviation of wavelet coefficients in the healthy case from faulty case shows that healthy rails can be classified from the faulty ones.

14. Non Destructive Testing (NDT)

Inspection and maintenance of existing infrastructure is increasingly becoming important because 40% of the total construction budget is spent on maintenance and repair work. Guided wave NDT is used for defect detection, classification and size analysis. During the inspection of materials, wavelet transform provides spectral or temporal signal decomposition. Thus wavelet transform is used to process acoustic emission signals to find failure faults. ND imaging of integrated circuits, medical echography, geo physical seismic imaging of earth's interior are some of the other applications of NDT.

14.1. Pipe Inspection

Pipe corrosion is a major problem in plant maintenance. NDT helps in detection and classification of faults in pipes. "LAMB" guided waves are used to inspect large pipes from a single location. Transmitter is made of piezoelectric or magneto-strictive transducer which is used to generate lamb guided waves in the pipes. Receiver located at a proper distance from the transmitter is another transducer. Receiver produces an electric signal when it detects the elastic wave travelling in the pipe. Defects in the path of guided wave change the scattered fields for a given frequency. Fault detection is done by analyzing the scattered field by wavelet transform. Wavelet transform also helps in the classification and size analysis. Different types of asymmetrical pipe defects are studied by wavelet transform matrix. Sensitivity of this wavelet transform matrix is used to detect the defects. Larger the defect volume, higher is the sensitivity of the wavelet matrix.

Depth-dependent properties of material are studied using wavelet transform [41]. Multi-scale structures of the spatial variations of acoustical properties of the material are studied using this step up. Wavelet transform effectively enhances the information about the abrupt changes present in a signal by showing conspicuous cone like patterns in the wavelet map. Thus a rapid and eventually automatic detection and characterization of depth is possible. Figure 5 shows the setup of rigid frame with ultrasonic transducers and the target is analyzed. The targets are polycarbonate plates with thickness 1, 2, 4 or 8 mm. Reflected waves produced by impinging of incident wavelets onto the reflectors present in the medium constitute the wavelet response. Thus, characterization and identification of reflectors present in the material is possible using wavelet transform. These reflectors produce conspicuous cone like patterns in wavelet transform with their apex pointing towards the homogeneity centers of the singularity. The application of wavelet response reveals that useful information is extracted from complex surfaces. Similar analysis can be done in many geophysical situations like marine seismic and solar probing.

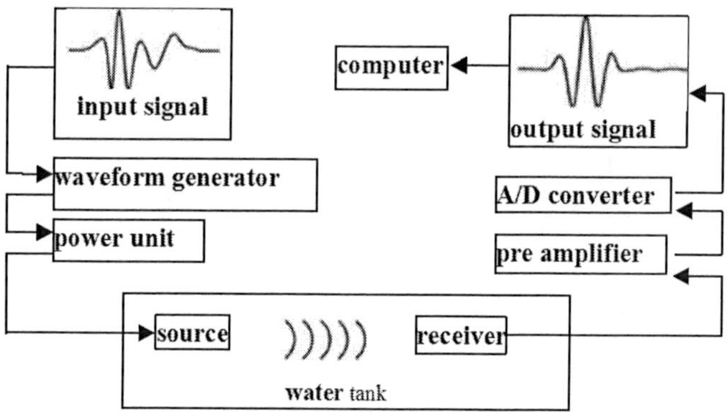

Figure 5. Set up for the experiment where numerical input is sent to the waveform generator to obtain the output signal [41].

14.2. Delamination

Lamb waves which are mechanical waves with wavelength proportional to the thickness of the plates, are used for non-destructive evaluation of planes for defect detection. Lamb waves are suitable for online structural health monitoring as they travel long distances and the sensors require very little power.

H. Sohn *et. al.* [42] used multi-resolution processing schemes based on wavelet analysis to extract a portion of the response signal that is suitable for defect detection. Two fundamental modes, S_0 and H_0 of Lamb waves and less dispersive frequency are used for damage detection. Energy is transferred back and forth between kinetic and elastic potential energy as the wave travels through a solid medium. Due to heating wave leaking, reflection and attenuation, the transfer is not perfect and the energy is wasted. Because of the presence of lamination, attenuation increases and energy splits over from driving frequency to the neighboring frequency values. Within the laminate, micro cracks cause reflection and dispersion of energy from the damaged area, resulting in excitation of high frequency local modes. Attenuation of energy is related to the amount of energy dissipated by damage which is given by damage index. Damage index is defined as a function of signal attenuation at a specific frequency.

15. Atmospheric Sciences

The time frequency decomposition of wavelet transform is used in atmospheric sciences as wavelet transform is an energy preserving transform. Scalogram is the squared modulus of the wavelet coefficients of CWT and cross scalogram is the product of two CWT of distinct functions. The scalogram describes that the analyzed signal has multi-scale characteristics and which scale is prominent in the signal. Scalogram reveals the information about the nature of non-stationary signal. It helps in the diagnosis of special events in the structural behavior during earth quake. Scalogram can identify any change in frequency content, e.g. start of stiffness degradation, occurrence of non ductile events and exchange of energy between modes. Wavelet spectrum of the zonal wind component is smooth as compared to its fourier transform. Because the wavelet coefficients are influenced by local events where as fourier transform coefficients are function of the frequency domain as a whole. Thus measurement of variance is better using wavelet spectrum as it is localized event. Another advantage is that it can be built even when there are gaps in the time series. Thus long period analysis is easier and feasible. The local maximum in this spectrum provides information on scales in which important character or coherent events contribute significantly.

Ground motion records contain high frequency energy which recites higher structural modes and is used to predict building response during earthquakes. The study of wave propagation in a structure can identify material properties, detection of crack, monitoring the health of structures via ultra sound, monitoring the characteristics of structural vibrations. Time - frequency analysis by wavelet transform is an ideal tool for scanning or mapping wave propagation at different frequencies. Wavelet analysis improves resolution and reduces time leakage. The ground motion record of 1940 El Centro earthquake contain energy at wide band of frequencies which excite higher modes in a building. These energy bursts at different times can be analyzed by scalogram and the events in the ground motion are understood better [13].

Wavelet analysis of Hurricane winds provides information about the distribution of energy as a function of time. The response of slender structure to wind contains contributions from the fundamental mode or higher modes. The contribution of each mode may vary significantly and building response may increase or reduce suddenly for the same wind speed because of the changes in the distribution of energy at different frequencies. Scalogram of different wind velocity at different times (two unrelated processes) show little correlation between the incident turbulence and pressure in separate regions, suggesting more complex interaction.

15.1. Solar Activity

Climate change is a very important issue and better understanding of natural climate changes is crucial for climate changes studies. Natural and anthropogenic influence on climate has to be distinguished to predict the future climate changes. Internal climate fluctuations are caused by interaction of atmosphere-ocean-land-sea-ice system which exhibits characteristics according to time scales. Influences from external forces (e.g. explosive volcanism or solar irradiance variations) are superposed on internal variations. These external forces disturb the climate on a consistent basis; the effect is seen both in time and magnitude. Some statistical properties of these forces can be studied by wavelet transform. Temporal scale of statistically significant relationship between solar forcing and climate time series can be accurately identified by wavelet analysis of solar activity data [44]. These major steps are undertaken:

1) Wavelet time scale decomposition is used to characterize the major components of solar or temperature variability in the different proxy records.
2) Dominant modes are extracted by decomposing different types of climate records.
3) Common temporal patterns between solar forcing or temperature variability are found at different scales.

Major components of solar variability in different proxy records are characterized by multi resolution decomposition. Temporal characteristics between solar forcing and surface temperature reconstruction can be evaluated using wavelet transform on temperature series [45].

15.2. Seismic Data Analysis

Wavelet techniques are used in atmospheric science for the analysis of seismic data for turbulence studies. Singularity detection and variance analysis are the two main analyses to be done. Atmosphere–ocean interaction, space conditions etc can be studied using wavelet transform. Torrence *et. al.* [46] elaborated the interaction between wind or ocean waves and used cross correlation to characterize the information. A fast numerical algorithm was developed to decompose the signal using bi-orthogonal spline wavelets. Wavelet transform reveals the points of energy transfer and the associated frequencies.

15.3. Wind Effects on Structures

Non-linear non stationary features of physical processes of civil engineering are studied using wavelet transform which permits the adaptation of a number of traditional system identification and analysis schemes. Time varying energy density can be estimated and intermittent correlated structures of the signals can be analyzed using wavelet based schemes. Wavelet transform is an ideal tool to accurately quantify the complex mechanism and non-linear characteristics of a dynamic parameter such as clamping. With reliable estimation of the behavior of dynamic parameters, satisfactory performance of the structures can be insured and state of art structural design can be achieved [47].

a) Time Domain Approach of System Identification

Scalogram representation of the localized wavelet transform is well suited to analyze non stationary signals. It reveals the frequency content of the signal at each time step to pin point the occurrence of transients while tracking the evolutionary phenomenon in both time and frequency. Ridge is defined by the scale at which the signal value is maximum. The real part of the wavelet coefficients along the ridge is proportional to the signal and imaginary part is proportional to the quadrature component. By calculating the phase of the wavelet skeletons for each ridge, time varying instantaneous frequency (IF) can be identified which is helpful in tracking amplitude dependent dynamic features. For a given mode, natural frequency can be estimated from the derivation of the wavelet phase of each skeleton. The slope of the skeleton's amplitude and phase term gives the damping and natural frequency respectively.

b) Frequency domain approach

From the wavelet scalogram, at a given time, wavelet coefficients are extracted across the range of frequencies. From the extracted frequencies, instantaneous power spectrum is produced which is focused at the instant frequency (IF) of the signal. Such analysis is done on wind generated random sea waves and scalogram of such data reflects several pockets of intense energy bursts associated with high amplitude events in the data. It is important to highlight that estimated wavelet IF when viewed along with the wavelet spectrum, acts as a microscope to study multiple harmonic components within the response [48].

The instantaneous wavelet spectrum has measurable spread about IF, which indicates instantaneous bandwidth of the signal. Morlet wavelet fits small localized sinusoidal waves at each point in time. The IF identified by the peak corresponds to the frequency of sinusoidal wave which is best fit at that instant of time. Additional neighboring frequencies are required to capture the deviations from pure sinusoid which accounts for the measurable spread around IF or the dominant frequency. Wavelet IF represents mean frequency and the bandwidth reflects the deviation of these frequencies from the mean as they evolve in time. The analysis of surface elevation of wind generated waves provides IF estimation and minor modulation reveal time-variance in the local mean frequency. These fluctuations about mean give instantaneous band width to characterize the non linear behavior.

16. Wavelet Applications in Power Systems

The most important wavelet transform applications in power systems are power system protection, power quality, power system transients, partial discharge, load forecasting and power system management. In analyzing power quality, wavelet transform helps in detection and location of disturbances to analyze interference, impulses, notches, glitches, interruptions, harmonics, flicker etc, of the power signal which is non-stationary.

Integer, non-integer harmonics and sub harmonics are identified by an algorithm where discrete wavelet packet transform using orthogonal high order Daubechies' functions is used to decompose the power waveform into two sub-bands. Then a continuous wavelet transform is applied to non-zero sub-bands to evaluate harmonics. Trapezoid complex wavelet function can also be used for harmonic and flicker analysis.

In power system distribution, wavelet transform is used to detect and locate various types of power quality disturbances. Multi-resolution analysis (MRA) is used to decompose the disturbance into wavelet coefficients. Squared wavelet coefficients act as an input to the neural network for classification of disturbances. Wavelet transform is also used to compress power quality disturbance signals. The partial discharge signals are very short duration, high frequency and low amplitude signals, which are very difficult to analyze. Wavelet transform can zoom in time to identify local variations of the signal. Thus partial discharges in transformer windings, cables and gas insulated sub stations can be detected using wavelet transform.

For efficient management of electrical power systems, demand forecasting is the most important issue. Wavelet transform and neural networks have been used for short term electrical load forecasting. At a particular time, electrical load is assumed to be a linear combination of different frequencies. The historical load waveform is decomposed into an approximation part, consisting low frequencies and detailed parts, consisting of higher frequencies. Wavelet coefficients are fed as input to neural network to forecast future load.

In the applications of power and rms measurements, wavelet transform is used to analyze the distribution of power and energy with respect to the individual frequency bands at each decomposition level. Wavelet packets are used to compose the waveform into uniform frequency bands which are used to measure the rms value of voltage or current and power of individual harmonic components. Thus, rms value of voltage or current and active power measurements can be done. Use of wavelet transform for improving the performance of protection relays for power distribution is very effective. Transient earth fault signals in a 20 kV resonant grounded network are generated by Electro Magnetic Transients Program (EMTP) is analyzed by wavelet transform. Significant features of the signals are extracted by wavelet transform which act as input to artificial neural network to diagnose faults. The fault generated traveling wave is processed by Wavelet transform to find out the travel time between the fault and the relay locations. The choice of wavelet family the number of decomposition levels, sampling rate strongly effect the arc fault time location.

Power system disturbances can be analyzed by wavelet transform for classifying them based on transient waveform signature. WT is applied on a transient signals to derive the waveform signature. Transient analysis of power systems can also be done by wavelet series expansion on integro-differential equations of power system to get system matrix. A set of algebraic equation are derived for the entire network. These equations can be solved in terms of wavelet expansion coefficients of voltages at the nodes of the network. The actual voltages can be reconstructed by wavelet series reconstruction.

16.1. Power Measurement

Time frequency domain analysis by complex wavelet transform helps in obtaining instantaneous amplitudes of voltage, current and instantaneous phase also. In electrical energy systems, because of a large number of non linear loads, voltages and currents are very irregular. The distortions like harmonics, rapid amplitude changes (flicker) and transients accounts for reduction in power quality. Thus complex wavelet transform deals with highly irregular and non stationary voltage and current. It yields phase information and instantaneous phase $\phi(t)$ can also be retrieved from polar representation of complex coefficients. Complex wavelet based power yields sliding average power, as the time window over which it averages is limited to the length of the wavelet function and is determined by dilation parameter. The average is weighted due to the decaying shape of the wavelet and the power can be localized in time. The frequency localization is broader than just a sharp single harmonic as the complex wavelet covers a frequency band. Wavelet transform is calculated using filters which yield thousands of coefficients over one time frame. This provides better time resolution and not so good frequency resolution especially for higher frequencies. We are mainly interested in sharp distinction of fundamental frequency behavior and good distinctions of low frequency harmonics which is achieved by wavelet transform. Higher frequency phenomenon can be predicted by low frequency components [49].

16.2. Failure Analysis

Power system quality deals with the problems caused by voltage sags, voltage swell and supply switching caused by auto reclose. Power system failure detection and identification of power quality problems both in transient and fixed signal is done using wavelet transform. Embedded system is used to capture instantaneous three phase voltage and current data for power system monitoring, protection and communication. The data is also used for fault failure analysis. Power quality is very important for electricity consumers as sensitive equipment and non linear loads are extensively used in industry and also in homes. DWT is applied to decompose the power signal into sub-trends. Any change in lower frequency of signal is detected at the finer resolution levels. The entropy contents of the signal are considered and a relationship between entropy content and one of the corresponding components is accomplished. The signals are analyzed by different wavelet functions and only high energy wavelet coefficients are retained. Signal is reconstructed back with the remaining coefficients [49].

16.3. Denoise

A power quality (PQ) monitoring system easily and correctly detects and localizes the disturbances in power systems by means of Wavelet transform, but the signal to be analyzed is corrupted by noise overlapping with high frequency spectrum of transient signal. To enhance the capability of PQ systems a denoising algorithm is essential to detect the transient disturbances in a noisy environment. To detect and localize the occurrence of power disturbances in noisy surroundings, H T Yang *et. al.* [50] suggested a technique based on Wavelet transform and de-noising. DWT is applied to the input signal (voltage and current waveforms) to decompose it into several scales at different levels of resolution through dilation and translation. Wavelet coefficients at different scales reveal the time organization information about the variation of signal from high to low frequency bands. Thus non periodic and high frequency signals arising from transient disturbances can be easily detected and localized in time. Different scales of Wavelet coefficients exactly pinpoint the occurrence of a transient disturbance in PQ monitoring system. Hard or soft thresholding can be used to denoise the wavelet coefficients. After denoising, if the disturbance still exists in the reconstructed signal, then the alarm is activated.

16.4. Transient Analysis

Power systems transients and harmonics can be analyzed using wavelet transform. System components such as resistors, capacitors and inductors are modeled for transient and steady state analysis in discrete wavelet domain. Orthogonal wavelet transform is used for this purpose. Resistor model is considered as scalar, inductor model as differential operator and the capacitor model as the integral operator. In discrete time domain, the linear differential and integral operations are first approximated. The initial condition is considered for transient analysis and periodic conditions are assumed for harmonic analysis. The wavelet domain representation is activated using DWT on the linear operator. To implement linear system analysis scheme in wavelet domain the steps are followed [51]:

1) First the mother wavelet is chosen. This is done by calculating the wavelet coefficients by different mother threshold are set to zero. The one with the fewest number of non-zero coefficients is chosen for analysis.
2) DWT is applied to obtain a multi resolution decomposition of input signal.
3) System component models are built in wavelet domain for both transient and harmonic analysis.
4) Different modeled system components and network topology is combined to build an equivalent circuit in wavelet domain.
5) Circuit is solved for transient harmonic analysis. Circuit is solved by impedance or admittance matrix method.
6) System transfer function is obtained in wavelet domain using time frequency response analysis.
7) Inverse DWT is applied to obtain the waveform in discrete time domain.

This scheme is suitable for transients and time varying harmonic analysis. This method makes it easier to study the linear systems because of using impedance or admittance matrix method.

16.5. Flicker

In power quality disturbances, flicker is an important parameter which is associated with lightening. The flicker phenomenon can be cyclic or random, caused by voltage fluctuations. Root-mean- square (RMS) voltage variations and frequency (or a band of frequencies) at which variations occurred have to be calculated and measured for voltage flickers. Characterization of flicker using inter harmonics is done using wavelet transform. The magnitude of the voltage can fluctuate and flicker might occur if voltage signal contains inter harmonics. With the help of wavelet transform, characterization of flicker is done in frequency domain and the source of flicker and inter harmonics can be identified. Inter harmonics are the frequencies which are not integral multiples of fundamental frequency and have frequencies between two harmonics. Wavelet based algorithm has the following steps [51]:

1) Mother wavelet is chosen.
2) Find fourier transforms of the mother wavelet.
3) Fourier transforms is applied to time series.
4) Starting from a minimum scale S_0, up to the higher scales do.

4.1 Compute daughter wavelet.

4.2 To preserve the variance of the time series normalization of the daughter wavelet is done.

4.3 Fourier transforms of the time series is multiplied to the normalized daughter Wavelet.

4.4 Inverse transform is taken.

4.5 A contour map is plotted.

Scalogram of the input signals waveforms can easily identify random flicker.

16.6. Diagnosis of Electric Motor

The direct current (dc), induction and synchronous machines are used extensively for industrial, commercial and household needs. Electric machines face various stresses including faults during different operating conditions. Thus, the condition monitoring, faults diagnostic, and protection become necessary in order to avoid catastrophic failures of electric machines. The use of comprehensive monitoring schemes with adequate warning for the continuous assessment of electrical machines is becoming increasingly important. In electrical machines, protection from over current, differential current, under and over voltage, Earth fault, over speed, field failure etc generally needed.

Wavelet packet transform (WPT) and neural network (NN) based algorithm is used for fault diagnostic and protection for electric motors. Features of fault currents of induction motors are extracted by using signature analysis technique [52]. Wavelet transform is used to decompose the stator current up to second level of resolution. dd^2 is the second level of high frequency details which are used to analyze the signatures of various faults. Generally, most of the induction motor fault current signals contain high frequency components along with the fundamental frequency. Significant features for fault detection are extracted based on the density of second level high frequency detail coefficients dd^2 of stator currents. These coefficients are passed to a three layer feed forward neural network with four inputs and one output. dd^2 co-efficients of faulted and normal currents are given as input to neural network. The neural network is trained for the faulted and normal current coefficients to indicate as 1 or 0 respectively at the output.

CONCLUSION

The objective of this chapter is to increase familiarity with basic wavelet applications in different areas and to provide reference sources and examples where wavelets can be potentially applied by researchers. A variety of wavelet based applications have been presented. It goes without saying that wavelet approaches will be growing importance in variety of fields. Potential for future advances seems quite good. Still there is scope for expanding into new areas and improving existing approaches with sophisticated techniques for better performance and reduced complexity. This is still largely incomplete review as Wavelet applications are numerous and invade nearly every domain.

REFERENCES

[1] J. M. Shapiro, *IEEE Trans. Signal Process.*, 41, 3445 (1993).

[2] Said and W.A. Pearlman, *IEEE Trans. Circuits Syst. Video Technol.*, 6, 243(1996).

[3] Skodras, C. Christopoulos and T Ebrahimi, *IEEE Signal Process. Mag.* 18, 36 (2001).

[4] S. Lawson and J. Zhu, *Electron. Comm. Eng* 14, 112 (2002).

[5] T.G. Lane, "JPEG software implementation", ver. 6b, Independent JPEG Group (1998).

[6] G. Sullivan and T. Wiegand, "Video compression - from concepts to the H.264 standard," *Proceedings of the IEEE*, 93, 18, (2005).

[7] P. Srinivasan and I.H. Jamieson, *IEEE Trans. Signal Process.,* 46, 1085 (1998).

[8] J.N. Bradley and C.M. Brislawn, "The Wavelet /Scalar Quantization Compression Standard for Digital Fingerprint Images", in *Proc. IEEE ISCAS*, London, pp 205-208 (1994).

[9] T. Hopper, C. Brislawn and J. Bradley, *WSQ Gray-Scale Fingerprint Image Compression Specification*, FBI, Tech. Rep. IAFIS-IC-O110v2, Fed. Bureau of Investigation, (1993).

[10] M. Saric, L.Bilicic and H.Dujmic, "White noise reduction of audio signal using wavelets transform with modified universal threshold", *Proceedings of the 4th WSEAS International Conference on Signal Processing, Robotics and Automation*, Salzburg (Austria) (2005).

[11] P.J. Wolfe and S.J. Godsill, "Audio signal processing using complex wavelets", Preprint 5829, presented at the *114th Convention of the Audio Engineering Society*, Amsterdam, 2003.

[12] H. Sheikhzadeh and H. R. Abutalebi, "An improved wavelet-based speech enhancement system", in *Proc. European Conf. on Speech Commun. and Technology (EuroSpeech)*, Aalborg, pp. 1855-1858 (2001).

[13] F.B Balcik and E. Sertel, "Wavelet based image fusion of landsat ETM images : A case study for different landscape categories of Istambul", *Conference on Information Extraction from SAR and Optical Data, with Emphasis on Developing Countries*, Istanbul, Turkey (2007).

[14] J.N. Ellinas, *Int.J. Comput Sci.*, 2, 206 (2007.)

[15] S. Lee, C.D. Yoo and T. Kalker, *IEEE Trans. Inf. Forensics Secur.* 2, 321 (2007).

[16] G. Fahmy, J. Black and S. Panchanathan, *IEEE Trans. Image Process.* 15, 1389 (2006).

[17] B.S. Manjunath and W.Y. Ma, *IEEE Trans. Pattern Anal. Mach. Intell.* 18, 837 (1996).

[18] Z. Hong, "Research and applications of electric power signal filter based on Wavelet multi-resolution technology", *International Conference on Environmental Science and Information Application Technology,* pp 367-370 (2010).

[19] N. Thakor, S. Yi-Chun, B. Gramatikov, H. Rix and P.Caminal, "Multiresolution wavelet analysis of ECG: Detection of ischemia and reperfusion in angioplasty", *Proceedings of the World Congress in Medical Physics and Biomedical Engineering,* Rio de Janeiro, pp. 392 (1994).

[20] O. Meste, H. Rix, P. Jane, P. Caminal and N.V. Thakor, *IEEE Trans. Biomed. Eng.*, 41 625 (1994).

[21] N. Thakor, B. Gramatikov and M. Mita "Multiresolution Wavelet Analysis of ECG during Ischemia and Reperfusion", *Proceedings of the 1993 Conf. on Computers in Cardiology*, London, pp. 895–898 (1993).
[22] A.Aldroubi and M. Unser, *Wavelets in Medicine and Biology*, CRC Press, Boca Raton, FL (1996).
[23] H. Adeli, Z. Zhou and N. Dadmehr, *J. Neurosci. Meth.*, 123, 69 (2003).
[24] J.M. Kates, *J. Acoust. Soc. Am.*, 97, 461 (1995).
[25] M. Penedo, W.A. Pearman, P.G.Tahoces, M. Souto and J.J.Vidal, *Proc. IEEE Int. Conf. Image Process.*, 2, 197 (2003).
[26] L. Shen and R.M. Rangayyan, *IEEE Trans. Med. Imaging*, 16, 301 (1997).
[27] A.J. Mendez, P.G. Tahoces, M.J. Lado, M.Souto, J.L. Correa and J.J. Vidal, *Comput. Methods Programs Biomed.*, 49, 253 (1996).
[28] K.S. Song, J.K. Kim, J.M. Park and H.W. Park, *IEEE Trans. Med. Imaging*, 16, 495 (1997).
[29] H.S. Own, *ICGST Int.J. Graphics, Vision and Image Processing*, 2, 9 (2005).
[30] J.N. Livingston and C.C. Tung, *IEEE Trans. Prof. Commun.*, 44, 1629 (1996).
[31] A.R.Lindsey, *IEEE Trans. Signal Process.*, 45, 1336 (1997).
[32] G.W. Wornell, *Proceedings of the IEEE*, 84, 586 (1996).
[33] L.Feng and W. Xinmei, *Electron. Lett.*, 33, 565 (1997).
[34] R. E. Learned, H. Krim, B. Claus, A.S. Willsky and W.C. Karl, "Wavelet-Packet-Based Multiple Access Communication," *Proceedings of the SPIE*, 2303, 246 (1994).
[35] K.Hetling, M. Medley, G, Saulnier and P. Das, "A PR-QMF (Wavelet) Based Spread Spectrum Communication System," *Proceedings of IEEE MILCOM*, 3, 760 (1994).
[36] C.C. Aggarwal, J.L. Wolf, P.S. Yu, C. Procopiuc and J.S. Park, "Fast algorithms for projected clustering ", in *Proceedings ACM SIGMOD Conference on Management of Data*, pp. 61–72 (1999).
[37] D. Keim and M. Heczko. Wavelets and their applications in databases. *Tutorial Notes of ICDE Conference* (2001)
[38] Bruce and H.-Y. Gao, *Stat. Sinica*, 4, 855 (1996).
[39] Shahabi, S. Chung, M. Safar and G. Hajj, "2D TSA-tree: A Wavelet-based approach to improve the efficiency of multi-level spatial data mining" in *Proceedings of SSDBM'2001*, pp. 55–68 (2001)
[40] H.A.Toliyat, K. Abbaszadeh, M.M. Rahimian and L.E. Olson, *IEEE Trans. Ind. Appl.*, 39, 5 (2003)
[41] Y.L Gonidec, F.Conil, D.Gibert, *Ultrasonics* 41 487 (2003).
[42] H. Sohn, J.R. Wait, N.P. Limback, C.R. Farrar, *Smart. Mater. Struct.*, 13, 153 (2004).
[43] O. Mendes, M. O. Domingues, A. Mendes da Costa and A. L. Clua de Gonzalez, "Wavelet analysis applied to magnetograms: singularity detections related to geomagnetic storms", *VI Latin American conference on space geophysics*, Sao José dos Campos, Brasil, pp 177 (2004).
[44] H. Lundstedt, L. Liszka and R. Lundin, *Ann. Geophys.*, 23, 1505 (2005).
[45] H S Oh, C. Ammann, P. Naveau, D. Nychka and B. Otto-Bliesner, *J. Atmosph. Sol-Terr. Phys.*, 165, 191 (2003).
[46] Torrence and G.P. Compo, *Bull. Am. Meteorol. Soc.*, 79, 61 (1998).
[47] K Gurley and A Kareem, *Eng. Struct.*, 21, 149 (2003).
[48] Kareem and T. Kijewski, *J. Wind Eng. Ind. Aerod.*, 90, 1435 (2002),.

[49] J. Driesen and R. Belmans, "Time-Frequency Analysis in Power Measurement using Complex Wavelets", *IEEE International Symposium on* Circuits and Systems, Scottsdale, Arizona, pp V681 - V684, (2002).
[50] H.T. Yang and C.C. Liao, *IEEE T Power Deliver.*, 16, 353 (2001).
[51] T. Zheng, E.B. Makram and A.A. Girgis, *IEEE Trans. Power Deliver.*, 14, 1461(1999).
[52] M.A.S.K. Khan and M.A. Rahman, Wavelet Based Diagnosis and Protection of Electric Motors, In W.Zhang (Ed.) *Fault Detection*, Ch. 12, InTech, Rijeka (Croatia), (2010).

In: Wavelets: Classification, Theory and Applications
Editors: Manel del Valle et al.
ISBN: 978-1-62100-252-9

Chapter 2

PROCESSING APPLICATIONS OF QUATERNION ATOMIC FUNCTION WAVELETS USING THE PHASE CONCEPT

E. Ulises Moya-Sánchez* and Eduardo Bayro-Corrochano
Department of Electrical Engineering and Computer Science,
CINVESTAV Campus Guadalajara, México

ABSTRACT

This chapter presents a novel quaternion wavelet transform using the atomic function. Atomic functions are compactly supported and infinitely differentiable solutions of differential equations. Consequently, they constitute a promising kernel for wavelet multi-resolution analysis. This work makes use of the phase concept for the detection of key features. For this goal, we steer the mother atomic function wavelet through a multi-resolution pyramid to detect quaternionic phase changes, which can be used for the feature detection of lines, edges, centroids, and orientation in geometric structures up to affine transformation. The experimental analysis gives more insight into our techniques and stimulates further applications.

INTRODUCTION

The atomic functions (*AFs*) were first developed in the 1970s, jointly by V. L. Rvachev and V. A. Rvachev. Since then, AFs have made important contributions in mathematical analysis, signal processing, pattern recognition, image processing, numerical methods, and so forth [1-3]. Atomic Functions have some useful properties, for instance, they are easy to derive (only a shift) in any order, they have has compact support in the space domain (good local characteristics), and they also have the possibility to represent any polynomial via its translations [1, 7]. Furthermore, complex, and hyper-complex algebras plays an important

* E-mail address: emoya@gdl.cinvestav.mx

role in signal and image processing [1, 2, 7, 9]. The quaternion algebra extends the complex domain into one real number and three imaginary numbers, which permits one to obtain more phase information [4, 6, 7].

We have extended the AF *up(x)* over the quaternion algebra (*Qup*). As a result, it has a real part and three imaginary parts (*i, j, k*).

The importance of wavelet transforms (real, complex, hypercomplex) has already been discussed in many references [1, 2, 4, 9]. We focus on the quaternion wavelet transform since we can easily obtain the phase information. The main contribution of this work has been the presentation of the atomic function QWT and its applications in phase to obtain a new geometric structure detector. The experimental results show that, this QWT can be used as an oriented direction filter of lines while the phase information of the image is not affected by illumination changes. We then show the phases profiles to detect lines and edges in a profile image. Also, we may use the convolution of the *Qup* with the image to obtain the phases, in order to, detect lines or edges in a specific orientation.

The structure of this chapter is as follows: In the first section, the AF *up(x)* and its properties are presented, and then we outline the quaternion algebra. We then present the quaternion *up(x,y)* and the QWT of *Qup*. After that, the applications using the phase concept are shown. We calculate the three phases of the filtered image, in a steerable pyramid, and then we use the quaternionic phase profile of the *QupWT* to detect lines and edges. We show how to use the 2D phase information from a pattern. This pattern can be useful when finding the center and orientation of different geometric structures or creating a new low- level structure detector in terms of *QupWT* phase information. Finally, the conclusions are presented.

ATOMIC FUNCTIONS AND *UP(X)*

The AFs are compactly supported and infinitely differentiable solutions of differential equations with a shifted argument [7, 1]; i.e.

$$Lf(x) = \lambda \sum_{k=1}^{M} c(k) f(ax - b(k)), |a| > 1, b \in \Re \tag{1}$$

where L is a linear differential operator with constant coefficients. In the AF class, the function *up(x)* is the simplest and, at the same time, the most useful primitive function to generate other kinds of atomic functions such as $fup_n(x)$, $\Xi_n(x)$ [1]. In this work we only use *up(x)*. In general, the atomic function *up(x)* is generated by infinite convolutions of rectangular impulses. The function *up(x)* has the following representation in terms of the Fourier transform:

$$up(x) = \frac{1}{2\pi} \int_{-\infty}^{\infty} e^{i\nu x} \prod_{k=1}^{\infty} \frac{\sin(\nu 2^{-k})}{\nu 2^{-k}} d\nu \,. \tag{2}$$

Another important characteristic of the function *up(x)* is that it is infinitely differentiable [1]:

$$d^{(n)}up(x) = 2^{n(n+1)/2} \sum_{k=1}^{2^n} \delta_k up(2^n x + 2^n + 1 - 2k), \tag{3}$$

where, $\delta_{2k} = -\delta_k, \delta_{2k-1} = \delta_k, \delta_{2k} = 1$. The first derivative has a simple expression:

$$dup(x) = 2[up(2x+1) + up(2x-1)] \tag{4}$$

and the function *dup(x)* provides a good window into the spatial frequency regions since, in this case (frequency), the side lobes have been completely eliminated [7]. Other characteristics are that *up(0) = 1*, it is even *up(−x) = up(x)*, it is analytical throughout the range, and it can represent any polynomial by means of its translations [7, 12]:.

$$X^N = \sum_k C(k) up(x - k2^N) \cdot \tag{5}$$

Figure 1 shows *up(x)* and the first derivative *dup(x)* in 1D. We can see a compact support in the space domain, and we see how *up(x)* and *dup(x)* are the same function with a shift in the argument.

Atomic functions were compared with classic ones [1, 7], in terms of their equivalent noise bandwidth, 50% overlapping region correlation, maximum conversion losses (in decibels), asymptotic decay rate of the side lobes (in decibels per octave), among other comparisons. Atomic windows exceed classic ones in terms of the asymptotic decay rate [1, 7].

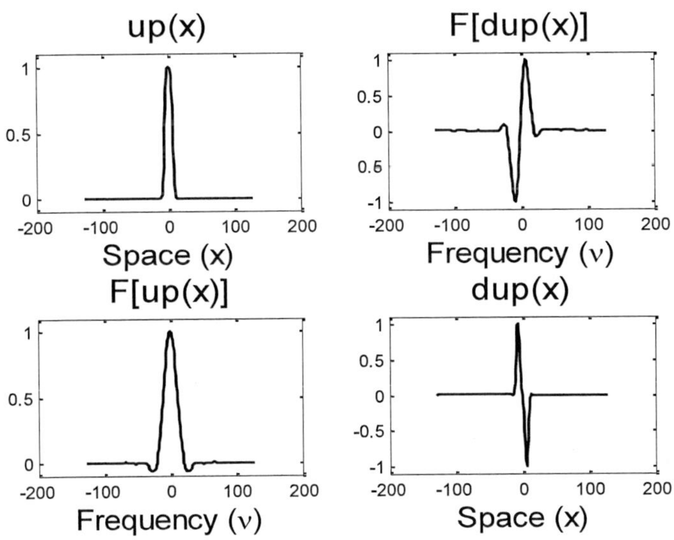

Figure 1. Left column *up(x)* and the Fourier transform *F[f(x)]* of *up(x)*, *F[up(x)]*. Right column *dup(x)* and *F[dup(x)]*.

Quaternion Algebra

The quaternion algebra H is geometrically inspired and is the basis of an even subalgebra [2] of the Geometric Algebra. H was invented by W. Hamilton in 1843 [2]. Specifically, the subalgebra G^{+}_{3} is isomorphic to the quaternion algebra, which is an associative non-commutative four-dimensional algebra that consists of one real scalar and three imaginary elements [2, 3]:.

$$q = a + ib + jc + kd, \tag{6}$$

where $a, b, c, d \in \Re$. The units i, j fulfill the relations $i^2 = j^2 = -1$, and $ij = -k$, and the imaginary components can be described in terms of the basis of $\Re^3$ (or bivector of G^{+}_{3}) space, $i \rightarrow e2e3$, $j \rightarrow e1e2$, $k \rightarrow e3e1$, where $\{e1, e2, e3\}$ is the basis of $\Re^3$. Another important property of H is the phase concept. The polar representation of q is (|q|, φ, θ, ψ), where $|q|$ is the magnitude and the angles (φ, θ, ψ) represent the three quaternionic phases [2, 4] $q = |q|e^{i\varphi}e^{k\psi}ej^{\theta}$. In image processing, the phase component contains the main part of the image information [5, 6] and is useful for encoding lines and edges that are invariant to brightness changes [2, 4].

Quaternion Atomic Function

The *up(x)* function is easily extendable to two dimensions. Since a 2D signal can be split into even *(e)* and odd *(o)* parts [3] as follows

$$f(x,y) = f_{ee}(x,y) + f_{oe}(x,y) + f_{eo}(x,y) + f_{oo}(x,y), \tag{7}$$

one can then separate the four components of Eq. (2) and represent them as a quaternion as follows:

$$\begin{aligned} Qup(x,y) &= up(x,y)[\cos(wx)\cos(wx) + i(\sin(wx)\cos(wy)) + ... \\ &+ j(\cos(wy)\sin(wx)) + k(\sin(wx)\sin(wy))] \\ Qup(x,y) &= Qup_{ee} + iQup_{oe} + jQup_{eo} + kQup_{oo} \end{aligned} \tag{8}$$

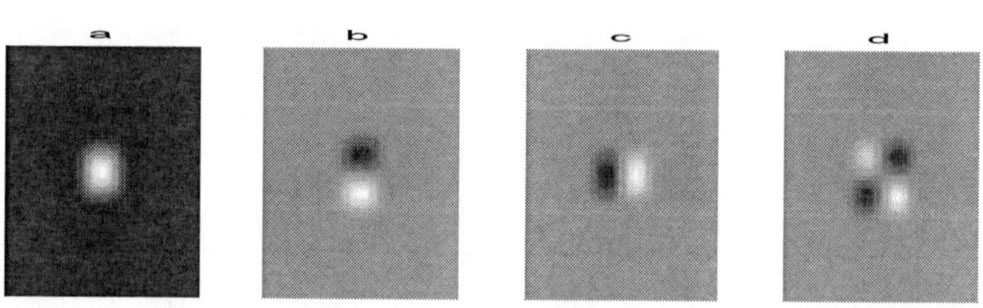

Figure 2. From left to right, *Qup(x,y)*: real part, *i-part, j-part, k-part.*

Figure 2 shows *Qup(x,y)* in the space domain with its four components: The real part Qup_{ee} is observed in part *a*, and the imaginary parts Qup_{eo}, $Qup_{o}e$ and Qup_{oo} are illustrated in parts b; c; and respectively. We can see clearly see the differences in each part of our filter.

QUATERNION ATOMIC FUNCTION FOR MULTI-RESOLUTION ANALYSIS

The real, complex, and recent Quaternion Wavelet Transforms (QWT), have been useful tools in different areas of applied mathematics and image processing [2, 4, 9]. In this work we only use the quaternion wavelet transform. The *QWT* can be seen as an extension of the complex wavelet transform (CWT) [9]. The multiresolution analysis applied in the real wavelet transform *(RW)T* and *CWT* can be straightforwardly extended to the *QWT* [9]. For the quaternionic versions of the wavelet pyramid, we use a scale function *h* and the wavelet function *g*, and we choose two *up(x)* filters in quadrature, i.e. [9]

$$h(x,y,\sigma,\omega) = up(x,y)\exp(i\frac{xw_1}{\sigma_1})\exp(j\frac{yw_2}{\sigma_2}) = h_{ee}(x,y) + h_{oe}(x,y) + h_{eo}(x,y) + h_{oo}(x,y), \quad (9)$$

$$g(x,y,\sigma,\omega) = up(x,y)\exp(i\frac{xw_1}{\sigma_1})\exp(j\frac{yw_2}{\sigma_2}) = g_{ee}(x,y) + g_{oe}(x,y) + g_{eo}(x,y) + g_{oo}(x,y), \quad (10)$$

where the parameters $\omega_1, \omega_2, \sigma_1, \sigma_2$ are selected for the fulfillment of the requirements of bandwidth and cut frequency. In this case the *x*-axis is related to *i* and the *y*-axis to *j*, both imaginary numbers of the quaternion algebra. For further explanation, see Refs. [2, 4, 9]. The procedure for quaternionic wavelet multiresolution analysis depicted partially in Figure 3 is as follows:

1. Convolve the image (2D signal) at level *n* and convolve it with the scale and wavelet filters *h* and *g* along the rows of the 2D signal. The latter filters are the discrete versions of those filters given in equations (9) and (10).
2. *h* and *g* are convolved with the columns of the previous responses of the filters *h* and *g*.
3. Subsample the responses of these filters by a factor of two (↓ 2).
4. The real part of the approximation at level *n+1* is taken as input at the next level. This process continues through all levels, repeating steps 1→4.

In particular, we have applied the *QWT* into images in a multiresolution analysis, because it can be more useful than *RWT* and *CWT* since we can obtain the three quaternionic phases, which is not possible with the real wavelet transform , and only one such phase exists in the *CWT* [2, 9]. We hence use the function *QupWT* in a steerable way, that is shown in Figure 4.

We applied the *Qup* on test images in four different ways: First, the image was convoluted with each part of the *Qup,* and then we calculated the quaternionic phases. Figure

4 shows the original image and the four resulting images (magnitude and three phases), which illustrates how the *Qup* filter works.

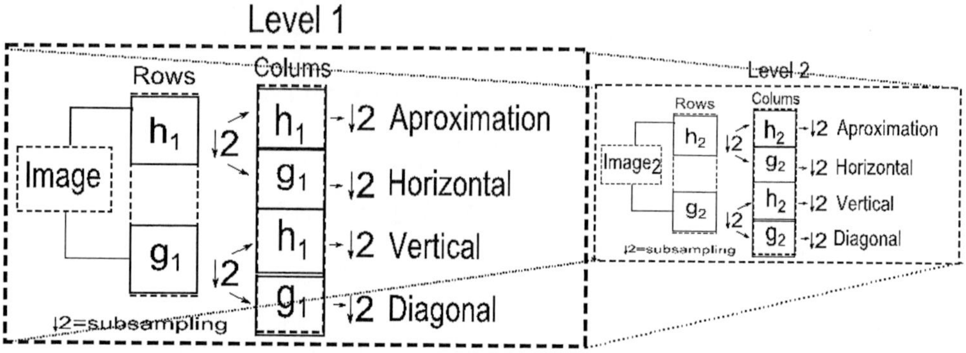

Figure 3. Scheme of two levels of the quaternionic wavelet pyramid.

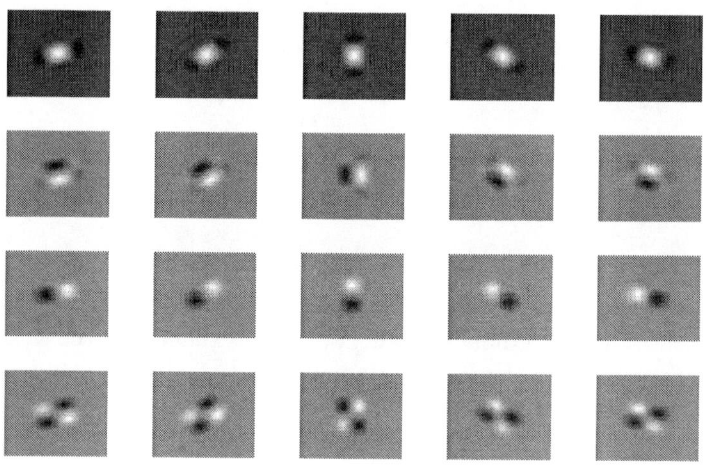

Figure 4. From left to right: *Qup* (15°), *Qup* (40°), *Qup* (90°), *Qup* (140°), and *Qup* (155°).

Applications in Image Processing

Additionally, we show the same result using a steerable *QupWT* at two levels 45° and 90° of the pyramid, respectively (second and third rows of Figure 5).

We can see how the phase information is immune to the illumination changes and how the phase-ψ can be used to detect two orthogonal fan lines' orientations, white lines in a particular direction and black lines in an orthogonal direction.

We use the quaternionic phases' information to detect lines and edges in different 2D geometric structures. Figure 6 shows the three quaternionic phases of the test image; this information is useful to detect phase changes along lines. Lines and edges are important in terms of low-level information for biological systems [5, 11].

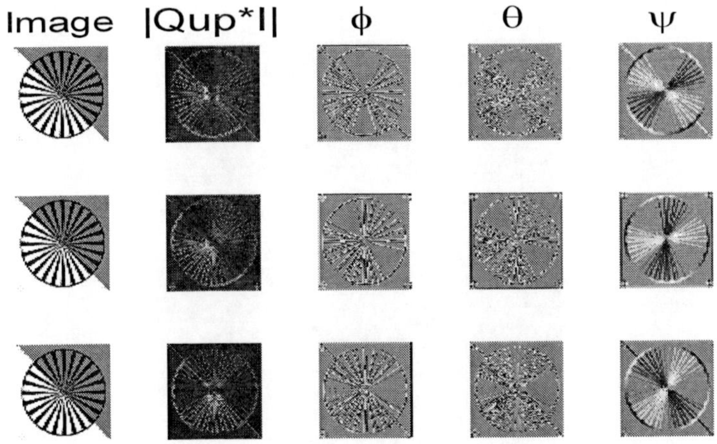

Figure 5. From left to right: Original image, magnitude, and each phase of *QupWT*. The first row shows 0° phases, then we present at 45°, and finally we present the 90° result.

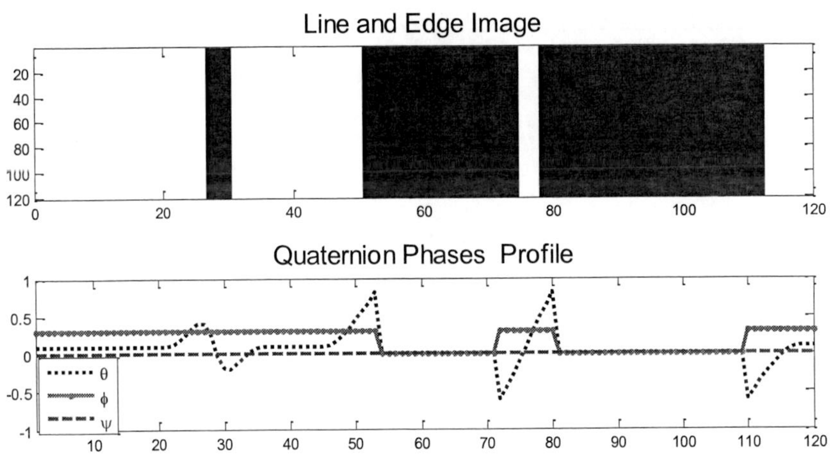

Figure 6. (From top to bottom). Line and edge image; quaternionic phases' profile.

The quaternionic phases-profiles show the different responses of each phase. In this case they are used to distinguish between lines and edges in the 1D profile of the image. The ϕ-phase′s only response results in edges and white lines. Another example of this appears is in Figures 7 and 8. We can see how the 2D phase is different in each phase profile. Briefly, the quaternionic phases can be used to obtain more structure information than would be possible by either the *RWT* or *CWT*.

Finally, we show other applications of the *QWT*. We compute the phase information to obtain the center of different objects or to obtain a useful detector of the centroid and orientation of geometric figures in terms of quaternionic phases. Figure 9 shows the 2D phases of some squares in different directions.

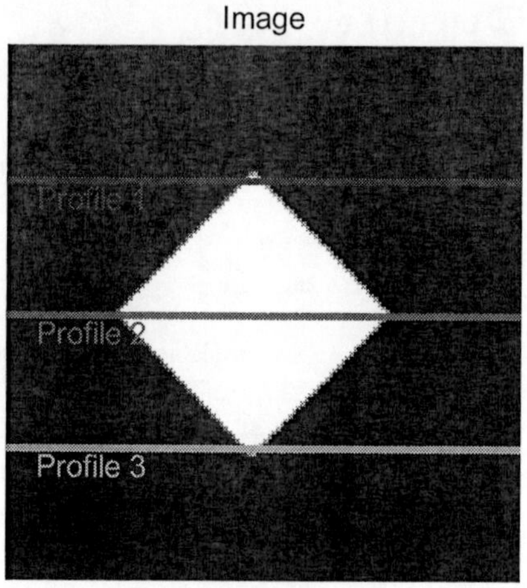

Figure 7. Test image profiles.

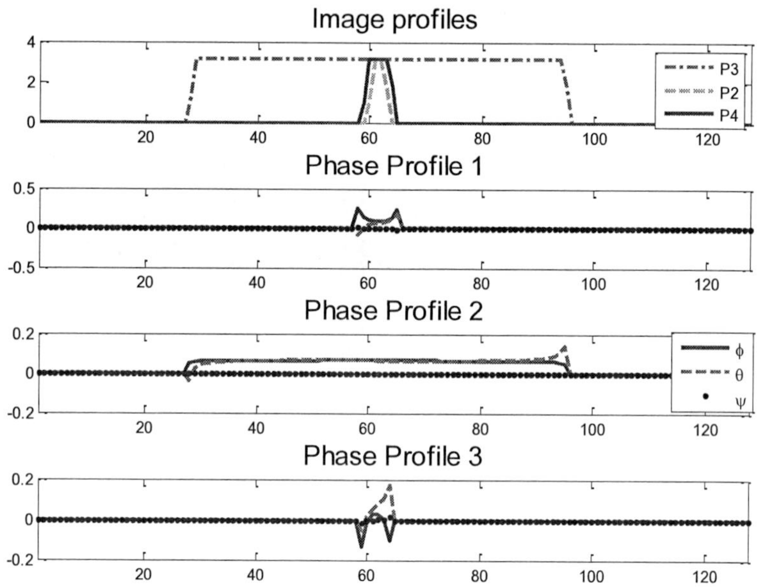

Figure 8. Plots of quaternionic phase profiles of Figure 7.

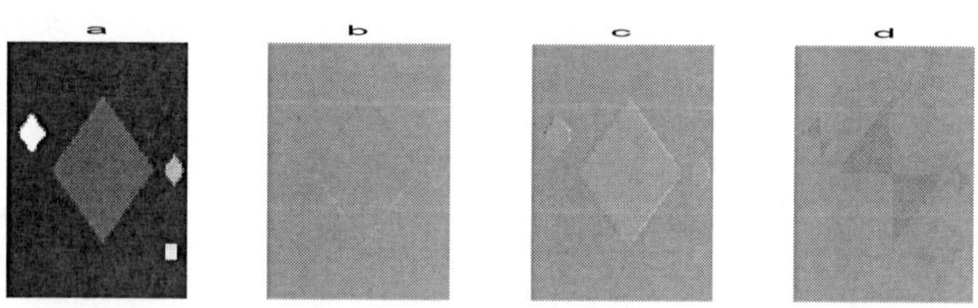

Figure 9. a) Magnitude and three quaternion phases (b) ϕ , (c) θ and (d) ψ .

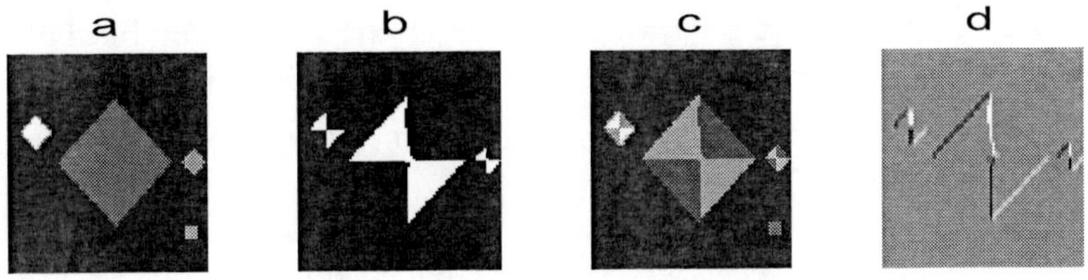

Figure 10. Plots of the quaternionic phase ψ to enhance image structures in different contrast.

Figure 10 shows the original image (a), the binarized quaternionic phase ψ (b), the superposition of the original image and the phase ψ (c) in (c) and the first derivative of part (b) in (d). We can see how this detector is such as preprocessing operator to detect the geometric structure up to an affine transform. This information can be useful in the future to obtain a new phase descriptor.

Conclusion

This work introduces a QWT multi-resolution analysis using the quaternion algebra and *up(x),* the most important atomic function. We have presented applications of the *QupWT* using the phase concept. Our work shows that *Qup,* can be more useful than the simple *up(x)* or the complex *up(x)*. *Qup* exploits the quarternionic phase concept and can be more useful to enhance different geometric structures than the complex phase. We have shown that the information in the three phases is independent of the illumination changes. The reader can see that the *Qup* wavelet filter disentangles the oriented structure of the image through the levels of a multiresolution steering pyramid. In fact, the *QupWT* can be used as a new preprocessing operator to detect geometric structures up to an affine transform.

Acknowledgments

We want to thank the financial support from SEP/CONACYT and CINVESTAV′s PhD program.

References

[1] V.F. Kravchenko, H.M. Perez-Meana, and V.I. Ponomaryov, *Adaptive Digital Processing of Multidimensional Signals with Applications*, Fizmatlit, Moscow (2009).

[2] E. Bayro-Corrochano, *Geometric Computing for Wavelet Transforms, Robot Vision, Learning, Control and Action*, Springer, London (2010).

[3] T. Bülow, *Hypercomplex Spectral signal representations for the processing and analysis of images*, Phd. Thesis, Christian Albrechts University, Kiel (1999).

[4] E.U. Moya-Sanchez and E. Bayro-Corrochano, Quaternion atomic function wavelet for applications in image processing, in I. Bloch and R.M. Cesar, Jr. (Eds.), *Progress in Pattern Recognition, Image Analysis, Computer Vision, and Applications,* LNCS Vol. 6419, Springer, Heidelberg (2010), pp. 346-353.
[5] J. Bigun, *Vision with Direction*, Springer, Heidelberg (2006).
[6] B. Jähne, *Digital Image Processing*, Springer, Heidelberg (2002).
[7] P. Kovesi, *Invariant measures of images features from phase information,* Ph.d Thesis, University of Western Australia, Perth (1996).
[8] Yu.V. Guyaev and V.F. Kravchenko, *Dokl. Math.* 75, 325 (2007).
[9] E. Bayro-Corrochano, *J. Math. Imaging Vis,* 24, 19 (2006).
[10] G.H. Granlund, and H. Knutsson, *Signal Processing for Computer Vision*, Kluwer Academic, Dordrecht, (2006).

In: Wavelets: Classification, Theory and Applications
Editors: Manel del Valle et al
ISBN 978-1-62100-252-9

Chapter 3

THE MULTIRESOLUTION WAVELET-BASED TRANSFORMATION OF 3D OBJECT SURFACE MESHES

*Agnieszka Szczesna**
Silesian University of Technology, Institute of Informatics, Poland

Abstract

3D meshes are a piecewise-linear surface approximation and are the most popular representation of real scanned and modeled object used in computer graphics, medicine, games, simulators, scientific visualisation, CAD and virtual reality systems. Such meshes can be set of millions of polygons, vertices and edges. Complex three-dimension objects must be processed, stored, transferred, animated and analysed which is very expensive. Multiresolution models arise a lot of interest because they make it possible to represent and process geometric data in different levels of detail (LOD) depending on application needs. This chapter presents a survey of the different wavelet-based multiresolution methods proposed for triangle surface meshes. First the preliminary background on triangle meshes and multiresolution analysis is presented. Then the different proposed methods for regular, semi-regular and irregular meshes are described. This survey consists also description of Author solution based on second generation wavelets for decomposition and reconstruction of irregular, triangle mesh. The generalisation from the first generation of wavelets to the second one gives an opportunity to construct wavelets on irregular meshes. This solution is based on lifting scheme as a wavelet construction tool. The conclusion is the different application of result multiresolution representation in algorithms of digital geometry processing.

Introduction

After processing signals, sounds, images and video, wavelets have been applied for digital geometry processing. Its goal is the research of mathematical and computational foundations of algorithms such as denoising, compression, transmission, enhancement, detection, analysis and editing. The basic framework for these tools is the multiresolution

*E-mail address: agnieszka.szczesna@polsl.pl

analysis that decomposes an initial data set into a sequence of approximations and details. Main advantages are reduction of data, quick access to data, hierarchical processing, LOD control and progressive transmission.

3D meshes are a piecewise-linear surface approximation and are the most popular representation of real scanned and modeled objects. Such meshes can be set of millions of polygons, vertices and edges. Multiresolution models arise a lot of interest because they make it possible to represent and process geometric data in different levels of detail (LOD) depending on application needs [1], [2], [3]. For given input original mesh the multiresolution representation is a decomposition to a sequence of topologically equivalent meshes with smaller geometric detail. Next approximations are next levels of detail of original mesh and consist of a lower number of vertices, edges and polygons. The wavelet-based multiresolution algorithms decompose an initial mesh into a sequence of mesh approximations and details - wavelet coefficients [4]. Result multiresolution representation can be base of different algorithms such as smoothing, enhancement, features detection, compression, progressive transmission, etc.

This chapter is a survey of different wavelet-based multiresolution methods proposed for triangle meshes (regular, semi-regular and irregular) [**?**]. The description of second generation wavelets for decomposition and reconstruction of irregular, triangle meshes is also presented. The conclusion is using of result multiresolution representation in the digital geometry processing applications.

Definition of Surface Mesh

There are three most popular representations of surfaces of the 3D objects: parametric surfaces (for example Bezier, B-spline surfaces), implicit surfaces and surface meshes. First two are continuous representation of surface and are commonly used for computer aided modeling (CAD). Mesh is discrete representation of surface. This is a piecewise-linear surface approximation and is the most popular representation of real scanned and modeled object used in computer graphics applications.

Because the triangle is the most popular graphics primitive so the triangles meshes are also most popular used. This chapter concerns only triangle surface meshes.

The triangle mesh is a pair $M = (P, K)$ where P is a set of n points $p_i = [x_i, y_i, z_i]$ with $1 \leq i \leq n$ and K is a simplicial complex which contains information about topology of a mesh. The complex is a set of three types of subsets called simplices: vertices $v_i = \{i\} \varepsilon V$, edges $e_{i,j} = \{i, j\} \varepsilon E$, faces $f_{i,j,k} = \{i, j, k\} \varepsilon F$. Two vertices $\{a\}$ and $\{b\}$ are neighbours if $\{a, b\} \varepsilon E$. The 1-ring neighbourhood of vertex $\{i\}$ is the set $N(v_i) = \{j | \{i, j\} \varepsilon E\}$. The valence of a vertex v_i is the number of edges meeting at this vertex $\#N(v_i)$. There are three types of triangular meshes shown on Figure 1:

- regular where every vertex has valence equals 6;
- semi-regular where the most vertices have valence 6, few isolated vertices have any valence;
- irregular where vertices can have any valence.

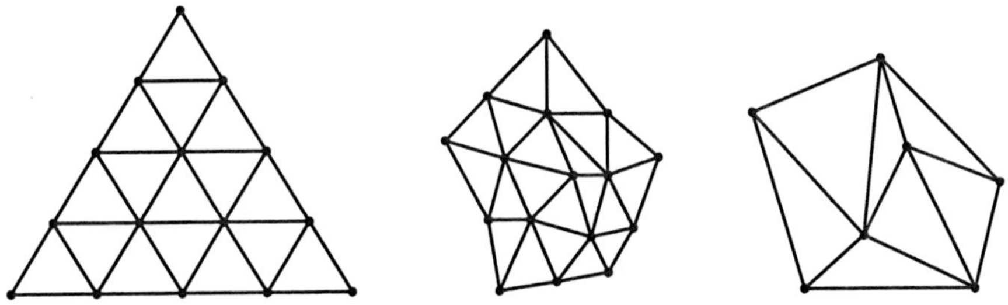

Figure 1. Types of triangle mesh: regular, semi-regular, irregular.

Many models have surface properties beyond simple geometry. In computer graphics, the most common is RGB color values. Vertices of surface mesh, in addition to its position in space, have additional values which describe other properties. Each vertex is treated as a extended vector $v \varepsilon R^n$. The first 3 components of v are spatial coordinates, and the remaining components are attributes values. Example vertex with RGB color is $[x_i, y_i, z_i, r_i, b_i, g_i]$.

The triangle mesh which make surface must meet the assumptions of the topological 2-manifold. This means that every vertex of mesh has to have a neighbourhood for which there exists a homeomorphism mapping that neighbourhood to R^2 disc.

The topological 2-manifold mesh M define the domain of wavelet multiresolution representation. Functions are defined based on set of vertices V, edges E and triangles F with topological information about mutual proper connections.

Level of Detail Representation

For given input triangle mesh $M = M^n$ the multiresolution representation is a decomposition to a sequence of topologically equivalent meshes with smaller geometric detail: $M^n, M^{n-1}, ..., M^0$. Where M^0 is mesh in lowers resolution (base mesh). Next approximations are next levels of detail of original mesh and consist of a lower number of vertices V, edges E and polygons F.

Generally methods to generation multiresolution representation can be divided to two groups: mesh decimation and mesh refinement. Some elements and assumptions of these base methods are often used as building block in wavelets transformation framework. So the short introduction is needed.

Mesh Decimation

Mesh decimation is coarsening of the mesh at full resolution through iterative application of local topological operator. These operators collapse vertex, edge, half-edge or triangle, the second group of operators remove vertex, edge or triangle and the result hole is triangulated. This operator simplifies mesh region chosen by local or global criteria which depends of algorithm. Resolution levels are constructed until the approximation error reaches a user-defined maximum.

Most popular simplification algorithms are vertex decimation [5], Quadric Error Metric (*QEM*) [6], [7], mesh optimization [8], progressive mesh (*PM*) [9], [10]. Full survey of simplification methods can be fund in [11], [12], [13], [14], [15].

Mesh Refinement

Mesh refinement is iterative algorithm which adds elements at each step to initial base mesh in coarser approximation. These algorithms need the input base mesh in lowest resolution. The main group of this algorithms are subdivision schemes.

Subdivision schemes were developed as generalizations of uniform B-spline knot insertion algorithms to settings with topologically irregular control meshes. Generally subdivision defines a smooth curve or surface as the limit of a sequence of successive refinements. A subdivision surface is created by iteratively refining a control mesh M^0 to produce a sequence on increasing detailed meshes that converge to a surface (1).

$$\sigma = \lim_{j \to \infty} M^j \tag{1}$$

The first step of algorithm is topological subdivision that splits each mesh face or edge. The second step, geometric positioning, computes positions for new vertices as a local affine combination of old vertices via subdivision relation. The subdivision process produces hierarchically sequence of meshes with increasing number of faces and vertices.

Example of semi-uniform subdivision schemes for triangle meshes are: approximating Loop [16] and $\sqrt{3}$ [17], interpolating Butterfly [18], [19]. To analysis of schemes and result surface the spectral analysis based on subdivision matrix can be used [20].

The non-uniform schemes (variational subdivision) can be used for irregular meshes [21], [22]. The subdivision operator computes positions of new vertex as a minimization of value of some function, for example: minimize of second order differences defined at every edge [23], [24], surface curvature minimization [21]. The disadvantage of that schemes are slower work, much more complex and not complete mathematical tool for analysis.

Wavelet-Based Multiresolution Transformation of Triangle Mesh

Background

Multiresolution analysis by using wavelets decomposes an initial mesh into a sequence of mesh approximations and details - wavelet coefficients (Figure 2). The approximations at different levels of resolution are computed by using scalar functions. Next approximations of original mesh consist of a lower number of vertices V, edges E and polygons F. The details are computed by using wavelet functions and represent the data lost between next approximations and consist information about vertices which were removed from mesh. The reconstruction process uses the dual scalar and wavelet functions.

The equation (2) is description of one step of wavelet transform. Wavelet coefficients D^j are set of difference vectors between meshes in $j+1$ and j resolution.

$$M^{j+1} \Delta M^j + D^j \tag{2}$$

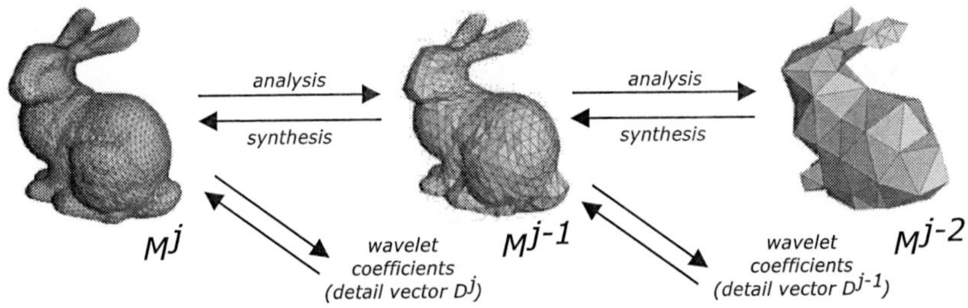

Figure 2. Wavelet based analysis of surface mesh.

The original mesh M^n is decomposed into base mesh in lowest resolution (M^0) and sequence of detail vectors. This is the result multiresolution representation (3).

$$M^n \,\Delta\, M^0 + \sum_{j=1}^{n} D^j \tag{3}$$

The general assumptions of wavelets methods for mesh transformation are:

- Meshes in lower resolutions should be a good approximation of the original surface mesh.
- The result of reconstruction process should be as close as possible to the original mesh. The calculations must be reversible.
- The detail vector should be small and capture local geometric details of represented surface.

Main problems in definition wavelet analysis and synthesis of surface mesh are no natural parametrization of mesh, irregularity of data and need for using non-separable wavelet bases.

Two groups of wavelets method can be singled out. First are methods for regular and semi-regular meshes with subdivision connectivity. These methods base on mathematical properties of uniform and semi-uniform subdivision methods. In this case input mesh first must be rebuild with remeshing algorithms [25], [26]. Another group of algorithms are methods to deal directly with irregular meshes with no need of remeshing process.

Lifting Scheme and Second Generation Wavelets

The second generation wavelets [27] are generalisation of biorthogonal classic wavelets, called first generation wavelets. The second generation wavelet is not necessarily constructed by translation and dilatation of one function (mother function). That wavelets can be built in more general settings where the Fourier can not be used as a construction tool. The second generation multiresolution representation may use many basic functions (wavelet and scaling functions), but must be possible to define nested spaces - must be meet the refinement relation of scaling functions.

The lifting scheme [28], [29], [30], [4] is a simple but powerful tool to construct the second generation wavelets. The main advantage of this solution is the possibility of building wavelet analysis on non-standard structures of data (irregular samples, bounded domains, curves, surfaces, manifolds) while keeping all powerful properties of the first generation wavelets such as speed and good ability of approximation.

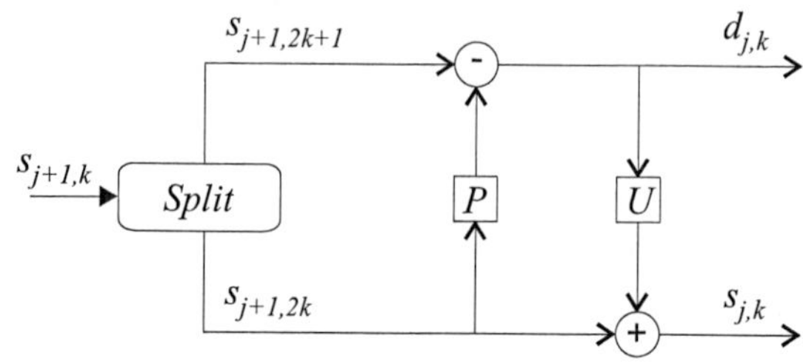

Figure 3. The lifting scheme.

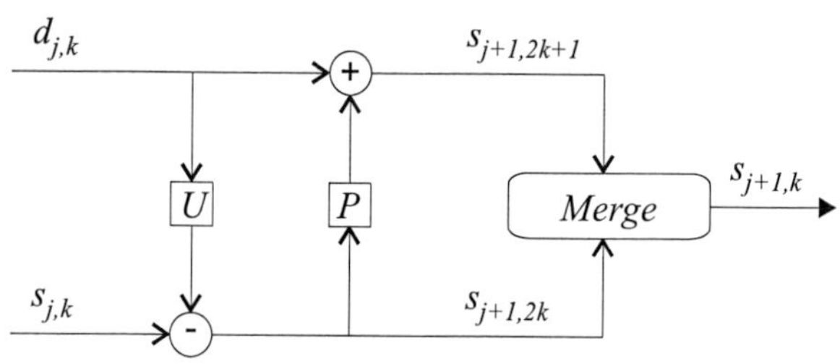

Figure 4. The reverse lifting scheme.

A general lifting scheme (Figure 3) consists of three types of operation:

- **Split**: splits input dataset into two disjoint sets of even and odd indexed samples (4). The definition of lifting scheme doesn't impose any restriction on how the data should be split nor on the relative size of each subsets.

$$(s_{j+1,2k+1}, s_{j+1,2k}) = S(s_{j+1,k}) \tag{4}$$

  Where first index denotes resolution level, higher index means more detailed resolution level. The second index denotes position of sample in data set.

- **Predict**: predicts odd indexed sample based on even samples. Next the odd indexed input value is replaced by the offset (difference) between the odd value and its prediction (5).

$$d_{j,k} = s_{j+1,2k+1} - P(s_{j+1,2k}) \tag{5}$$

- **Update**: updates the output, so that coarse-scale coefficients have the same average value as the input samples (6). This step is necessary for stable wavelet transform [27].

$$s_{j,k} = s_{j+1,2k} + U(d_{j,k}) \tag{6}$$

This calculations can be performed in-place. In all stages input samples can be overwritten by output of that step. Inverse transform (Figure 4) is easy to find by reversing the order of operations and flipping the signs.

Method for Regular and Semi-regular Meshes

The first connection between wavelets and subdivision (semi-uniform Loop subdivision) which allows to define multiresolution surface representation for semi-regular mesh with subdivision connectivity, was presented in [31], [25]. This is extension of multiresolution analysis based on tensor-product, which can be used only for functions parametrized on R^2, to arbitrary topological domains but with subdivision connectivity (semi-regular meshes). This solution is based on creating refinable scaling functions using recursive surface subdivision.

The recursive subdivision methods applied to surfaces meshes leads to a collection of refinable scaling functions and to a sequence of nested linear spaces required by multiresolution analysis. Only surfaces generated through subdivision can be hierarchically decomposed using that wavelets methods. The domain of scaling function is initial control mesh M^0. This mesh has the same topological type as the limit surface. A chain of nested linear space can be defined as:

$$V^0 \subset V^1 \subset \dots$$

where V^j is a set of mesh vertices in resolution j. The inner product is defined for functions which domain is M^0:

$$f, g \varepsilon V^j(M^0) :< f, g >= \sum_{\tau \varepsilon \Delta(M^0)} \frac{1}{area(\tau)} \int_{x \varepsilon \tau} f(x) g(x) dx \tag{7}$$

Where τ denote a triangle in set of triangles of M^0 mesh ($\Delta(M^0)$), $area(\tau)$ is area of triangle and x is the point of M^j corresponding to base mesh.

Result multiresolution transformation is computed by two analysis filters $\widetilde{H}^j$ and $\widetilde{G}^j$ for each resolution level j (8, 9).

$$V^j = \widetilde{H}^{j+1} V^{j+1} \tag{8}$$

$$W^j = \widetilde{G}^{j+1} V^{j+1} \tag{9}$$

The reconstruction is done with two synthesis filters H^j and G^j (10). Where W^j is the corresponding wavelet coefficients.

$$V^{j+1} = H^j V^j + G^j W^j \tag{10}$$

The analysis and synthesis filters make following assumption (11):

$$\left[\begin{array}{c} \widetilde{H}^j \\ \widetilde{G}^j \end{array} \right] = \left[\begin{array}{cc} H^j & G^j \end{array} \right]^{-1} \tag{11}$$

In [32] this method was extended to processing meshes with additional attributes.

This biorthogonal wavelet basis can be build with lifting scheme notation. This *k-disc* wavelet construction begin with lazy wavelets and then is improved with use of lifting scheme [4].

Based on semi-uniform subdivision methods the prediction and update blocks of lifting scheme as a masks of subdivision schemes where also defined. In [33] the Loop scheme was used.

In [34], [35] the second generation wavelets on sphere were constructed as lifting scheme blocks. The Lazy, Linear, Quadratic, and Butterfly bases are lifted to assure that the wavelet has at least one vanishing moment.

The Loop wavelet transform with MAPS (Multiresolution Adaptive Parameterization of Surfaces, [26]) parametrisation was also introduced in [36]. The MAPS algorithm uses hierarchical simplification to efficiently induce a parameterization of the original mesh over a base domain consisting of a small number of triangles (base mesh M^0). In [37] a new biorthogonal wavelet decomposition using the $\sqrt{3}$ subdivision was described.

Another possibility, in construction of surface wavelets, is adaptation of Burt-Adelson pyramid scheme (Figure 5). This is combination of subdivision (upsampling), half-edge collapse topological operator (downsampling) and smoothing steps. Each step of this schema produces detail vector as a difference between meshes in two next resolution levels. In [38] the Loop subdivision and Taubin smoothing operator as elements of pyramid algorithm were used.

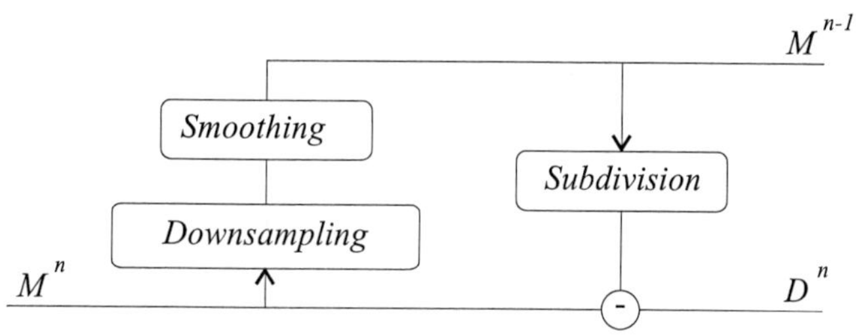

Figure 5. The Burt-Adelson piramid for decomposition of surface mesh.

Method for Irregular Meshes

The BLaC-wavelets (Blending of Linear and Constant) [39], [40] define local decomposition which is generalisation of Haar wavelets for non-nested spaces. The connection of hierarchical Delaunay triangulation allows to build multiresolution model of planar or spherical, irregular surface mesh.

The next proposition in literature is wavelet-based multiresolution analysis of irregular meshes using a new irregular subdivision scheme [41], [42]. This is expansion of the solution presented in [31], [25]. The Authors proposed a new irregular subdivision scheme based on *subdivision cookbook* which defined different combinations of split and merge steps. Analysis and synthesis filters are defined in this same way as for semi-regular meshes but the new inner product was specified because the new subdivision scheme splits triangle on different count of new triangles (from 1 to 4) (12).

$$< f,g >= \sum_{\tau \varepsilon \Delta(M^j)} \frac{K^j(\tau)}{pow(\tau)} \int_{x \varepsilon \tau} f(x)g(x)dx \tag{12}$$

Where $K^j(\tau)$ is count of new triangles from τ triangle of M^j mesh.

Another direction in research is the use of non-uniform subdivision schemes to build wavelet analysis. In [23], [24] the central ingredient of multiresolution analysis is a non-uniform relaxation operator which minimises divided differences. This operator together with the mesh simplification method (progressive mesh [9]) and pyramid algorithm (Burt-Adelson pyramid) allows to define signal processing tools for irregular connected triangle meshes (Figure 5).

In [43] the progressive mesh algorithm is used to build the coarser resolution mesh and semi-uniform discrete Laplacian smoothing operator to estimate the higher resolution mesh. This method is not a wavelet transform but allows define multiresolution representation of surface mesh.

Lifting scheme is used for multiresolution analysis in [44] where a prediction block is non-uniform relaxation operator based on surface curvature minimisation (Meyer smoothing operator). This transformation is not a complete lifting scheme because it does not have update block. This framework also allows processing irregular meshes with attributes [45].

In [46], [47] the lifting scheme framework to processing irregular surface mesh was defined. The proposition of split, predict and update block with influence on result multiresolution representation (scaling and wavelet function) was described. In [48] the extension of lifting scheme for the multiresolution decomposition and reconstruction of irregular triangle surface meshes with additional attributes was presented.

General Lifting Scheme Framework to Multiresolution Transformation of Irregular Triangle Mesh

This section summarises proposition of general transform based on the lifting scheme for multiresolution analysis of irregular, triangle closed mesh (the 2-manifold). Main operations of lifting scheme (Figure 7 and 6) can be singled out:

- Select from input mesh M^{j+1} (where j is the resolution level) the even and odd vertices (split block, S, definition in 13). The odd vertex (v_n) will be removed from mesh. The even vertices are 1-ring neighbourhood of odd vertex ($N(v_n)$).

$$S(M^{j+1}) = (v_n^{j+1}, N(v_n^{j+1})) \tag{13}$$

One step of lifting scheme removes only one vertex (this is "one coefficient at a time" lifting scheme). One of proposition is vertex selection based on the value of criterion which is vertex distance to the regression plane defined by 1-ring neighbourhood of the odd vertex ($S_d(v_n)$). This criterion reflected the curvature of surface very well. So the vertices from plain area of mesh are removed first.

The selected vertex is removed by half-edge collapse. Also the edge of vertex to collapse must be chosen. This is very important and has big influence on quality of result meshes. For every edge of selected vertex (v_n) the sum of differences between triangles normal before and after edge collapse is computed (14). Edge with minimal value β_n will be chose to collapse. This criteria minimize the changes in the curvature of surface in lower resolution.

$$\beta_n(v_n) = \min_{v_c \varepsilon N(v_n)} \sum_{m \varepsilon \Delta(v_n)} (1 - \cos(\alpha_{\Delta_m, \Delta'_m})) \tag{14}$$

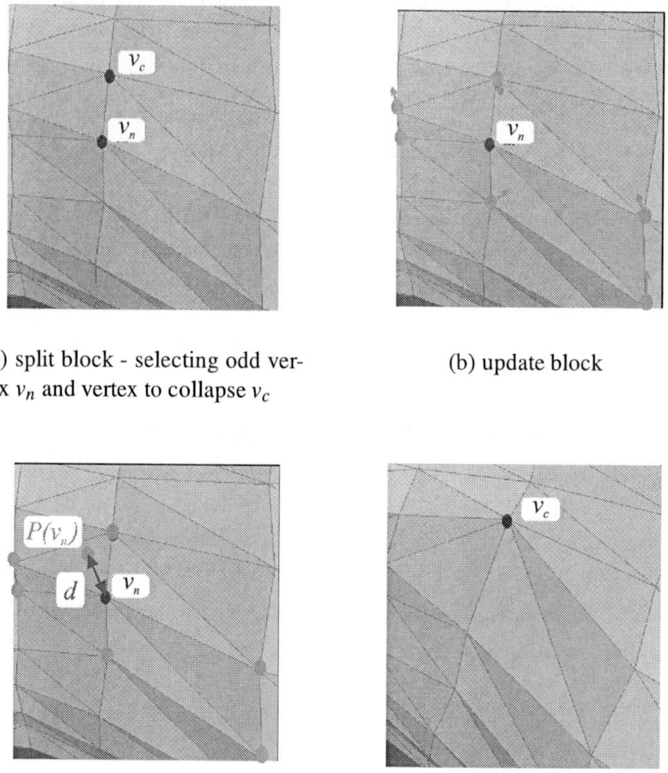

(a) split block - selecting odd vertex v_n and vertex to collapse v_c

(b) update block

(c) predict block and detail computation d

(d) odd vertex remove by half-edge collapse $(v_n, v_c) \rightarrow v_c$

Figure 6. Decomposition lifting scheme steps.

Where $\alpha_{\Delta_m, \Delta'_m}$ is the angle between normal of triangle Δ_m and triangle Δ'_m after half-edge collapse $(v_n, v_c)\ \Delta\ v_c$.

- Update even vertices.

Update block is a simple assurance that the values of the input and output vertices in coarser resolution have the same average. If this step is missing, the even indexed vertices proceed unchanged to the next level. This block smoothes mesh in lower resolution, which influences on better approximation by this mesh and in consequence smaller wavelet coefficients. The update value is added to every vertex in neighbourhood of processed vertex in this lifting scheme step (15).

$$\bigwedge_{v_k \varepsilon N(v_n^{j+1})} v_k^j = v_k^{j+1} + U(v_n^{j+1}, N(v_n^{j+1})) \tag{15}$$

The suggested update block is: equal average value of single coordinates (AVV) before and after odd vertex removal. Odd and even vertices are used in computing so as to avoid allocation memory for additional coefficients in multiresolution model, update block is before prediction block ("update first lifting scheme", [49]).

- Prediction odd vertex based on even ones (predict block, *P*).

  The proposed prediction blocks take advantage of deterministic spatial interpolation methods such as natural neighbours (NNI) and thin plate splines interpolation based on first (TPS1) and second ring neighbourhoods (TPS2). To calculate prediction value the local parameterization of odd vertex and its neighbours should be determined, so the proposed prediction block consists of two parts: parameterization and interpolation. The searched parametric plane is the regression plane (total least squares plane) used also in split block. This is the best-fitting plane to a given set of vertices in 1-ring neighbourhood by minimizing the sum of the squares of the perpendicular distances of the points from the plane.

- Compute detail vector (wavelet coefficient, *d*) as difference between predicted and original vertex (16). Detail is stored in multiresolution model.

$$d^j = v_n^{j+1} - P(N(v_n^{j+1})) \tag{16}$$

  The detail vector is difference between odd vertex and its prediction. This encode the data lost by the next approximations and can be seen as local frequency spectrum of the mesh (output of high pass filter).

- Remove the odd vertex, selected in Split block, from the mesh by an half-edge collapse. The edge of odd vertex to collapse was also chosen in Split block. The output is the coarser mesh M^j that is the approximation of initial mesh.

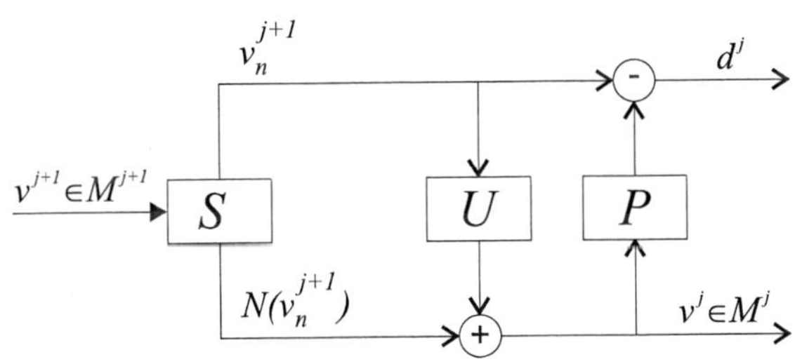

Figure 7. The lifting scheme for decomposition of surface mesh.

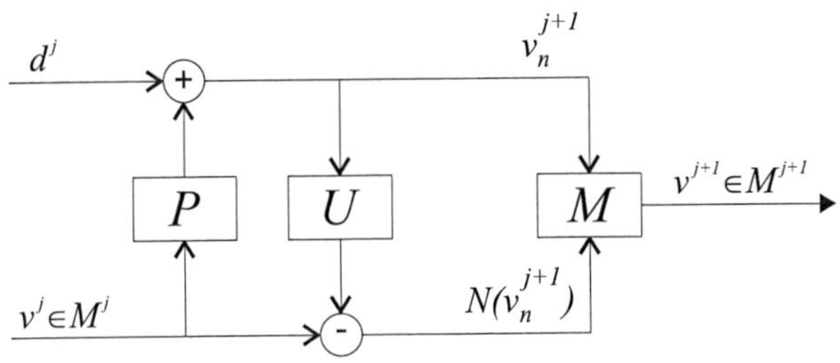

Figure 8. The reverse lifting scheme for reconstruction of surface mesh.

The inverse lifting scheme consists of following operations (Figure 8):

- Prediction odd vertex based on even vertices from mesh M^j.
- Compute original value of odd vertex as sum of detail vector remembered in multiresolution model and prediction vertex.
- Update even vertices.
- Insertion computed vertex to mesh (merge block, M). Reconstruction of topological connections between odd vertex and its neighbors (even vertices). The output is the mesh M^{j+1}.

If functions in split, predict and update blocks are defined with respect to additional attributes in vertices this general scheme will be able to process meshes with additional attributes.

Wavelet and Scaling Function

The lifting scheme is a tool for constructing second generation wavelets, which are no longer dilates and translates of one single function. For irregular settings scaling and wavelets functions depend on local surroundings. For proposed scheme these functions depend on valence of odd vertex. Example scaling function for scheme with this same prediction block but for vertices with different valence are on Figure 9 and Figure 10. The basic idea behind lifting is to start with simple multiresolution analysis and modify it by prediction and update block. This starting analysis is called lazy wavelet, which split the input date into even and odd samples. In filter bank algorithm starting multiresolution analysis is implemented by analysis ($\widetilde{H}^{old}$ and $\widetilde{G}^{old}$) and synthesis filters (H^{old} and G^{old}). To find equations of scaling and wavelet functions, first the modification of input filters by update and prediction blocks of one step of reverse lifting scheme should be defined as in (18 and 18).

$$G^j = G^{j,old} - H^{j,old}U^j \tag{17}$$

$$H^j = G^{j,old}P^j + (1 - U^jP^j)H^{j,old} \tag{18}$$

Modification of scaling function of even vertices are defined in (20 and 20).

$$\Phi^j = \Phi^{j+1}H^j \tag{19}$$

$$\Phi^j_{N(v_n)} = \Phi^j_{N(v_n)} + (\phi^{j+1}_{v_n} - \Phi^j_{N(v_n)}U^j)P^j \tag{20}$$

After removal of odd vertex (v_n) from mesh the area taken by its scaling function $\phi^{j+1}_{v_n}$ is occupied by scaling functions of neighbours vertices $\Phi^j_{N(v_n)}$. Because proposed lifting scheme is update first lifting scheme so on scaling functions influence update and prediction blocks. This can be seen on Figure 9 and Figure 10 for scheme with prediction block NNI and on Figure 11 with $TPS1$ and $TPS2$. In classic scheme the scaling functions are modified only by prediction block.

Similarly the result wavelet functions can be analysed (22 and 22):

$$\Psi^j = \Phi^{j+1}G^j \tag{21}$$

$$\psi^j_{v_n} = \phi^{j+1}_{v_n} - \Phi^j_{N(v_n)}U^j \tag{22}$$

The update block modifies wavelet function to add vanishing moments. If scheme doesn't have update block wavelet function is this same as scaling function of next level (Figure 12).

To find shape of result scaling function the reverse lifting scheme can be used. This scaling function is defined as follows: set values of input data to Kronecker delta and wavelet coefficients equal to 0. After running reverse lifting scheme starting from level j ad infinitum we obtain scaling function $\phi^j_{v_n}$. For wavelet function the detail values are equal Kronecker delta and data set to 0.

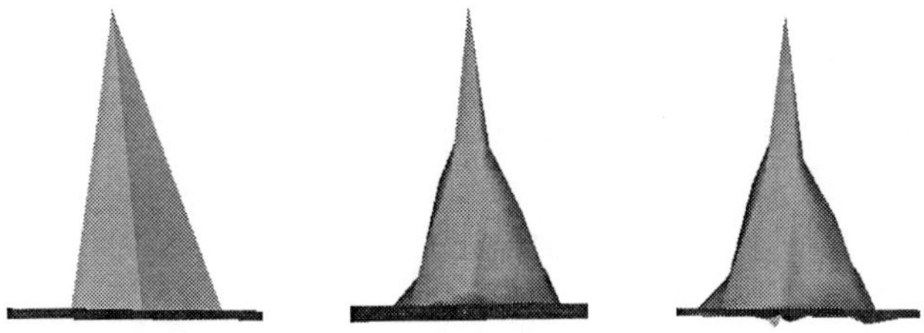

(a) (b) (c)

Figure 9. The scaling function of vertex with valence equals 6 after 268 steps of reverse scheme with prediction block *NNI*: (a) input scaling function, (b) result scaling function for scheme without update block, (c) result scaling function for scheme with update block.

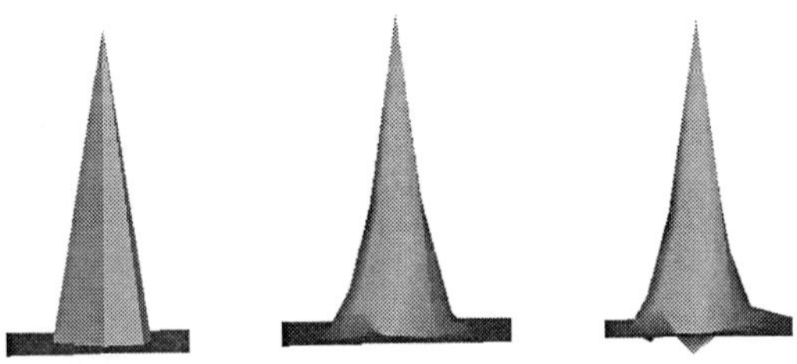

(a) (b) (c)

Figure 10. The scaling function of vertex with valence equals 4 after 268 steps of reverse scheme with prediction block *NNI*: (a) input scaling function, (b) result scaling function for scheme without update block, (c) result scaling function for scheme with update block.

Implementation

The input mesh is processing in main loop until archive requested level of resolution or no more verities can be removed from mesh. The main loop of lifting scheme is written in pseudocode on listing 1. First the split criteria for every vertex is computed. Next is the while loop with calling update, predict function, detail computing, removing one vertex and split criteria updating. The output is the continuous multiresolution model with base mesh and set of detail vectors. Such model supports LOD (for mesh in defined level of resolution) and ROI (for region of mesh in defined level of resolution) questions.

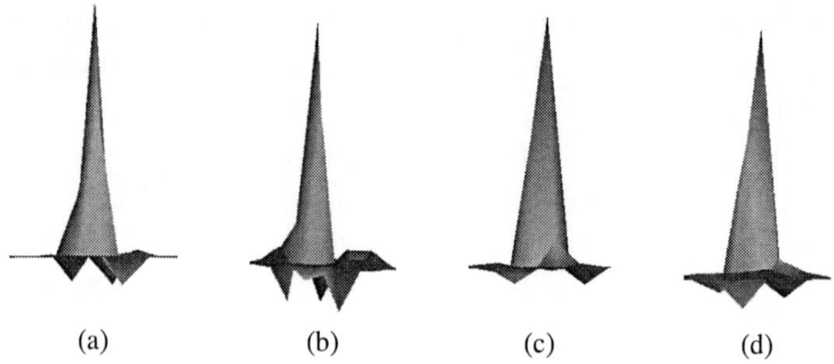

Figure 11. The scaling function of vertex with valence equals 4 after 268 steps of reverse scheme: (a) result scaling function for scheme with prediction block $TPS1$ without update block, (b) result scaling function for scheme with prediction block $TPS1$ with update block, (c) result scaling function for scheme with prediction block $TPS2$ without update block, (d) result scaling function for scheme with prediction block $TPS2$ with update block.

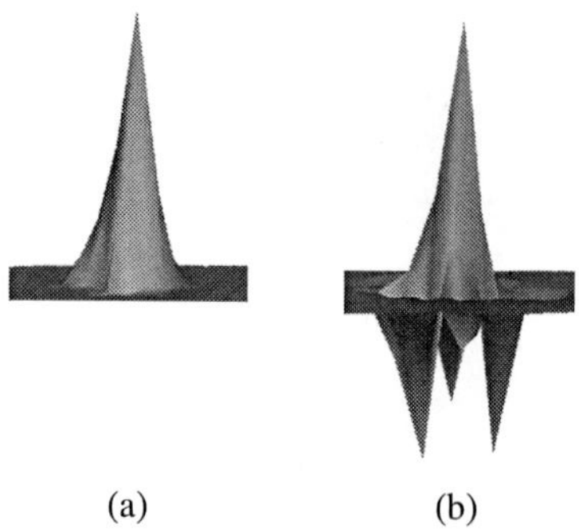

Figure 12. The wavelet function for scheme with prediction block NNI: (a) result wavelet function for scheme without update block, (b) result wavelet function for scheme with update block.

The reverse lifting scheme reconstructs the mesh to defined level of resolution from multiresolution model. Loop is written on listing 2 and consists of calling predict, undo update function and inserting computing vertex to mesh by reconstruction the topological connections.

Application of Multiresolution Representation

In this section the most common applications of multiresolution representation are presented. That algorithms need multiresolution model which store base mesh and detail informations - wavelet coefficients.

Listing 1 Main loop of lifting scheme DOLIFTINGSCHEME

Input:	M	–	triangle mesh
	$MultiM$	–	multiresolution model
	$vCnt$	–	count of vertices to remove
Output:			mesh M after removing $vCnt$ vertices and multiresolution model $MultiM$

function DOLIFTINGSCHEME(M, $vCnt$, $MultiM$):*void*
    for $v_i \in M$ **do** ▷ for every vertex from mesh M
        DOSPLIT(v_i) ▷ set the criteria value
    end for
    while ($vCnt > 0$) and (is possible to remove another vertex from M) **do**
        $v_n \leftarrow$ vertex from M with smallest value of split criteria
        DOUPDATE(v_n, $false$)
        $v_p \leftarrow$DOPREDICT(v_n)
        $v_n \leftarrow v_n - v_p$
        $MultiM \leftarrow MultiM^{-} v_n$ ▷ save detail vector in multiresolution model
        $M \leftarrow M\ ^{\circ}\ v_n$ ▷ remove vertex v_n by half-edge collapse
        $v2Near \leftarrow$ second ring neighbourhoods of v_n
        for $v_i \in v2Near$ **do**
            DOSPLIT(v_i) ▷ update the split criteria value
        end for
        $vCnt \leftarrow vCnt - 1$
    end while
end function

Mesh Compression

Generally meshes obtained by 3D scanners are very complex and big. They usually demand a huge amount of storage space and/or transmission bandwidth. So the most popular kind of application based of wavelets multiresolution representations is compression. In these methods wavelets coefficients of the analysed mesh are encoded using zerotree and entropy coding. More informations about multiresolution meshs compression can be found in [50], [51], [52], [36], [53], [54], [55].

Figure 13 is the example of mesh after the hard threshold, where 50% of smallest wavelets coefficients of lifting scheme multiresolution representation with *NNI* prediction block, were set to 0.

Feature Extraction

Important features (like sharp edges) can be efficiently found by using detail coefficients through different scales or verities from base mash. This can be base of mesh segmentations algorithms [56] or base of classification algorithms to compare and recognise 3D objects [57], [58].

Listing 2 Loop of reverse lifting scheme DOREVERSELIFTINGSCHEME

Input: *MultiM* – multiresolution model
vCnt – count of vertices to process
Output: mesh *MultiM* after
inserting *vCnt* verities (mesh in higher resolution)

function DOREVERSELIFTINGSCHEME(*vCnt*, *MultiM*):*void*
  while ($vCnt > 0$) **do**
    $d \leftarrow$ detail vector from *MultiM*
    $v_p \leftarrow$DOPREDICT(d)
    $d \leftarrow d + v_p$
    DOUPDATE(d, *true*)
    $MultiM \leftarrow MultiM^{-} d$ ▷ insert vertex to mesh
    $vCnt \leftarrow vCnt - 1$
  end while
end function

Figure 13. Mesh of Stanford Bunny after the hard threshold of 50% smallest wavelets coefficients.

Filtering

Most of traditional multiresolution signal techniques can be defined for wavelet multresolution representation of surface mesh.

Example filter is a denosing algorithm using soft-thresholding of wavelet coefficients. First, the initial noisy mesh is decomposed using wavelet analysis. In this case the large coefficients are mainly the shape of surface and the smaller ones represent the noise. By the shrink the detail coefficients and next reconstruction the mesh, the noise can be removed [59].

Another examples are smoothing and enhancement filters. Here also the multiresolution representation is a base of these algorithms. The wavelet coefficients are the detail representations (high frequencies) and next meshes are the approximations of original surface (low frequencies). The smoothing filter sets to zero wavelets coefficients from determined levels of resolutions. Enhancement filter highlights some characteristics features in opposite to smoothing filter. This is available by scaling detail coefficients by multiply then by factor higher than 1. On the Figure 14 we can see the results of applying the smoothing and enhancement filter to multiresolution representation. This representation is the result of lifting scheme algorithm with *NNI* prediction block. This same filters can be applying to multiresolution representation of mesh with additional attributes in verities.

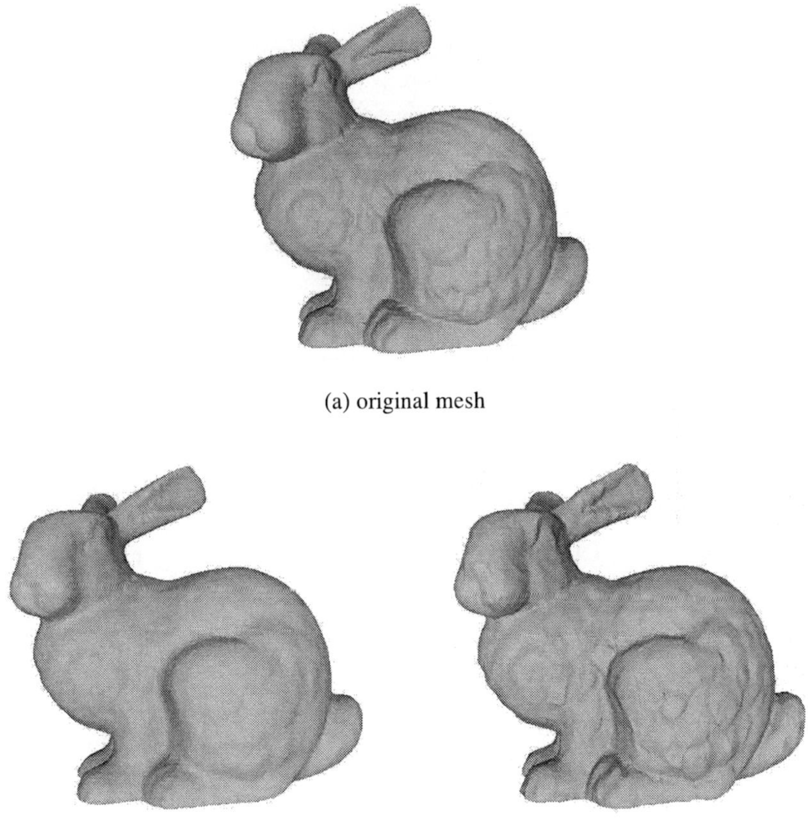

(a) original mesh

(b) smoothing mesh (c) enhancement mesh

Figure 14. Applying of smoothing and enhancement filters to mesh of the Stanford Bunny.

Level of Detail Visualization

When the visualizing complex object is far away from the object or is moving very fast, it is unnecessary and inefficient to draw a highly detailed representation [1], [60]. We can use level of detail (LOD) adaptive visualization where the information about the view determine the complexity of rendering model (Figure 15).

In LOD visualisation all object is in lower resolution. The second possibility is using the region of interest (ROI) visualisation where in higher resolution is only the part of object - pick by user or actually in the viewport. Rest of mesh can be in lower resolution. Example is on Figure 16.

Figure 15. Level of detail (LOD) visualisation dependent of viewer distance.

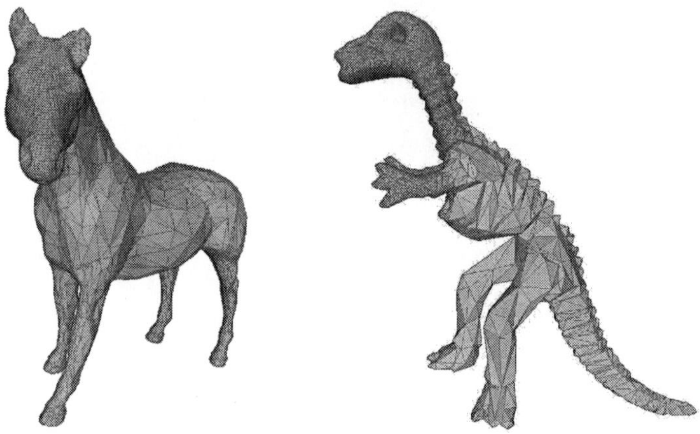

(a) mesh of the Horse model (b) mesh of the Dinozaur model

Figure 16. Region of interest visualisation (ROI).

For this kind of visualization the results of wavelet decomposition must be stored in the continuous multiresolution model. This is a hierarchical structure with base mesh and details from all levels of resolution.

Multiresolution Editing

The wavelet multiresolution representation can be used also to edit shapes at variety of resolutions. The idea is to edit the mesh at the lower resolution and use the reconstruction algorithm to obtain the high resolution mesh based on its edited version. The main advantage of these metod is the ability to edit very large meshes [38], [43].

Progressive Transmission

The complex geometric models are becoming very common in the Internet. The distribution of these geometric object motivates the need for efficient transmissions methods of models across relatively low-bandwidth networks.

The progressive transmission allows gradually send complex model so the user does not have to wait until the entire model is received. First, the base mesh in lowers resolution is send. It is received and displayed very quickly. Next, the wavelets coefficients are transmitted in order of decreasing magnitude [50], [36], [53].

Watermarking

The digital watermarking allows copyright protection of various multimedia contents. This technique hides some secret information in the cover content [61]. Example algorithms for meshes can use wavelets coefficient vectors as authentication primitives [62], [63], [64].

Conclusions

The multiresolution methods for 3D surfaces meshes presented in this chapter do not always respect the strict definitions of the wavelet transform. In lots of examples there is lack of mathematical framework to define and proof of wavelets properties such as coverage, smoothness and stability. There are no mathematical utility to uniformly writing all this methods.

By the examples of using the result multiresolution representation we can see the ability of the multriresolution transformation to efficiently represent and process surface meshes data. This is base of efficients digital geometry processing algorithms and can by used in many 3D graphics applications such as compression, progressive transmission, filtering, feature extractions, LOD visualisations and watermarking.

References

[1] D. Luebke, M. Reddy, J. Cohen, A. Varshney, B. Watson, and R. Huebner. *Level of Detail for 3D Graphics*. Computer Graphics and Geometric Modeling. Morgan Kaufmann, 2002.

[2] L. De Floriani, L. Kobbelt, and E. Puppo. A survey on data structures for level-of-detail models. In N.Dodgson, M.Floater, and M.Sabin, editors, *Advances in Multiresolution for Geometric Modelling*. Springer, 2004.

[3] L. De Floriani and P. Magillo. Multiresolution modeling of three-dimensional shapes. In M. Abdelguerfi, editor, *3D Synthetic Environment Reconstruction*. Kluwer Academic Publishers, 2001.

[4] E.J. Stollnitz, T. DeRose, and D.H. Salesin. *Wavelets for Computer Graphics Theory and Applications*. Morgan Kaufmann, 1996.

[5] P. Schröder, J. Zarge, and W. Lorensen. Decimation of triangle meshes. *Computer Graphics (Proc. SIGGRAPH)*, **26**(2), 1992.

[6] M. Garland and P. Heckbert. Surface simplification using quadric error metrics. *SIGGRAPH*, 1997.

[7] M. Garland and P. Heckbert. Simplifying surfaces with color and texture using quadric error metrics. *IEEE Visualization*, 1998.

[8] H. Hoppe, T. DeRose, T. Duchamp, J. McDonald, and W. Stuetzle. Mesh optimization. *ACM SIGGRAPH*, 1993.

[9] H. Hoppe. Progressive meshes. *ACM SIGGRAPH*, 1996.

[10] H. Hoppe. Efficient implementation of progressive meshes. *Computers and Graphics*, **22**(1), 1998.

[11] E. Puppo and R. Scopigno. Simplification, LOD and Multiresolution Principles and Applications. *EUROGRAPHICS*, 1997.

[12] D.P. Luebke. A Developer's Survey of Polygonal Simplification Algorithms. *IEEE Computer Graphics and Applications*, 2001.

[13] P.S. Hackbert and M. Garland. Survey of Polygonal Surface Simplification Algorithms. *SIGGRAPH 97 Course Notes, ACM Press, New York*, 1997.

[14] C. Gostman, S. Gumhold, and L. Kobbelt. Simplification and compression of 3D meshes. In M.S. Floater, A. Iske, and E. Quak, editors, *Tutorials on Multiresolution in Geometric Modelling*, L. De Floriani. Springer, 2002.

[15] M. Garland. Multiresolution modeling: Survey N future opportunities. *Eurographics, State of the Art Report*, 1999.

[16] C. Loop. Smooth subdivision surfaces based on triangles. Master's thesis, University of Utah, Department of Mathematics, 1987.

[17] L. Kobbelt. $\sqrt{3}$ subdivision. *Computer Graphics Proceedings, Annual Conference Series*, 2000.

[18] D. Zorin, P. Schröder, and W. Sweldens. Interpolating subdivision for meshes with arbitrary topology. *Computer Graphics Proceedings, SIGGRAPH*, 1996.

[19] N. Dyn, D. Levin, and J.A. Gregory. A butterfly subdivision scheme for surface interpolation with tension control. *ACM Trans.*, 1990.

[20] J. Warren and H. Weimer. *Subdivision Methods for Geometric Design: A Constructive Approach*. Morgan Kaufmann, 2001.

[21] L. Kobbelt. Discrete fairing and variational subdivision for freeform surface design. *Visual Comput. J.*, 2000.

[22] L. Kobbelt. A variational approach to subdivision. *Comput. Aid. Geomet. Design*, (13), 1996.

[23] I. Daubechies, I. Guskov, P. Schröder, and W. Sweldens. Wavelets on irregular point sets. *Royal Society*, 1999.

[24] I. Guskov, W. Sweldens, and P. Schröder. Multiresolution signal processing for meshes. *Computer Graphics Proceedings*, 1999.

[25] M. Eck, T. DeRose, T. Duchamp, H. Hoppe, M. Lounsbery, and W. Stuetzle. Multiresolution analysis of arbitrary meshes. *Proceedings of SIGGRAPH*, 1995.

[26] A. Lee, W. Sweldens, P. Schröder, L. Cowsar, and D. Dobkin. MAPS: Multiresolution Adaptive Parameterization of Surfaces. *Proceedings of SIGGRAPH*, 1998.

[27] M. Jansen and P. Ooninck. *Second Generation Wavelets and Applications.* Springer, 2005.

[28] W. Sweldens. The Lifting Scheme: A New Philosophy in Biorthogonal Wavelet Constructions. In A.F. Laine and M. Unser, editors, *Wavelet Applications in Signal and Image Processing III*, pages 68–79. Proc. SPIE 2569, 1995.

[29] W. Sweldens. The lifting scheme: A custom-design construction of biorthogonal wavelets. *Appl. Comput. Harmon. Anal.*, **3**(2):186–200, 1996.

[30] W. Sweldens. The lifting scheme: A construction of second generation wavelets. *SIAM J. Math. Anal.*, **29**(2):511–546, 1997.

[31] J.M. Lounsbery. *Multiresolution analysis for surfaces of arbitrary topological type.* PhD thesis, Department of Mathematics, University of Washington, 1994.

[32] A. Certain, J. Popovic, T. DeRose, T. Duchamp, D. Salesin, and W. Stuetzle. Interactive multiresolution surface viewing. *Proceedings of SIGGRAPH*, 1996.

[33] M. Bertram. Biorthogonal Loop-subdivision wavelets. *Computing*, **72**(1-2), 2003.

[34] P. Schröder and W. Sweldens. Spherical wavelets: Texture processing. In P. Hanrahan and W. Purgathofer, editors, *Rendering Techniques '95*, pages 252–263. Springer Verlag, Wien, New York, August 1995.

[35] P. Schröder and W. Sweldens. Spherical wavelets: Efficiently representing functions on the sphere. *Computer Graphics Proceedings (SIGGRAPH 95)*, pages 161–172, 1995.

[36] A.Khodakovsky, P. Schröder, and W. Sweldens. Progressive geometry compression. In *Proceedings of the 27th annual conference on Computer graphics and interactive techniques*, SIGGRAPH '00, pages 271–278, 2000.

[37] W. Huawei, Q. Kaihuai, and S. Hanqiu. $\sqrt{3}$-subdivision-based biorthogonal wavelets. *IEEE Trans. Vis. Comput. Graph.*, **13**(5):914–925, 2007.

[38] D. Zorin, P. Schröder, and W. Sweldens. Interactive multiresolution mesh editing. Technical Report CS-TR-97-06, Department of Computer Science, Caltech, 1997.

[39] G.P. Bonneau. Multiresolution analysis on irregular surface meshes. *IEEE Trans. Visualiz. Comput. Graph.*, **4**(4), 1998.

[40] G.-P. Bonneau and A. Gerussi. Hierarchical decomposition of datasets on irregular surface meshes. *Proceedings CGI*, 1998.

[41] S. Valette, Y.S. Kim, H.J. Jung, I. Magnin, and R. Prost. A multiresolution wavelet scheme for irregularly subdivided 3D triangular mesh. *IEEE Int. Conf. on Image Processing ICIP99, Japan*, 1999.

[42] S. Valette and R. Prost. Wavelet based multiresolution analysis of irregular surface meshes. *IEEE Trans. Visualiz. Comput. Graph.*, 2004.

[43] L. Kobbelt, S. Campagna, J. Vorsatz, and H.-P. Seidel. Interactive multi-resolution modeling on arbitrary meshes. *ACM SIGGRAPH*, 1998.

[44] M. Roy, S. Foufou, A. Koschan, F. Truchetet, and M. Abidi. Multiresolution analysis for irregular meshes. *Proc SPIE Photonics East, Wavelet Applications in Industrial Processing*, 5266, 2003.

[45] M. Roy, S. Foufou, and F. Truchetet. Multiresolution analysis for irregular meshes with appearance attributes. *In Proceedings of International Conference on Computer Vision and Graphics*, 2004.

[46] A. Szczesna. The multiresolution analysis of triangle surface meshes with lifting scheme. In A. Gagalowicz and W. Philips, editors, *Computer Vision/Computer Graphics Collaboration Techniques, Proceedings of MIRAGE, LNCS 4418*. Springer, 2007.

[47] A. Szczesna. Designing lifting scheme for second generation wavelet- based multiresolution of irregular surface meshes. In *Proceedings of IADIS International Conference on Computer Graphics and Visualization*, 2008.

[48] A. Szczesna. The lifting scheme for multiresolution wavelet-based transformation of surface meshes with additional attributes. In L. Bolc, J.L. Kulikowski, and K.W. Wojciechowski, editors, *ICCVG*, volume 5337 of *Lecture Notes in Computer Science*, pages 487–495. Springer, 2008.

[49] M. Nunes, M. Knight, and G. Nason. Adaptive lifting for nonparametric regression. *Statistics and Computing*, **16**(2), 2006.

[50] G. Taubin. 3D geometry compression and progressive transmission. *Eurographics - State of the Art Report*, 1999.

[51] S. Valette and R. Prost. A wavelet-based progressive compression scheme for triangle meshes : Wavemesh. *IEEE Trans. Visualizat. Comput. Graph.*, **10**(2), 2004.

[52] S. Valette, A. Gouaillard, and R. Prost. Compression of 3D triangular meshes with progressive precision. *Computers and Graphics*, **28**(1), 2004.

[53] C. Roudet, F. Dupont, and A. A. Baskurt. Semi-Regular 3D Mesh Progressive Compression and Transmission based on an Adaptive Wavelet Decomposition. In *Proceedings of SPIE Volume 7248 - Wavelet Applications in Industrial Processing VI*, 2009.

[54] J. Peng, Ch.S. Kim, and C.C.Jay Kuo. Technologies for 3d mesh compression: A survey. *J. Vis. Comun. Image Represent.*, 16:688–733, December 2005.

[55] A. Khodakovsky and I. Guskov. Compression of normal meshes. In *IN GEOMETRIC MODELING FOR SCIENTIFIC VISUALIZATION*, pages 189–206. Springer-Verlag, 2003.

[56] C. Roudet, F. Dupont F., and A. Baskurt. Multiresolution mesh segmentation based on surface roughness and wavelet analysis. volume 6508, page 65082E. SPIE, 2007.

[57] A. Hubeli and M. Gross. Multiresolution feature extraction for unstructured meshes. In *VIS '01: Proceedings of the conference on Visualization '01*. IEEE Computer Society, 2001.

[58] D. Nain, S. Haker, A.F. Bobick, and A. Tannenbaum. Shape-driven 3d segmentation using spherical wavelets. In R. Larsen, M. Nielsen, and J. Sporring, editors, *MICCAI (1)*, volume 4190 of *Lecture Notes in Computer Science*, pages 66–74. Springer, 2006.

[59] M. Roy, S. Foufou, and F. Truchetet. Denoising 3D models with attributes using soft thresholding. *Wavelet Applications in Industrial Processing II*, 2004.

[60] L. De Floriani, E. Puppo, and P. Magillo. A formal approach to multiresolution modeling. In W. Strasser, R. Klein, and R. Rau, editors, *Geometric Modeling: Theory and Practice*. Springer, 1997.

[61] K. Wang, G. Lavoué, F. Denis, and A. Baskurt. A comprehensive survey on three-dimensional mesh watermarking. *IEEE Transactions on Multimedia*, **10**(8):1513–1527, 2008.

[62] K. Wang, G. Lavoué, F. Denis, and A. Baskurt. Hierarchical watermarking of semiregular meshes based on wavelet transform. *IEEE Transactions on Information Forensics and Security*, **3**(4):620–634, 2008.

[63] M.-S. Kim, S. Valette, H.-Y. Jung, and R. Prost. Watermarking of 3D irregular meshes based on wavelet multiresolution analysis. *Lecture Notes on Computer Science, proceedings of IWDW 05*, 2005.

[64] M. Kim, J. Cho, R. Prost, and H. Jung. Wavelet analysis based blind watermarking for 3-d surface meshes. In Y.-Q. Shi and B. Jeon, editors, *IWDW*, volume 4283 of *Lecture Notes in Computer Science*, pages 123–137. Springer, 2006.

In: Wavelets: Classification, Theory and Applications
Editors: Manel del Valle et al.
ISBN: 978-1-62100-252-9

Chapter 4

AUTOCORRELATION REMOVAL FOR SPATIAL STATISTICAL MODELLING AND ITS APPLICATION TO BIOGEOGRAPHICAL DATA

Gudrun Carl*[*] *and Ingolf Kühn
UFZ - Helmholtz Centre for Environmental Research,
Department Community Ecology (BZF), Germany

ABSTRACT

Biogeographical studies are often based on a statistical analysis of data sampled in a spatial context. If data sampled at adjacent locations are more likely to be similar to each other than distant ones, this is called spatial autocorrelation. However, in many cases standard analyses such as regression models may yield incorrect results, because the presence of autocorrelation violates the assumption of independently and identically distributed errors.

First, we show that the theory of wavelets provides a method to remove autocorrelation in generalized linear models. Autocorrelation can be described by smooth wavelet coefficients at small scales. Therefore, data can be decomposed into uncorrelated and correlated parts. Using an appropriate linear transformation, we are able to extend generalized linear models to autocorrelated data. We illustrate our new method, called the wavelet-revised model, by applying it to multiple regressions with response variables conforming to various distributions. We recommend wavelet-revised models as a method for lattice data and applicable without any prior knowledge of the real autocorrelation structure.

Second, we deal with the problem that statistical models may suffer not only from autocorrelation, but also from collinearity, i.e., predictor variables in a multiple linear regression can be highly correlated with each other. In the presence of collinearity among predictors, methods modelling, e.g., with latent variables such as partial least squares regression are recommended in literature. We introduce a novel approach coping simultaneously with both problems: collinearity and autocorrelation. Therefore, we develop an extension to partial least squares based on wavelet-revised modelling. We

[*] E-mail address: gudrun.carl@ufz.de

illustrate our new method, called wavelet-revised model for partial least squares (WRMPLS), by applying it to multiple regressions with responses of both normal and binary distribution. Its predictive performance is evaluated for simulated data.

INTRODUCTION

Biogeographical studies are often based on a statistical analysis of data sampled in a spatial context. The analysis of such spatial data is complicated due to spatial autocorrelation [1, 2, 3]. Positive spatial autocorrelation occurs when adjacent data points (i.e., data sampled at adjacent locations) are more likely to be similar than distant ones. Standard statistical methods such as Generalized Linear Models (GLMs), however, may yield incorrect results, because the presence of autocorrelation violates the basic assumption of independently and identically distributed (i.i.d.) errors.

The aim of this chapter is to present a wavelet analysis removing the effect of spatial autocorrelation in such models. To this end, a pre-filtering process will be carried out by means of wavelets and a procedure of *parametric* regression, i.e., a method for estimating a finite number of parameters, will follow this filter. Note that this concept is quite different from *nonparametric* regression, i.e., denoising and adapting to unknown smoothness, see, e.g., [4]. It is well-known that wavelet techniques provide an effective tool for nonparametric regression. There denoising is done by wavelet shrinkage; i.e., by thresholding techniques in the wavelet domain. Flandrin shows the decorrelation property of wavelets [5]. He derives formulas for the correlation structure of wavelet coefficients; i.e., in the wavelet domain. Furthermore, Johnstone and Silverman provide an extension of thresholding methods to deal with correlated data [6]. They recommend level dependent thresholding for denoising. Our case of parametric regression, however, is different in various aspects: First, denoising by means of wavelets is not what we want to do here. Instead, the amount of error or noise is estimated by the parametric regression itself. Second, thresholding for wavelet coefficients is not what we are allowed to do here. Thresholding would alter estimates for parameters. Instead, projection into an orthogonal subspace specified by a linear wavelet transform ensures that estimates are unaltered. Third, decorrelation for the wavelet coefficients is not what we need here. Instead, errors should be uncorrelated in the data domain.

A variety of methods have been developed to correct effects of *spatial autocorrelation* in normal linear models, but only a few are applicable for non-Gaussian distributed response variables. Dormann et al. presented an overview of such modelling approaches [7]. Our wavelet analysis is also applicable to a regression in which a response variable has a non-normal distribution [8, 9]. Therefore, it is an extension of the GLM. To the best of our knowledge, wavelets have not been used to remove spatial autocorrelation in GLMs. Though Fadili and Bullmore provide a wavelet-based method for linear regressions in the context of autocorrelated errors [10], it is applicable only for normal linear models.

Our new technique, called the Wavelet-Revised Model (WRM), is developed for (regular gridded) lattice data. It is applicable without any prior knowledge of the real autocorrelation structure. The WRM is illustrated by its application to a real spatial dataset comprising the geographic distribution of plants in Germany. Moreover, for the purpose of assessing model performance, we apply our method to simulated datasets and analyze the validity of their hypothesis tests. While previously we have tested our model against other methods and proved to be among the best performing methods [9, 11], we here investigate how the choice of different types of wavelets and different resolution levels make an impact on correlograms and error calibration curves.

In the second part of this chapter, we deal with the problem that statistical models may additionally suffer from *collinearity*. This means that predictor variables in multiple regressions can be highly correlated with each other [12, 13]. For instance, in species distribution models (or habitat models), the abundance or occurrence of species is regressed on environmental and/or geographical predictors. In such spatial analyses, two undesirable effects frequently occur in multiple linear regressions: spatial autocorrelation and collinearity [14]. While autocorrelation in the strict sense is a small scale process, predictors are highly collinear in case that they are strongly related to each other across their full range. This is often the case when the environment exhibits strong spatial trends, e.g. temperature is strongly correlated with latitude and both are used as predictors. Both challenging aspects of modelling can have a substantial impact on the predictive performance of such models. However, the alternatives proposed to generalized linear models so far are different in both cases and methods tackling both problems are missing.

Several approaches to deal with the problem of collinearity have been established. Among them, methods modelling with latent variables are recommended in literature. A frequently used and appropriate method is Partial Least Squares (PLS) regression [15, 16].

We introduce a novel approach coping simultaneously with both problems. Therefore, we develop an extension to partial least squares based on wavelet-revised modelling. We illustrate our new method, called wavelet-revised model for partial least squares (WRMPLS), by applying it to multiple regressions with responses of both normal and binary distribution. In order to assess its predictive performance for simulated data, we investigate whether the accuracy of predictions can be influenced by different degrees of collinearity and autocorrelation in training and test datasets.

Notation: Throughout this article, matrices will be denoted by boldface uppercase letters and column vectors by boldface lowercase letters. Transposition of a matrix $\mathbf{A}$ is shown by $\mathbf{A}'$.

1. Wavelets

1.1. The Wavelet Approximation

Wavelets are analyzing functions fulfilling certain conditions, e.g., being orthogonal basis functions. For an orthogonal wavelet basis denoted by

$$((\psi_{j,k})_{1\le j\le J,k\in Z},(\phi_{J,k})_{k\in Z}), \tag{1}$$

where ψ is a compactly supported "mother" wavelet, which integrates to zero $\int \psi(x)dx = 0$, ϕ is the associated scaling function or "father" wavelet, which integrates to one

$\int \phi(x)dx = 1$, and J a suitable chosen decomposition level, the functions $\phi_{j,k}(x)$ and $\psi_{j,k}(x)$ are generated from ψ and ϕ via scaling and translation as follows:

$$g_{j,k}(x) = 2^{-j/2} g(2^{-j} x - k) \tag{2}$$

Scaling and translation produce wavelets to different levels j and shifts k, i.e., dilations and locations. For basics on wavelets, we refer to [17, 18, 19].

The mother wavelets are used to describe the detail and high-frequency parts of given data, whereas the father wavelets are used to describe the smooth and low-frequency parts. The oldest and simplest example of a function ψ is the Haar function

$$\psi(x) = \begin{cases} 1 & 0 \le x < 1/2 \\ -1 & 1/2 \le x < 1 \\ 0 & \text{otherwise.} \end{cases} \tag{3}$$

(However, wavelets do not need to have an analytic form in general.) Using such a wavelet basis (1), orthogonal series representations of signals can be performed.

The orthogonal wavelet series approximation [20] for a function $f(x)$ is given by

$$f(x) = \sum_{j=1}^{J} \sum_{k} d_{j,k} \psi_{j,k}(x) + \sum_{k} s_{J,k} \phi_{J,k}(x). \tag{4}$$

This wavelet transform calculates the wavelet transform coefficients $d_{1,k}, \ldots, d_{J,k}, s_{J,k}$ of the wavelet expansion (4). High magnitude of a wavelet transform coefficient means high similarity between the function $f(x)$ and the corresponding wavelet. The information of function $f(x)$ is completely preserved in a minimal set of wavelet coefficients if the used wavelet family is an orthogonal one.

Due to the character of wavelets, the coefficients $d_{j,k}$ (coefficients d for short) of equation (4) describe the detail parts of the shape of function $f(x)$, while the coefficients $s_{J,k}$ (coefficients s for short) represent its smooth part. In general, detail coefficients d have to be calculated at different levels j in contrast to smooth coefficients s.

Thus, changes in magnitude of f are captured in detail coefficients for different levels $j = 1, \ldots, J$. Parts of f which are not included so far are mapped in smooth coefficients s of last level J.

Note that higher dilation levels mean lower resolution. The level of lowest resolution J can arbitrarily be chosen. Since the full information of f is captured in the coefficients d and

s, it is possible to reconstruct the original function f by applying the inverse wavelet transform.

1.2. The Two-Dimensional Wavelet Approach

In order to make wavelet analysis applicable to spatial statistics, we have to take into account the spatial, two-dimensional structure of lattice data and their statistical origin, i.e. the discrete structure of sampled data.

Therefore, first we have to use the *two-dimensional* (2-D) wavelet approximation of a function $F(x, y)$:

$$F(x,y)=\sum_{m=1}^{3}\sum_{j=1}^{J}\sum_{k_x,k_y} d^{m}_{j,k_x,k_y}\Psi^{m}_{j,k_x,k_y}(x,y)+\sum_{k_x,k_y} s_{J,k_x,k_y}\Phi_{J,k_x,k_y}(x,y), \quad (5)$$

where the functions Ψ^{m}_{j,k_x,k_y} and Φ_{J,k_x,k_y} are given orthogonal 2-D wavelet functions and m corresponds to 3 different spatial directions, i.e., to wavelets operating horizontally, vertically or diagonally [21]. Similarly, as described above, the wavelets Ψ are used to detect the detail and high-frequency parts of given data, whereas the wavelets Φ are used to detect the smooth and low-frequency parts. Accordingly, the wavelet coefficients d represent the detail part, while the coefficients s represent the smooth part of F [22]. Multiresolution analysis is a decomposition of equation (5) into orthogonal image components:

$$F^{i}(x,y)=D^{m}_{j}(x,y)=\sum_{k_x,k_y} d^{m}_{j.k_x,k_y}\Psi^{m}_{j,k_x,k_y}(x,y)\ ,\ i=1,\ldots,3J \quad (6)$$

$$F^{3J+1}(x,y)=S_{J}(x,y)=\sum_{k_x,k_y} s_{J,k_x,k_y}\Phi_{J,k_x,k_y}(x,y) \quad (7)$$

Thus the function F can be reconstructed by a sum of all $3J+1$ image components.

Second, the *discrete* wavelet transform calculates the coefficients for a finite set of discrete data points. It is equivalent to a matrix multiplication. Thus the 2-D discrete wavelet transform enables us to transform discrete image data, i.e., a matrix $\mathbf{F}$, into a matrix consisting of all wavelet coefficients in a specific hierarchical order. Without loss of generality, we can assume that both are $(2^{J}\times 2^{J})$ matrices. This decomposition allows us to analyze 2-D data such as a matrix or a geographical pattern of an ecological or an environmental variable. Here again, decomposition and reconstruction is possible by means of $3J+1$ orthogonal lattice components:

$$\mathbf{F}=\sum_{i=1}^{3J+1}\mathbf{F}^{i}=\sum_{m=1}^{3}\sum_{j=1}^{J}\mathbf{D}^{m}_{j}+\mathbf{S}_{J}\ , \quad (8)$$

where m, j, and i are summation indices and J is the resolution level [23].

1.3. Autocorrelation Removal via Haar Wavelets

If locations close to each other are more similar than those further apart, this is represented in the smoothness as a specific feature of a "signal value" of these data. Thus it represents autocorrelation at a certain scale. In spatial statistics it is desirable to revise data by removing such smooth components, i.e. removing spatial autocorrelation. For this purpose, wavelets can be used as a feature extraction tool.

If this data preparation is carried out by means of, for example, 2-D Haar wavelets (for 1-D: see Eq. (3)) then it has a very intuitive meaning. This procedure basically averages data values within squared subareas and subtracts this average from the data values to remove autocorrelation [11]. Let us demonstrate the Haar wavelet analysis of the following data matrix $\mathbf{F}$:

$$\mathbf{F} = \begin{pmatrix} f_{11} & f_{12} & f_{13} & f_{14} \\ f_{21} & f_{22} & f_{23} & f_{24} \\ f_{31} & f_{32} & f_{33} & f_{34} \\ f_{41} & f_{42} & f_{43} & f_{44} \end{pmatrix}$$

In the first step the matrix is simplified by calculating the average value of the resulting non-overlapping 2×2 blocks of cells. Therefore, the so-called smooth matrix $\mathbf{S}_1$ is:

$$\mathbf{S}_1 = \begin{pmatrix} a & a & b & b \\ a & a & b & b \\ c & c & d & d \\ c & c & d & d \end{pmatrix}$$

where $a = average(f_{11}, f_{12}, f_{21}, f_{22})$, $b = average(f_{13}, f_{14}, f_{23}, f_{24})$, etc.

The sum of the detail matrices $\sum_{m=1}^{3} \mathbf{D}_1^m$ gives the difference to the original matrix $\mathbf{F}$:

$$\sum_{i=1}^{3} \mathbf{F}^i = \sum_{m=1}^{3} \mathbf{D}_1^m = \begin{pmatrix} f_{11}-a & f_{12}-a & f_{13}-b & f_{14}-b \\ f_{21}-a & f_{22}-a & f_{23}-b & f_{24}-b \\ f_{31}-c & f_{32}-c & f_{33}-d & f_{34}-d \\ f_{41}-c & f_{42}-c & f_{43}-d & f_{44}-d \end{pmatrix}$$

In the second step we proceed analogously, now averaging 4×4 blocks. Therefore, the smooth matrix $\mathbf{S}_2$ on level 2

$$\mathbf{S}_2 = \begin{pmatrix} e & e & e & e \\ e & e & e & e \\ e & e & e & e \\ e & e & e & e \end{pmatrix}$$

is associated with the sum of the detail matrices $\sum_{m=1}^{3}\sum_{j=1}^{2}\mathbf{D}_j^m$

$$\sum_{i=1}^{6}\mathbf{F}^i = \sum_{m=1}^{3}\sum_{j=1}^{2}\mathbf{D}_j^m = \begin{pmatrix} f_{11}-e & f_{12}-e & f_{13}-e & f_{14}-e \\ f_{21}-e & f_{22}-e & f_{23}-e & f_{24}-e \\ f_{31}-e & f_{32}-e & f_{33}-e & f_{34}-e \\ f_{41}-e & f_{42}-e & f_{43}-e & f_{44}-e \end{pmatrix}$$

More formally this procedure can be explained as follows: A two-dimensional discrete Haar wavelet analysis can be performed, provided that data are given as a $2^n \times 2^n$ matrix. On level J we divide this original matrix $\mathbf{F}$ in $2^J \times 2^J$ submatrices and construct a smooth $2^n \times 2^n$ matrix $\mathbf{S}_J$ by assigning the mean value within each submatrix to each element of the submatrix. Then the sum of the detail matrices $\sum_{i=1}^{3J}\mathbf{F}^i$ is the difference between original matrix and smooth matrix

$$\sum_{i=1}^{3J}\mathbf{F}^i = \mathbf{F} - \mathbf{S}_J \tag{9}$$

1.4. Autocorrelation Removal via Wavelets

Generally speaking, the full information of a signal can be captured in coefficients which modify the amplitudes of wavelets, and vice versa: one can reconstruct the original signal using all coefficients and the appropriate set of wavelets. Moreover, the information on signal properties can be manipulated by selecting a subset of coefficients and neglecting the other part. Such data manipulation can be executed by the following three steps: wavelet transform, wavelet coefficient selection, and back transform.

We will show that we can detect and extract the feature of autocorrelation by means of wavelet analysis. Because the original data can be regarded as a sum of detail and smooth components on a certain resolution level, the smoothness as data feature is captured separately. Smoothness implies that data values of locations close to each other are more similar than those further apart. Thus it represents autocorrelation at a certain scale. Detail components, however, are data adjusted for autocorrelation. In order to extract short-range autocorrelation in its strict sense, the subareas should be small, i.e., the resolution level J should be low. Thus we remove the smooth component $\mathbf{S}_1$ or $\mathbf{S}_2$. This means that J is 1 or 2 in our applications.

Therefore, changing from equation (8) to equation (9), i.e., transform

$$\mathbf{F} \rightarrow \mathbf{PF} = \sum_{i}^{3J} \mathbf{F}^i \tag{10}$$

provides a tool to remove autocorrelation, at least in a good approximation. Using this feature extraction tool we obtain manipulated lattice data below referred to as wavelet-revised data. Because removing the smooth matrix $\mathbf{S}_J$ and keeping only detail matrices $\mathbf{D}_j^m$ can be carried out by the three steps, namely 2-D discrete wavelet transform, wavelet coefficient selection, and 2-D discrete back transform, $\mathbf{PF}$ is mathematically equivalent to a matrix multiplication. Hence, transform $\mathbf{P}$ is a linear transformation.

Because the aim of this chapter is to present a wavelet analysis removing the effects of spatial autocorrelation in multiple regression, it remains to be clarified how this data revision can be embedded into the framework of multiple regressions, such as generalized linear models and partial least squares regressions.

2. Spatial Autocorrelation in Linear Regression

2.1. Generalized Least Squares

Suppose we have the normal linear model usually written in the form:

$$\mathbf{y} = \mathbf{X}\boldsymbol{\beta} + \boldsymbol{\varepsilon} \tag{11}$$

where $\mathbf{y}$ is an $(n \times 1)$ vector of responses, $\mathbf{X}$ is an $(n \times p)$ matrix of predictors, $\boldsymbol{\beta}$ is a $(p \times 1)$ vector of unknown parameters, and $\boldsymbol{\varepsilon}$ is an $(n \times 1)$ vector of normally distributed errors. Assume further that the expected value of $\boldsymbol{\varepsilon}$ is $E(\boldsymbol{\varepsilon}) = \mathbf{0}$ and the variance-covariance-matrix is $\mathrm{var}(\boldsymbol{\varepsilon}) = \mathbf{V}$. If errors are correlated and matrix $\mathbf{V}$ has nonzero off-diagonal elements, then the standard least squares assumptions are not valid and the Ordinary Least Squares (OLS) estimate of $\boldsymbol{\beta}$

$$\mathbf{b} = (\mathbf{X}'\mathbf{X})^{-1}\mathbf{X}'\mathbf{y} \tag{12}$$

and the corresponding vector of fitted values

$$\hat{\mathbf{y}} = \mathbf{X}(\mathbf{X}'\mathbf{X})^{-1}\mathbf{X}'\mathbf{y} \tag{13}$$

are no longer optimal.

Instead of that, a transformed model $\mathbf{Py} = \mathbf{PX\beta} + \mathbf{P\varepsilon}$, where $\mathbf{P}$ is an $(n \times n)$ matrix, $E(\mathbf{P\varepsilon}) = \mathbf{0}$ and $\mathrm{var}(\mathbf{P\varepsilon}) = \mathbf{P}E(\mathbf{\varepsilon\varepsilon}')\mathbf{P}' = \mathbf{PVP}'$, can help to satisfy the standard least squares assumptions. If $\mathbf{P}$ is chosen to decorrelate the errors in the model, then we have approximately $\mathbf{PVP}' = \sigma^2\mathbf{I}$. Therefore, fitting by OLS leads to the following normal equations for $\mathbf{b}$, i.e. the estimate of the original $\mathbf{\beta}$:

$$(\mathbf{PX})' \cdot (\mathbf{Py} - \mathbf{PX}\,\mathbf{b}) = \mathbf{0}. \tag{14}$$

If $\mathbf{X}$ and $\mathbf{PX}$ have full rank, then this equation has the unique solution

$$\mathbf{b} = ((\mathbf{PX})'(\mathbf{PX}))^{-1}(\mathbf{PX})'\mathbf{Py} \tag{15}$$

Therefore, the Generalized Least Squares (GLS) estimator $\mathbf{b}$ can be obtained using OLS on the transformed response $\mathbf{Py}$ and the transformed matrix of predictors $\mathbf{PX}$.

2.2. Generalized Linear Models

In order to extend this well-known concept of GLS [24] to Generalized Linear Models (GLM), we have to emphasize the crucial steps for GLM. In GLM, the response variables can have distributions other than the normal (or Gaussian) distribution. First, consider the method of maximum likelihood as basis for parameter estimation. Second, use the (canonical) link $g(\mu_i) = \mathbf{x}_i'\mathbf{\beta} = \eta_i$, $i = 1, 2, \ldots, n$ with the expected value of response $E(y_i) = \mu_i$ and sample size n. For parameter estimation, we need the following partial derivative of the log likelihood ℓ_{GLM}:

$$\frac{\partial \ell_{GLM}}{\partial \mathbf{\mu}} = [\mathbf{V}(\mathbf{\mu})]^{-1}(\mathbf{y} - \mathbf{\mu}) \tag{16}$$

Since the responses are uncorrelated by assumption in GLM, the off-diagonal elements of the variance-covariance matrix $\mathbf{V}$ are approximately zero. Thus $\mathbf{V}$ can be substituted by a diagonal matrix $\mathbf{A} = diag\{v_{ii}\} = diag\{\mathrm{var}(y_i)\}$. The score $\mathbf{U}_{GLM}$ can be obtained by differentiation of equation (16) with respect to $\mathbf{\beta}$. Hence the score equation is

$$\mathbf{U}_{GLM} = (\mathbf{MX})' \cdot (\mathbf{A}^{-1}(\mathbf{y} - \mathbf{\mu})) = \mathbf{0}, \tag{17}$$

where $\mathbf{M}$ is a diagonal matrix $\mathbf{M} = diag\{\partial\mu_i / \partial\eta_i\}$. Let $\mathbf{W}$ denote a diagonal weight matrix $\mathbf{W} = diag\{v_{ii}^{-1}(\partial\mu_i / \partial\eta_i)^2\}$. The matrix $E(\mathbf{UU}')$ is called information matrix $\mathbf{i}$. This matrix is given by

$$\mathbf{i}_{GLM} = \mathbf{X}'\mathbf{WX}. \tag{18}$$

Using $\mathbf{U}_{GLM}$ (see Eq. (17)) and $\mathbf{i}_{GLM}$ (see Eq. (18)) we obtain the iterative solution

$$\mathbf{b}^{(m)} = (\mathbf{X}'\mathbf{WX})^{-1}\mathbf{X}'\mathbf{Wz}, \tag{19}$$

where $\mathbf{b}^{(m)}$ is the vector of estimates at the mth iteration [24, 25]. The vector $\mathbf{z}$ can be written as

$$\mathbf{z} = \mathbf{Xb}^{(m-1)} + \mathbf{W}^{-1/2}\mathbf{A}^{-1/2}(\mathbf{y} - \boldsymbol{\mu}). \tag{20}$$

Therefore, the GLM estimator $\mathbf{b}$ of equation (19) can be obtained iteratively using OLS (see Eq. (12)) on the adjusted matrix of predictors $\mathbf{W}^{1/2}\mathbf{X}$ and the adjusted dependent variable $\mathbf{W}^{1/2}\mathbf{z}$, where $\mathbf{z}$ is given by equation (20).

2.3. Wavelet-Revised Models

In case of correlated observations, the score equation (17) can be extended to the generalized estimating equation (GEE) $\mathbf{U}_{GEE} = (\mathbf{MX})'\mathbf{V}^{-1}(\mathbf{y} - \boldsymbol{\mu}) = \mathbf{0}$, see, e.g., in [25]. The matrix $\mathbf{V}$ can be written as $\mathbf{V} = \mathbf{A}^{1/2}\mathbf{RA}^{1/2}$, where $\mathbf{R}$ is the correlation matrix $\mathbf{R} = E[\mathbf{A}^{-1/2}(\mathbf{y} - \boldsymbol{\mu})(\mathbf{y} - \boldsymbol{\mu})'\mathbf{A}^{-1/2}]$. Therefore, the score is also given by $\mathbf{U}_{GEE} = (\mathbf{W}^{1/2}\mathbf{X})'\mathbf{R}^{-1}\mathbf{A}^{-1/2}(\mathbf{y} - \boldsymbol{\mu})$. Generalizing this score according to the linear transformation $\mathbf{P}$ of GLS we are led to transformations $\mathbf{A}^{-1/2}(\mathbf{y} - \boldsymbol{\mu}) \rightarrow \mathbf{PA}^{-1/2}(\mathbf{y} - \boldsymbol{\mu})$ and $\mathbf{W}^{1/2}\mathbf{X} \rightarrow \mathbf{PW}^{1/2}\mathbf{X}$. Here the correlation matrix is transformed as follows $\mathbf{R} \rightarrow E[\mathbf{PA}^{-1/2}(\mathbf{y} - \boldsymbol{\mu})(\mathbf{y} - \boldsymbol{\mu})'\mathbf{A}^{-1/2}\mathbf{P}'] = \mathbf{PRP}'$. If this transformed correlation matrix is approximately the identity matrix $\mathbf{I}$, then the transformed score equation is given by

$$\mathbf{U} = (\mathbf{PW}^{1/2}\mathbf{X})' \cdot \mathbf{PA}^{-1/2}(\mathbf{y} - \boldsymbol{\mu})) = \mathbf{0}. \tag{21}$$

A vector $\mathbf{U}$ is identified as 'quasi-score function' provided that the following three properties hold [26]. (i) The vector $\mathbf{U}$ has a multivariate normal distribution, at least asymptotically. (ii) Its expected value is $E(\mathbf{U}) = \mathbf{0}$. (iii) Its variance-covariance matrix is equivalent to the negative expected value of the partial derivatives of $\mathbf{U}$ with respect to $\boldsymbol{\beta}$, i.e. $E(\mathbf{UU}') = -E[\partial \mathbf{U}(\boldsymbol{\beta})/\partial \boldsymbol{\beta}]$. Condition (ii) is fulfilled because $E(\mathbf{y}) = \boldsymbol{\mu}$. If we assume that $\mathbf{PRP}' = \mathbf{I}$, then the information matrix is

$$\mathbf{i} = E(\mathbf{UU}') = (\mathbf{PW}^{1/2}\mathbf{X})' \cdot (\mathbf{PW}^{1/2}\mathbf{X}) \tag{22}$$

and condition (iii) is fulfilled. In order to verify condition (i), we may assume that there is only one parameter β. Then $U_{GLM}/\sqrt{i_{GLM}}$ is normally distributed and $U/\sqrt{i}$ is a linear combination of $U_{GLM}/\sqrt{i_{GLM}}$. Thus $U/\sqrt{i}$ is normally distributed as well. Moreover, if $\mathbf{PRP}' = \mathbf{I}$, then $U/\sqrt{i} \sim N(0,1)$, at least asymptotically.

Then it follows from the above that $\mathbf{U}$ given by equation (21) is a quasi-score function. Thus $\mathbf{U}$ and $\mathbf{i}$ (see Eq. (22)) can be used for the method of scoring. Therefore, we obtain the iterative solution

$$\mathbf{b}^{(m)} = ((\mathbf{PW}^{1/2}\mathbf{X})'\ \mathbf{PW}^{1/2}\mathbf{X})^{-1}\ (\mathbf{PW}^{1/2}\mathbf{X})'\ \mathbf{PW}^{1/2}\mathbf{z}\ , \tag{23}$$

where

$$\mathbf{PW}^{1/2}\mathbf{z} = \mathbf{PW}^{1/2}\mathbf{X}\ \mathbf{b}^{(m-1)} + \mathbf{PA}^{-1/2}(\mathbf{y} - \boldsymbol{\mu}). \tag{24}$$

From equation (22), the asymptotic variance-covariance matrix of $\mathbf{b}$ is given by $\text{var}(\mathbf{b}) = \mathbf{i}^{-1} = ((\mathbf{PW}^{1/2}\mathbf{X})'\ \mathbf{PW}^{1/2}\mathbf{X})^{-1}$. If $\mathbf{X}$ and $\mathbf{PW}^{1/2}\mathbf{X}$ have full rank, then equation (23) has a unique solution.

Summarizing the above, the estimator $\mathbf{b}$ of equation (23) can be obtained iteratively using OLS (see Eq. (12)) on the adjusted matrix of predictors $\mathbf{PW}^{1/2}\mathbf{X}$ and the adjusted dependent variable $\mathbf{PW}^{1/2}\mathbf{z}$ given by equation (24). If $\mathbf{P}$ is equivalent to the identity matrix $\mathbf{I}$, then our quasi-score concept is equivalent to a standard GLM result. Recall that in standard GLM theory, the responses are assumed to be independent, i.e., the variance-covariance matrix of $\mathbf{A}^{-1/2}(\mathbf{y} - \boldsymbol{\mu})$ are assumed to be equivalent to the identity matrix $\mathbf{I}$. In general, however, responses as well as predictors and errors are autocorrelated rather than independent random variables. Therefore, matrix $\mathbf{P}$ has to ensure that the variance-covariance matrix of the variables $\mathbf{PA}^{-1/2}(\mathbf{y} - \boldsymbol{\mu})$ is close to the identity matrix $\mathbf{I}$. As we showed in subsection 1.4., one can achieve this goal by using the concept of wavelet decomposition, this means, in particular, by using transform $\mathbf{P}$ given by expression (10).

Therefore, we derive a spatial extension of the non-spatial GLM, including tools for parameter estimation and statistical inference. The equations (23) and (24) in conjunction with transform (10) form our new method. If observations are originally autocorrelated, the use of these equations should lead to improved parameter estimates compared to standard GLM results.

Our method is named Wavelet-Revised Model (WRM), since it is based on a wavelet-supported revision of data. For further details, we refer the reader to [9].

2.4. WRM Algorithm

Summarizing the content of the previous sections we give the following algorithm for calculating regression parameters via WRM:

Step 1. Find an initial estimate $\mathbf{b}^{(0)}$.

Step 2. Use $\mathbf{b}^{(0)}$ to obtain $\boldsymbol{\mu}_0$.

Step 3. Calculate $\mathbf{A}_0$, $\mathbf{W}_0$, $\mathbf{X}_{\text{new}} = \mathbf{W}_0^{1/2}\mathbf{X}$, $\mathbf{z}_{\text{new}} = \mathbf{W}_0^{1/2}\mathbf{z}_0$.

Step 4. Create matrices for $\mathbf{z}_{\text{new}}$ and all columns of $\mathbf{X}_{\text{new}}$ according to the spatial structure.

Step 5. Perform multiresolution analysis on these matrices.

Step 6. Sum all orthogonal image components except the last one to pick up only all detail components (and no smooth ones). This yields $\mathbf{P}\,\mathbf{z}_{\text{new}}$ and $\mathbf{P}\,\mathbf{X}_{\text{new}}$.

Step 7. Transform matrices into vectors.

Step 8. Use equation (23) to estimate the regression model and to obtain the new estimator $\mathbf{b}^{(1)}$.

Step 9. Substitute $\mathbf{b}^{(1)}$ for $\mathbf{b}^{(0)}$ and return to step 2.

Step 10. Continue iterating until convergence is achieved.

3. Application for WRM

3.1. Implementing the WRM

Our computations are based on software packages housed in the R language and environment for statistical computing [27]. The tools for calculating wavelet transforms are available in package *waveslim* [28]. We used either the functions *dwt.2d* and *idwt.2d* for discrete wavelet transform and inverse discrete wavelet transform, respectively, or the function *mra.2d* for multiresolution analysis.

Because of the truncation to finite sets in discrete wavelet transforms, boundary treatment rules must be provided. In the 2-D discrete wavelet transform, type *periodic* is implemented for boundary conditions. This causes a restriction on the sample size. The number of rows and columns must be divisible by 2^j in order to perform dilation and location of wavelets (see Eq. (2)). In general, however, one wishes to analyze samples of arbitrary size. For this reason, we decide to pad data with zeros until a quadratic matrix of required size is reached.

3.2. An Application to Biogeographical Data

In this subsection we apply the wavelet-revised method to a real biogeographical dataset. We relate two environmental variables to the distribution of plant species richness in Germany.

Information on plant species distribution is available from FLORKART (see www.floraweb.de) which contains species location in a grid of 2995 grid cells. The cells of this lattice are $10'$ longitude $\times 6'$ latitude, i.e. about $11 \times 11\,\text{km}^2$, and therefore almost square cells. We additionally use the following variables per grid cell: mean annual temperature and average altitude. The values for mean annual temperature are based on data provided by the "Deutscher Wetterdienst, Department Klima und Umwelt" (recording period for temperature data was 1951-1980). Altitude data are based on the ARCDeutschland500 dataset provided by ESRI.

To visualize the degree of residual autocorrelation, the plant species richness, i.e., normally distributed number of all plant species found per grid cell, is regressed on these environmental variables. Figure 1(a) shows the distribution of species richness in Germany with light gray for species poor cells and black for species rich cells. Apart from these raw response data Figure 1 displays residual maps for (b) GLM and (c) WRM. Our WRM model and the non-spatial model GLM applied to such data differ obviously in the spatial signature of their residuals. On the GLM residual maps, areas of equal colour and large pixel size detect a high amount of correlation. Therefore, we are able to determine already visually whether there are any spatial patterns of regression residuals. The maps demonstrate that these areas of equal colour are essentially reduced for WRM residuals compared to GLM residuals. Therefore, when response values are clearly correlated, regression results such as regression coefficients and hypothesis testing will be more reliable for WRM than for GLM.

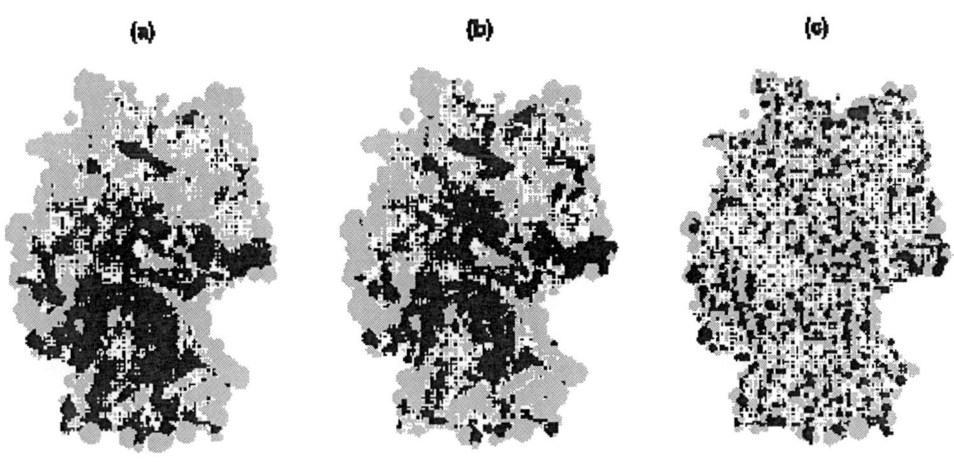

Figure 1. Maps for distributional data and regression residuals across Germany. For the models (a) plant species richness is regressed on environmental variables, and (b) GLM residual maps, and (c) WRM residual maps are given for comparison. On the residual maps, pixel sizes indicate sizes of residuals. The two colours black and grey represent the positive and negative signs of residuals, and areas of equal colour indicate autocorrelation.

3.3. Assessing Model Performance for WRM

The simulated datasets used here were prepared to check the models for autocorrelation effects [29, 30, 31]. For this purpose values for two normally distributed predictors and normally distributed errors were randomly generated. Moreover, these data are lattice data, i.e. they simulate a sampling on a lattice containing 30×30 regular grid cells. In this way we have introduced an isotropic spatial autocorrelation structure by an exponential function. Both the vector of errors and the vectors of predictors were multiplied by a matrix related to the correlation matrix. This procedure creates correlated normal random errors and predictors, enabling the calculation of correlated responses.

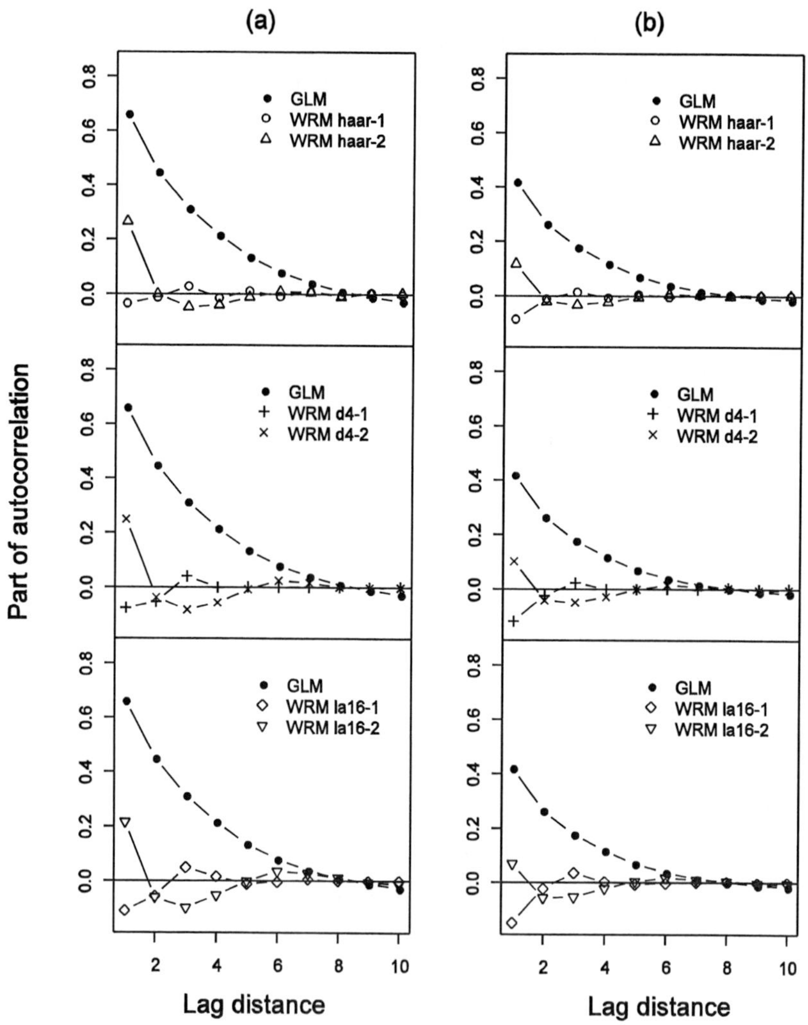

Figure 2. Correlograms for residual autocorrelation. The averages are calculated based on 100 randomly generated datasets for: (a) a normal distribution and (b) a binomial distribution. The compared methods are GLM and WRM characterized by three different wavelet families and two different resolution levels.

Figure 2 shows that residual autocorrelation calculated by correlograms and obtained as an average of the results for 100 simulation runs. We consider the following models: GLM and WRM. In order to assess the WRM, we utilize three different families of orthogonal wavelets: haar, d4, and la16 [32].

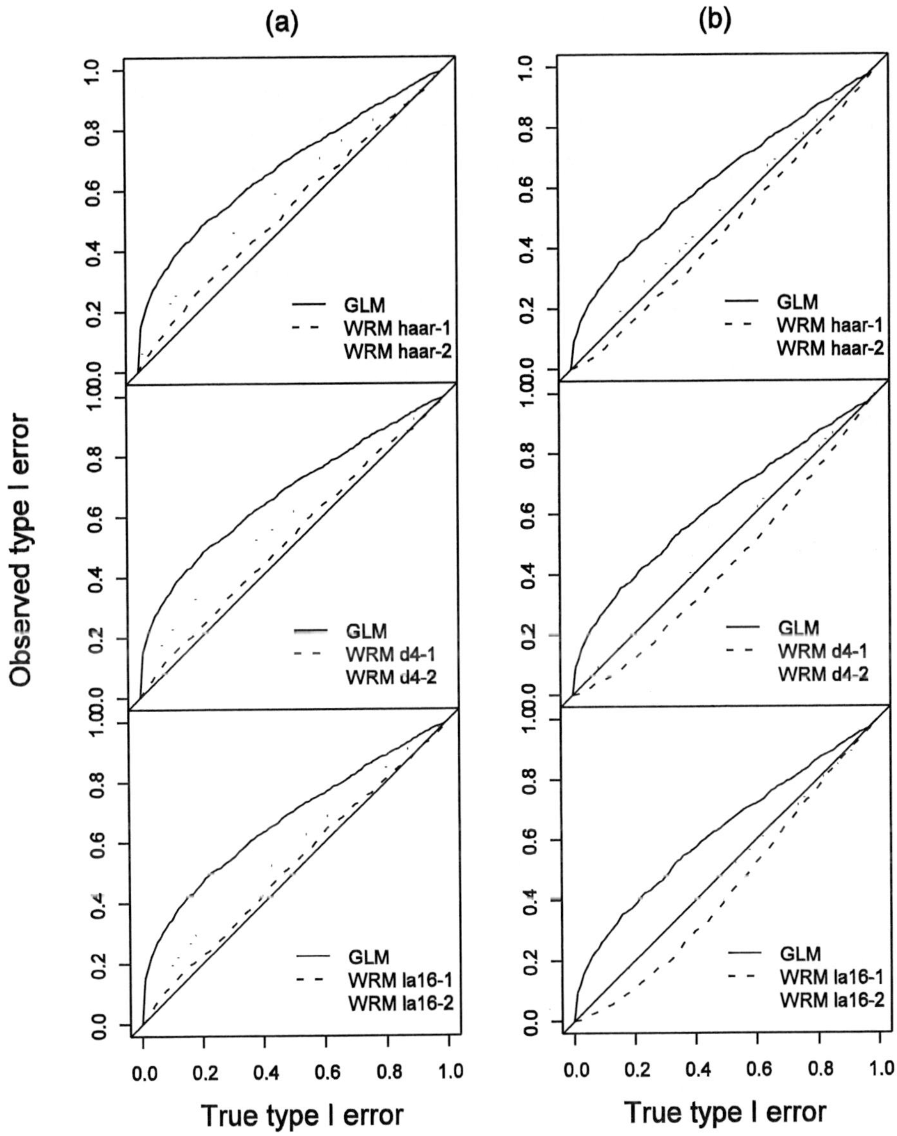

Figure 3. Type I error calibration curves. The probabilities are calculated based on 1,000 randomly generated datasets for: (a) a normal distribution and (b) a binomial distribution. The compared methods are GLM and WRM characterized by three different wavelet families and two different resolution levels.

This choice stands for a mixed assortment of samples for the following wavelet characteristics in the spatial domain: Haar wavelets are symmetric, block-shaped (i.e., non-smooth, see Eq. (3)), and very compact; d4 wavelets are asymmetric, non-smooth, and compact; la16 wavelets are nearly symmetric (i.e., least asymmetric), smoother and wider. On the one hand, the application of Haar wavelets is justified because of their capability of detecting edges in the spatial domain [8]. This is an important aspect for minimizing border effects. On the other hand, smoother wavelets such as d4 and, in particular, la16 have a higher number of vanishing moments and better frequency localization properties. Therefore, they work better as high-pass filters, which is essential for transform $\mathbf{P}$. Each of the three wavelet families is combined with smooth coefficients removal at the lowest resolution levels 1 and 2. Therefore, we label the WRM models as follows: haar-1, haar-2, d4-1, d4-2, la16-1, la16-2. Figure 2 shows that in all WRM models autocorrelation is considerably reduced compared to GLM. For long-range autocorrelation in GLM the corresponding values for WRM models are approximately zero. For nearest neighbours, i.e., at lag distance = 1, WRM level 1 models over-compensate the effect of residual autocorrelation, whereas WRM level 2 models under-compensate it. This holds true for all three types of wavelets.

To assess the performance of parameter estimates in terms of type I errors (i.e., the probability α of falsely rejecting the null hypothesis $H_0 : \beta = 0$), we calculated type I error calibration curves of the methods [10]. For this purpose we generated 1,000 datasets for both distributions. The real parameters of our simulations are $\beta_1 = 1$ and $\beta_2 = 0$; i.e., the first predictor is a significant one, whereas the second predictor does not have any explanatory power. It is, however, included as a dummy variable. We recorded how often the probability value of the dummy variable was falsely estimated to be less than a theoretical significance level α. Then the observed number of such results per 1,000 simulation runs is given by $n_r/1000$, where n_r is the number of rejections. All plots in Figure 3 show $n_r/1000$, i.e., the observed type I error, versus the expected number α, i.e., the true type I error, in the full range possible for probability α. Methods work best when the observed number equals the predicted one, i.e. when the calibration curve coincide with the line of identity given as straight line in all plots. It is shown that GLM considerably overestimates the true type I error, whereas WRM level 1 models yield good error calibration curves for normal distributions and WRM level 2 models yield good error calibration curves for binomial distributions. For binomial distributed responses, for instance, we consider it best to utilize la16 wavelets at resolution level 2.

4. Collinearity in Multiple Regression

4.1. Collinearity and Autocorrelation

If two or more predictor variables in multiple linear regressions (see Eq. (11)) are highly correlated with each other then this is called collinearity (synonym for multicollinearity). The effect arises since the columns of the design matrix $\mathbf{X}$ are nearly linearly dependent. Hence the matrix $\mathbf{X'X}$ is nearly singular and at least one of the eigenvalues of the correlation matrix of $\mathbf{X}$ is small (i.e., nearly zero). Moreover, at least some of the diagonal elements of the inverse matrix $(\mathbf{X'X})^{-1}$ will be large and, as a result, the variances of the corresponding elements of $\mathbf{b}$ will be large. This means that parameter estimates $\mathbf{b}$ themselves are unbiased. However, parameter estimates may be unstable in a sense that small changes in the data may cause large changes in regression parameters [25, 33]. That is why predictions, i.e., values for the response variable at values of the predictor variables that we did not measure, need not but may be seriously flawed.

Spatial structure can act at two scales on model performance: (i) at large scale, spatial trends may be highly collinear to environmental predictors such as temperature°ii) at small scale, spatial structure may cause autocorrelation. Hence data where predictor variables are collinear can show the feature of spatial autocorrelation as well. That means if data are sampled as lattice data in a 2-D domain, autocorrelation may arise in both response and predictor variables. Moreover, the spatial structure can affect the errors when relevant variables are either not included or inadequately mapped onto the response. Linear regression models affected by spatial autocorrelation in the errors work less efficient than others, i.e. the mean squares errors of parameter estimates increase [9]. Thus prediction can be flawed as well as in case of collinearity.

In summary, both challenging aspects of modelling can have a substantial impact on the predictive performance of such models. It can be assumed that, on the one hand, methods accounting only for autocorrelation yield bad results in case of collinearity. On the other hand, methods accounting only for collinearity probably fail to yield high prediction accuracy in case of spatial autocorrelation. Obviously, methods that are capable of handling autocorrelation as well as collinearity would be of advantage.

Several approaches to deal with the problem of collinearity have been established. Only few methods are dimension reduction methods such as principal component regression. Among them, one frequently used and appropriate method is Partial Least Squares (PLS) regression [15, 16]. But we are currently not aware of any approach coping simultaneously with both problems. Therefore, the purpose of this section is to generalize and combine features from PLS regression and WRM modelling. Furthermore, we evaluate the combined method by using simulated data.

4.2. Partial Least Squares (PLS) Regression

PLS regression consists of two steps. In the first step, so-called latent variables are calculated. These variables are linear combinations of the original predictors. The appropriate weights for this linear combination have to be calculated under certain additional constraints. In the second step, a multiple regression is performed. Both steps are incorporated in an iterative procedure. PLS regression belongs to the so-called dimension reduction techniques. Therefore, an important matter is to specify the number of latent variables to be included in the regression. If the number of latent variables introduced in the model is not reduced, then there is no difference between modelling with latent variables and standard multiple regression. If the removal of latent variables is optimal, then the estimates of regression coefficients are stable. In case of collinearity, therefore, PLS predictions have higher potential than those of standard methods.

Once more we consider the case of normal linear models written in the form (11): $\mathbf{y} = \mathbf{X}\boldsymbol{\beta} + \boldsymbol{\varepsilon}$. Let us assume that $\mathbf{X}$ and $\mathbf{y}$ are scaled. In PLS, there are different variants of computation. We utilize an algorithm which is called power method [16]. The goal is: Find a set of weights $\mathbf{w}$ in order to create a linear combination of the columns of $\mathbf{X}$, i.e. $\mathbf{t} = \mathbf{Xw}$, such that their covariance to $\mathbf{y}$ is maximum. This goal can be attained by maximization of $\mathbf{w'X'yy'Xw}$ with respect to $\mathbf{w}$ under the constraints that weight vector $\mathbf{w}$ is normalized and the matrix $\mathbf{T} = (\mathbf{t}_1, \mathbf{t}_2, \ldots)$ is orthogonal. This problem leads to an eigenanalysis solution and is related to the singular value decomposition. The vector $\mathbf{w}$ is the first eigenvector of $\mathbf{X'yy'X}$, and moreover, the first right singular vector of the matrix $\mathbf{X'y}$. Furthermore, a deflation of $\mathbf{X}$ ensures that a given component $\mathbf{t}_i$ is orthogonal to all others. That is, when the first vector $\mathbf{t}_1$ is found, its projection onto $\mathbf{X}$ is subtracted from $\mathbf{X}$. Then, the procedure is re-iterated until $\mathbf{X}$ becomes a null matrix. For regression the first m vectors of the so-called latent variables: $\mathbf{T}_m = (\mathbf{t}_1, \mathbf{t}_2, \ldots, \mathbf{t}_m)$ are used as an optimal set. Therefore, response $\mathbf{y}$ can be regressed on $\mathbf{T}_m$ and OLS prediction (see Eq. (13) for $\mathbf{y}$ is:

$$\hat{\mathbf{y}}_{PLS}{}^{m} = \mathbf{T}_m (\mathbf{T}_m{}'\mathbf{T}_m)^{-1} \mathbf{T}_m{}'\mathbf{y} \tag{25}$$

Further calculations and substitutions yield the PLS estimation of $\boldsymbol{\beta}$ [34].

PLS regression can be extended to GLM [35]. This so-called iteratively reweighted partial least squares estimation can be applied to data with non-normal response variable. Starting from an initial value for the regression parameters, one has to calculate and to update a weight matrix and the response vector because it is replaced by an adjusted dependent variable. According to equations in subsection 2.2., it can be regarded as an iterative method applying the PLS procedure described above to the adjusted matrix of predictors $\mathbf{W}^{1/2}\mathbf{X}$ and the adjusted dependent variable $\mathbf{W}^{1/2}\mathbf{z}$, where $\mathbf{z}$ is given by equation (20).

4.3. Wavelet-Revised Model for Partial Least Squares (WRMPLS)

We give the following algorithm for calculating regression parameters via WRMPLS:

Let us consider the response vector $\mathbf{y}$ and the predictors arranged as columns in the design matrix $\mathbf{X}$.

Step 1. Create matrices for $\mathbf{y}$ and all columns of $\mathbf{X}$ according to the spatial structure.

Step 2. Perform multiresolution analysis on these matrices.

Step 3. Sum all orthogonal image components except the last one to pick up only all detail components (and no smooth ones). This yields $\mathbf{P\,y}$ and $\mathbf{P\,X}$.

Step 4. Transform matrices into vectors.

Step 5. Find for this wavelet-revised response vector and this wavelet-revised design matrix m latent variables $\mathbf{T}_m$ as described in subsection 4.2.

Step 6. Regress the revised response vector on $\mathbf{T}_m$ to obtain the OLS prediction for $\mathbf{y}$ (see Eq. (25)) and the PLS estimation of $\boldsymbol{\beta}$.

Note that these 6 steps are sufficient if the response variables are normally distributed. For non-normal responses, however, the WRMPLS algorithm has to be incorporated into the framework of GLM. For this purpose, the approach embeds these steps into the iterative reweighted least squares procedure. So we have two iterative procedures: the outer one of WRM as extension of GLM and the inner one of PLS. The two nested loops are iterated until a stopping criterion for the accuracy of parameter estimate $\mathbf{b}$ is satisfied.

Summarizing the content of the previous sections, the Wavelet-Revised Model for Partial Least Squares (WRMPLS) can be regarded as an iterative method applying the PLS procedure described in subsection 4.2. to the adjusted matrix of predictors $\mathbf{PW}^{1/2}\mathbf{X}$ and the adjusted dependent variable $\mathbf{PW}^{1/2}\mathbf{z}$ of equation (24), where the meaning of transform $\mathbf{P}$ is given by expression (10).

5. Application for WRMPLS

5.1. Simulation

Simulations were performed to demonstrate the application and accuracy of the proposed method. For this purpose, datasets for both normal and logistic regression models were generated.

First, values for 5 normally distributed predictors were randomly generated. The number of values is $n = 900$. In order to produce groups of highly correlated covariates, i.e. collinearity, 4 of the predictors were used to establish clusters, while the last one remains independent. Each of the 4 predictors provided a basis required for generating another 4 predictors differing only to a certain extent from its origin. Therefore, we have 4 clusters each with 5 predictors and an individual predictor.

The clusters exhibit different degrees of collinearity. A hierarchical cluster analysis can visualize the cluster structure (see Figure 4).

Second, a regular grid of 30×30 square grid cells was generated. Spatial autocorrelation is assumed to be equal for each pair of equal distance. In this way, we have introduced an isotropic autocorrelation structure AC by an exponential function: $AC = \rho^{d_{kl}}$. Here d_{kl} is the Euclidean distance between centre points of grid cells and ρ is the correlation parameter for nearest neighbours.

Having incorporated collinearity in the design matrix $\mathbf{X}$ and autocorrelation in its columns, one can calculate the response vector using the linear component and autocorrelated noise, for details see [8]. All regression parameters of our simulations are set to 1. For a systematic investigation, we provide diverse datasets which are affected by the harmful effects of collinearity and spatial autocorrelation to an increasing degree. Moreover, each dataset consists of a training set, along with an independent test set. Both sets are of same size.

The tools for calculating PLS regressions are available in the R packages *pls* for Gaussian response variables [15] and *gpls* for binary response variables [35].

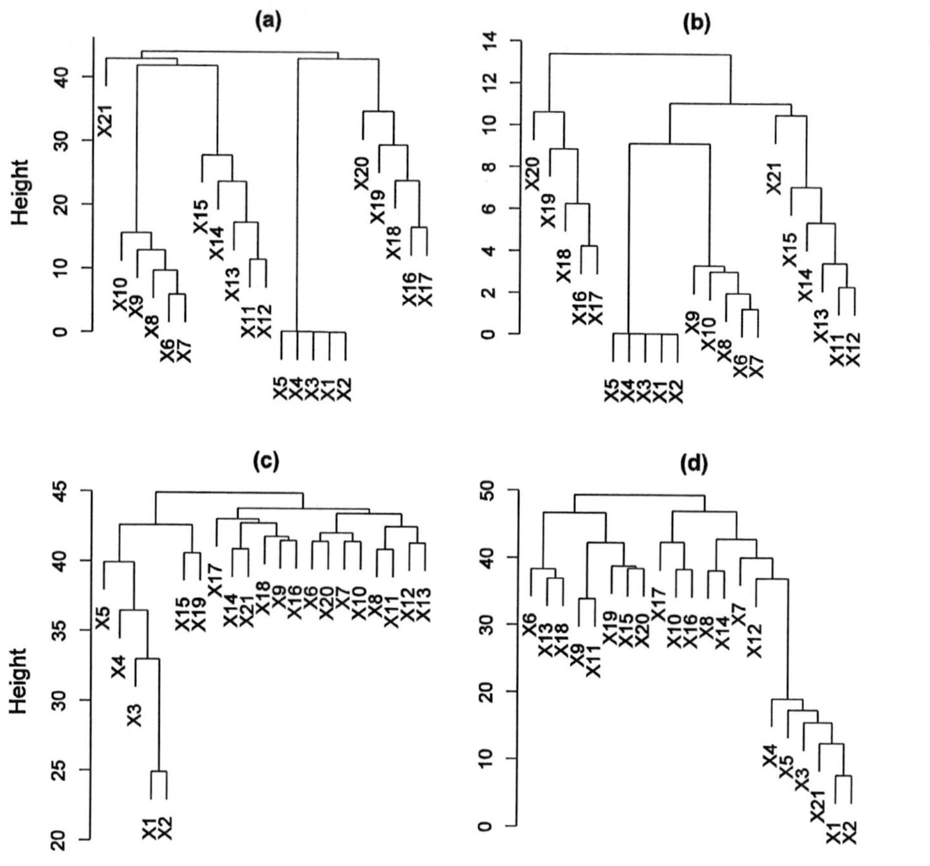

Figure 4. Hierarchical cluster analysis for the design matrix. The cluster setting is characterized by (a) high collinearity, no autocorrelation, (b) high collinearity and additional autocorrelation, (c) low collinearity, no autocorrelation, (d) low collinearity and additional autocorrelation.

5.2. Relationship Between Collinearity and Autocorrelation

Our first aim is to investigate the relationship between collinearity and autocorrelation in the design matrix $\mathbf{X}$. Therefore, we analyze simulated data, which are different regarding incorporated collinearity in matrix $\mathbf{X}$ and incorporated autocorrelation in its columns. Hierarchical cluster analysis helps to demonstrate that (i) autocorrelation can aggravate the effect of collinearity, however, (ii) autocorrelation is not completely absorbed in collinearity. For this purpose, we compare a cluster structure of data resulting from incorporated collinearity of a certain extent but without any autocorrelation to a cluster structure of unchanged collinearity and additionally incorporated autocorrelation. (i) If a cluster setting with high collinearity, i.e. clearly detectable cluster structure, but without any autocorrelation is given (Figure 4a), then its cluster structure becomes even more evident in case of additional autocorrelation (Fig 4b). This is characterized in that the complete structure arises on lower levels (see axis). (ii) If a cluster setting with low collinearity, e.g. with only one cluster, and without any autocorrelation is given (Figure 4c), then this cluster arises earlier in case of additional autocorrelation (Fig 4d). Parts without any hierarchical structure, however, remain in the background. As a result, the distance between cluster and arbitrary tree structure is increasing, i.e., parts without any hierarchical structure are not able to reflect autocorrelation. This means that autocorrelation as a small-scale phenomenon can intensify but not cause collinearity as a full-range phenomenon in matrix $\mathbf{X}$.

5.3. Assessing Predictive Performance for WRMPLS

In all of the following, we consider 4 models for comparison. These are GLM, PLS, WRM and WRMPLS. Since two of them, PLS and WRMPLS, are dimension reduction techniques, an important step in dealing with these methods is to find the appropriate number of latent variables to be included in the regression. Despite several textbooks proposing methods such as the Kaiser rule or the scree procedure, these criteria have been criticized by several investigations, see, e.g., in [36]. Therefore, we always determine the most appropriate number of retained latent variables by validation.

In order to systematically investigate how the accuracy of predictions can be influenced by the degree of collinearity and autocorrelation, we used 16 data setups with different combinations for presence/absence of collinearity and autocorrelation in the training set as well as in the test set. In Tab. 1, the numbers 1 and 0 of the first 4 columns give information about which combination is considered. Here 1 and 0 mean presence and absence, respectively. The accuracy of prediction is measured by the correlation of $\mathbf{y} = \mathbf{y}_{test}$ and $\hat{\mathbf{y}} = \mathbf{y}_{predict}$ for the 4 methods. For autocorrelation, we used correlation parameter $\rho = 0.8$ as parameter for nearest neighbours. The extent of collinearity can be measured by the mean value of off-diagonal elements of the covariance matrix. It depends on presence or absence of autocorrelation as discussed above. Here the amount of collinearity is 0.04 if autocorrelation is absent and 0.23 if autocorrelation is present, i.e. the collinearity if present is rather low. As can be seen in Tab. 1, model performances can noticeably differ from line 9 downwards where collinearity is present in the training data. The difference is significant for the setups in line 13 and 14, where (i) collinearity intensified by autocorrelation increases, and (ii) the training set is combined with a test set without any collinearity. It is worth mentioning that WRMPLS yields by far the best performance in these cases. Similarly, results for high collinearity are given (Tab. 2). The parameter settings are as follows: autocorrelation parameter ρ is 0.8, collinearity is 0.14 if autocorrelation is absent, and 0.91 if autocorrelation is present.

Here all cases where collinearity in the training set is combined with no collinearity in the test set are relevant. WRMPLS and PLS provide significantly better performance than GLM and WRM in such cases.

The aim of the following analysis is to investigate more precisely the 2 setups given in the lines 13 and 14 of Tab. 1. Therefore, the estimated coefficients based on a training set with both collinearity and autocorrelation are used to predict $\mathbf{y}$ for a test set without collinearity, i.e. without any precast cluster structure. The difference of these 2 setups herein after referred to as setup1100 and setup1101 is absence and presence of autocorrelation in the test set, respectively.

Table 1. Performance of different models in terms of different setups for training and test dataset with low collinearity. Results for correlation are given for datasets with normal responses, high autocorrelation (ρ = 0.8) and low collinearity (collinearity without autocorrelation = 0.04, collinearity with autocorrelation = 0.23). 1/0 indicates presence/absence of collinearity (coll) and autocorrelation (autocorr) in data setup

	Training data		Test data		Correlation (Ytest,Ypredict)			
	coll	autocorr	coll	autocorr	GLM	PLS	WRM	WRMPLS
1	0	0	0	0	0.71	0.71	0.71	0.71
2	0	0	0	1	0.63	0.64	0.64	0.66
3	0	0	1	0	0.80	0.80	0.80	0.80
4	0	0	1	1	0.91	0.91	0.91	0.91
5	0	1	0	0	0.67	0.67	0.68	0.68
6	0	1	0	1	0.60	0.60	0.64	0.64
7	0	1	1	0	0.77	0.77	0.78	0.78
8	0	1	1	1	0.90	0.90	0.91	0.91
9	1	0	0	0	0.64	0.70	0.59	0.70
10	1	0	0	1	0.58	0.63	0.57	0.61
11	1	0	1	0	0.79	0.80	0.79	0.80
12	1	0	1	1	0.91	0.91	0.91	0.91
13	1	1	0	0	0.00	0.59	0.07	0.68
14	1	1	0	1	0.01	0.48	0.03	0.58
15	1	1	1	0	0.62	0.72	0.69	0.77
16	1	1	1	1	0.88	0.88	0.91	0.91

For the purpose of assessing model performances, the simulated datasets were generated for increasing collinearity in the training set and for increasing autocorrelation parameters: ρ = 0.2, 0.5, and 0.8. Having added weight on collinearity initiated by autocorrelation, we calculate the actual degree of collinearity. Our goal is to examine the performances of GLM, PLS, WRM, and WRMPLS in terms of collinearity and autocorrelation. The main criterion used to check the consistency between simulated record and prediction is correlation. High correlations, however, do not guarantee adequate predictive capacity. Thus a second quality criterion based on errors should be considered. Therefore, we use the following 2 criteria regarding $\mathbf{y}_{test}$ and $\mathbf{y}_{predict}$: (i) the correlation calculated as a Pearson correlation coefficient, and (ii) the mean squares of residuals (MSR). Because MSR can have unlimited growth, we discuss the reciprocal instead, i.e. 1/MSR.

Table 2. Performance of different models in terms of different setups for training and test dataset with high collinearity. Results for correlation are given for datasets with normal responses, high autocorrelation ($\rho = 0.8$) and high collinearity (collinearity without autocorrelation = 0.14, collinearity with autocorrelation = 0.91). 1/0 indicates presence/absence of collinearity (coll) and autocorrelation (autocorr) in data setup

	Training data		Test data		Correlation (Ytest,Ypredict)			
	coll	autocorr	coll	autocorr	GLM	PLS	WRM	WRMPLS
1	0	0	0	0	0.71	0.71	0.71	0.71
2	0	0	0	1	0.63	0.64	0.64	0.66
3	0	0	1	0	0.89	0.89	0.89	0.89
4	0	0	1	1	0.98	0.98	0.98	0.98
5	0	1	0	0	0.67	0.67	0.68	0.68
6	0	1	0	1	0.60	0.60	0.64	0.64
7	0	1	1	0	0.88	0.88	0.88	0.88
8	0	1	1	1	0.98	0.98	0.98	0.98
9	1	0	0	0	0.08	0.70	0.03	0.71
10	1	0	0	1	0.09	0.63	0.09	0.63
11	1	0	1	0	0.89	0.89	0.89	0.89
12	1	0	1	1	0.98	0.98	0.98	0.98
13	1	1	0	0	-0.04	0.71	-0.02	0.71
14	1	1	0	1	-0.02	0.64	-0.05	0.63
15	1	1	1	0	0.55	0.89	0.72	0.89
16	1	1	1	1	0.96	0.98	0.98	0.98

Figure 5 depicts correlation for datasets with normal responses. Line plots in the same column are characterized by the same autocorrelation parameter. Plots in the same row are characterized by the same data setup. Ignoring the effect of collinearity, accuracy of GLM results largely decreases under collinearity (e.g. Fig 5a-f). Reducing the set of latent variables, PLS achieves an optimal predictive performance for high collinearity (Fig 5a-f), but has lower performance comparable to GLM for low collineariy in the presence of autocorrelation (Figures 5c, f). Removing the effect of autocorrelation instead of collinearity, WRM yields an increased model fit for low collinearity caused by autocorrelation, whereas high collinearity caused by cluster structures are not detected (Figures 5c, f). Only the combined model WRMPLS is able to optimize the performance not only for low, but also for high collinearity in the presence of autocorrelation. It becomes more obvious for increasing autocorrelation (Figures 5a-c). Additional autocorrelation in the test set can further increase the differences between model performances (Figures 5d-f).

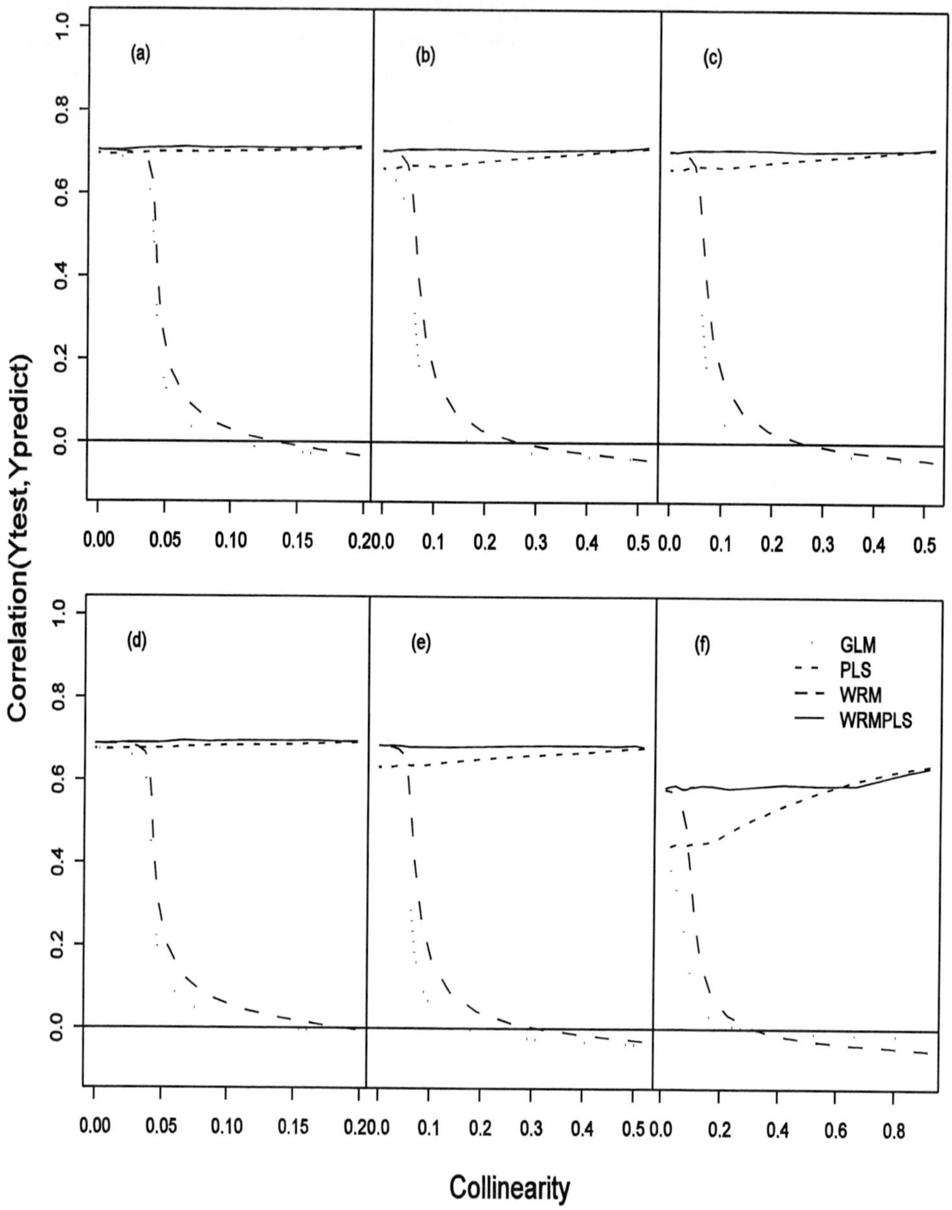

Figure 5. Correlation between predicted and real Gaussian responses of different models in terms of collinearity and autocorrelation. The y-axis depicts the correlation in terms of collinearity using datasets with normal responses for (a) setup1100, autocorrelation parameter $\rho = 0.2$; (b) setup1100, $\rho = 0.5$; (c) setup1100, $\rho = 0.8$; (d) setup1101, $\rho = 0.2$; (e) setup1101, $\rho = 0.5$; (f) setup1101, $\rho = 0.8$.

Figure 6 shows correlation for datasets with binary responses. Here WRMPLS seems to be even more superior to PLS than for models with normal responses (Figure 6f vs. Figure 5f). If the reciprocal of mean squared residuals is considered as criterion, the best models should have the highest values for 1/MSR. As a matter of fact, the same general trends as for correlation are evident for 1/MSR (Figure 7 and 8).

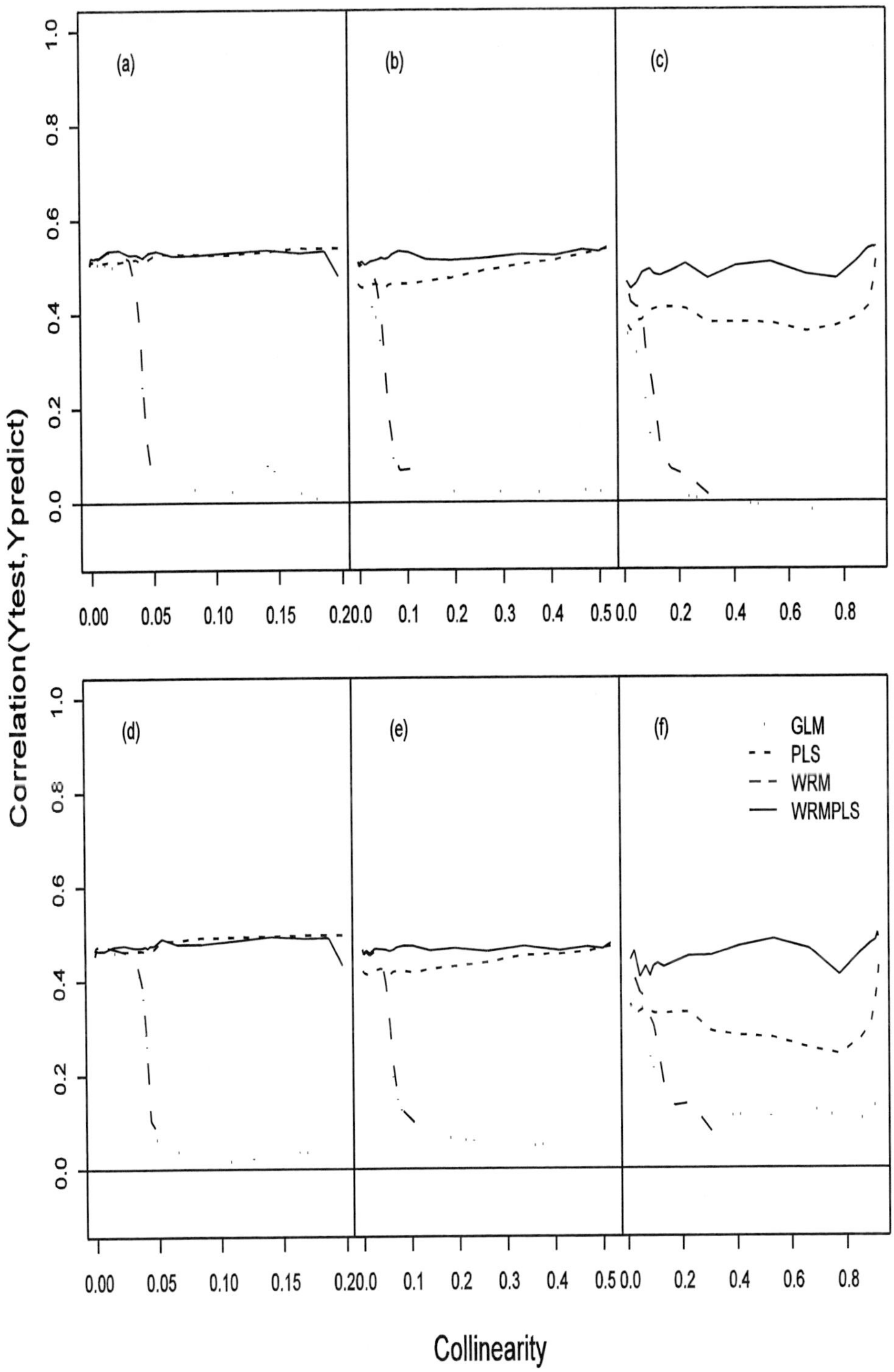

Figure 6. Correlation between predicted and real binary responses of different models in terms of collinearity and autocorrelation. The y-axis depicts the correlation in terms of collinearity using datasets with binary responses for (a) setup1100, autocorrelation parameter $\rho = 0.2$; (b) setup1100, $\rho = 0.5$; (c) setup1100, $\rho = 0.8$; (d) setup1101, $\rho = 0.2$; (e) setup1101, $\rho = 0.5$; (f) setup1101, $\rho = 0.8$.

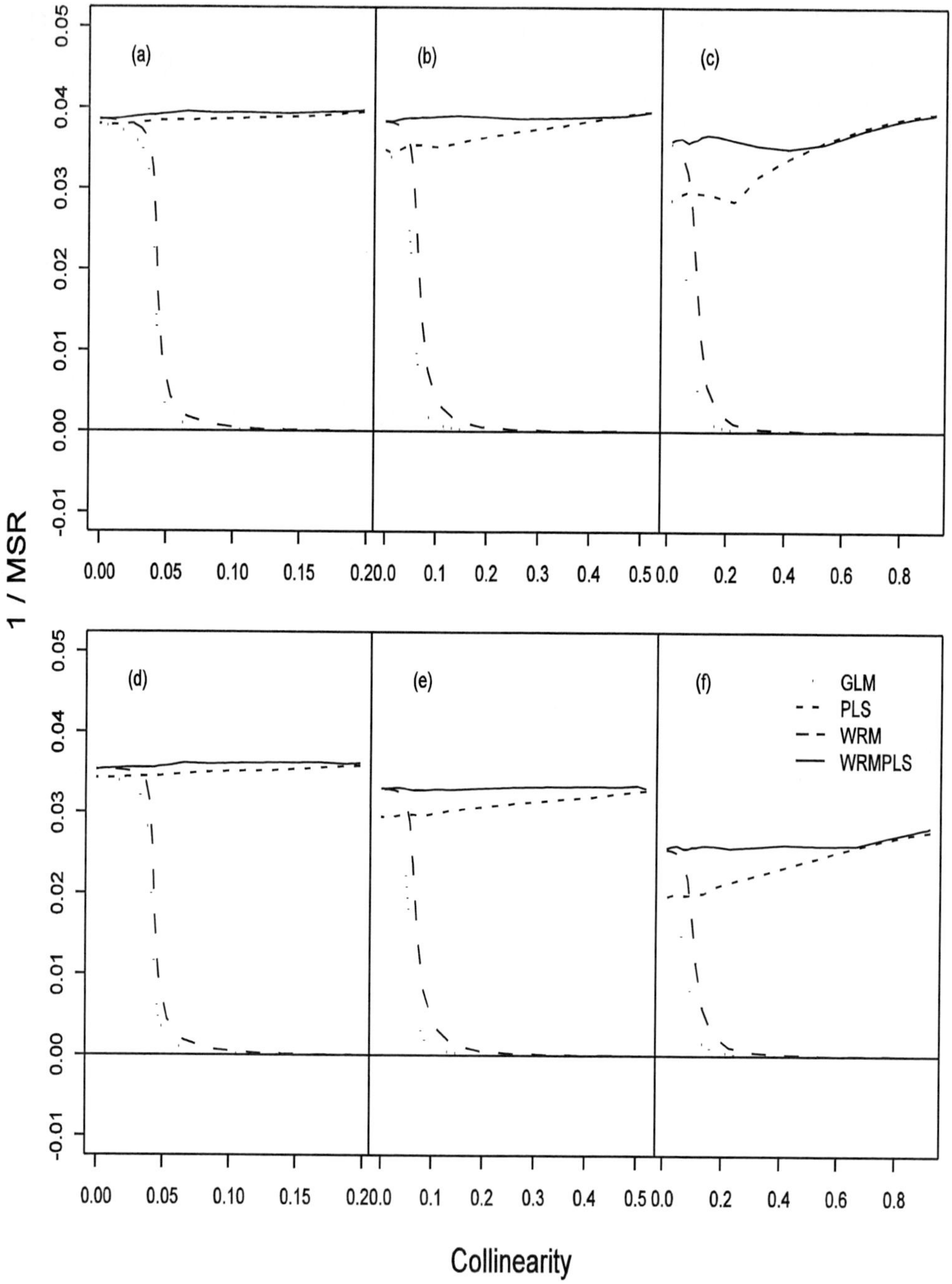

Figure 7. Mean squares of residuals between predicted and real Gaussian responses of different models in terms of collinearity and autocorrelation. The y-axis depicts the reciprocal mean squared residuals in terms of collinearity using datasets with normal responses for (a) setup1100, autocorrelation parameter $\rho = 0.2$; (b) setup1100, $\rho = 0.5$; (c) setup1100, $\rho = 0.8$; (d) setup1101, $\rho = 0.2$; (e) setup1101, $\rho = 0.5$; (f) setup1101, $\rho = 0.8$.

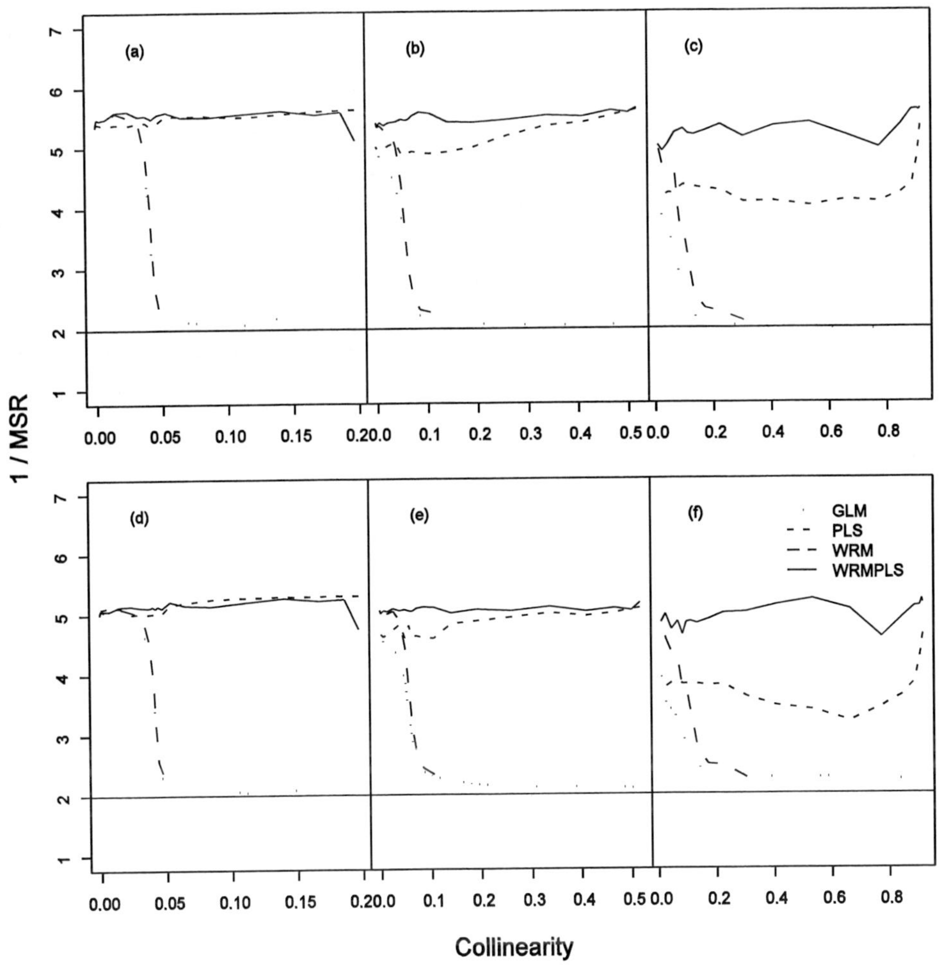

Figure 8. Mean squares of residuals between predicted and real binary responses of different models in terms of collinearity and autocorrelation. The y-axis depicts the reciprocal mean squared residuals in terms of collinearity using datasets with binary responses for (a) setup1100, autocorrelation parameter $\rho = 0.2$; (b) setup1100, $\rho = 0.5$; (c) setup1100, $\rho = 0.8$; (d) setup1101, $\rho = 0.2$; (e) setup1101, $\rho = 0.5$; (f) setup1101, $\rho = 0.8$.

Conclusion

The main contribution of this work is to show that the theory and application of wavelets provides a method to remove undesired effects of spatial autocorrelation in spatial statistics. Autocorrelation can be described by smooth wavelet coefficients at small scales. Therefore, data can be decomposed into uncorrelated and correlated parts. In order to make wavelet analysis applicable to spatial statistics, one has to take into account the spatial, two-dimensional structure of lattice data and their statistical origin, i.e. the discrete structure of sampled data. Therefore, our extension of relevant statistical modelling is based on discrete wavelet transforms, and is carried out as two-dimensional analysis. Thus multiresolution analysis gives the background for autocorrelation extraction from 2-D (image) data. Standard statistical analyses such as regression models yield incorrect results, because the presence of autocorrelation violates the assumption of independently and identically distributed errors. Using an appropriate linear transformation, we are able to extend standard regression models, in particular, generalized linear models (GLM) to autocorrelated data.

After describing the properties of wavelets and developing an extension of GLM, we present an algorithm to estimate regression parameters, and we calculate the information matrix as a tool for hypothesis testing. For an application of our new wavelet-revised method (WRM) to a real biogeographical dataset, we present maps representing regression data for Germany and demonstrating that areas characterized by autocorrelation are essentially reduced for WRM residuals compared to GLM residuals. Moreover, we use correlograms to show that autocorrelation is considerably reduced in WRM models.

The main difficulty arising from the GLM applied to such data is that autocorrelation affects its performance in terms of type I errors (i.e., the probability of falsely rejecting the null hypothesis). Therefore, we analyze whether the WRM test statistic is valid, finding that its empirical significance levels are very close to their theoretical counterparts. For binomial distributed responses, for instance, we consider it best to utilize la16 wavelets at resolution level 2. In that case, the WRM test statistic is valid even for strong and long ranged autocorrelation.

We therefore suggest the use of WRM, especially when analyzing data observed on a regular two-dimensional lattice and characterized by large sample size. WRM is a powerful method for removing autocorrelation without any prior knowledge of the underlying correlation structure. It is a computationally very fast and efficient procedure.

Spatial analyses, in particular, species distribution models can suffer not only from autocorrelation, but also from collinearity. Therefore, predictions need not but may be seriously flawed. Using different combinations for presence/absence of collinearity and autocorrelation in training and test datasets, we investigate whether the accuracy of predictions can be influenced by the strength of these effects. Predictions for GLM fail if collinearity is present in the training set, however, absent in the test set. Predictions for partial least squares (PLS) regression indicate severe problems in the special case of low collinearity and additional autocorrelation in the training set. Here autocorrelation is not completely absorbed in collinearity.

Therefore, methods seeking for latent variables such as PLS have lower predictive potential when the data additionally suffer from autocorrelation. For this reason, we propose an extension to PLS, i.e., a combination of WRM and PLS, for correcting data with respect to autocorrelation. Our studies show that our novel wavelet-revised method for partial least squares (WRMPLS) obtains good results in applications. Its outcome is demonstrated in simulation studies. WRMPLS has high predictive performance as can be measured by both correlation and mean squares of residuals in terms of "true" response and predicted response variables. It clearly outperforms PLS for data setups of high autocorrelation combined with different collinearity in training and test set.

ACKNOWLEDGMENTS

We acknowledge support by the Integrated Project "ALARM: Assessing LArge scale environmental Risks to biodiversity with tested Methods" (GOCE-CT-2003-506675) from the European Commission within Framework Programme 6 and EU infrastructure project "Distributed Infrastructure for EXPErimentation in Ecosystem Research – EXPEER" (project reference 262060) funded under 7th Framework Programme.

REFERENCES

[1] N.A.C. Cressie, *Statistics for spatial data*, Wiley, New York (1993).
[2] P. Legendre, *Ecology* 74, 1659 (1993).
[3] J.W. Lichstein, T.R. Simons, S.A. Shriner and K.E. Franzreb, *Ecol. Monogr.* 72, 445 (2002).
[4] G.P. Nason and B.W. Silverman, The stationary wavelet transform and some statistical applications, in A. Antoniadis and G. Oppenheim (Eds.), *Wavelets and Statistics*, Springer, Heidelberg (1995) pp. 281-299.
[5] P. Flandrin, *IEEE T. Inform. Theory* 38, 910 (1992).
[6] I.M. Johnstone and B.W. Silverman, *J. Roy. Stat. Soc.* 59, 319 (1997).
[7] C.F. Dormann, J.M. McPherson, M.B. Araújo, R. Bivand, J. Bolliger, G. Carl, R.G. Davies, A. Hirzel, W. Jetz, W.D. Kissling, I. Kühn, R. Ohlemüller, P.R. Peres-Neto, B. Reineking, B. Schröder, F.M. Schurr, and R. Wilson, *Ecography* 30, 609 (2007).
[8] G. Carl and I. Kühn, *Stoch. Env. Res. Risk A.* 22, 315 (2008).
[9] G. Carl and I. Kühn, *Geogr. Anal.* 42, 323 (2010).
[10] M.J. Fadili and E.T. Bullmore, *NeuroImage* 15, 217 (2002).
[11] G. Carl, C.F. Dormann, and I. Kühn, *Web Ecology* 8, 22 (2008).
[12] S. Chatterjee, A.S. Hadi, and B. Price, *Regression analysis by example*, Wiley, New York (2000).
[13] M.H. Graham, *Ecology* 84, 2809 (2003).
[14] R.P. Freckleton, *Behav. Ecol. Sociobiol.* 65, 91 (2011).
[15] H. Martens and T. Naes, *Multivariate calibration*, Wiley, Chichester (1989).
[16] H. Abdi, Partial least squares regression (PLS-regression). In M. Lewis-Beck, A. Bryman and T. Futing (Eds.), *Encyclopedia for research methods for the social sciences,* Sage, Thousand Oaks, California (2003) pp. 792-795.
[17] Y. Meyer, *Wavelets and Operators*, Cambridge University Press, Cambridge (1992).
[18] Daubechies, *Ten lectures on wavelets*. CSBM-NSF series application mathematics, vol. 61, SIAM Publications, Philadelphia, Pennsylvania (1992).
[19] C.K. Chui, *An introduction to wavelets*, Academic Press, San Diego (1992).
[20] R.H. Shumway and D.S. Stoffer, *Time Series Analysis and Its Applications*, Springer Texts in Statistics, Springer, New York (2000).
[21] Bruce and H.Y. Gao, *Applied Wavelet Analysis with S-Plus*, Springer, New York (1996).
[22] K. Müller, G. Lohmann, S. Zysset and D.Y. von Cramon, *J. Magn. Reson. Imag.* 17, 20 (2003).
[23] S.G. Mallat, *Trans. Am. Math. Soc.* 315, 69 (1989).

[24] R.H. Myers, D.C. Montgomery and G.G. Vining, *Generalized linear models*, Wiley, New York (2002).
[25] A.J. Dobson, *An introduction to generalized linear models* (2nd ed.), Chapman and Hall/CRC, Boca Raton, Florida (2002).
[26] P. McCullagh and J.A. Nelder, *Generalized linear models* (2nd ed.), Chapman and Hall, London (1989).
[27] R Development Core Team. *R: A language and environment for statistical computing*, R Foundation for Statistical Computing, Vienna, Austria (2010).
[28] B. Whitcher, Waveslim: Basic wavelet routines for one-, two- and three-dimensional signal processing. *R package,* version 1.5, (2005).
[29] S.M. Ross, *Simulation* (2nd ed.), Academic Press, San Diego (1997).
[30] R. Haining, D.A. Griffith, and R. Bennet, *Geogr. Anal.* 15, 247 (1983).
[31] P.J. Heagerty and S.R. Lele, *J. Am. Stat. Assoc.* 93, 1099 (1998).
[32] D.B. Percival and A.T. Walden, *Wavelet Methods for Time Series Analysis*, Cambridge University Press, Cambridge (2000).
[33] J.H. Jinn, C.C. Song, and W.B. Mou, An Application of the Generalized Shrunken Least Squares Estimator on Principal Component Regression (2007). Available from: http://interstat.statjournals.net/YEAR/2007/articles/0702005.pdf
[34] Phatak and S. de Jong, *J. Chemometr.* 11, 311 (1997).
[35] B.D. Marx, *Technometrics* 38, 374 (1996).
[36] W.F. Velicer and D.N. Jackson, *Multivar. Behav. Res.* 25, 1 (1990).

In: Wavelets: Classification, Theory and Applications
Editors: Manel del Valle et al.
ISBN: 978-1-62100-252-9

Chapter 5

WAVELETS AND CLUSTERING: METHODS TO ASSESS SYNCHRONIZATION

Irene L. Hudson[1,*], Marie R. Keatley[2] and In Kang[3]
[1]School of Mathematical and Physical Sciences, University of Newcastle, Callaghan, NSW, Australia
[2]Dept of Forest and Ecosystem Science, University of Melbourne, Creswick, Victoria, Australia
[3]Department of Mathematics and Statistics, University of Canterbury, Christchurch, New Zealand

ABSTRACT

In this chapter the discrete wavelet transform (DWT) is shown to accurately enumerate and visually demonstrate synchronisation of a flowering time series. The DWT is applied to flowering intensity records (covering 1940 – 1970) of *Eucalyptus leucoxylon*, *E. microcarpa*, *E. polyanthemos* and *E. tricarpa*. The DWT is used as a correlational method, and proposed as a new metric for synchronisation and visualisation for a multidimensional time series. The wavelet based results are compared to Moran's traditional synchronisation test and to a self organising map time series clustering approach. The Maximal Overlap DWT analysis successfully identify four subcomponents in each flowering series characterised as a (i) non-flowering phase (d1), (ii) duration (d2), (iii) annual (d3) and (iv) intensity (d4) cycles. The wavelet correlation (WCORR) signatures are shown to exhibit a common pattern for synchronous species pairs, for example d2, d3 and d4 WCORR values being significant and positive for *E. leucoxylon* and *E. tricarpa* whereas for species pairs that are asynchronous the d2 and d3 specific WCORR values are significant and negative for *E. tricarpa* and *E. polyanthemos*.

The profiles of wavelet cross correlation (WCCORR) enable us to further understand how synchrony may relate to the subcomponents of flowering (peak annual cycles, start, cessation, duration, overlap) and climate. The positive periods of the WCCORR profiles for d2 reflect flowering overlap between species pairs. The d3 WCCORR's peak at the lag which is the difference between the species pair's peak flowering months. Distinctly

* E-mail addresses: Irene.Hudson@newcastle,edu.au; irenelena.hudson@gmail.com;

different patterns for d3 WCCORR are associated with asynchronous and synchronous time series. The sinusoidal and cycling of the d3 specific WCCORR profiles mirror the cyclic 6 monthly climatic influence on species flowering, where for 6 months if temperature is positively correlated with flowering, rainfall is negative and vice versa.

1. Introduction

Phenology: the timing of life stages (e.g. flowering, fruiting, bird arrival) underpins or influences many different ecological processes [1, 2]. These processes also have a significant role in influencing what society values (e.g. human health, biodiversity, forestry, agriculture and tourism [3-5]). Since the 1990s primarily because of climate change [6] phenological time series have been used to determine and report the impacts of global warming in both natural and managed systems [7-9]. Determining trends in relation to long-term climate, however, is not easy as they can be confounded by short-term interannual trends hence not only are long term records required but also analytic techniques methods which can accommodate this confounding factor [10-12]. Wavelets have been extensively applied to many arenas (e.g. to the study of change in European spring temperatures [13] and rainfall [14], cycles in Antarctic seabird population [12], changes in vegetation cover [15], and to brain imaging [16, 17]. It is the capacity of wavelets to handle non-stationary data: to deconstruct a time series into its subcomponents and remove noise; to accommodate multi-scale information, and to minimize correlation and time-dependency in data [18-21] that have added to their popularity. As phenological time series are usually non-stationary and noisy, wavelets presents as a useful analytic method [22, 23] for examining phenological records and for the determination of possibly changing climatic impacts on flowering, at an annual and across years basis.

In this chapter the discrete wavelet transform (DWT) [19, 20] is tested as a tool by which to accurately enumerate and visually demonstrate synchronisation of time series. The DWT, follows the development of the maximal overlap DWT (MODWT) of Percival and Walden [20] and the rationale for its use is that, given the resultant MODWT coefficients, the original (flowering) time series can be reconstructed as an additive decomposition - a multi-resolution analysis (MRA) [24]; and that subsequently the individual detail (subcomponent) series could be studied via wavelet (cross) correlation methods. The DWT is applied to flowering intensity records of four Eucalyptus species - namely, *E. leucoxylon, E. microcarpa, E. polyanthemos* and *E. tricarpa*, from a rare long-term 31 year record of flowering (1940 – 1971) of over 350 monthly time points [24, 25]. The DWT is used as a correlation method, and proposed as a new metric for synchronisation, for visualisation and assessment of synchronisation (of flowering) for multidimensional (phenological) time series. The utility of wavelets for examining the flowering time series of Australian eucalypt tree species in relation to climate was recently demonstrated [22]. This work builds, in part on an initial study by Kang et al. [26] and on the early premise by Hudson et al. [23] that wavelets per se could add integrity to the use of phenological records to detect climate change. This was recently confirmed by a study of the four Eucalypt species studied in this chapter by Hudson et al. [22] (see also [23, 27]) and on a recent study expanding wavelets analysis to 8 Eucalypt species [28].

The wavelet based results in this chapter are compared to: (i) Moran's traditional synchronisation test [29, 30] (as a benchmark [31-33]) and to (ii) a self organising map (SOM) time series clustering approach [34]).

The Moran effect [35] states that the correlation between two populations equals the correlation between the environmental factors influencing their dynamics if the intrinsic structure of the dynamics is the same. Moran suggested an autoregressive integrated moving average (ARIMA) [36] model with lagged dependency to test for synchrony and a correlation approach to examine synchrony between pairs of species. The afore-mentioned SOM [37] approach adapted for time series [38, 39], follows the development proposed recently Hudson et al 2010 [34]. SOMs [40] create an imagined geography and the SOM algorithm converts complex, nonlinear relationships between high-dimensional data into simple networks and a map - a kind of topological map which organises itself based on the most likely patterns in the time series that it trains [34]. As far as the authors are aware, this is the first study to date to use wavelet signatures as a procedure to test synchronization of phenological time series.

2. METHODS

2.1. Maximal Overlap DWT (MODWT) and Multiresolution Analysis (MRA)

The discrete wavelet transform (DWT) as applied here, following Percival and Walden [20], requires a discretization of the continuous time variable, Y_t. For most practical applications, the DWT, which analyzes signals over a discrete set of scales, that are usually sampled at dyadic sequence ($\lambda_j = 2^{j-1}, j = 1, 2, 3, \ldots$), is sufficiently accurate and can recover signals perfectly [41]. The maximal overlap DWT (MODWT) is one discretization choice which gives N wavelet coefficients for each scale [42]. The MODWT is a non-decimated variation of the DWT [43]. It is also equivalent to the original time series, in the sense that, given the MODWT coefficients, Y, can be constructed as an additive decomposition, which is known as a multiresolution analysis (MRA) [19, 24, 44]. The MRA decomposition is as follows:

$$Y = \sum_{j=1}^{J_0} \bar{d}_j + \bar{s}_{j_0} \tag{1}$$

where $\bar{d}_j$ are the "detail series", ($j = 1, 2, \ldots, J_0$) for a pre-specified J_0, and are part of the MRA of Y that can be attributed to variations on a scale of λ_j [19, 20]. Recall that scale $\lambda_2 = 2$. In equation (1) $\bar{s}_{j_0}$ is a N dimensional vector, which depends on the scaling coefficients. The vector, $\bar{s}_{j_0}$, is called the "smooth series" since it is associated with averages over scales $2\lambda_{j_0}$ and longer. Note that $\bar{s}_{j_0}$ captures the slowly varying portion of Y, which is often considered to be the overall trend.

2.2. Wavelet Correlation and Cross-Correlation and Confidence Limits

Section 2.1 considers decomposing the original time series into J_0+1 subcomponents. However, the scale λ_j MODWT coefficients may also be used to examine the wavelet correlation and wavelet cross-correlation of bivariate time series X_t and Y_t [19, 45, 46] as described belowover.

The wavelet correlation (WCORR) of (X_t, Y_t) at scale $\lambda_j=2^{j-1}$ is defined as

$$\rho_{XY,\tau=0}(\lambda_j)=\frac{\text{Cov}\left\{\overline{W}^X_{j,t},\overline{W}^Y_{j,t}\right\}}{\sigma_X(\lambda_j)\cdot\sigma_Y(\lambda_j)}=\frac{\gamma_{XY,\tau=0}(\lambda_j)}{\sigma_X(\lambda_j)\cdot\sigma_Y(\lambda_j)} \tag{2}$$

where $\sigma^2_X(\lambda_j)=\text{var}\left\{\overline{W}^X_{j,t}\right\}$ is the wavelet variance with scale λ_j. $\overline{W}^X_{j,t}$ and $\overline{W}^Y_{j,t}$ are the scale λ_j MODWT coefficients for X_t and Y_t, respectively [47]. Note that Equation 2 is a theoretical quantity that is well-defined under an assumption of stationarity. For a time lag τ, between the two series, X_t and Y_t, the wavelet cross-covariance and wavelet cross-correlation (denoted by WCCORR) [19], for scale $\lambda_j=2^{j-1}$ and time lag τ is

$$\rho_{XY,\tau}(\lambda_j)=\frac{Cov\left\{\overline{W}^X_{j,t},\overline{W}^Y_{j,t}\right\}}{\sigma_X(\lambda_j).\sigma_Y(\lambda_j)}=\frac{\gamma_{XY,\tau}(\lambda_j)}{\sigma_X(\lambda_j).\sigma_Y(\lambda_j)} \tag{3}$$

By setting $\tau = 0$ $\gamma_{XY,0}(\lambda_j)$ reduces to the wavelet variance for X_t or Y_t denoted by $\sigma^2_X(\lambda_j)$ or $\sigma^2_Y(\lambda_j)$, respectively. The wavelet cross-correlation in equation 3 is thus able to provide the lead or lag relationship on a scale-by-scale basis [19], just as is the case for conventional cross-correlations, which can determine lead or lag relationships between two series. When the time lag $\tau = 0$, the corresponding MODWT estimator of the wavelet correlation (WCORR) between the two series X_t and Y_t [20] is derived.

Construction of the 95% two-sided confidence interval (CI) of WCORR and WCCORR follows the development of Gencay et al. [19]. To produce CIs for the cross-correlation coefficient, the Fisher's nonlinear z-transformation $h(\rho)$ [20], as follows, $h(\rho)=1/2\log(1+\rho/1-\rho)=\tanh^{-1}(\rho)$ is required. For $\hat{\rho}_x$, an unbiased estimator of WCORR based on the MODWT, the following asymptotic normal distribution holds

$$\sqrt{N-3}\left[h(\hat{\rho})-h(\rho)\right]\sim N(0,1). \tag{4}$$

Applying the transformation tanh maps the confidence interval back to [-1,1] to produce an approximate 95% CI for $\rho_X(\lambda j)$ as follows [19, 46]

$$\tanh\left\{h\left[\hat{\rho}_X\left(\lambda_j\right)\right]\pm 1.96\left(\frac{1}{\overline{N}_j-3}\right)^{1/2}\right\} \qquad (5)$$

The quantity $\overline{N}_j$ in equation (5) is the number of DWT coefficients associated with scale λ_j.

2.2. Phenological Data and Data Analysis

The phenological data were collected between 1940 to 1970 and comprises the flowering intensity profiles of the four eucalypts species growing in the region of Havelock, Victoria, Australia. The four eucalypt species examined in this chapter are: *E. leucoxylon*, *E. microcarpa*, *E. polyanthemos* and *E. tricarpa* (Table 1).

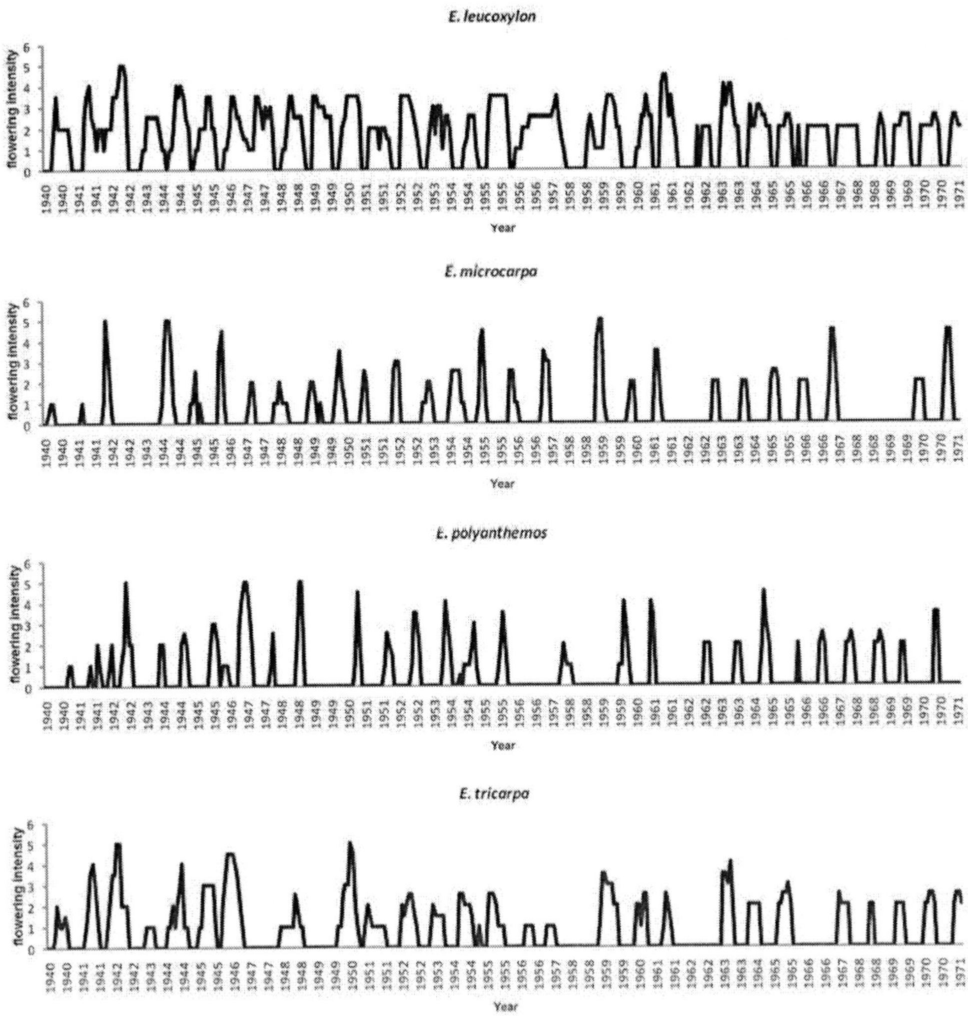

Figure 1. Raw flowering intensity time series for the four eucalypt species.

Table 1. Four eucalypt species studied mean (months + 1 SD) and median flowering duration and months in which main phenophases occur. Terms describing flowering intensity and their assigned value [49]. (Table from Hudson et al. [34])

Species	Common name and taxonomic authority	Mean duration	Median duration	Most probable month$^{\phi}$	Peak Flowering$^{\#}$	Flowering start month	Flowering cessation months
E. leucoxylon	Yellowgum F. Muell	9.7 ± 4.2	9	October Spring	September Spring	May Autumn	December Summer
E. microcarpa	Grey box Maiden	4.0± 1.0	4.0	March Autumn	March Autumn	February Summer	May Autumn
E. polyanthemos	Red box L. A. S. Johnson and K. D Hill	3.6 ± 1.5	3.5	November Spring	November Spring	October Spring	December Summer
E. tricarpa	Red ironbark L. A. S. Johnson and K.D Hill	6.2 ± 2.1	6.0	July Winter	July Winter	April Autumn	September Spring

$^{\phi}$ Based on actual likelihood/probabilities for Havelock [25].

$^{\#}$The month with the highest mean intensity within a flowering year.

Phenological phases are determined by the observer and are now commonly categorized according to the BBCH code [48]. Observations on the timing, quantity and distribution of flowering of these species were collected on a monthly basis [25] and placed into categories which had been pre-determined by the Forest authorities [49]. Flowering intensity of each species was quantified by assigning a rank value (ranging from 0 to 5) [25] and producing a discrete time series (Figure 1).

A score of 0 indicates that no flowering was observed in the given month, whilst a score of 5 indicates that flowering was heavy and distributed throughout the observation area. More specifically the categories for quantity: no flowering (0), very scattered or isolated (0.5); light flowering (1); medium flowering (2), and heavy flowering (3): and for distribution: isolated (0.5), scattered (1), fairly general (1.5) and general (2). DWT methodology [19, 20] was applied to the afore-mentioned species flowering intensity scores [25]. DWT, MODWT and MODWT-MRA analyses were performed using code based on the wavelet methods developed by Percival and Walden [20] and Gencay et al. [19].

3. Results

3.1. MODWT-MRA Subcomponents

The maximal overlap discrete wavelet transform – multi-resolution analysis (MODWT-MRA) with $J_0 = 4$ provided subcomponents of the flowering signal within each species (see Figure 2 for *E. tricarpa*).

The raw series is given by X, d1 to d4 are the different subcomponents (detail) of the original series X. Traditionally, d1 to d4 (also derived were d5 and d6) are associated with changes in averages over 1, 2, 4, 8, 16 and 32 months (Figures 2 — 76, 8). The s4 profile is the smoothed series associated with averages over 32 months, and is also provided by MODWT-MRA (Figure 2).

First The first two subcomponents reflect the duration and the overall pattern of the original flowering data. d1 and d2 appear similar in their profiles (Figure 3), however, d1 contains subcycles per year (in a flowering year) but it is unclear as to what these cycles are related. The annual cycle of flowering is delineated by d3, and also reveals (by its peaks), in the majority of cases, the month of peak flowering in a flowering year (or within a month on either side) (Figure 4). Note that for *E. leucoxylon* 73% of the peak months are thus selected, for *E. microcarpa* 72%, for *E. polyanthemos* 69% and for *E. tricarpa* 65%.

Table 2. Species and their years of non-flowering

Species	Non-flowering years
E. leucoxylon	Always flowered
E. microcarpa	1943, 1958, 1962, 1968 and 1969
E. polyanthemos	1949/50, 1956/57, 1958/59 and 1961/62
E. tricarpa	1947, 1949, 1958, 1962 and 1966

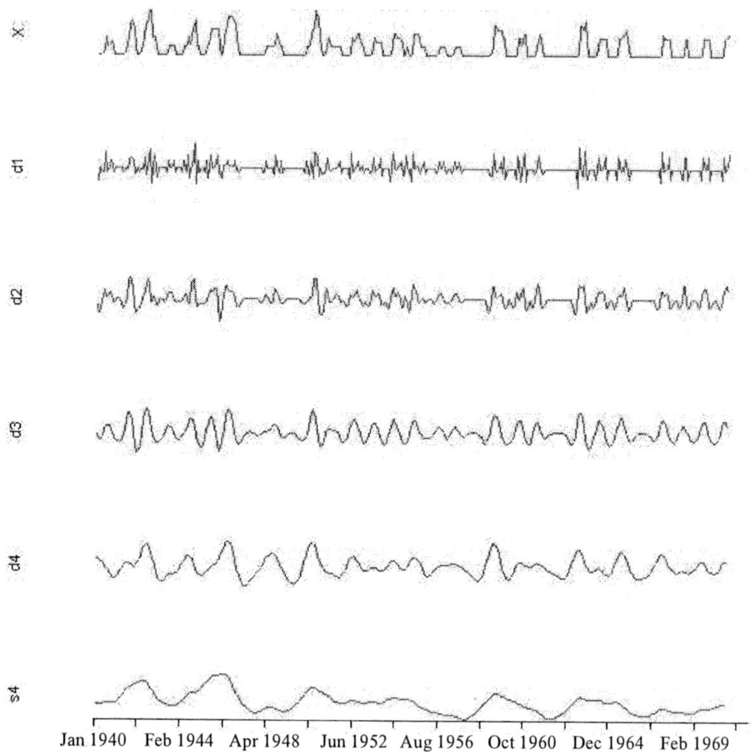

Figure 2. Example of MODWT-MRA output using *E. tricarpa*. X = the raw data, d1 to d4 different subcomponents (details) of the original series X, and s4 = the smooth series.

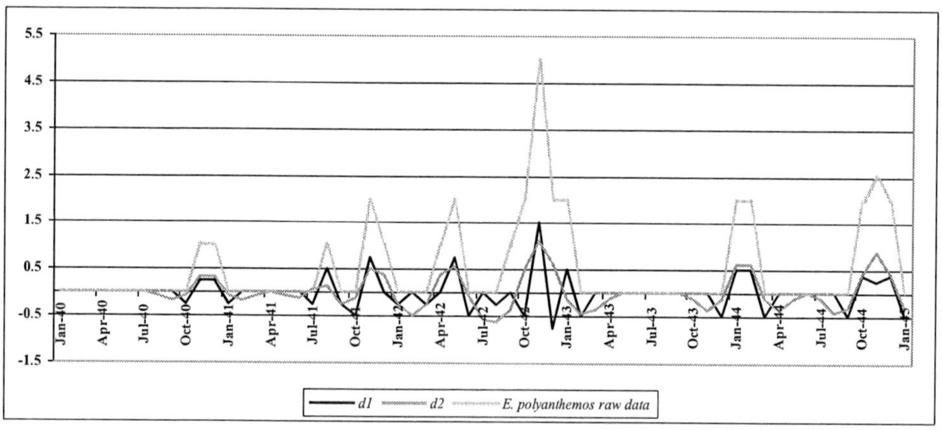

Figure 3. d1, d2 and raw data of *E. polyanthemos* between January 1940 and March 1945.

Additionally, d3 highlights years when flowering does not occur – for profiles of d3 near zero, true in all species except *E. leucoxylon* which always flowered (Figure 4). Years of non-flowering are given per species in Table 2. The flowering period of *E. polyanthemos* generally includes the months of December and January so two years are given (Table 2).

Note that the years 1958 or 1958/59 as a non-flowering was common to three species: *E. microcarpa*, *E. polyanthemos* and *E. tricarpa*.

d4 outlines annual and biennial cycles of flowering intensity for each species (Figure 5). For three of the species, *E. microcarpa, E. polyanthemos* and *E. tricarpa*, the biennial cycle is associated with years of low flowering intensity (< 2) and also includes years of no flowering (i.e. flowering intensity = 0). This is particularly influenced by the large number of non-flowering years (Table 2). The species with the smallest average period between peaks was *E. leucoxylon* 19.1 months, then followed by *E. polyanthemos* 20.3 months, *E. tricarpa* 21.3 months and *E. microcarpa* 22.5 months (Figure 5). All species had at least one year between 1952 and 1958 when the number of months between peaks was between 11 to 14 months (Figure 5).

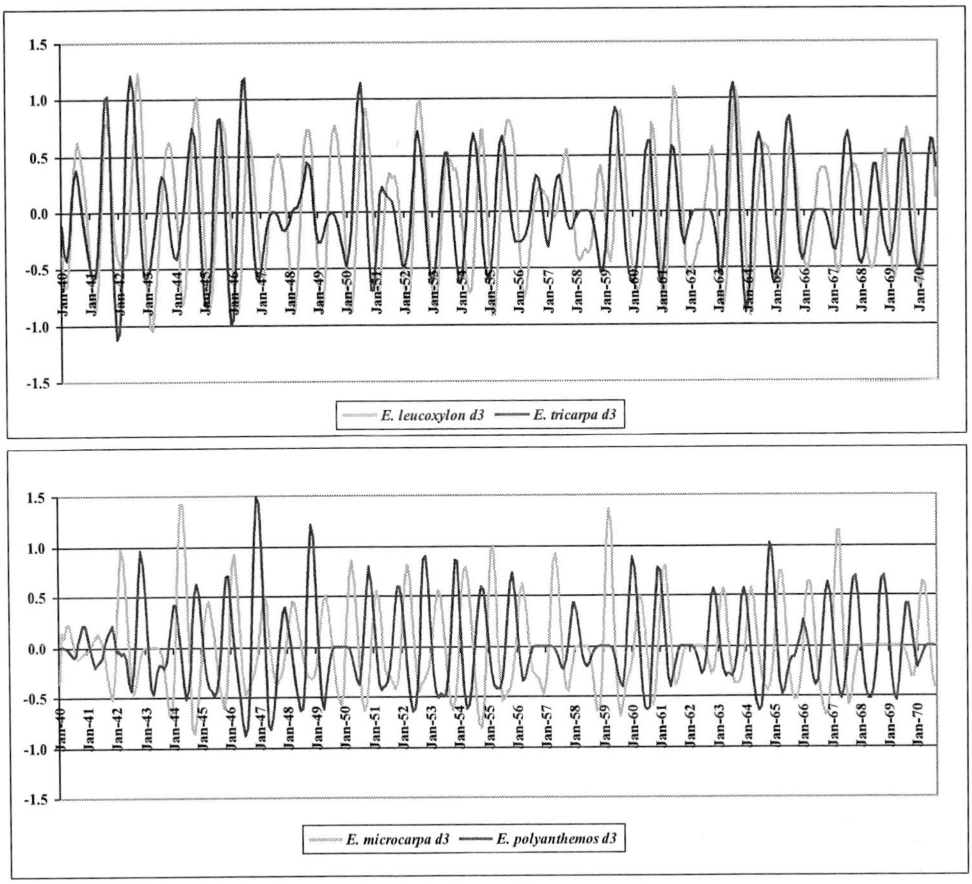

Figure 4. The d3 subcomponent for each of the species.

d5 for *E. tricarpa*, is a quasi-four year cycle (Figure 6). The average number of months between peaks is 47.9 with a range of 30 to 55 months. *Eucalyptus tricarpa* is also the species with the largest amplitude for d5 followed by *E. leucoxylon* and then *E. microcarpa* and *E. polyanthemos*.

A four year cycle for *E. tricarpa* was also found within the reconstructed 6 and 7th SSA series by Hudson and Keatley et al. [27] (Figure 7B). Clearly comparing A and B in Figure 7 shows that wavelets is detecting what SSA has previously indicated, namely a four year cycling of flowering for *E. tricarpa*. In terms of the DWT analysis we refer to *E. microcarpa* as having a quasi-four average cycle (48.1 month) (range 28 to 69 months). Figure 8 gives the profile of d6 for all four species, but this subcomponent was not easily interpreted.

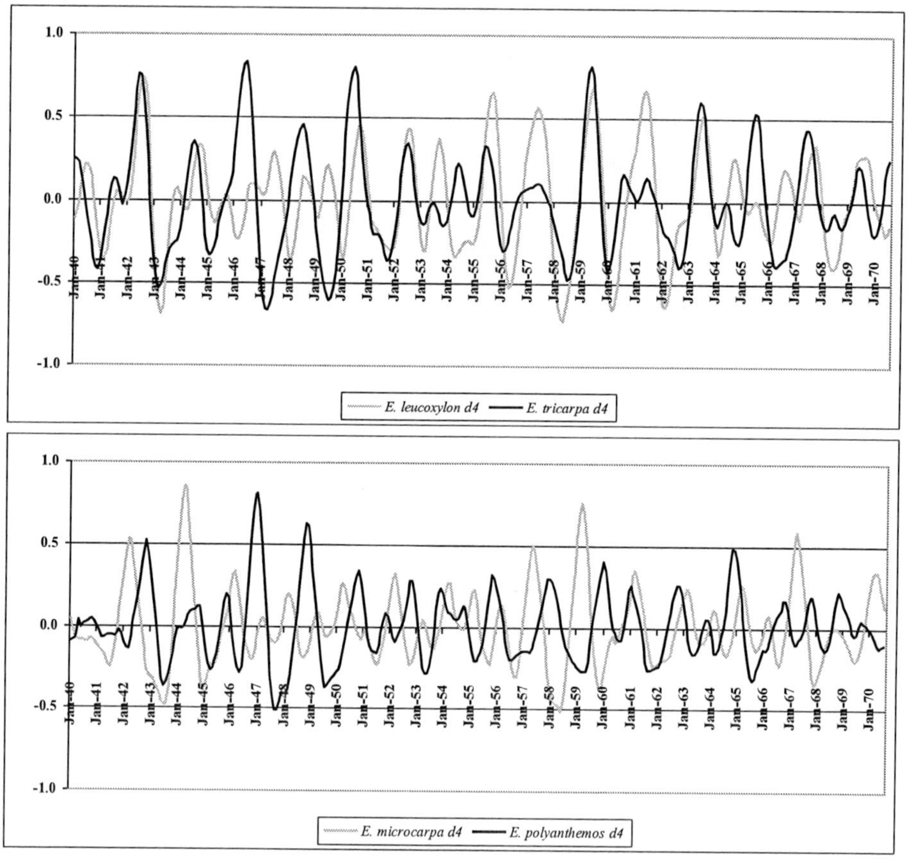

Figure 5. The d4 subcomponent of each species.

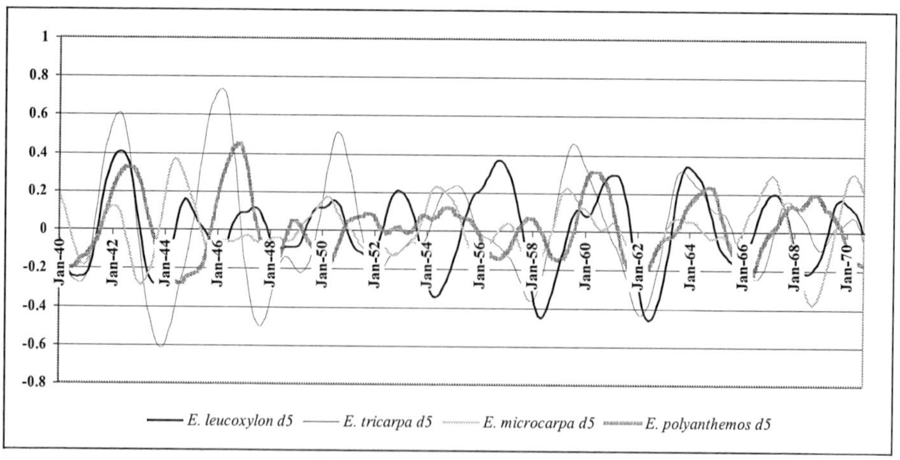

Figure 6. d5 subcomponents for each species.

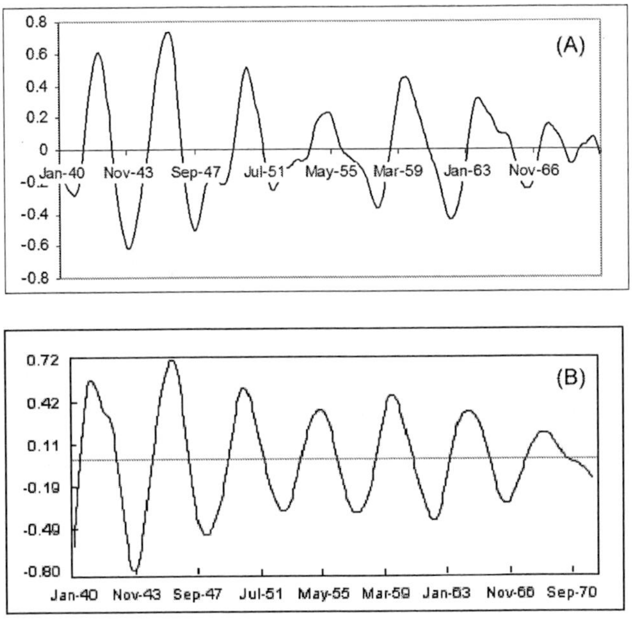

Figure 7. d5 for *E. tricarpa* (A) and its SSA reconstructed RS 6-7 series (B).

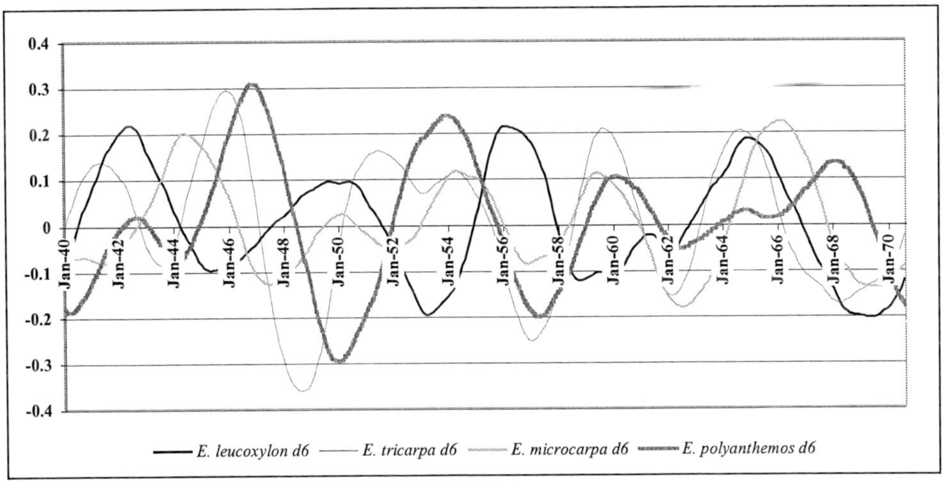

Figure 8. d6 subcomponents for each species.

Although as stated earlier, s4 is the smoothed series it is also reflective of the trend and variation in flowering intensity. s4 indicates that flowering has become less intense from 1940 to 1970 in all four species (Figure 9). *Eucalyptus leucoxylon* on average flowers (mean intensity 1.63) more intensely than the other species except for a brief period, between June 1945 and July 1946, when *E. tricarpa* flowered most (Figure 9). *Eucalyptus tricarpa* is the next most intense flowering species (mean intensity = 0.89). The remaining two species have flowered as intensely as each other (*E. microcarpa* mean = 0.59 and *E. polyanthemos* mean =0.57) over time, but alternate between periods of heightened intensity (Figure 6). The most intense flowering period for all species occurred prior to 1958. Hudson et al. [23] noted wetter years than average years in the mid 1940's to the end of the 1950's; and maximum temperature at Maryborough increasing until late 1967/early 1968 (also reflected by the Southern Oscillation Index [50, 51]).

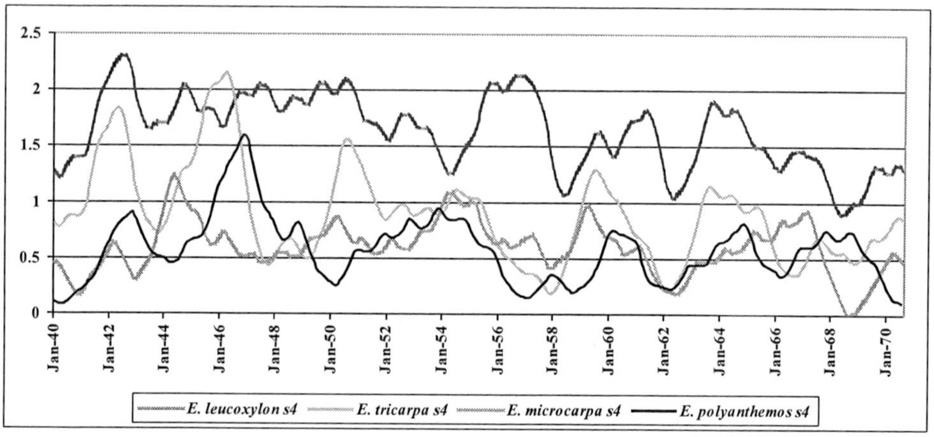

Figure 9. The s_4 subcomponent profiles for each species.

3.2. Correlation between Species Pairings

The wavelet correlation (WCORR) between all six species pairings at different wavelet levels or scales (1, 2, 4, 8, 16 and 32 months) were examined for the 4 species. Table 3 reports these WCORR values and associated 95% confidence lower and upper confidence interval limits (L.CI and U.CI, respectively). Significant ($P < 0.05$) wavelet correlations are those, where the upper (U) and lower (L) confidence limits are on the same side of the horizontal zero line (Figure 10). This is illustrated by the WCORR profiles of the flowering of *E. microcarpa* and *E. leucoxylon* (top), where wavelet scales 1 (d1), 2 (d2) and 4 (d3) are significant and negatively so, and *E. leucoxylon* with *E. tricarpa* (bottom) where wavelet scales 2 (d2), 4 (d3) and 8 (d4) are significant and positive. Table 4 compares the corresponding autoregressive (order 2) (AR (2)) pair-wise correlations based on the Moran synchrony test [31], and the corresponding SOM based network correlations [34], along with the d1 to d6 specific WCORR values. The a/synchronous species pairings are listed in decreasing order of Moran-based synchrony correlation in Table 4. The following abbreviations for species names are used in Table 4 and Table 6: *E. leucoxylon* (leu), *E. tricarpa* (tri), *E. polyanthemos* (pol) and *E. microcarpa* (mic).

The WCORR signatures exhibit a common pattern for synchronous species pairs, with generally positive WCORR values associated with d2 and d3. Specifically, d2 WCORR values are significant and positive for the species pair, *E. leucoxylon* and *E. tricarpa* (WCORR = 0.272; 95% CI (0.071, 0.451)), d3 (WCORR = 0.410; 95% CI (0.136, 0.626) and d4 (WCORR = 0441; 95% CI (0.035, 0.722) (Table 3, Figure 10). Likewise WCORR values are significant and positive in *E. leucoxylon* and *E. polyanthemos* for d1 (WCORR = 0.198; 95% CI (0.055, 0.333)) and d2 (WCORR = 0.221; 95% CI (0.017, 0.408)) (Table 3). *Eucalyptus leucoxylon* is the only species to synchronise flowering with *E. tricarpa;* and the flowering of *E. leucoxylon* is synchronous with *E. polyanthemos*. This is in agreement with the results from SOM networks and correlations and with Moran synchrony tests (Table 4) [34]; and is in accord with Keatley et al. (2004) and also with results based on mixed transition distribution (MTD) based Moran residuals [32-34]. Note the flowering period of *E. polyanthemos* is entirely encompassed by that of *E. leucoxylon*. It is also noteworthy that all these species have been found to overlap – to varying degrees [52]

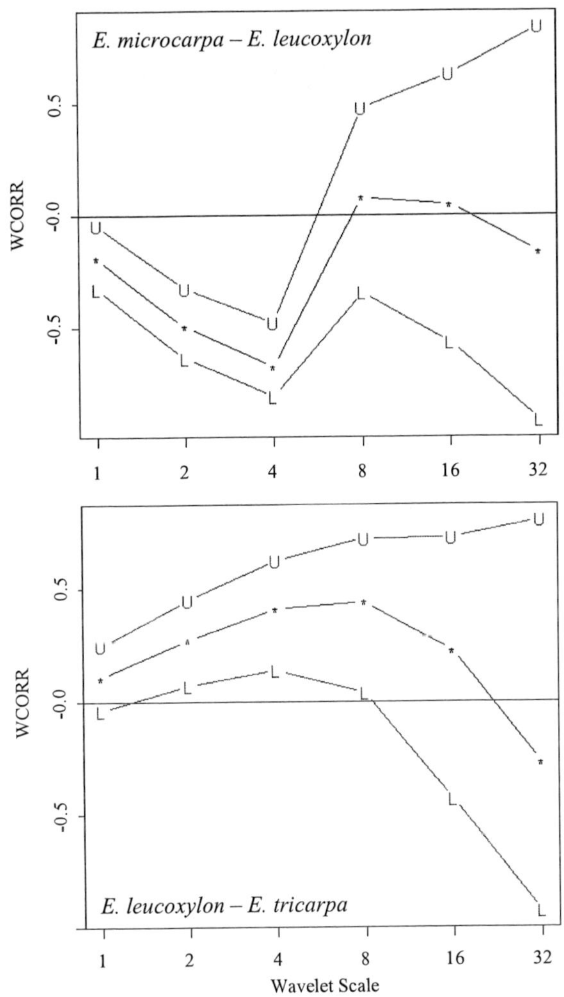

Figure 10. Wavelet correlations (WCORR) of the flowering of E. microcarpa and E leucoxylon (top) and of *E. leucoxylon* with *E. tricarpa* (bottom) (95% confidence intervals indicated by U and L).

Distinctly different patterns and negative values for WCORR are associated with asynchronous species (Tables 3 and 4). The d2 (WCORR = -0.277; 95% CI (-0.413, -0.024)) and d3 (WCORR = -0.521; 95% CI (-0.705, -0.724)) specific WCORR values are significant and negative for *E. tricarpa* and *E. polyanthemos* (Table 3).

Table 3. Wavelet correlation (WCORR) of the 6 pairs of species peak flowering intensities (* signifies P< 0.05)

	E. leucoxylon – E. tricarpa			*E. microcarpa – E. tricarpa*			*E. polyanthemos - E. tricarpa*		
	L.CI	WCORR	U.CI	L.CI	WCORR	U.CI	L.CI	WCORR	U.CI
d1	-0.041	0.105	0.246	-0.120	0.025	0.169	-0.201	-0.058	0.087
d2	0.071	0.272*	0.451	-0.249	-0.046	0.160	-0.413	-0.227*	-0.024
d3	0.136	0.410*	0.626	-0.300	-0.01	0.281	-0.705	-0.521*	-0.272
d4	0.035	0.441*	0.722	-0.052	0.368	0.677	-0.546	-0.172	0.258
d5	-0.435	0.223	0.726	-0.396	0.267	0.747	-0.476	0.174	0.700
d6	-0.931	-0.271	0.803	-0.663	0.528	0.962	-0.730	0.428	0.951
	E. microcarpa – E. leucoxylon			*E. polyanthemos - E. leucoxylon*			*E. polyanthemos - E. microcarpa*		
	L.CI	WCORR	U.CI	L.CI	WCORR	U.CI	L.CI	WCORR	U.CI
d1	-0.323	-0.187*	-0.044	0.055	0.198*	0.333	-0.273	-0.134	0.011
d2	-0.637	-0.497*	-0.325	0.017	0.221*	0.408	-0.420	-0.236*	-0.032
d3	-0.806	-0.674*	-0.477	-0.222	0.073	0.355	-0.527	-0.279	0.012
d4	-0.344	0.079	0.476	-0.301	0.127	0.512	-0.564	-0.198	0.233
d5	-0.568	0.048	0.630	-0.614	-0.023	0.585	-0.732	-0.235	0.425
d6	-0.914	-0.164	0.840	-0.953	-0.445	0.72	-0.807	0.262	0.929

Table 4. Results from the AR (2) and SOM based tests (in decreasing order of AR (2) r values) for pairs synchronous at zero lag (* signifies P< 0.05) (NS denotes not significant at the 5% level of significance) ($^{\Re}$ signifies significant at 10%). The following abbreviations for species names are used: *E. leucoxylon* (leu), *E. tricarpa* (tri), *E. polyanthemos* (pol) and *E. microcarpa* (mic)

Model species	Fitted Species	AR(2) Correlation (r)	AR (2) P value	SOM Correlation (r)	SOM P value	WCORR values					
Synchronous											
Leu	Pol	0.185	0.001	0.137	<0.015	0.198*	0.221*	NS	NS	NS	NS
Pol	Leu	0.171	0.002	-	-	-	-	-	-	-	-
Leu	Tri	0.171	0.002	0.361	<0.0005	NS	0.272*	0.410*	0.441*	NS	NS
Tri	Leu	0.142	0.010	-	-	-	-	-	-	-	-
Asynchronous											
Tri	Pol	-0.121	0.028	-0.217	<0.0005	NS	-0.227*	-0.521*	NS	NS	NS
Pol	Tri	-0.129	0.020	-	-	-	-	-	-	-	-
Mic	Pol	-0.132	0.017	-0.213	<0.0005	NS	-0.236*	-0.279$^{\Re}$	NS	NS	NS
Pol	Mic	-0.147	0.008	-	-	-	-	-	-	-	-
Leu	Mic	-0.226	<0.0005	-0.372	<0.0005	-0.187*	-0.497*	-0.674*	NS	NS	NS
Mic	Leu	-0.169	0.002	-	-	-	-	-	-	-	-
No relation											
Tri	Mic	0.023	NS	0.090	0.090$^{\Re}$	NS	NS	NS	NS	NS	NS
Mic	Tri	0.060	NS	-		-	-	-	-	-	-

For *E. leucoxylon* and *E. microcarpa* their d2 (WCORR = -0.497; 95% CI (-0.637, -0.325)) and d3 (WCORR = -0.674; 95% CI (-0.806, -0.477)) specific WCORR values are significant and negative (along with d1 WCCORR values (WCORR = -0.187; 95% CI (-0.323, -0.044)). *Eucalyptus polyanthemos* and *E. microcarpa* are shown to have different flowering durations (Table 1), as evidenced by d2 specific WCORR values which are significant and negative, though they flower at the same time of year (Table 1). The species pairs identified as asynchronous by wavelet analysis, namely, *E. microcarpa and E. polyanthemos, E. tricarpa* and *E. polyanthemos*, *E. leucoxylon* and *E. microcarpa,* were those which had low overlap, (0.06, 0.13 and 0.29, respectively) as determined by Keatley and Hudson [52].

3.3. Wavelet Cross-Correlation between Species Pairings

Wavelet cross-correlations (WCCORRs) between all six species pairings were examined for all wavelet scales (1 to 32 months, d1 to d6). Wavelet cross-correlations between flowering intensity of all species pairings are lagged from 0 and ± 36 months.

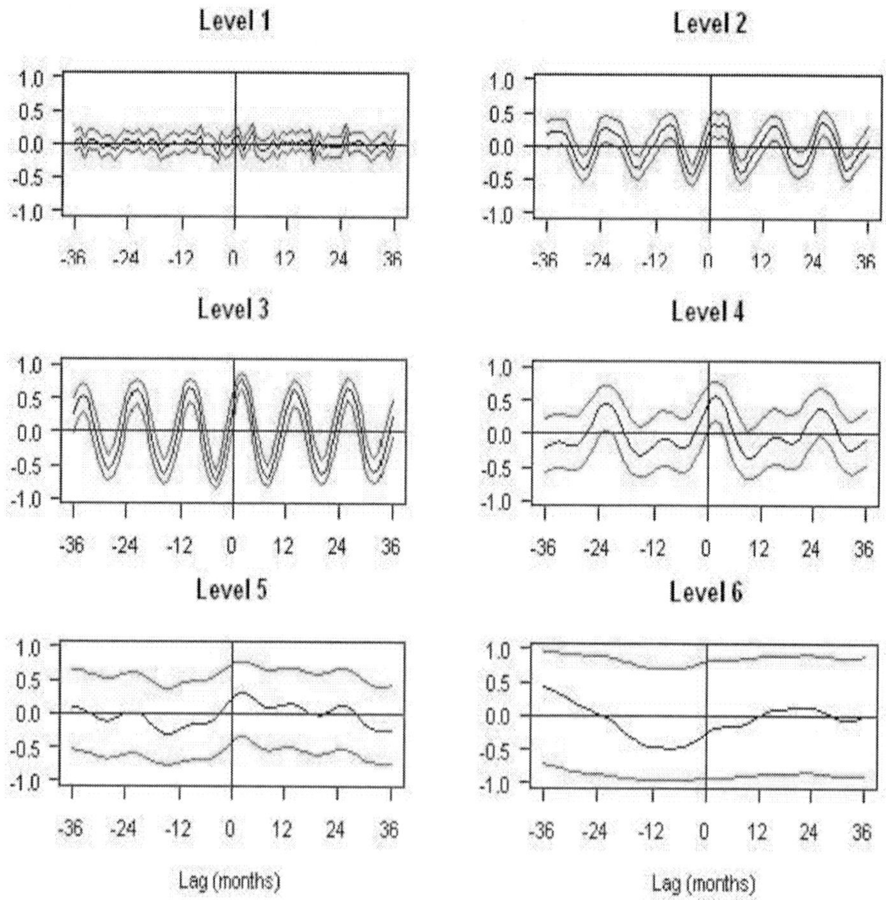

Figure 11. Wavelet cross-correlations (WCCORR) plots of d1, d2, d3, d4, d5 and d6 for the flowering of *E. leucoxylon* and *E. tricarpa* (95% confidence intervals indicated by the red lines).

Figure 11 gives wavelet cross-correlation profiles for each of d1 (level 1), d2 to d6 between flowering of *E. leucoxylon* and *E. tricarpa*, and analogously Figure 12 for *E. microcarpa* and *E. leucoxylon*, respectively (95% confidence intervals indicated by U and L). Note the positive cross-correlations at zero lag for level 1, 2, 3, 4 and 5 for *E. leucoxylon* and *E. tricarpa* (Figure 11). This species pair was highly synchronous by wavelet correlation (WCORR) analysis (Figure 10, Table 3). In contrast the WCCORR profiles of d1, d2, d3, d4 and d5 are negative for *E. microcarpa* and *E. leucoxylon* (Figure 12), this species pairing were shown to be highly asynchronous by WCORR analysis of their flowering intensities (Figure 10, Table 4).

The first two subcomponents reflect, as reported above, the duration and the overall pattern of the original flowering intensity profiles for each species, and that of the annual cycle of flowering is delineated by d3 (level 3). The median flowering duration for *E. tricarpa* and *E. leucoxylon* are 6 and 9 months, respectively (Table 1); peak flowering occurs in July and September, respectively; flowering commencement in April and May, and flowering finishes in September and December, respectively – their d2 specific period of positive WCCORR (Figure 13) is thus the overlap, namely 6 months. The overlap in flowering of a given species pairing is generally reflected in the periods of the WCCORR profiles of d2 (Figures 11 and 12) per species (see also Figure 13 for a selection of four of the six species pairings).

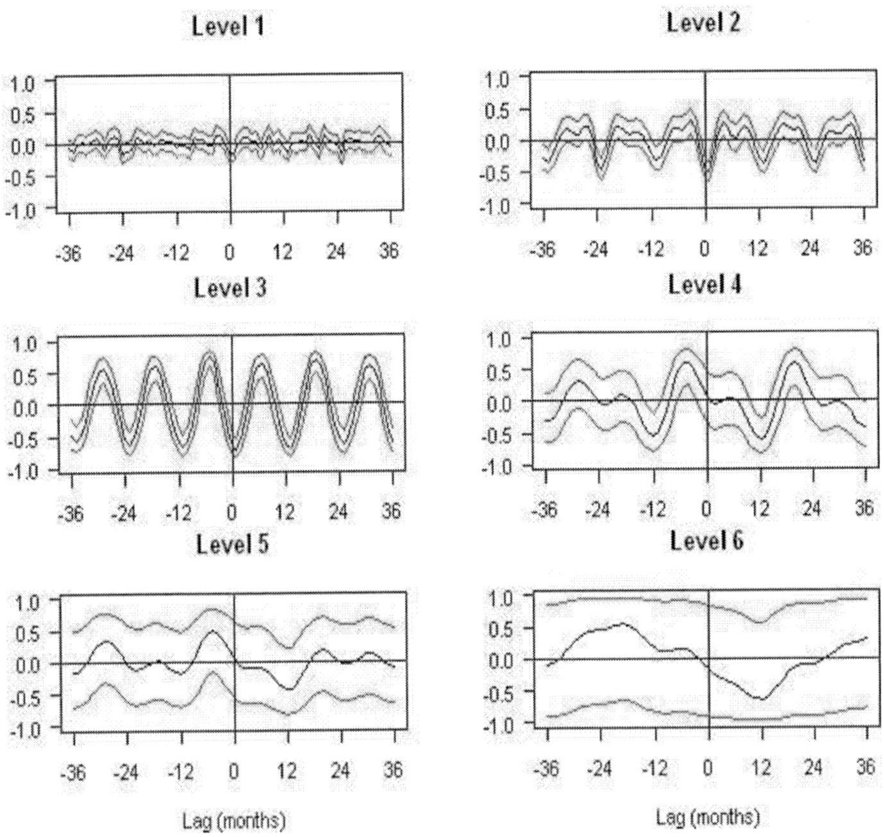

Figure 12. Wavelet cross-correlations (WCCORR) plots of d1, d2, d3, d4, d5 and d6 for the flowering of *E. microcarpa* and *E. leucoxylon* (95% confidence intervals indicated by the red line).

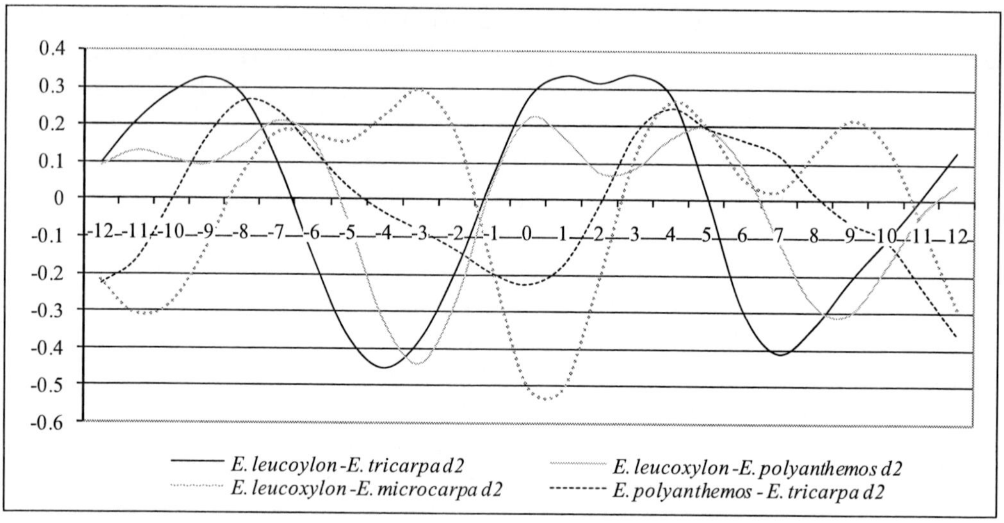

Figure 13. Individual wavelet cross-correlation plots for d2 WCCORR for the 4 species pairings.

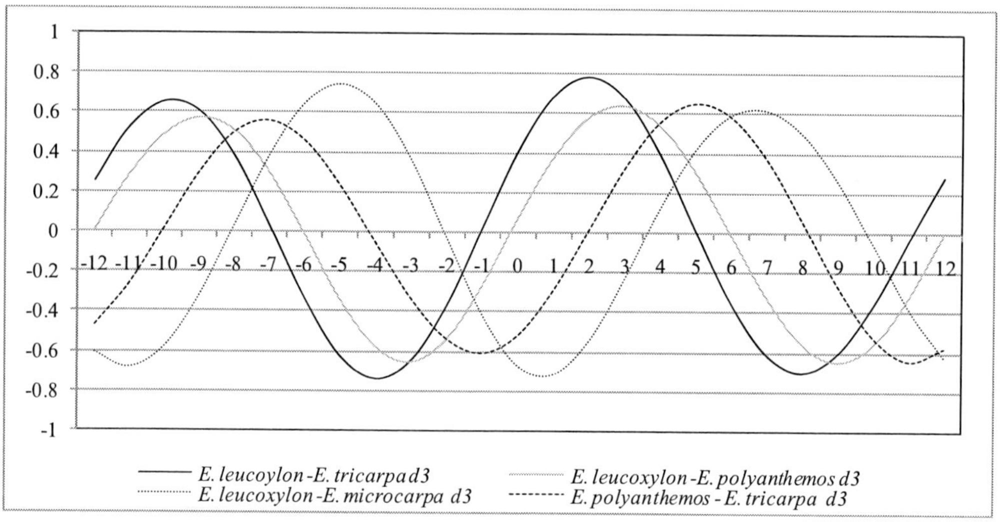

Figure 14. Individual wavelet cross-correlation plots for d3 WCCORR for the 4 species pairings.

From Figures 11 and 12 there is a distinct sinusoidal cycling in d3 for all species. From Figure 14 *E. leucoxylon* and *E. tricarpa* has d3 specific WCCORR values which are positive from lag -1 month, zero lag up to 5 months (i.e. positive for a period of 6 months); and negative from lag -7 to -2 months (i.e. negative for a period of 6 months). Thus there is a six month period of negative correlations from -7 months, followed by six months of positive WCCORR values to lag 5 months. Likewise *E. polyanthemos* and *E. leucoxylon* has positive d3 specific WCCORR values from lag 0 to 6 months, and negative d3 specific WCCORR from lag -5 to lag -1 months (Figure 14). Thus there is a six month period of negative correlations from -5 months, followed by six months of positive WCCORR values to lag 6 months for *E. polyanthemos* and *E. leucoxylon*. It is noteworthy that there are positive d3

WCCORR values for lags at zero to 5 or 6 months for these two synchronous pairs, *E. leucoxylon* and *E. tricarpa;* and *E. polyanthemos* and *E. leucoxylon.*

Table 5. Species specific change points (month, season) d3 for the species

Mean temperature	Positive to negative		Negative to positive	
Species	Month	Season	Month	Season
E. leucoxylon	August	Winter	February	Summer
E. tricarpa	April	Autumn	October	Spring
E. polyanthemos	December	Summer	July	Winter
E. microcarpa	September	Spring	March	Autumn

In contrast for the asynchronous pair *E. polyanthemos* and *E. tricarpa* the wavelet WCCORR values for subcomponent d3 are negative from -4 lag to zero lag and up to 2 months, and positive from lag -10 to -5 months (Figure 14). Thus there is a six month period of positive correlations from lag -10 up to lag -5 months, followed by six months of negative WCCORR values to lag 2 months for *E. polyanthemos* and *E. tricarpa*. Similarly, the WCCORR values for wavelet d3 for the asynchronous pair *E. leucoxylon* and *E. microcarpa* are negative from -2 lag to zero lag and up to 3 months, and preceded by positive wavelet cross-correlations for earlier lags from -8 to -3 months (Figure 14). Thus there is a six month period of positive correlations from lag -8 up to lag -3 months, followed by six months of negative WCCORR values to lag 3 months for *E. leucoxylon* and *E. microcarpa*; again a six month period of positive d3 cross-correlations at zero lag (up to 2-3 months), preceded by negative WCCORRs for the asynchronous species' d3 earlier (from -10 or -8 months). The cycling of the d3 specific WCCORR profiles mirror changing climatic influence on species flowering [28]., but However, the profile of positive cross-correlations at earlier lags followed 6 months later by negative cross-correlations, is opposite to the profile of negative WCCORR at earlier lags followed by positive cross-correlations for the synchronous pair; the signs of earlier positive to negative WCCORRs, and earlier negative to positive profiles of WCCORR, thus mirror asynchronous or synchronous pairings, respectively. These sinusoidal profiles of d3 specific WCCORR show that there are positive and negative cross-correlational relationships between pairs of species' flowering of consistently 6 months duration. This agrees with earlier results [28, 34] which established seasonal change points from positive to negative cross-correlation and vice-versa with regard to a given species' flowering and climate predictors (temperature or rainfall) (see Table 5, where the season specific to each change point month is given per species, as found by SOMs [34]. Table 1 also details the species' phenophases and their seasons.

Distinctly different patterns for d3 WCCORR are associated with asynchronous and synchronous time series. d3 wavelet cross correlations and their changing sign are indicative of changing climatic impact on flowering, synchrony *per se* and peak at the lag which indicates the difference between the species pair's peak flowering months. Specifically from Figure 14 the positive peaks of d3-specific WCCORR profiles align with the number of months difference between the peak flowering months of the two species in the pair under consideration. For example, for the pair *E. leucoxylon* – *E. tricarpa* the highest d3 WCCORR peak occurs at two months, which is the difference between their respective peak flowering months, September and July, respectively (Δ= 2 months); for *E. leucoxylon* –

E. polyanthemos the highest peak occurs at three months, which is the difference between their respective peak flowering months, September and November, respectively (Δ= 2 months); for the pair *E. leucoxylon – E. microcarpa* the highest peak occurs at seven months, the difference between their flowering peaks, September and March, respectively, is 6 months; for *E. polyanthemos – E. tricarpa* the highest d3 WCCORR peak is at five months, the difference between flowering peaks November and July, respectively, is 4 months, (Figure 14, Table 1).

Table 6 gives the lags that maximize the absolute value of the d3 WCCORR profiles for the species pairs that are asynchronous at lag zero (Table 3). Maximum synchrony (positive absolute value for WCCORR for the d3 annual cycling subcomponent is achieved via appropriate lagging for most previously discordant pairs (Table 6); in agreement with Kim et al.[32] and Hudson et al. [31]. For the species pairs that can be synchronised, the amount of lag needed to synchronise the two species, in any given pair, denoted by Δ, is often, *but not always*, the difference between the peak flowering months of each species (peak months are given in Tables 1). For example, the peak flowering months for *E. microcarpa* and *E. polyanthemos* are March and November, respectively (8 months between the peaks) and the lag to synchronise *E. microcarpa* and *E. polyanthemos* is $\pm\Delta = \pm 8$. Likewise *E. tricarpa*'s peak flowering month is July and that of *E. polyanthemos* is November, thus, Δ = 4 (Table 6). The only species pair that cannot be synchronised by lagging, with a negative maximum d3 WCCORR value of -0.712, (Table 6), and which is also asynchronous at lag zero (d3 WCORR = -0.674, Table 4) is *E. leucoxylon* and *E. microcarpa* (Table 6). If synchronisation occurred only by matching of the peak flowering months then all species should be able to be made synchronous by lagging. Synchronisation is thus a complex mechanism that incorporates, apart from peak flowering times, the duration, timing of flowering commencement and cessation, along with climatic impacts [28, 34].

Table 6. AR (2) and SOM correlations (r) (in decreasing order of AR (2) *r*) with lags to maximize |*r*/ for lag zero asynchronous pairs. (* P value <0.0005). The following abbreviations for species names are used: *E. leucoxylon* (leu), *E. tricarpa* (tri), *E. polyanthemos* (pol) and *E. microcarpa* (mic)

Model species	Fitted Species	AR (2) (r)	SOM (r)	Peak month of fitted species	AR(2) Lag to synchronise	Max WCORR d3	Lag to maximise |WCORR| of d3
Lagged							
Pol	Mic	0.296*	0.501*	March	8 =12-4	0.675	8
Mic	Pol	0.255*	0.490*	Nov	-8	-	-
Pol	Tri	0.269*	0.443*	July	4	0.663	5
Tri	Pol	0.234*	0.271*	Nov	4	-	-
Mic	Tri	0.237*	0.480*	July	-4	0.324	-4
Tri	Mic	0.208*	0.497*	March	4	-	-
Mic	Leu	-0.232*	-0.521*	Sept	11	-0.712	11
Leu	Mic	-0.255*	-0.525*	March	-11	-	-

4. Discussion

Wavelets Multiresolution multiresolution analysis (MRA) has identified five subcomponents in each flowering series: these are a non-flowering phase, duration, annual cycle, flowering intensity, a quasi-four year cycle (for *E. tricarpa* only); as well as the overall trend for each species. More specifically: d1 is the non-flowering phase per species. These so-called "off phases" coincide with other reproductive phases (e.g. budding, seeding) in eucalypts [53, 54]. This phenomenon has also been reported specifically in *E. tricarpa* [55]. The second subcomponent, d2, reflects both the duration and the pattern of the original flowering time series record, for a given species (see also Hudson et al. [22]). The annual cycle has been clearly delineated by subcomponent d3, with the mean month of peak intensity, so identified by peaks in d3. For all the species this is in agreement with results based on different approaches (the mean flowering intensity in a flowering year [25] and/or by singular spectrum analysis (SSA) [23, 27]. d4 relates to the annual or biennial cycling of the intensity of flowering years. In all species, except *E. leucoxylon,* the biennial cycle is related to years of low flowering intensity (< 2); where low flowering intensity includes in addition years of no flowering (i.e. flowering intensity = 0). In *E. leucoxylon* flowering alternates between a quasi-biennial (1950 to 1952, 1955 until 1963) and an annual cycle. It is the only species, where the biennial cycle is associated solely with low (< 2) intensity, and/or late commencement with short duration (commencing in September or October with a most probable commencement month of April [25].

Variability in flowering intensity, in general, as well as its cyclic behaviour has been long reported in eucalypts [56]. Two yearly flowering cycles in these species has also been previously discussed, as based on the judgment and expertise of apiarists [57-59]; the collection of reproductive components [55] and confirmed by autocorrelogram analysis [60], and by singular spectrum analysis [27]. SSA is a global analysis [23, 27], in contrast to the wavelets approach in this chapter, which affords a localized analysis. Yiou and colleagues [61] have discussed methods by which SSA can be made more wavelet-like specifically via the use of windowing techniques, not however, used in the SSA applications of Hudson et al. [23] and Hudson and Keatley [27]. d5 represents a quasi-four year cycle. A four year cycle was also previously detected by SSA for *E. tricarpa* [27]; and was shown via autocorrelogram analysis by Wells [60], who with Somerville and Campbell [59] also identified a four year flowering cycle in *E. microcarpa.* s4 is the smoothed series that illustrates the slowly varying portion of the flowering signal, and is taken to be an estimate of the overall trend. The results of wavelet analysis, namely a decreasing trend in flowering intensity over the period of observation, also agree with SSA results [22]. Note that the so-called trend is usually the first primary reconstructed component in SSA and mirrors the smoothed overall trend s4 discussed in this chapter.

The wavelet correlation (WCORR) signatures exhibit a common pattern for synchronous species pairs, with d2 , d3 and d4 WCORR values being significant and positive for the species pair, *E. leucoxylon* and *E. tricarpa*; and with d1 and d2 WCORR values being significant and positive for species pair *E. leucoxylon* and *E. polyanthemos.* These species pairings were likewise found to be synchronous by SOM and Moran tests [34]. *Eucalyptus leucoxylon* is the only species to synchronise flowering with *E. tricarpa.* This species pair was also shown to be synchronous by Keatley et al. [52]. It is noteworthy that recent work by

Hudson et al. [22, 28] showed that the DWT wavelet based *climate* signature of *E. leucoxylon* and *E. tricarpa* is similar with respect to temperature, both exhibit decreased flowering with increased temperature. Separation of flowering time is one mechanism that species employ to avoid hybridization [62, 63], hence synchrony between *E. leucoxylon* and *E. tricarpa* is unexpected, as they are also within the same taxonomic series [64] and have similar sized flowers [65] (increases the likelihood of cross pollination). Production of *E. leucoxylon* flowers at Havelock are negatively skewed, not so for *E. tricarpa*. This results in the temporal separation of their months of peak flowering (September and July, respectively) [25] therefore reducing the likelihood of hybridization. Given the flowering overlap between these species, as well as their similarly shaped flowers, there may also be a facilitation effect that is the pollination of *E. leucoxylon* is enhanced as the earlier flowering of *E. tricarpa* has already attracted suitable pollinators. A strong synchrony leads to a greater chance of hybridization; a low synchrony less so; overlap does not necessarily imply result in hybridization. Griffin et al. [63] gives details about what is known in regard to hybridization between some of these species.

According to the MODWT-MRA approach in this chapter the flowering of *E. leucoxylon* is also synchronous with *E. polyanthemos,* in agreement with the results from SOM and Moran synchrony tests [34]; in accord with Keatley et al. [52] and also with results based on mixed transition distribution (MTD) based Moran residuals [32-34]. Indeed the flowering period of *E. polyanthemos* is entirely encompassed by that of *E. leucoxylon*. These two species both cease flowering in December and share the same season for peak flowering (Spring), though their peaks are separated by two months [25]. These two species can occur in the same geographical area and have flowering periods that overlap significantly [34, 52]. Differentiation of these two species is based on their differing months of peak flowering as well as their separated months of most probable flowering; October and November, respectively. Likewise their flowering commencement months differ, May and October, respectively [25].

Distinctly different patterns for WCORR were shown to be associated with asynchronous species. The d2 and d3 specific WCORR values are significant and negative for *E. tricarpa* and *E. polyanthemos* (also shown to be asynchronous by SOMs, in agreement with previous work [52] and results based on Moran residuals [31, 34]); and for *E. leucoxylon* with *E. microcarpa* (along with d1). *Eucalyptus polyanthemos* and *E. microcarpa* are shown to have different flowering periods, as evidenced by d2 specific WCORR values which are significant and negative. These species may occur together. For example, *E. tricarpa* may be found with *E. leucoxylon* and *E. polyanthemos*. Separation of flowering times is just one mechanism to avoid competition for pollinators and reduce hybridization. This is particularly the case for four species examined which are all in the same informal subgenus – Symphomyrtus and hence hybridization is more likely. *Eucalyptus tricarpa's* mean overlap with *E. polyanthemos* and *E. microcarpa* is low (0.13 and 0.36, respectively). This is because the flowering of *E. tricarpa* and *E. polyanthemos* did not overlap in 18 years of 30 years [52], in addition they have differing months of peak flowering July and November, respectively. Their cessation of flowering also occurs in different months, September and December, respectively. Overlap did not occur between *E. tricarpa* and *E. microcarpa* in 10 of the 30 years. They also have different commencement, peak and cessation months. However, in some years their flowering period can largely overlap (0.53 and 0.78, respectively) [52].

Wavelets correlation and cross-correlation, SOM and Moran synchrony tests generally agree showing that: *E. leucoxylon* flowers asynchronously to *E. microcarpa*; *E. polyanthemos* is asynchronous to *E. tricarpa* and *E. microcarpa*; *E. tricarpa* is asynchronous to *E. polyanthemos*; *E. leucoxylon* cannot be synchronised by lagging with respect to *E. microcarpa. Eucalyptus leucoxylon* is the only species to significantly synchronise flowering with *E. tricarpa.* The highest degree of asynchrony occurred between *E. leucoxylon* and *E. microcarpa* (for d2 and d3). This result agrees with Keatley et al. [52] who established that a cross between *E. leucoxylon* and *E. microcarpa* is impossible; this species pair has 10 years (of 30 years recording) of no overlap (with a long term mean synchrony value of 0.29). Note *E. leucoxylon's* long duration of flowering (median of 9.0 months) versus the shorter duration of 4.0 months for *E. microcarpa*; as well the different average flowering periods: May to December for *E. leucoxylon* and February to May for *E. microcarpa.* This species pair remains asynchronous and cannot be synchronised by lagging the flowering records of one versus the other. DWT based results in this chapter, along with SOM based clustering and Moran AR (2) tests also found that *E. polyanthemos* was asynchronous to both *E. microcarpa* and *E. tricarpa,* in agreement with the extended Kalman filter (EKF) [66, 67] based synchrony measures of Kim et al. [32]. Keatley et al. [52] demonstrated that *E. polyanthemos* and *E. microcarpa* have 25 years with no overlap (with a long term mean synchrony value of 0.29). *Eucalyptus microcarpa* and *E. tricarpa* are not significantly correlated via DWT WCORR methods as reported in this chapter, nor by SOM network nor Moran based synchrony statistics [34]. *Eucalyptus microcarpa* and *E. tricarpa* have a long term mean synchrony value of 0.36 which was considered low by Keatley et al. [52]. These two species have separated months of peak flowering (March and July, respectively) and of flowering cessation (March and September, respectively). As these species are in the same taxonomic section temporal separation of flowering peaks makes sense in terms of avoiding competition. However, given that they overlap the production of hybrids is a possibility.

The wavelet cross-correlational profiles for d2 reflect flowering overlap between species pairs. The sinusoidal shaped d3 wavelet cross correlations and their changing sign are indicative of differences in climate, synchrony *per se* and peak at the lag which indicates the difference between the species pair's peak flowering months. Distinctly different patterns for d3 WCCORR are associated with asynchronous and synchronous time series. The cycling of the d3 specific WCCORR profiles mirror changing climatic influence on species flowering as established by Hudson et al [28]. The profile of positive cross-correlations at earlier lags followed 6 months later by negative cross-correlations for asynchronous species is opposite to the profile of negative WCCORR at earlier lags followed by positive cross-correlations for the synchronous pairings. The ordering of earlier positive to negative WCCORRs, versus earlier negative to positive profiles of WCCORR, thus mirror asynchronous or synchronous pairings, respectively. These sinusoidal profiles show that there are positive and negative cross-correlational relationships between pairs of species' flowering of 6 months duration. This agrees with earlier results by Hudson et al. [28] which established seasonal change points from positive to negative cross-correlation and vice-versa with regard to a given species' flowering and climate (temperature or rainfall). Furthermore, the profiles of wavelet cross-correlation (WCCORR) enable us to further understand how synchrony may relate to patterns (subcomponents) of flowering (peak annual cycles, start, cessation, duration, overlap), adding to recent insights based on SOMs by Hudson et al [34]. The MODWT-MRA results mirror those of SOM [34], and of Moran synchrony statistics (whether based on ARIMA, transitional state or extended Kalman filter residuals [32, 33].

There appears to be at a very broad level a seasonal influence which points to a similar combination of climatic drivers (e.g. temperature or more precisely growing degree days or temperature and rainfall) [68]. The species studied in this chapter (*E. leucoxylon, E. microcarpa, E. polyanthemos* and *E. tricarpa)* have had their flowering relationship with temperature and rainfall examined [22, 69, 70]. Flowering of each of these species has been shown to be significantly influenced by temperature two months prior to flowering commencement [22, 33, 70]. This is a known influence for species which commence flowering in Spring and Summer [7, 71-74]. *Eucalyptus microcarpa* commences flowering in Spring and *E. polyanthemos* in Summer. The remaining two species commenced flowering in Autumn. The drivers for autumn phenology are less clear [75] – and for the most part the phases examined have been leaf colouring, leaf fall and fruit ripening [7, 75]. However, other drivers such as day length, wind, and frost [75, 76] have been suggested. It is generally rare to find studies informing on multiple flowering characteristics, and for such a long time period of observation; as in the investigation reported in this chapter. The relationships between phenophases and climatic variables, may be considered to be physiologically and/or statistically based [77, 78]. In an earlier study [79], it was suggested that even though flowering time seems to be under the control of climatic conditions, it is during bud formation and development that there may be genetic control of the actual flowering dates in local (sub-) populations. It was advocated that heavy and light flowering seasons appear to occur in cycles in many species (also evidenced in some species in this chapter) and are related to resource availability.

Conclusion

The utility of wavelets for examining synchronisation of flowering of Australian eucalypts species is demonstrated in this chapter. The MODWT-MRA approach is shown to accurately enumerate and visually demonstrate synchronisation of multidimensional time series. Synchronisation of eucalypt flowering is shown to be a complex mechanism that incorporates all the flowering elements—flowering duration, timing of peak flowering, and the timing of start and finishing of flowering, as well as possibly specific climate drivers for flowering. A similar, but criterion-based approach has been presented by Kang et al. [80] as a diagnostic tool by which to assess the validity of agitation-sedation protocols of patients in ICU. The utility of wavelets for examining the relationship of flowering to *climate* was also recently established for these 4 species by Hudson et al. [22] (see also [23, 27]); and recently expanded to a study of 8 species to find wavelet-based flowering-climate signatures [28].

References

[1] M. Dunlop and P. R. Brown, Implications of climate change for Australia's National Reserve System: A preliminary assessment. Report to the Department of Climate Change., *Department of Climate Change*, Canberra, Australia (2008).

[2] J. Forrest and A. J. Miller-Rushing, *Proc. Roy. Soc. London B.* 365, 3101 (2010).
[3] P. J. Beggs, *Clin. Exp. Allergy*. 34, 1507 (2004).
[4] H. Fitter and R. S. R. Fitter, *Science*. 296, 1689 (2002).
[5] J. H. van Vliet, Societal adaptation options to changes in phenology, in Phenological Research: Methods for environmental and climate change analysis, I. L. Hudson and M. R. Keatley, Eds., Springer, Berlin, (2010) p. 75.
[6] M. R. Keatley and I. L. Hudson, Introduction and overview, in Phenological Research: Methods for environmental and climate change analysis, I. L. Hudson and M. R. Keatley, Eds., Springer, Berslin, (2010) p. 1.
[7] Menzel, T. H. Sparks, N. Estrella, E. Koch, A. Aasa, R. Ahas, K. Alm-Kubler, P. Bissolli, O. Brasavská, A. Briede, F.-M. Chmielewski, Z. Crepinsek, Y. Curnel, A. Dahl, C. Defila, A. Donnelly, Y. Filella, K. Jatczak, F. Mage, A. Mestre, Ø. Nordli, J. Peñuelas, P. Pirinen, V. Remišová, H. Scheifinger, M. Striz, A. Susnik, A. J. H. van vliet, F.-E. Wielgolaski, S. Zach and A. Zust, *Global Change Biol.* 12, 1969 (2006).
[8] Rosenzweig, D. Karoly, M. Vicarelli, P. Neofotis, Q. Wu, G. Casassa, A. Menzel, T. L. Root, N. Estrella, B. Seguin, P. Tryjanowski, C. Liu, S. Rawlins and A. Imeson, *Nature*. 453, 353 (2008).
[9] T. H. Sparks, P. J. Croxton, N. Collinson and P. W. Taylor, *Ann. Appl. Biol.* 146, 531 (2005).
[10] L. Hudson, Meta-analysis and its application in phenological research: a review and new statistical approaches, in Phenological Research: Methods for environmental and climate change analysis, I. L. Hudson and M. R. Keatley, Eds., Springer, Berlin, (2010) p. 463.
[11] L. Hudson, Meta analysis, in Encyclopedia of Climate and Weather, S. H. Schneider, M. Mastrandrea and T. L. Root, Eds., Oxford University Press, New York. (2011).
[12] S. Jenouvrier, H. Weimerskirch, C. Barbraud, Y.-H. Park and B. Cazelles, *Proc. Roy. Soc. London B. 272*, 887 (2005).
[13] M. Paluš, D. Novotná and P. Tichavský, *Geophys. Res. Lett.* 32, L12805 (2005).
[14] M. Koch and D. Marković, "Evidences for Climate Change in Germany over the 20th Century from the Stochastic Analysis of Hydro-Meteorological Time-Series", MODSIM 2007 International Congress on Modelling and Simulation, (Christchurch, New Zealand) pp 596.
[15] X. Lu, R. Liu, J. Liu and S. Liang, Photogram. *Eng. Remote Sensing*. 73, 1129 (2007).
[16] E. Bullmore, J. Fadili, M. Breakspear, R. Salvador, J. Suckling and M. Brammer, *Stat. Methods Med. Res.* 12, 375 (2003).
[17] L. Sendur, J. Suckling, B. Whitcher and E. Bullmore, *Neuroimage*. 37, 1186 (2007).
[18] R. Cornish, C. S. Bretherton and D. B. Percival, Bound-lay. *Meteorol.* 119, 339 (2006).
[19] R. Gencay, F. Selcuk and B. Whitcher, An Introduction to Wavelets and Other Filtering Methods in Finance and Economics, Academic Press, San Diego, (2001).
[20] Percival and A. Walden, Wavelet Methods for Time Series Analysis, Cambridge University Press, Cambridge, UK (2000).
[21] B. Vidakovic, Statistical modelling by wavelets, John Wiley and Sons, New York (1999).
[22] L. Hudson, M. R. Keatley and I. Kang, *Environ. Ecol. Stat.* (2010). DOI: 10.1007/s10651-010-0149-5.

[23] L. Hudson, M. R. Keatley and A. M. I. Roberts, "Statistical Methods in Phenological Research", Statistical Solutions to Modern Problems. Proceedings of the 20th International Workshop on Statistical Modelling, July 10-15 (Sydney, Australia) pp 259.

[24] O. Bratteli and P. Jorgensen, Wavelets through a looking glass. The world of the spectrum, Birkhäuser, Boston, MA (2002).

[25] M. R. Keatley and I. L. Hudson, *Environ. Model. Assess.* 12, 279 (2007).

[26] Kang, Hudson I.L, Keatley M.R and A. T. Walden, "Wavelets analysis in phenological research- the study of flowering in relation to climate in four Eucalypt species (1940-1971).", IBC 2004 XXIInd International Biometric Conference, (Cairns, Australia).

[27] L. Hudson and M. R. Keatley, Singular spectrum analysis: climatic niche identification, in Phenological Research: Methods for environmental and climate change analysis, I. L. Hudson and M. R. Keatley, Eds., Springer, Berlin (2010) p. 393.

[28] L. Hudson, M. R. Keatley and I. Kang, Wavelet signatures of climate and flowering: identification of species groupings, in Discrete Wavelet Transforms InTech, Vienna, Austria. (2011).

[29] P. A. P. Moran, *Aust. J. Zool.* 1, 291 (1953).

[30] P. A. P. Moran, *Aust. J. Zool.* 1, 163 (1953).

[31] L. Hudson, M. R. Keatley, S. K. Kim and I. Kang, "Synchroncity in Phenology: from PAP Moran to now", Australian Statistical Association and New Zealand Statistical Association Conference, 3-6 July, 2006 (Auckland, New Zealand)

[32] S. K. Kim, I. L. Hudson, M. Agrawal and M. R. Keatley, "Modelling and synchronization of four Eucalyptus species via mixed transition distribution MTD and extended kalman filter EKF", Proceedings of the 23d International Workshop on Statistical Modelling, July 7-11, 2008 (Utrecht, The Netherlands) pp 287.

[33] S. W. Kim, I. L. Hudson and M. R. Keatley, "Modelling the flowering of four eucalypts species via MTDg with interactions", World Congress - MODSIM09 International Congress on Modelling and Simulation, 13–17 July, 2009 (Cairns, Australia) pp 2625.

[34] L. Hudson, M. R. Keatley and S. Y. Lee, *Int. J. Biometeorol.* (2011). DOI: 10.1007/s00484-011-0427-4.

[35] T. Royama, *Ecol. Monogr.* 75, 277 (2005).

[36] P. J. Brockwell and R. A. Davis, Time Series: Theory and Methods, New York (1991).

[37] T. Kohonen, Self-Organizing Maps, Springer-Verlag, Berlin (2001).

[38] B. H. Junker, C. Klukas and F. Schreiber, *BMC Bioinformatics*. 7, 109 (2006).

[39] King, J. Wang, L. Chan, D. Wang, M. Martín-Merino and J. Román, A New SOM Algorithm for Electricity Load Forecasting, in Neural Information Processing, Vol 4232. Lecture Notes in Computer Science. Springer-Verlag, Berlin. (2006) p. 995.

[40] B. Reusch, R. B. Alley and B. C. Hewitson, *J. Geophys. Res.* 112, (D2):D02104. doi:10.1029/2006jd007460 (2007).

[41] S. Mallat, *IEEE T. Pattern Anal.* 11, 674 (1989).

[42] B. Percival and P. Guttorp, Long-memory processes, the Allan variance and wavelets, in Wavelets in Geophysics, E. Foufoula-Georgiou and P. Kumar, Eds., Academic Press, New York. (1994) p. 325.

[43] D. B. Percival and O. Mofjeld, *J. Am. Stat. Assoc.* 92, 868 (1997).

[44] Hernández and G. Weiss, A first course on wavelets, CRC-Press, Boca Raton, FL (1996).

[45] Serroukh and A. T. Walden, *J. Nonparametr. Stat.* 13, 1 (2000).
[46] J. Whitcher, P. Guttorp and D. B. Percival, *J. Geophys. Res.* 105, 941 (2000).
[47] B. Percival, *Biometrika* 82, 619 (1995).
[48] Bruns, F.-M. Chmielewski and A. J. H. van Vliet, The Global Phenological Monitoring Concept, in Phenology: An integrative environmental science, M. D. Schwartz, Ed., Kluwer Academic Publishers, Dordrecht. (2003) p. 93.
[49] M. R. Keatley, I. L. Hudson and T. D. Fletcher, The use of long-term records for describing flowering behaviour: A case-study in Victorian Box-Ironbark Forests., in Australia's Ever-changing Forests IV, J. Dargavel and B. Wasser, Eds., Australian University Press, Canberra (1999) p. 311.
[50] M. Ghil, M. R. Allen, M. D. Dettinger, K. Ide, D. Kondrashov, M. E. Mann, A. W. Robertson, A. Saunders, Y. Tian, F. Varadi and P. Yiou, *Rev. Geophys.* 40, 1 (2002).
[51] E. Trenberth and T. J. Hoar, *Geophys. Res. Lett.* 23, 57 (1996).
[52] M. R. Keatley, I. L. Hudson and T. D. Fletcher, *Aust. J. Bot.* 52, 47 (2004).
[53] D. H. Ashton, *Proc. Ecol. Soc. Aust.* 11, 75 (1981).
[54] O. D. Bassett, *Aust. For.* 65, 237 (2002).
[55] M. R. Keatley and M. D. Murray, Flowering and seed crop development of Eucalyptus tricarpa in East Gippsland, *Dept. of Sustainability and Environment, Melbourne, Victoria* (2011).
[56] T. Chambers, The Colonial Beekeeper, J.C. Stephens, Melbourne (1893).
[57] R. D. Goodman, Honey Flora of Victoria, *Department of Agriculture*, Melbourne (1973).
[58] D. C. Paton, E. L. Crossfield, B. Hurrell and D. J. Rogers, Floral Resources used by the South Australian Apiary Industry, *Rural Industries Research and Development Corporation*, Barton, ACT (2004).
[59] D. Somerville and S. Campbell, Beekeeping in the Narrandera State Forests, N.S.W Agriculture, Goulburn, N.S.W (1997).
[60] Wells, Long term cyclic and environmentally induced effects on flowering of four box-ironbark eucalypts, University of Melbourne, Parkville (2000).
[61] P. Yiou, D. Sornette and M. Ghil, Physica D. 142, 254 (2000).
[62] Eldridge, J. Davidson, C. Harwood and G. van Wyk, Eucalypt Domestication and Breeding, Oxford University Press, New York (1993).
[63] R. Griffin, I. P. Burgess and L. Wolf, *Aust. J. Bot.* 36, 41 (1988).
[64] D. Pryor and L. A. S. Johnson, A Classification of the Eucalypts, Australian National University, Canberra (1971).
[65] L. Delaporte, J. G. Conran and M. Sedgley, *Int. J. Plant Sci.* 162, 1317 (2001).
[66] R. E. Kalman, Trans. ASME. Ser. D. *J. of Basic Eng.* 82, 34 (1960).
[67] R. Van der Merwe, Quick-start guide for ReBel toolkit, Oregon Health and Science University, (2004).
[68] P. R. Martin, F. Bonier, I. T. Moore and J. J. Tewksbury, *Ideas Ecol. Evol.* 2, 9 (2009).
[69] L. Hudson, S. W. Kim and M. R. Keatley, Climatic influences on the flowering phenology of four Eucalypts: a GAMLSS approach, in Phenological Research: Methods for environmental and climate change analysis, I. L. Hudson and M. R. Keatley, Eds., Springer, Berlin, (2010) p. 209.
[70] R. Keatley and I. L. Hudson, Influences on the flowering phenology of three eucalypts, in Biometeorology and Urban Climatology at the Turn of the Century. Selected Papers

from the Conference ICB-ICUC '99, R. J. de Dear, J. D. Kalma, T. R. Oke and A. Aucliems, Eds., World Meteorological Organisation, Geneva, Switzerland. (2000) p. 191.

[71] S. Abu-Asab, P. M. Peterson, S. G. Shelter and S. S. *Orli, Biodivers. Conserv.* 10, 597 (2001).

[72] Menzel and P. Fabian, *Nature.* 397, 659 (1999).

[73] S. D. Johnson, *J. Ecol.* 82, 567 (1993).

[74] T. H. Sparks, E. P. Jeffree and C. E. Jeffree, Int. J. Biometeorol. 44, 82 (2000).

[75] T. H. Sparks, M. Górska-Zajączkowska, W. Wójtowicz and P. Tryjanowski, *Int. J. Biometeorol.* (2010).

[76] Menzel and T. Sparks, Temperature and plant development: phenology and seasonality, in Plant Growth and Climate Change, J. I. L. Morison and M. D. Morecroft, Eds., Blackwell Publishing Ltd, (2006) p. 70.

[77] D. Spano, C. Cesaraccio, P. Duce and R. L. Synder, *Int. J. Biometeorol.* 42, 124 (1999).

[78] S. Yang, J. Logan and D. L. Coffey, *Agr. Forest. Meteorol.* 74, 61 (1995).

[79] S. M. House, Reproductive biology of Eucalypts, in Eucalypt Ecology: Individuals to Ecosystems, J. E. Williams and J. Woinarski, Eds., Cambridge University Press, Cambridge. (1997) p. 30.

[80] Kang, I. L. Hudson, A. Rudge and Chase J.G, Wavelet signatures and diagnostics for the assessment of ICU Agitation-Sedation protocols, in *Discrete Wavelet Transforms. InTech*, Vienna, Austria (2011).

In: Wavelets: Classification, Theory and Applications ISBN 978-1-62100-252-9
Editors: Manel del Valle et al

Chapter 6

EXTENDED WAVELET TRANSFORM FOR DISCRETELY SAMPLED DATA

Robert W. Johnson*
Alphawave Research, Atlanta, GA, US

Abstract

The theory of the wavelet transform using the complex Morlet basis in the context of discrete signal sampling is thoroughly explored. A sequence of steps is developed relating the Fourier transform in the continuum to the wavelet transform on a lattice, including remarks on the admissibility criteria. The wavelet domain is extended to include all coefficients touched by convolution with the data. Nearly perfect reconstruction is obtained for bandwidth limited test signals with the inclusion of a correction factor on the norm of the basis wavelets. The Gibbs effect is observed in the residual near the edges of a finite duration of data, and the conditions required for perfect reconstruction are evaluated. An analysis of a coastal sea level time series is performed to demonstrate the utility of the extended wavelet transform.

Keywords: Fourier analysis, Integral transforms, Spectral methods, Wavelets and other special systems.

PACS: 02.30.Nw, 02.30.Uu, 02.70.Hm.

AMS Subject Classification: 42C40, 65T60.

1. Introduction

Research is what one is doing when one does not know what one is doing. While we would like to present a complete and coherent theory of spectral analysis using the Morlet basis [1], what we have is a summary of explorations down various avenues of inquiry. The motivation for this chapter is the attainment of perfect reconstruction for discretely sampled data using a variation of the wavelet transform whose relation to the Fourier transform is

*E-mail address: robjohnson@alphawaveresearch.com

transparent. By perfect reconstruction, what we mean is that the inverse transform of the wavelet coefficients recreates the original data down to the level of machine precision, not simply that it resembles the data in form but not detail. While techniques exist using the filter bank implementation of bi-orthogonal discrete wavelet bases [2–4], a methodology similar in spirit to the spectral analysis of data using the Fourier transform has yet to be developed fully. This chapter comes close: by extending the wavelet domain and including the reconstruction factor in the wavelet normalization, the integrated spectral energy is nearly equal to the signal energy, and the reconstruction residual is orders of magnitude lower than previous techniques over the majority of the signal duration. We will compare this wavelet method to the short-time Fourier transform (STFT) in terms of its temporal and frequency localization; a similar comparison of spectrograms and scalograms may be found in Reference [5]. Various examples of wavelet methodology as applied to data analysis are given in References [6–18].

We begin the analysis with the definition of the Fourier transform in the continuum, followed by its implementation for discrete signal sampling of finite duration. Some subtleties arise not commonly appreciated when one goes from continuous, infinite transform axes to those which are discrete and finite. We then characterize the response of the Fourier transform with an arbitrary Gaussian window in both the continuum and lattice formulations. The Morlet family of wavelet bases is recognized as a Gaussian window for which the parameters are related to the central frequency rather than remaining fixed. Consideration of the effects of aliasing indicate that the zero mean condition should not be applied in light of discrete sampling, so that the transform response remains smooth as it crosses the zero frequency axis. The domain of calculation is extended well beyond the duration of the data, including every coefficient which contributes to the reconstruction of the signal. The empirical reconstruction factor is absorbed into the wavelet normalization, giving an integrated spectral energy within one percent of the signal energy and a reconstruction residual vanishing to ten decimal places for the majority of a bandwidth limited test signal. The residual displays oscillations resembling the Gibbs effect at locations of jump discontinuities, such as those found at the beginning and end of the measurement duration, whose characterization is explored. The quality of reconstruction is found to depend on the amount of high frequency content contained within the signal and diminishes for the case of a low sample rate. Nonetheless, in its extended form, the wavelet transform provides a useful tool for the spectral analysis of discretely sampled data satisfying Parseval's theorem to within some small factor and with nearly perfect reconstruction, as demonstrated by an analysis of coastal sea level measurements sampled hourly over 16 weeks.

2. Various Fourier Transforms

The starting point of virtually all approaches to spectral estimation is the Fourier transform defined in the continuum. The problem is that we live in a discrete world: regardless of how one feels about the quantization of space-time, the fact remains that practical measurements are given in terms of discrete, finite sampling, whose influence on the nature of the analysis should not be underestimated. The basic dichotomy is that, in the continuum derivatives are easy but integrals are (relatively) complicated, whereas on the lattice differences are complicated but sums generally are not. One cannot avoid the effect of the window function in any practical implementation, as there always exists the window defining the measurement duration. (Someone trying to circumvent that window would still be measuring their data and not reading this book!) Only by getting a firm grip on the window can one feel confident in the results of one's spectral analysis of data.

Let us briefly discuss the matter of units in signal analysis, which must be real on physical grounds. Supposing a continuum signal $y(t)$ with units of u_y, the signal energy is defined to be $E_y \equiv \int_{-\infty}^{\infty} |y(t)|^2 dt$ with units of $u_y^2 u_t$. This energy's units may be related to those of physical energy, mass times velocity squared or joules in the SI system, upon consideration of the load driven by the signal, $E_J = E_y/Z_L$, where Z_L is the impedance of the load. The most familiar example is of a sequence of readings from a voltmeter, so that $u_y = V$ and E_y has units of $V^2 s$. Accounting for the load impedance in ohms, E_J has units of $V^2 s/\Omega = (m^2 l^4/t^4 q^2)t/(ml^2/tq^2)$, which is equivalent to joules. Power, whether signal or physical, is then defined as energy per duration, and the power spectral density has units of power per pixel unit, so that all the integrations come out with the correct units. On the lattice, the time unit is assumed to be given by the sampling rate $u_t = \Delta_t \equiv 1$, with frequency then in units of $u_f = \Delta_t^{-1}$.

2.1. Continuous Infinite

In the continuum, the basis functions θ are of infinite duration $t \in (-\infty, \infty)$ and domain $f \in (-\infty, \infty)$,

$$\theta(t, f) = \exp(i2\pi f t) \,. \tag{1}$$

The forward transform

$$\hat{y}(f) = \int_{-\infty}^{\infty} \theta^*(t, f) y(t) dt \tag{2}$$

satisfies Parseval's theorem, so that the spectral energy equals the signal energy,

$$E_{\hat{y}} \equiv \int_{-\infty}^{\infty} |\hat{y}(f)|^2 df = E_y \,. \tag{3}$$

The inverse transform

$$\hat{\hat{y}}(t) = \int_{-\infty}^{\infty} \theta(t, f) \hat{y}(f) df \tag{4}$$

has perfect reconstruction $y(t) - \hat{\hat{y}}(t) = 0$ for signals of finite energy without jump discontinuities. For a real signal $y(t)$ the Fourier transform is symmetric $\hat{y}(-f) = \hat{y}^*(f)$, so that the coefficients along the negative frequency axis are the complex conjugate of those on the positive frequency axis. Consequently, one also encounters the one-sided transform over the non-negative frequencies $f \to f \in [0, \infty)$ whose normalization is given by $\theta \to \sqrt{2}\theta$.

2.2. Discrete Finite

We next consider discrete regular sampling $t \to t\Delta_t$ and $f \to f\Delta_f$ for integers t and f, where $t \in [1, T]$ and $f \in [-F, F]$ for $\Delta_f \equiv (F2\Delta_t)^{-1}$. The number of strictly positive frequencies is F, and in this case the window duration T equals the number of data points D. The basis functions may be represented by a matrix $\Theta \equiv \Theta_{tf} = \theta(t, f)$, as can the integration measures $dt \to \mathbf{D}_t \equiv \mathbf{I}_T \Delta_t$ and $df \to \mathbf{D}_f \equiv \mathbf{I}'_F \Delta_f$, where $\mathbf{I}_T$ is the identity matrix

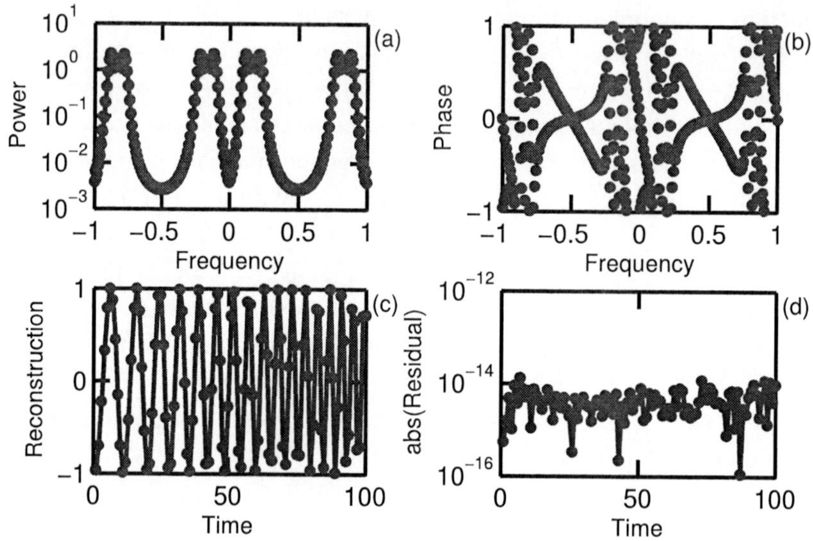

Figure 1. The Fourier spectrum of a discretely sampled signal is periodic over the frequency axis with period Δ_t^{-1} given by the sample rate, as seen in (a) and (b). Here the signal is a real sinusoid whose frequency increases linearly with time, reconstructed in (c). The residual (d) is zero to double precision for the entire signal duration.

of rank T and $\mathbf{I}'_F \equiv \mathbf{I}_F - (| -F\rangle\langle -F | + | F\rangle\langle F |)/2$. The measure $\mathbf{D}_f$ accounts for the half width of the edge-most pixels within one period of the frequency axis. Then, the forward and inverse transforms may be written in terms of matrix multiplication

$$\hat{\mathbf{y}} = \Theta^\dagger \mathbf{D}_t \mathbf{y} \,, \qquad \hat{\hat{\mathbf{y}}} = \Theta \mathbf{D}_f \hat{\mathbf{y}} \,, \tag{5}$$

as can the expression for the signal energy $E_y = \mathbf{y}^\dagger \mathbf{D}_t \mathbf{y}$. The resolution along the frequency axis is arbitrary, as long as there are a sufficient number of degrees of freedom to recreate the data. When $F = T/2$ for even T, this approach is equivalent to the discrete Fourier transform [19] up to the treatment of the domain boundaries.

The effect of aliasing is often misunderstood. One commonly finds the statement, made even by authoritative figures, that the power carried by frequencies outside of the Nyquist interval $f \in [-1/2, 1/2]\Delta_t^{-1}$ has been shifted to within that range, so that the remainder of the power spectrum is zero. Such is not the case. What really happens is that the Fourier spectrum of the data becomes periodic over the frequency axis with period Δ_t^{-1}, such that $\hat{y}(f) = \hat{y}(f + k\Delta_t^{-1})$ for integer $k \in [-\infty, \infty]$, as may be verified by explicit calculation shown in Figure 1. Parseval's theorem still holds, as the normalization of the spectral energy is now defined over one period, $E_{\hat{y}} = \hat{\mathbf{y}}^\dagger \mathbf{D}_f \hat{\mathbf{y}}$. Our experience with the Fourier transform indicates that perfect reconstruction and satisfaction of Parseval's theorem to machine precision rely on a uniform (linear) sampling of the frequency axis. While expressions for a logarithmic frequency axis transform may be found in the continuum [20], the periodicity induced by the discrete sampling wrecks havoc with their implementation for data analysis. Only the linear sampling maintains uniformity over the entire infinite, periodic frequency axis, so that aliasing effects respect the pixel width.

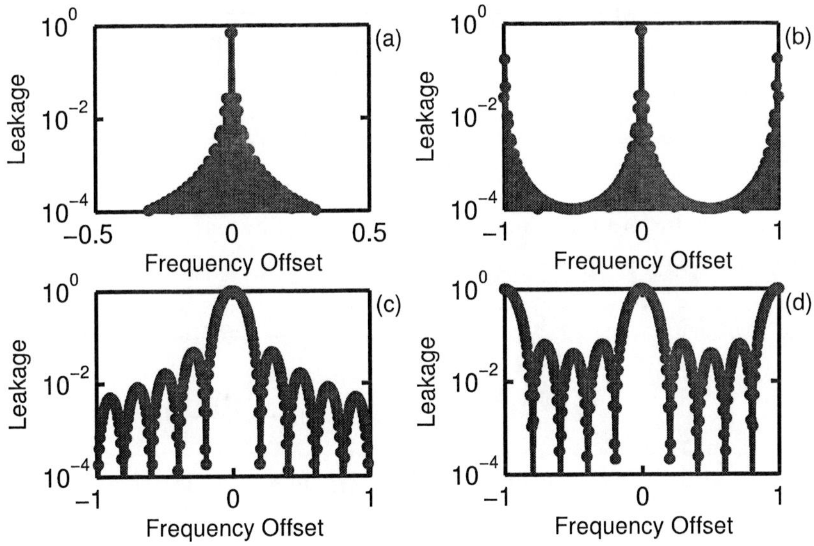

Figure 2. The square window defined by the signal duration introduces sidelobes to the leakage function, shown in (a) for a duration $T = 100$. The effect of discrete sampling is to turn the sidelobes into a periodic function (b). To better see the sidelobes, (c) and (d) were calculated for a duration $T = 5$.

The finite duration of the data introduces the first window which we must address. Looking initially at the continuum response to a sinusoidal signal of frequency f', one finds

$$\int_{c-T/2}^{c+T/2} \exp[-i2\pi(f-f')t]dt = \exp[-i2\pi(f-f')c]\frac{\sin[T\pi(f-f')]}{\pi(f-f')}, \tag{6}$$

whose power distribution $\{\sin[T\pi(f-f')]/\pi(f-f')\}^2$ gives the spectral leakage displaying the expected oscillatory lobes. On the lattice, however, one has (omitting the phase)

$$\sum_{t=1}^{T} \exp[-i2\pi(f-f')t]\Delta_t \propto \frac{\sin[T\pi(f-f')]}{\sin[\pi(f-f')]}, \tag{7}$$

where the effect of the discrete sampling turns the formerly linear denominator into one that is periodic in f. In Figure 2 we compare the leakage functions for the continuous and discrete cases for durations $T = 100$ and $T = 5$, where the number of lobes in one frequency period equals $T-1$. As $\sin(x) \approx x$ for $|x| \ll 1$, one sees that $\sin(x)$ is the periodic analogue of the linear function $y = x$. Here is our first example of how one must be cautious when implementing discretely equations derived in the continuum.

2.3. Fixed Gaussian Window

Next let us consider a windowed Fourier transform whose window function Φ is a Gaussian of fixed temporal bandwidth, sometimes called the short-time Fourier transform (STFT). The basis functions now acquire a variable for their centroid c as well as a parameter h defining the decay of the normal distribution,

$$\begin{aligned}\psi(t,f,c;h) &\equiv C_\infty^{-1}\Phi(t,c;h)\,\theta(t,f,c) &(8)\\ &\propto \exp[-h^2(t-c)^2]\exp[i2\pi f(t-c)]\,, &(9)\end{aligned}$$

with normalization $C_\infty^2 = \int_{-\infty}^{\infty}\Phi^2 dt = \sqrt{2\pi}/2h$ so that $\|\psi\|^2 = 1$ for $f \in (-\infty,\infty)$. In the continuum, the window duration may be of infinite extent, allowing one to verify the integrals for the Fourier transform of ψ,

$$\begin{aligned}\hat{\psi}(f,f',c;h) &= C_\infty^{-1}\int_{-\infty}^{\infty}\exp\{-h^2(t-c)^2 - i2\pi[ft - f'(t-c)]\}dt &(10)\\ &= C_\infty^{-1}\frac{\sqrt{\pi}}{h}\exp(-i2\pi fc)\exp[-\pi^2(f-f')^2/h^2]\,, &(11)\end{aligned}$$

and for the first and second moments of time and frequency, giving their central and bandwidth values,

$$\begin{aligned}t_c &\equiv \int_{-\infty}^{\infty} t|\psi|^2 dt = c\,, &(12)\\ t_b^2 &\equiv \int_{-\infty}^{\infty}(t-t_c)^2|\psi|^2 dt = 1/4h^2\,, &(13)\\ f_c &\equiv \int_{-\infty}^{\infty} f|\hat{\psi}|^2 df = f'\,, &(14)\\ f_b^2 &\equiv \int_{-\infty}^{\infty}(f-f_c)^2|\hat{\psi}|^2 df = h^2/4\pi^2\,, &(15)\end{aligned}$$

such that the bandwidth product $t_b f_b = 1/4\pi$ satisfies the limit of the Fourier uncertainty principle [21] (related to Heisenberg's by a factor of Planck's constant).

Introducing now a parameter for the window duration $T \ll D$, assumed to be significantly shorter than the signal duration, the continuum response to a sinusoidal signal $y(t) = \exp(i2\pi f't)$ is

$$\begin{aligned}\check{y}(f,c;h,T) &\equiv C_T^{-1}\int_{c-T/2}^{c+T/2}\exp\{-h^2(t-c)^2 - i2\pi[f(t-c) - f't]\}dt\,, &(16)\\ &= C_T^{-1}\frac{\sqrt{\pi}}{2h}\exp(i2\pi f'c)\exp[-\pi^2(f-f')^2/h^2][\text{erf}(z) - \text{erf}(-z^*)]\,, &(17)\end{aligned}$$

where $z = hT/2 - i\pi(f-f')/h$ and $C_T^2 \equiv \int_{-T/2}^{T/2}\exp(-2h^2t^2) = C_\infty^2\,\text{erf}(hT/\sqrt{2})$. The final factor in Equation (17) is real because of symmetry, $\text{erf}(-z^*) = -[\text{erf}(z)]^*$, equaling $2\,\text{Re}\,\text{erf}(z)$. The lattice response to a sinusoid (for $c = T/2 + 1/2$) is

$$\check{y}(f,c;h,T) \propto \sum_{t=1}^{T}\exp\{-h^2(t-c)^2 - i2\pi[f(t-c) - f't]\}\Delta_t\,, \quad (18)$$

for which we cannot find a closed form expression; however, it is easily evaluated numerically, so that the power spectral density of the Fourier transform of the basis functions gives the spectral leakage of the Gaussian windowed Fourier transform. In Figure 3 we show the real parts of ψ for central frequencies 1/10 and 1/2 and parameters $h = 1/5$ and

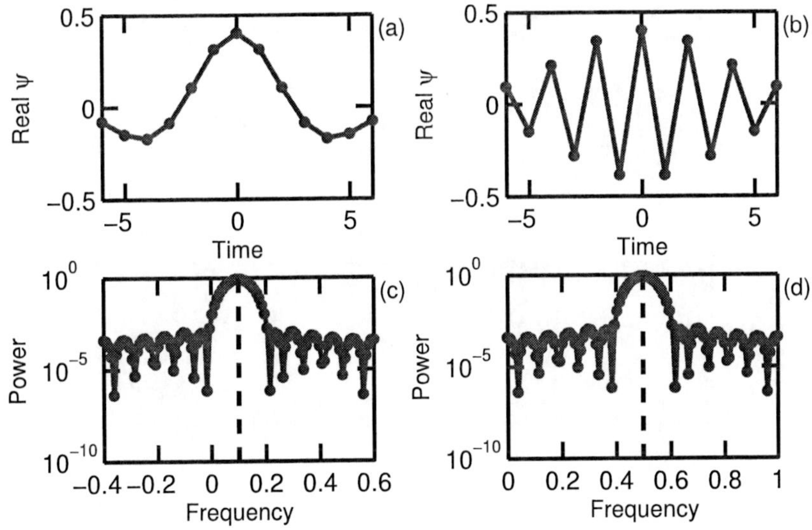

Figure 3. The real parts of Gaussian windowed basis functions ψ with decay $h = 0.2$ and duration $T = 13$ are shown for central frequency 0.1 in (a) and 0.5 in (b). Their spectral leakage $|\hat{\psi}|^2$ in (c) and (d) displays a convolution of the Gaussian distribution from the decay with the sidelobes from the window duration.

$T = 13$ and their corresponding leakage functions $|\hat{\psi}|^2$, which display a convolution of the Gaussian distribution with the sidelobes and differ only by a shift along the frequency axis. In Table 1 are the central and bandwidth values for time and frequency, evaluated on axes symmetric about the central values.

Considering now a real test signal comprised of four unit amplitude sinusoids of duration $D = 100$ with the mean subtracted, the results of a one-sided ($C_T^{-1} \to \sqrt{2}C_T^{-1}$) Gaussian window Fourier transform using the above parameters are shown in Figure 4. The power spectral density in (a) displays an intermittent response whose strength corresponds to the amplitude of the signal rather than the uniform response one would expect from steady sinusoids. The phase spectral density in (b) displays regularity on the order of the window duration $T = 13$. Note that the transform coefficients extend to either side of the signal duration by an amount $\lfloor T/2$, where the first data location is at $t = 1$. The reconstruction in (c) accurately represents the data, as the residual in (d) is on the order of 10^{-16}, and the spectral energy $E_{\check{y}}$ equals the signal energy E_y to machine precision. The time integrated power spectral density in (e) shows a broad response over the region of the signal frequencies, but

Table 1. Central and bandwidth values for time and frequency of the fixed Gaussian window Fourier basis functions ψ with decay $h = 0.2$ and duration $T = 13$

f_c	f_b	f_b/f_c	t_c	t_b	ht_b	$\sqrt{2}t_bf_c$	$4\pi t_bf_b$
0.1	0.038175	0.381750	0.0	2.4116	0.48232	0.34105	1.1569
0.5	0.038175	0.076351	0.0	2.4116	0.48232	1.70530	1.1569

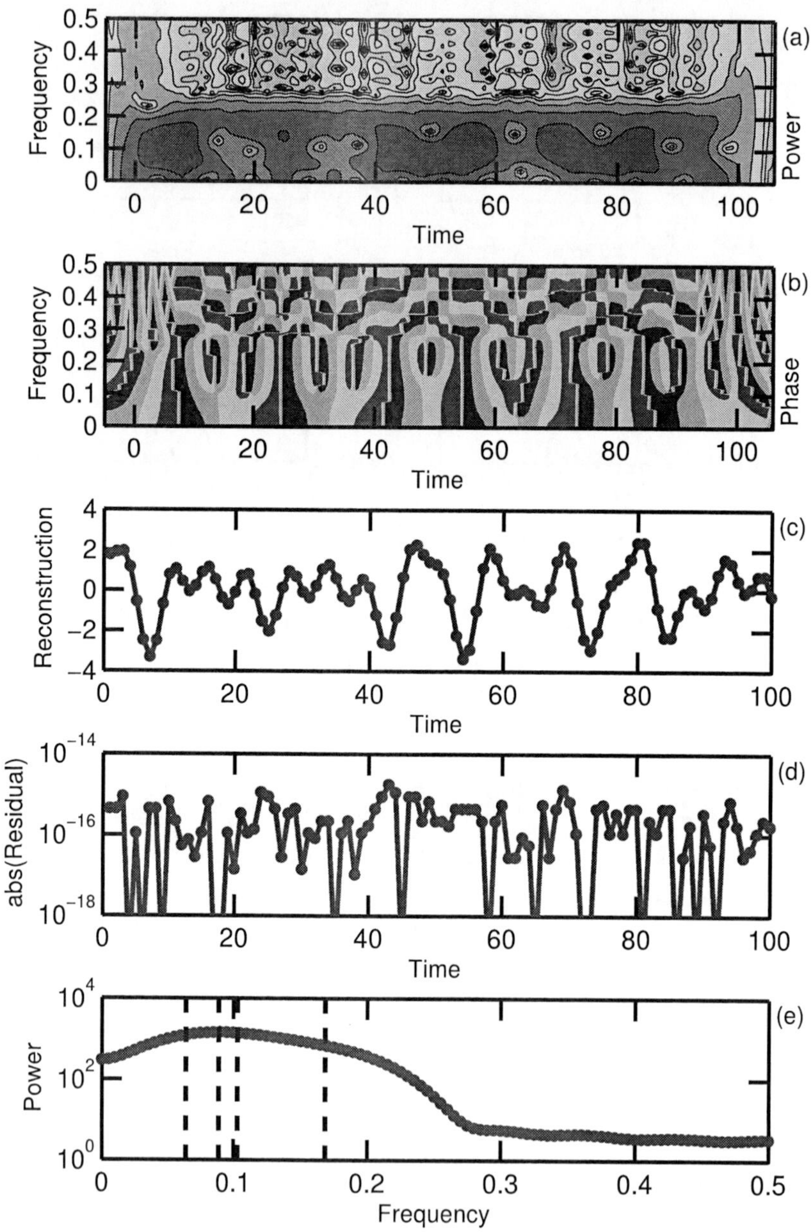

Figure 4. The power and phase spectral densities for a fixed Gaussian window Fourier transform of a test signal with four unit amplitude components are shown in (a) and (b) respectively. The reconstruction (c) accurately represents the original signal, and the residual (d) is on the order of 10^{-16}. The time integrated power spectral density (e) shows a broad response over the component frequencies, indicated by dashed lines, but does not resolve the individual peaks.

the individual peaks are not resolved. While the fixed Gaussian window Fourier transform provides an accurate representation of the signal's information, its utility for tracking signal components through time is questionable.

3. Extended Wavelet Transform

The essential feature of wavelet basis functions is their scale invariance; that is to say, the function at frequency f should look the same as one at frequency f', which is not the case for the fixed window Fourier transform. If one adopts the view that the duration of the wavelet is fixed in terms of physical units, then the wavelet basis is given by varying the sampling frequency. Alternately, when the sampling frequency is considered fixed, the duration of the wavelet depends upon its central frequency, as does its temporal bandwidth. There are a menagerie of wavelet families available in the literature [22–24], with various properties of symmetry and support. As we are staying close in spirit to the Fourier transform, we will restrict consideration to the Morlet family of wavelets given by a Gaussian window times an exponential oscillation.

Following the example of the fixed window Fourier transform, for discretely sampled data we will calculate every coefficient which contributes to the reconstruction of the signal. As the duration of the wavelet is inversely related to its central frequency, the low frequency content requires expansion of the wavelet domain to several times the signal duration. Admissibility is not an issue, as will be discussed later. Various other approaches have been suggested to address the finite duration of the signal, including truncating the wavelet domain to those spanned entirely by the data (within the cone-of-influence) [25], reflecting the signal about its endpoints [26], or attempting to adjust the basis functions to account for edge effects [27, 28]. Here we have taken a different point of view, evaluating instead the coefficient for every basis function which overlaps the data, even if by only one sample, without alteration of the basis beyond the reconstruction factor mentioned later. This approach supposes that the true signal and wavelet domain are of infinite duration, of which we have only a finite contribution.

3.1. Morlet Basis

The Morlet family of wavelet bases is recognized as a Gaussian window whose decay and duration are related to the central frequency $f_c = f$ of the Fourier component. In the continuum, the decay is given by $h = \alpha f$ and the duration by $T = 2\beta/f + 1$ for some parameters $\alpha, \beta > 0$. Substitution into Equations (13) and (15) yields $T = \infty$ bandwidths of $t_b = 1/2\alpha f$ and $f_b = \alpha f/2\pi$ satisfying the uncertainty limit, and the ratio of frequency bandwidth to central frequency $f_b/f_c = \alpha/2\pi$ is a constant. When $\alpha = 1/\sqrt{2}$ one recovers the familiar Morlet basis written in terms of $\eta \equiv (t-c)f$. In Figure 5 we display the continuum leakage function evaluated from Equation (17) for central frequencies 1/4 and 1/2 and duration parameter β of 2 and 4; note these leakage functions are not periodic but of infinite extent.

Considering now the discrete case, the window duration $T = 2\lfloor\beta/f + 1$ must be an

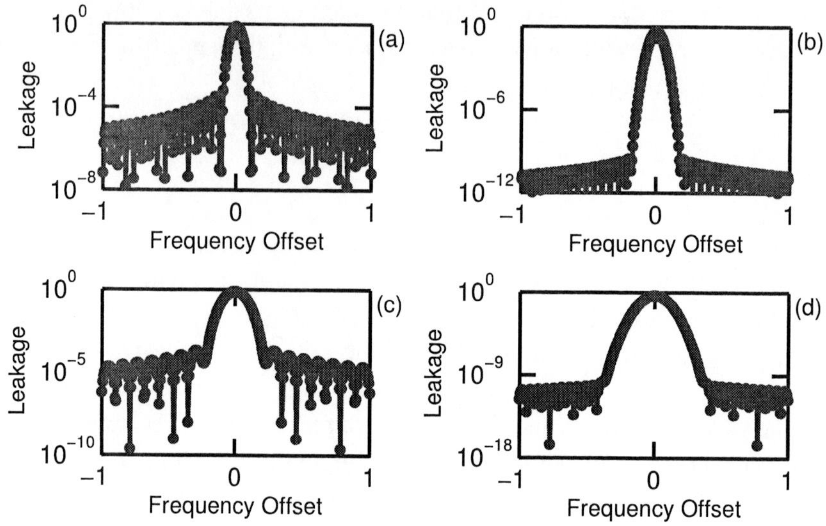

Figure 5. Continuum leakage functions for the Morlet basis with decay parameter $\alpha = 1/\sqrt{2}$ and parameters $f_c = 0.25$, $\beta = 2$ in (a), $f_c = 0.25$, $\beta = 4$ in (b), $f_c = 0.5$, $\beta = 2$ in (c), and $f_c = 0.5$, $\beta = 4$ in (d).

integer chosen to be odd. The discrete Morlet wavelet basis functions are then written

$$\psi(t, f, c; \alpha, \beta) \equiv C_T^{-1} \exp[-\alpha^2 (t-c)^2 f^2] \exp[i2\pi f(t-c)] \,, \tag{19}$$

where $C_T^2 \equiv \sum_1^T \exp[-2\alpha^2(t-c)^2 f^2]$ for $c = \lfloor \beta/f + 1$ is evaluated numerically so that $\|\psi\|^2 = 1$. The leakage function given by the Fourier transform of ψ is now periodic because of the discrete sampling. Taking $\alpha = 1/\sqrt{2}$ and $\beta = 3$, in Figure 6 we display $\mathrm{Re}\,\psi$ for central frequencies 1/10 and 1/2 as well as their Fourier power spectra over one period of the frequency axis. The frequency bandwidth scales with the central frequency so that the wavelet transform has superior low frequency resolution compared to that for the fixed Gaussian window. The trade-off is that the high frequency resolution is poor, with a significant broadening of spectral features near the Nyquist limit of 1/2. In Table 2 are their bandwidth values, as well as those for the case of $\beta = 6$.

Note that we have imposed no admissibility criteria other than that of finite energy. In particular, we have not enforced the zero-mean condition, $\sum_t \psi \neq 0$, so that the wavelet basis

Table 2. Central and bandwidth values for time and frequency of the Morlet basis functions ψ with parameter $\alpha = 1/\sqrt{2}$

β	f_c	f_b	f_b/f_c	t_c	t_b	$\alpha f_c t_b$	$\sqrt{2} t_b f_c$	$4\pi t_b f_b$
3	0.1	0.011272	0.11272	0.0	7.0700	0.49992	0.99985	1.0014
3	0.5	0.056275	0.11255	0.0	1.4142	0.49998	0.99997	1.0001
6	0.1	0.011254	0.11254	0.0	7.0711	0.50000	1.00000	1.0000
6	0.5	0.056270	0.11254	0.0	1.4142	0.50000	1.00000	1.0000

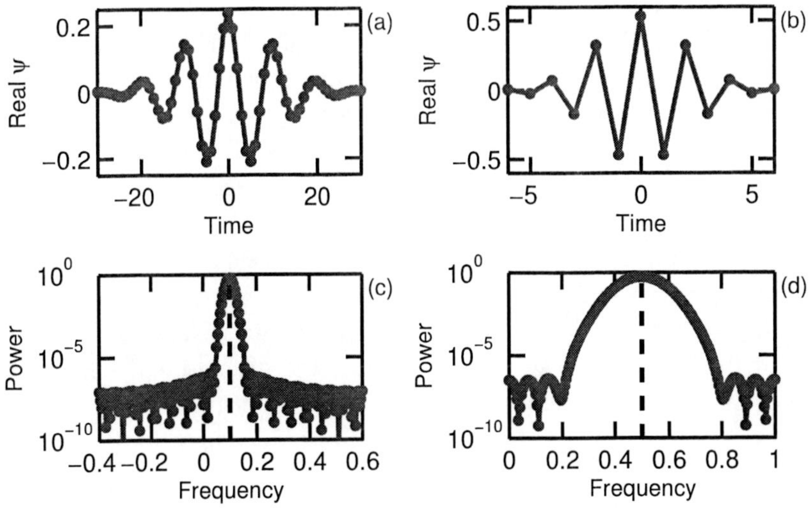

Figure 6. The real part of the Morlet basis functions with parameters $\alpha = 1/\sqrt{2}$ and $\beta = 3$ for central frequencies 0.1 in (a) and 0.5 in (b) indicated by dashed lines. Their Fourier power spectral densities in (c) and (d) are periodic over the frequency axis. The bandwidth of the wavelet response scales with the central frequency.

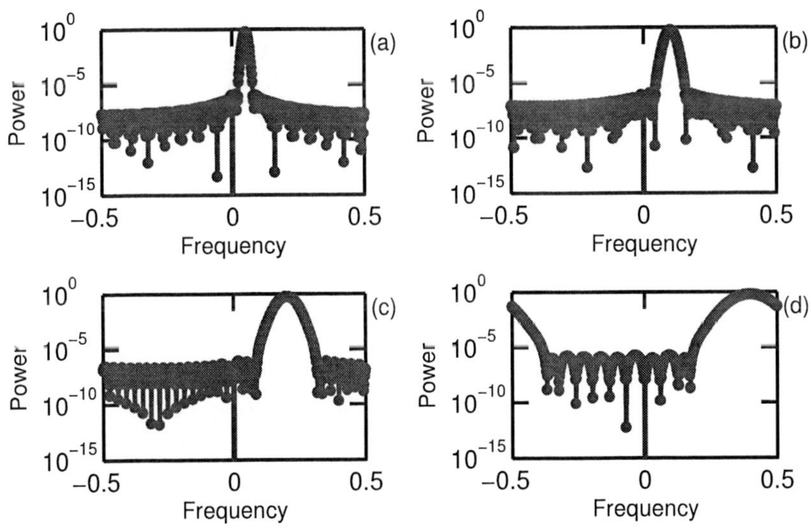

Figure 7. When the zero-mean condition $\sum_t \psi = 0$ is enforced, an artificial asymmetry appears in the Fourier transform of the wavelet basis functions. Shown are the Fourier power spectral densities of discrete Morlet wavelets at central frequencies 0.05 in (a), 0.1 in (b), 0.2 in (c), and 0.4 in (d).

functions maintain a zero frequency response, $\hat{\psi}(0) \neq 0$. The reason is because, in light of the discrete sampling as well as the window response, that pixel contains contributions from frequencies other than zero. Removing the mean of the discrete wavelet via the imposition

$\theta \to \theta - \sum_t \Phi\theta / \sum_t \Phi$ introduces an artificial asymmetry to the leakage function, as shown in Figure 7 for $\alpha = 1/\sqrt{2}$ and $\beta = 3$ at several values of central frequency.

The Morlet basis is notable for its lack of a father function [29]; that is, if the wavelet basis functions are recognized as a collection of bandpass filters, there seems to be no corresponding low-pass filter available to capture the zero frequency content of the signal data. Our attempts to introduce a father function implemented as the Gaussian window for either edge of the frequency axis have met with limited success. We have also explored extending the range of central frequencies considered beyond the Nyquist value of $f = 1/2$. The algorithm for the discrete wavelet basis functions remains well-defined for $f_c > 1/2$; as the central frequency increases, the wavelet becomes progressively shorter, up until $f_c = \beta$ which defines the last wavelet which differs from the discrete unit impulse $\psi(t) = \delta_{0t}$, beyond which the transform coefficients would simply replicate the data. With β an integer that frequency corresponds to an alias analogue of zero frequency. In Figure 8 we show the basis functions ψ for central frequencies $f_c \in [1/2, 3/4, 1, 4/3, 2]$ as well as their Fourier power spectral densities. Recalling that the lobe structure comes from the finite window, the effect of the discrete sampling is to turn the otherwise normal distribution into a wrapped normal distribution [30], *i.e.* $p(x;\mu,\sigma) \propto \exp[-(x-\mu)^2/2\sigma^2]$ for $x \in [-\infty, \infty]$ becomes $p(x;\mu,\sigma) \propto \sum_{k=-\infty}^{\infty} \exp[-(2\pi k + x - \mu)^2/2\sigma^2]$ for $x \in [-\pi, \pi]$, whose limits are the normal distribution for small σ and the uniform distribution for large σ. (We mitigated the effects of wrapping in the evaluation of the bandwidths above [31] by centering the frequency axis on the mean value.) We have looked at the transform response for $f_c \in [1/2, \beta]$ but not seen anything useful; nonetheless, the utility of that region remains to be thoroughly explored.

So far our discussion pertains to the two-sided wavelet transform (with appropriate inclusion of signs); upon restriction to real signals $y(t)$ the one-sided transform for $f > 0$ is normalized by $C_T^{-1} \to \sqrt{2}C_T^{-1}$ so that $\|\psi\|^2 = 2$. The frequency axis resolution is chosen to correspond to that for the Fourier transform, evaluated at integers $f \in [1, F]$. Given the absence of a father function for $f = 0$, the mean of the signal is subtracted $y(t) \to y(t) - D^{-1}\sum_t y(t)$ so that the wavelet response measures variation about the signal mean. The signal is then zero-padded for $t < 1$ and $t > D$ when convoluted with the basis functions to evaluate the transform coefficients—one is supposing here that in the absence of measurements the most likely value of the signal is given by the mean of the available measurements. Explicitly, the forward transform is given by

$$\tilde{y}(f,t) = \sum_{t'} \psi^*(t', f, t) y(t') \Delta_t \,, \tag{20}$$

where the transform domain is $t \in [1 - \tau_1, D + \tau_1]$ for $\tau_1 = \lfloor \beta/\Delta_f$. The nonzero coefficients for a given f have a domain of $t \in [1 - \tau_f, D + \tau_f]$ for $\tau_f = \lfloor \beta/(f\Delta_f)$. The power spectral density is $P_y(f,t) = D^{-1}|\tilde{y}|^2$, and the phase spectral density is $A_y(f,t) = \arg \tilde{y}$. (Recall power is energy divided by duration.) The inverse transform is given by

$$\tilde{\tilde{y}}(t) = \sum_f \sum_{t'} \psi^*(t', f, t) \tilde{y}(f, t') \Delta_t \Delta_f \,, \tag{21}$$

where we have used $\psi(-t', f, t) = \psi^*(t', f, t)$ and the measure for the pixels at $f = 1/2$ is treated as per the discussion above, *ie* their weight is $\Delta_f/2$.

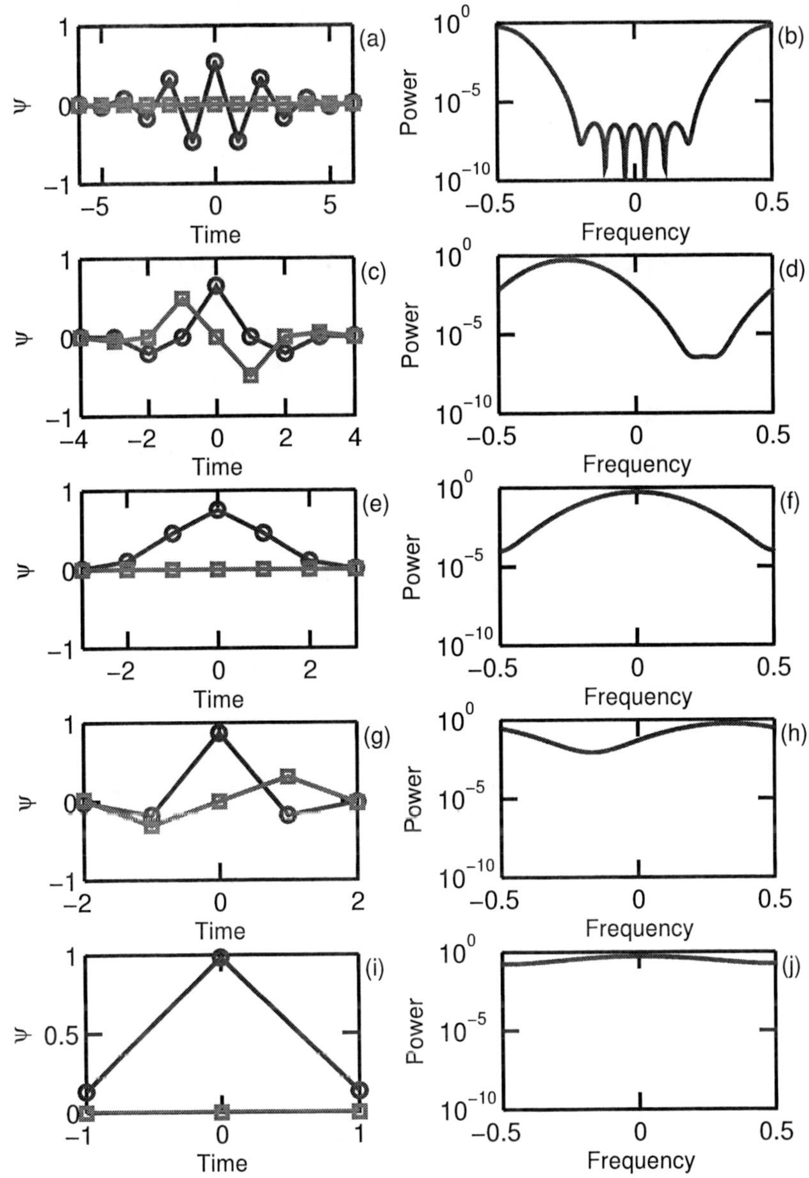

Figure 8. The discrete Morlet wavelet basis function algorithm remains well-defined for central frequencies beyond the Nyquist limit. Shown are the real ○ and imaginary □ parts of ψ for central frequencies 0.5 in (a), 0.75 in (c), 1 in (e), $1.\bar{3}$ in (g), and 2 in (i). Their Fourier power spectral densities proceed from a wrapped normal distribution with lobes to a uniform distribution as f_c increases.

Considering the same four frequency test signal as before, the results of its extended wavelet transform analysis using parameters $\alpha = 1/\sqrt{2}$ and $\beta = 6$ are shown in Figure 9. In (a) and (b) the power and phase spectral densities are displayed over their entire domain; one can see how the domain extends considerably to either side of the signal duration ($t \in [1, D]$) for the low frequency region. The reconstruction in (c) has a residual on the order of 10^{-2},

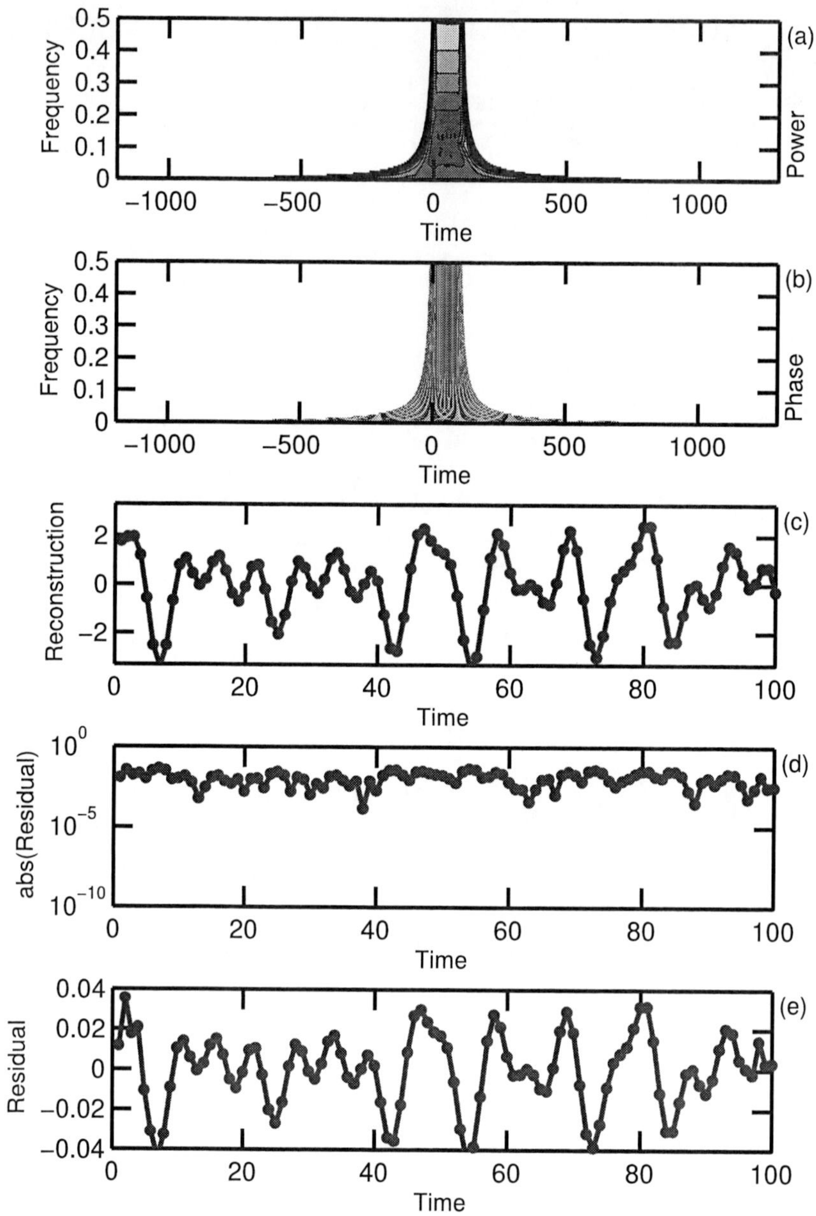

Figure 9. Extended wavelet analysis of a real four frequency test signal. The power (a) and phase (b) spectral densities have a domain considerably larger than the signal duration. The reconstruction (c) reproduces the original signal to an accuracy of one percent, as shown in (d). The residual (e) is proportional to the signal up to some discrepancies at the edges of the signal duration.

as shown in (d). Plotting the residual in (e), one sees that, except for some business at the edges, the residual is proportional to the original signal. That proportionality is the motivation for the reconstruction factor discussed next.

Table 3. Signal, spectral, and reconstruction energies for the extended wavelet transform of opposite polarity unit impulses with separation Δ_δ

Δ_δ	E_y	$E_{\tilde{y}}$	$E_{\tilde{\tilde{y}}}$	$E_{\tilde{y}}/E_y$	$E_{\tilde{\tilde{y}}}/E_{\tilde{y}}$
40	2	2.000000	2.001372	1.000000	1.000686
1	2	1.975278	1.952942	0.987639	0.988692

3.2. Reconstruction Analysis

Torrence and Compo [23] discuss the Farge method of reconstruction [32], in which the reconstruction basis is chosen to be the discrete delta function δ. They introduce a factor C_δ determined empirically from the reconstruction of a unit impulse from its transform coefficients which renormalizes the inverse basis. We maintain the use of the Morlet basis for reconstruction so that the forward and inverse transforms are symmetric. Let us consider a signal comprised of two unit impulses of opposite sign (so that the signal mean is zero) and separated by a span of Δ_δ. The reconstruction analysis of two such signals, with $\Delta_\delta = 40$ and $\Delta_\delta = 1$, is shown in Figure 10. Visually, the reconstructions appear to reproduce the original signal quite well; it is only when we plot the residual that we can see how close (or far) the reconstruction is to being perfect. For comparison, the magnitude of the residual for Erickson's implementation[1] of the Farge method using the package by Torrence and Compo is also displayed, remaining several orders of magnitude above that for the extended wavelet transform for the majority of the signal duration. (The Fourier transform has no problems with accurately reconstructing these signals.) When Δ_δ is large, surrounding each reconstructed impulse is a series of oscillations whose magnitude diminishes on a characteristic scale of $\lesssim 20$ samples to either side. When Δ_δ is small, these oscillations overlap, apparently from an interference effect in the wavelet basis. Looking at the values for the signal, spectral, and reconstruction energies in Table 3, we see that somehow the interference diminishes the wavelet response by a factor of approximately 0.987, which is also close to the ratio of the spectral and reconstruction energies.

In light of the preceding analysis, as well as the observation of proportionality between the residual and original data for more complicated signals, let us introduce an empirical reconstruction factor C_f. This factor is absorbed into the wavelet normalization via $C_T^{-1} \to C_f C_T^{-1}$ and appears in both the forward and inverse transforms. We know the value of C_f^2 will be close to 0.987, determining its precise value by a numerical minimization of the magnitude of the residual over the region $t \in [21, D-20]$. For now we will determine its value independently for each test signal so that we may discuss its relation to the quality of reconstruction. For our four frequency test signal, the results of the renormalized extended wavelet transform are shown in Figure 11. The magnitude of the residual is now on the order of 10^{-9} for the majority of the signal duration, compared to 10^{-1} for the Farge method. The value for C_f will be given later in comparison to other signals' reconstruction factors. We see that the residual remains significant at either edge of the signal duration, displaying oscillations reminiscent of the Gibbs effect with the same characteristic width as found for the delta function reconstruction.

[1] http://www.mathworks.com/matlabcentral/fileexchange/20821

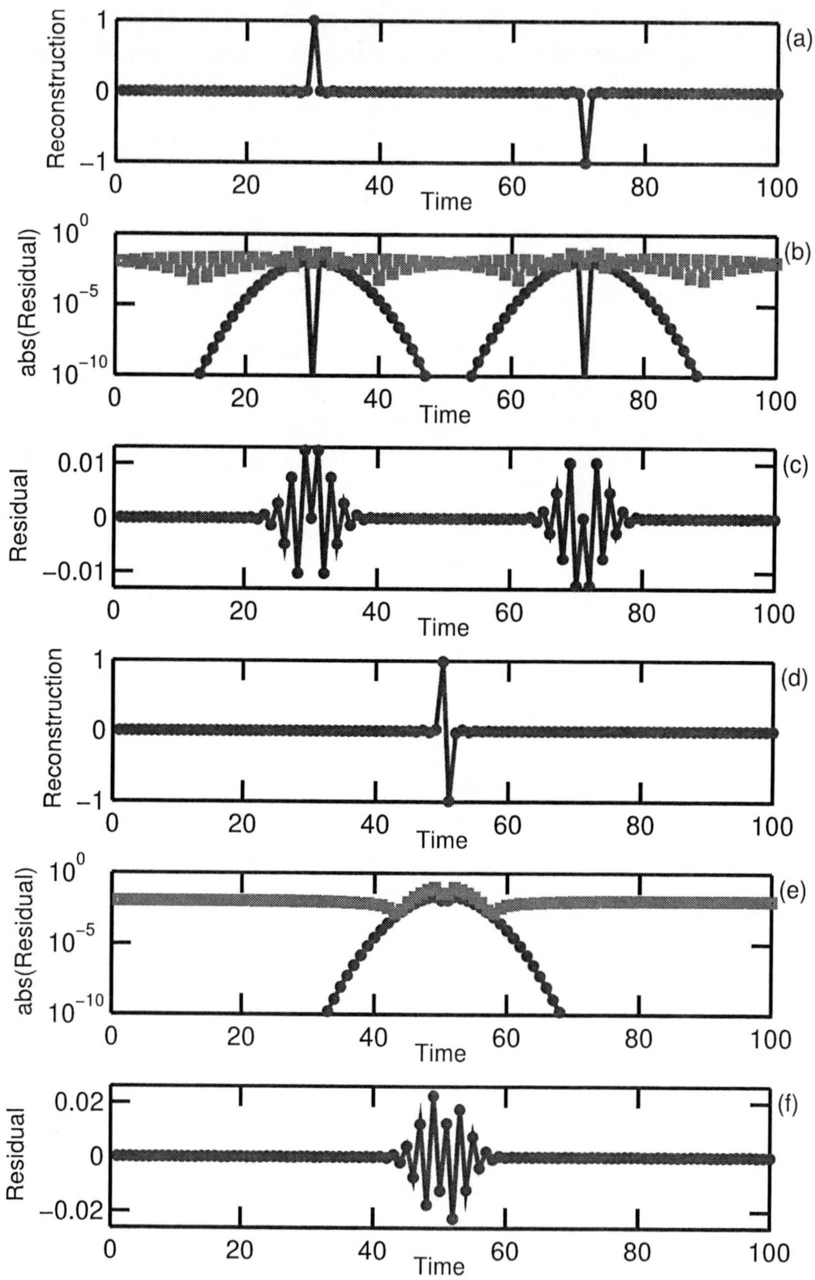

Figure 10. Reconstruction analysis of test signals comprised of two unit impulses, with a separation $\Delta_\delta = 40$ for (a)-(c) and $\Delta_\delta = 1$ for (d)-(f). The magnitude of the residual for the Farge method is indicated by □ in (b) and (e) and remains orders of magnitude above that for the extended wavelet transform. The residual displays oscillations to either side of the impulse locations which diminish on a scale of $\lesssim 20$ samples.

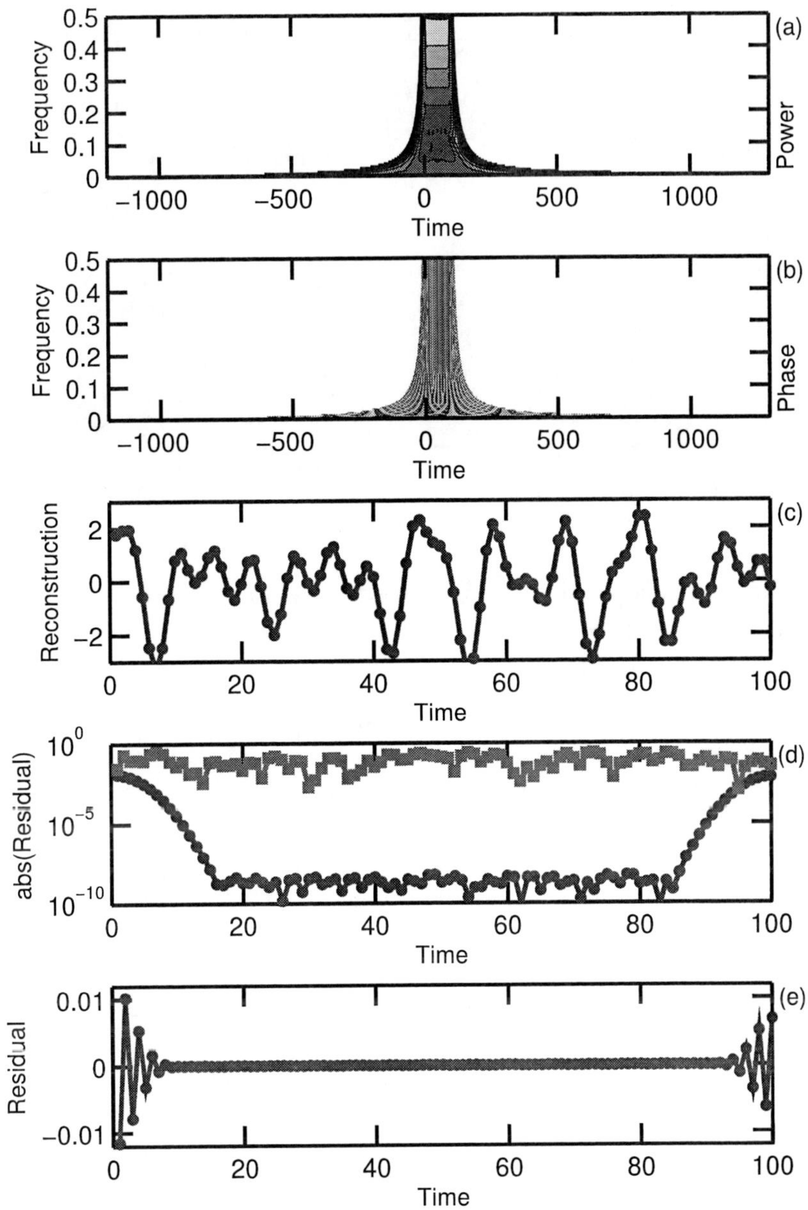

Figure 11. Renormalized extended wavelet analysis of a real four frequency test signal. While the reconstruction appears nearly the same as for the unrenormalized transform, its residual is many orders of magnitude smaller, as seen in (d). For comparison, the magnitude of the residual for the Farge method is indicated by □. The residual (e) remains significant at the edges of the signal duration, where oscillations appear surrounding the locations of the jump discontinuities.

The extended domain of the time axis makes it hard to see the behavior of the transform coefficients over the signal duration. In Figure 12 we show the power and phase spectra for our four frequency test signal over the signal duration extended by 20 samples to either side.

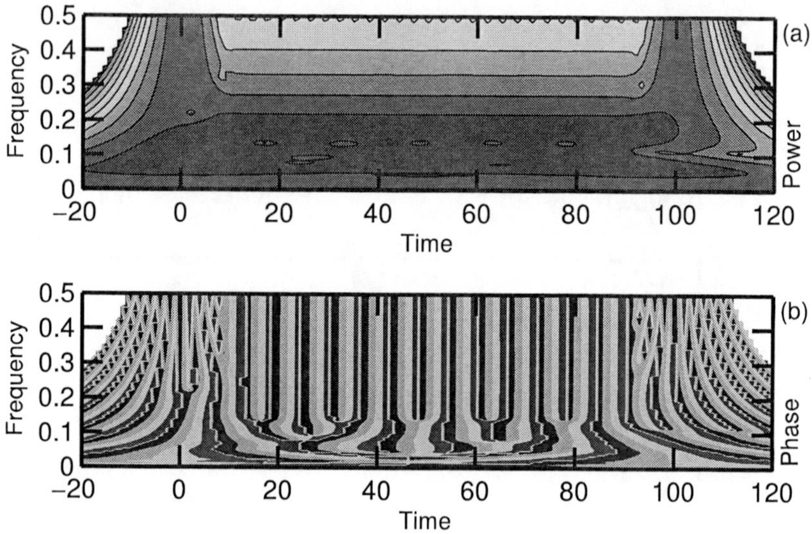

Figure 12. The power spectrum (a) of the extended wavelet transform of the four frequency test signal displays the regularity expected from constant amplitude sinusoidal components. The phase spectrum (b) displays structure on the scale of the period of the basis functions.

We now see the expected regularity in the power spectrum rather than the blockiness of the fixed Gaussian window Fourier transform, and similarly the phase spectrum displays structure on the scale of the period of the basis functions. The wavelet response at high frequency remains strong at the locations of the jump discontinuities defining the data duration, indicating that the remaining residual is indeed a high frequency Gibbs effect and motivating the (so far unsuccessful) incorporation of the wavelet domain beyond the Nyquist limit. In Figure 13 we display the time-integrated wavelet power spectral density, which is simply the sum over the entire time domain of the squared amplitude of the forward transform coefficients divided by the signal duration. The power spectrum is in excellent accord with that given by the Fourier transform, also shown. The wavelet power spectral density displays peaks at the locations of the frequencies of the signal components, where the middle two have been merged into a single peak, in contrast to the behavior of the fixed Gaussian window Fourier transform seen in Figure 4 (e). The Fourier power spectrum resolves the two middle peaks but also displays the inherent spikiness induced by the duration window. This time integrated wavelet power spectrum satisfies the expectation of being a smoothed version of the Fourier power spectrum. The price of having compact support in time is the spectral leakage associated with the Gaussian window function.

Let us inspect the remaining residual in more detail. In Figure 14 we show the residual for a region of ± 20 samples surrounding each jump discontinuity defining the the width of the data window. These subsignals have been normalized to unit energy, and the time axis has been relabeled for the right edge so that $t = 0$ locates the first sample beyond the data. The mean of these subsignals is on the order of 10^{-9}. Their Fourier power and phase spectra are also displayed in the figure. These oscillations resemble the Gibbs effect well known in the theory of Fourier analysis, which arises from trying to represent the sharp transition of a

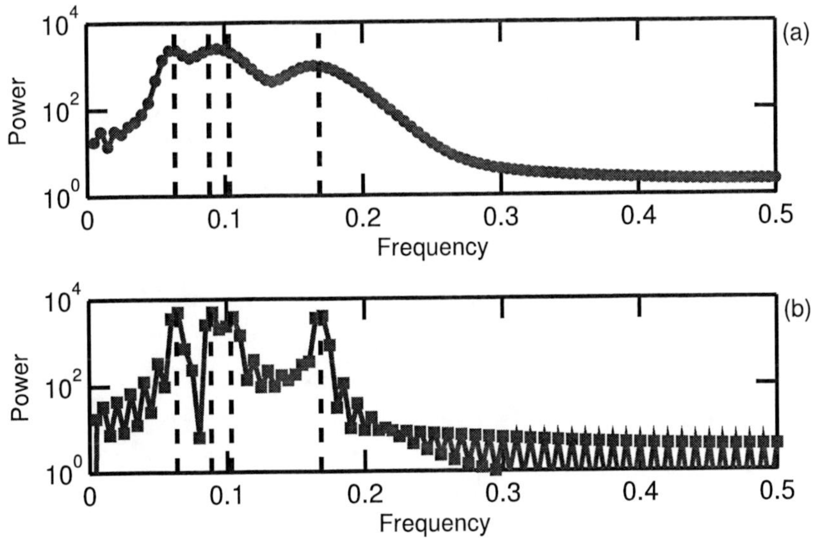

Figure 13. The time integrated wavelet power spectrum in (a) is in agreement with the Fourier power spectrum in (b). The wavelet spectrum merges the response to the two signal components around $f = 0.1$ into a single peak, whereas the Fourier spectrum resolves them independently. The frequencies of the signal components are indicated by vertical dashes.

jump discontinuity with a finite Fourier series. Their bandwidth characterization is given in Table 4, from which we see that they themselves appear to be wavelets, but not of the same parameters as used in the analysis. In general, the two edges need not be as symmetric as displayed here—this particular test signal happens to have a high degree of correspondence between its two edges. These remaining residual oscillations have been a source of much consternation in our quest to achieve truly perfect reconstruction, for which we have found no amelioration.

In order to further characterize the reconstruction factor C_f, let us now consider some variations to the test signal. In Figure 15 we display a linear signal, a quadratic signal, our four frequency signal, and their sum, all with duration $D = 100$. In each case the mean is subtracted before entering the wavelet analysis. The magnitude of the residuals is shown in Figure 16, all of which display the Gibbs effect associated with the discontinuities at the edges of the data duration. Discounting the edges, the polynomial signals have a residual on the order of 10^{-10}, while their sum with the four frequency signal has a residual around 10^{-8}. Their values for C_f^2 as well as their signal, spectral, and reconstruction energies are

Table 4. Central and bandwidth values for time and frequency of the residual subsignals in Figure 14 at either edge of the data window

Edge	f_c	f_b	f_b/f_c	t_c	t_b	$\sqrt{2}t_b f_c$	$4\pi t_b f_b$
left	0.5	0.040892	0.081785	0.53636	1.9460	1.3760	1.000
right	0.5	0.041610	0.083220	−0.93699	1.9125	1.3523	1.000

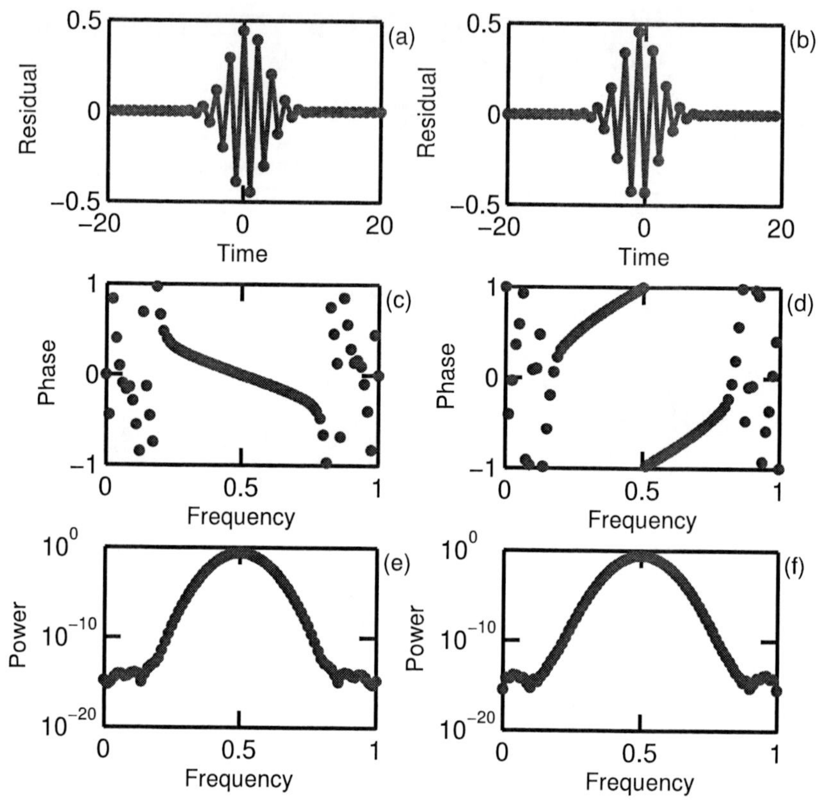

Figure 14. The remaining residual oscillations at either edge of the signal duration have been normalized to unit energy in (a) and (b). The time axis for the right edge (b) has been relabeled so that $D+1 \to 0$. Their Fourier power spectra in (c) and (d) resemble that found for a Morlet wavelet at $f_c = 0.5$. Their phase spectra in (e) and (f) display the mirror symmetry of the jump discontinuities.

given in Table 5. We see that Parseval's theorem is very close to being satisfied, $E_y \approx E_{\tilde{y}} \approx E_{\tilde{\tilde{y}}}$, as their relative ratios are near unity.

It appears that the reconstruction factor, as well as the quality of reconstruction, has some dependence on the frequency content of the signal. To investigate that dependence, let us compare the reconstruction of a low frequency signal to that of a high frequency signal (or equivalently a high sample rate to a low sample rate). In Figure 17 we show the reconstruction and residual magnitude of two such signals, with their reconstruction factors and energies given in Table 6. We see that the reconstruction of the low frequency signal has a residual (discounting the Gibbs effect) of less that 10^{-10}, whereas the reconstruction of the high frequency signal has a residual magnitude around 10^{-2} so that the Gibbs effect is not visible. From these considerations, it seems the reconstruction factor C_f should not be a constant but rather should include some slight frequency dependence. Without a theoretical justification for this factor, it is hard to determine what that dependence should be, and so we leave the renormalization C_f as a constant in what follows.

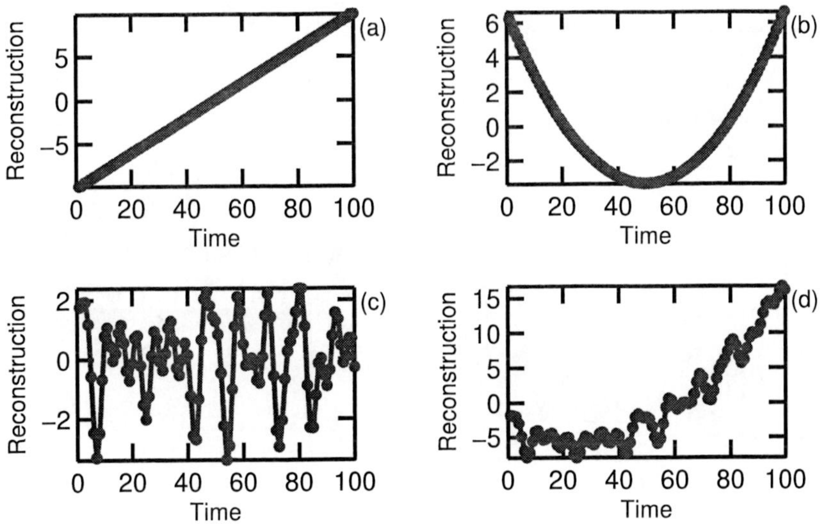

Figure 15. Reconstructions of a linear (a), quadratic (b), and oscillatory signal (c), as well as their sum in (d).

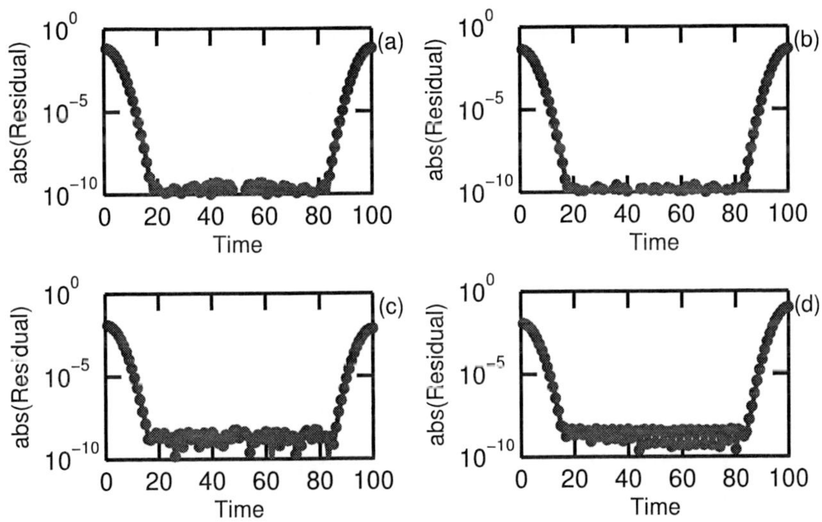

Figure 16. Residual magnitudes from the reconstructions found in Figure 15 of a linear (a), quadratic (b), and oscillatory signal (c), as well as their sum in (d).

4. Real Data

Having explored thoroughly the properties of the extended continuous wavelet transform using the Morlet basis in the context of discrete sampling, it is time to put it to practical use. The physical signal we analyze here is a record of coastal sea level height measurements sampled hourly at the NOAA/NOS St. Petersburg station, with an unbroken set of measurements spanning 16 years. Because of computational constraints, we will consider

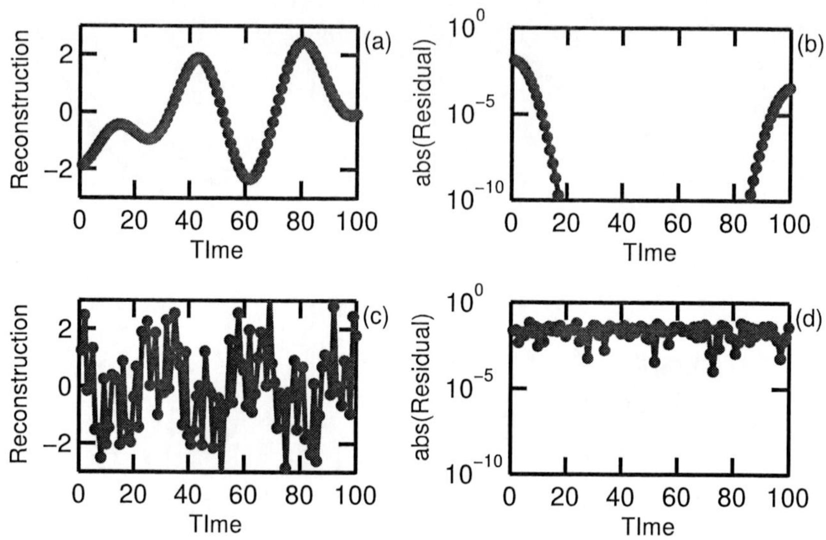

Figure 17. The reconstruction and residual magnitude of a low frequency signal are shown in (a) and (b), compared to those for a high frequency signal in (c) and (d).

only the first 16 weeks of data, comprising $D = 2688$ samples, as shown in Figure 18. Real data necessarily come with measurement errors, which the forward (or inverse) transform assumes to be negligible—their proper treatment requires the use of a conditional probability analysis beyond the scope of this chapter. For this data, the quantal unit is 10^{-3} m, giving an approximate variance of $2.5 \times 10^{-7}\,\mathrm{m}^2$, which we will neglect.

Let us start with the one-sided Fourier transform of the data with resolution $F = D$ along the frequency axis. In Figure 19 we show the Fourier power spectral density, displayed on a logarithmic frequency axis but calculated using linear sampling. The spectrum has two prominent peaks at about 0.04 and 0.08 hr^{-1}, corresponding to approximately 24 and 12 hour cycles in the data. Looking more closely, we see that these peaks appear to be double

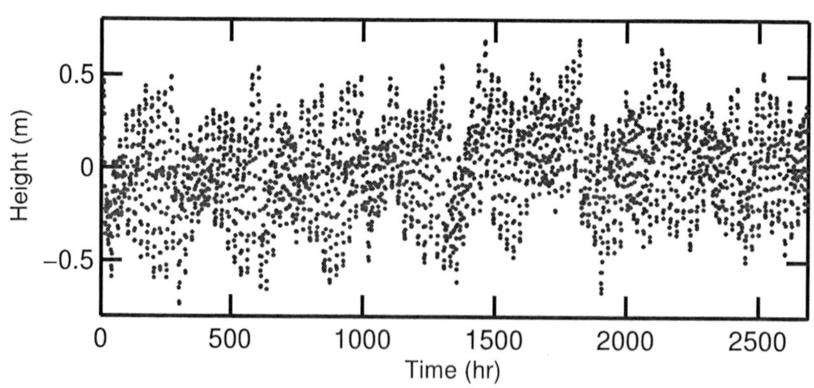

Figure 18. Coastal sea level measurements sampled hourly over 16 weeks at the NOAA/NOS St. Petersburg station.

Table 5. Reconstruction factor and signal, spectral, and reconstruction energies for the extended wavelet transform of various test signals

Signal	C_f^2	E_y	$E_{\tilde{y}}$	$E_{\tilde{\tilde{y}}}$	$E_{\tilde{y}}/E_y$	$E_{\tilde{\tilde{y}}}/E_{\tilde{y}}$
linear	0.986991549423	3333.00	3332.33	3331.68	0.999798	0.999805
quadratic	0.986991549428	889.78	889.48	889.19	0.999664	0.999674
oscillatory	0.986991550458	191.05	191.03	191.02	0.999926	0.999929
summed	0.986991549475	4624.17	4623.31	4622.48	0.999815	0.999821

Table 6. Reconstruction factor and signal, spectral, and reconstruction energies for the extended wavelet transform of a low frequency and a high frequency test signal

Signal	C_f^2	E_y	$E_{\tilde{y}}$	$E_{\tilde{\tilde{y}}}$	$E_{\tilde{y}}/E_y$	$E_{\tilde{\tilde{y}}}/E_{\tilde{y}}$
low frequency	0.986991549427	178.23	178.22	178.21	0.999929	0.999932
high frequency	0.998963608640	200.96	201.15	201.41	1.000908	1.001318

peaks, reflecting the epicycles of the earth, moon, and sun. No information is available on their temporal variation, as the Fourier basis functions span the entire data duration.

The results of the extended wavelet transform of this data using a reconstruction factor of $C_f^2 = 0.98699155$ are shown in Figure 20. The power spectral density is displayed only for the duration of the data and reveals the temporal variation of signal components expected from wavelet analysis with basis functions of compact temporal support. The residual magnitude of about 10^{-4} is less than the quantal error of the data, indicating that the reconstruction is as faithful as it can be. The time integrated power spectral density has two prominent peaks at around 12 and 24 hour cycles, and some additional structure is seen in the low frequency region. Recalling the smoothing action of the wavelet transform, any peaks which appear in the time integrated power spectral density are likely to represent significant signal components in the data.

The values for the signal and spectral energies of the Fourier and wavelet transforms are given in Table 7, where we see that Parseval's theorem is satisfied. The interesting part of the spectrum lies below a frequency of 0.1 hr^{-1}. Comparing the wavelet and Fourier power spectra on linear axes for $f \in [0, 0.1]$ hr^{-1} in Figure 21, we see that the wavelet transform has a stronger response in the low frequency region compared to the higher frequency peaks at 0.04 and 0.08 hr^{-1}, relative to the response given by the Fourier transform. The spectral resolution of the Fourier transform is superior, as is to be expected, at the cost of less low frequency response as well as a lack of temporal resolution.

Table 7. Signal and spectral energies for the Fourier and wavelet transforms of the coastal sea level data

E_y	$E_{\hat{y}}$	$E_{\tilde{y}}$	$E_{\tilde{\tilde{y}}}$	$E_{\tilde{y}}/E_y$	$E_{\tilde{\tilde{y}}}/E_{\tilde{y}}$
167.848407	167.848407	167.845441	167.842650	0.999982	0.999983

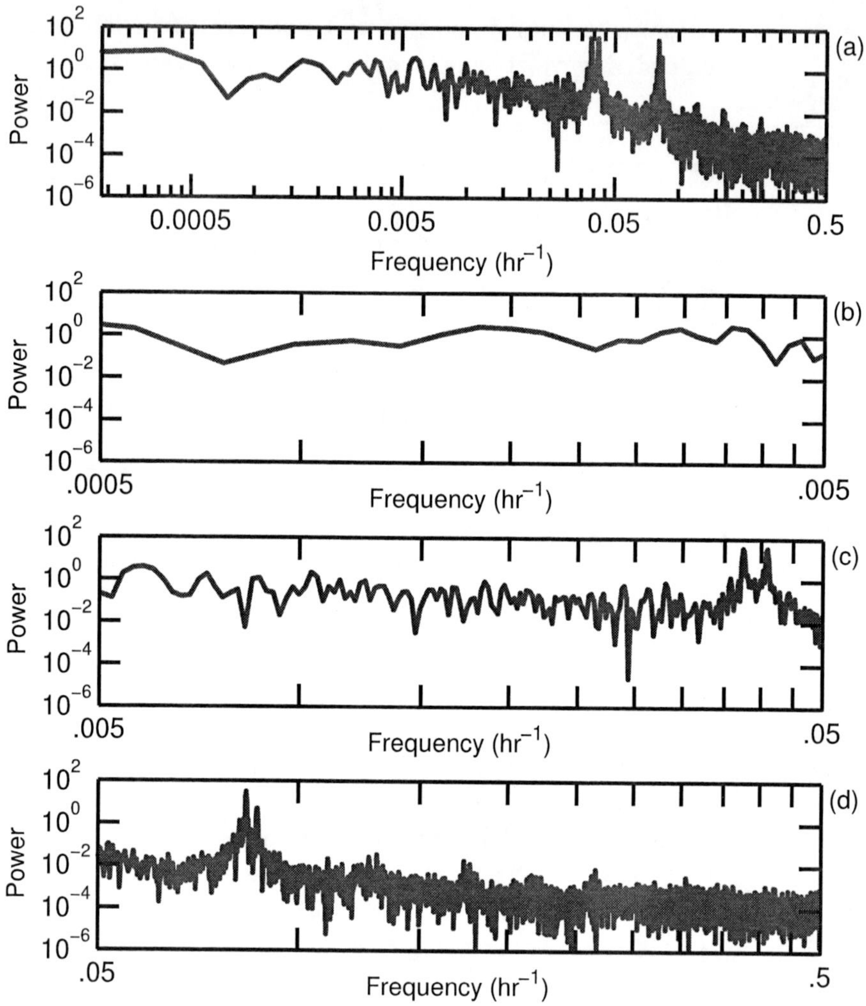

Figure 19. Fourier power spectral density of the coastal sea level measurements in Figure 18, displayed along a logarithmic frequency axis. The two peaks in (a) are at roughly 0.04 and 0.08 hr^{-1}. No low frequency structure is apparent in (b). The peak at 0.04 hr^{-1} in (c) is seen to be a double peak, as is the one at 0.08 hr^{-1} in (d).

Looking now at the wavelet and Fourier power spectra on logarithmic axes, shown in Figure 22 for $f \in [10^{-3}, 0.5]$ hr^{-1}, we see that the wavelet transform supports the significance of structure in the low frequency region. In Table 8 we give the locations of the peaks we can identify in the wavelet spectrum in units of both frequency and period. The bumps in the spectrum at 0.01674 and 0.3 hr^{-1} do not quite satisfy the definition of a peak, in that the gradient never goes positive, but their presence indicates some component amplitude at those frequencies. The peak periods are also given in units of 1 day = 24 hr; it would be interesting to identify physical processes at the periods of the low frequency peaks in the spectrum, which are not quite harmonically spaced.

Finally, let us look at the temporal variation in the wavelet power spectrum. Starting

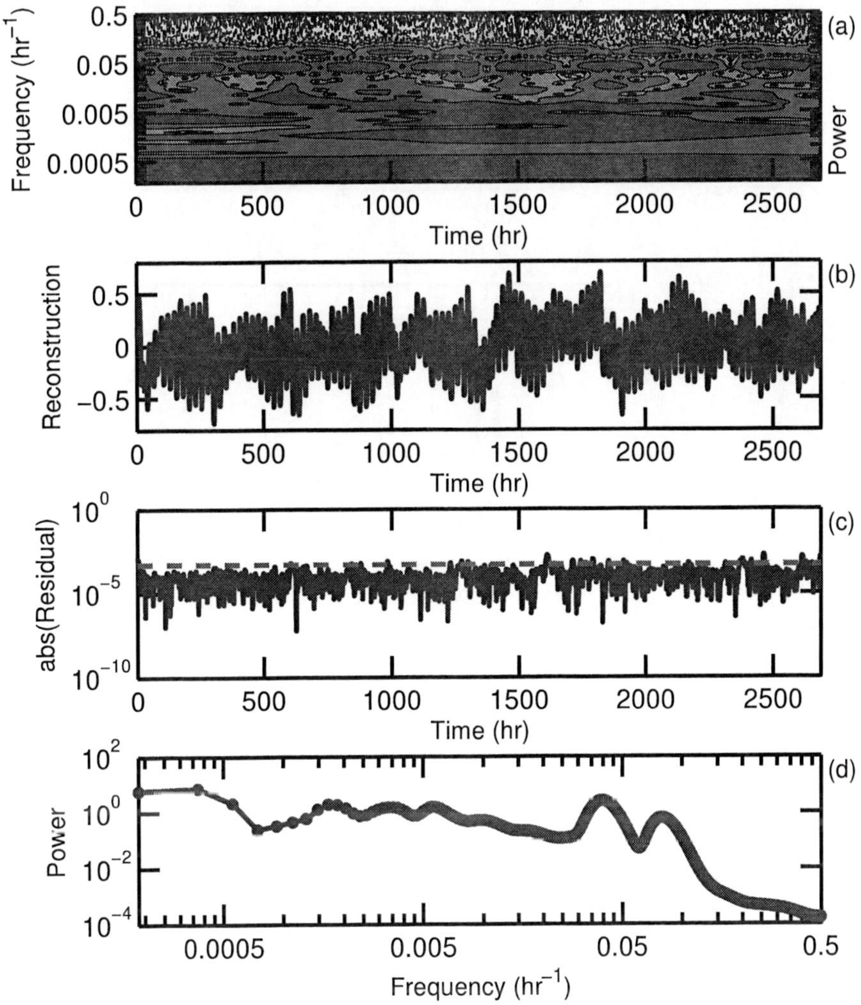

Figure 20. The wavelet power spectral density in (a) displays the temporal variation of the signal components. The reconstruction in (b) reproduces the data with a residual generally less than the quantal error, indicated by the dashed line in (c). The time integrated power spectrum in (d) shows two prominent peaks at roughly 0.04 and 0.08 hr^{-1}, as well as some low frequency structure between 0.001 and 0.01 hr^{-1}.

with the higher frequency peaks, the wavelet power spectral density over the duration of the data is displayed in Figure 23. Here one sees a clear alternation in amplitude of the 0.04 and 0.08 hr^{-1} peaks, itself with a period of approximately 300 – 350 hr^{-1} or 12-14 days, indicating when the moon is aligned with the sun. In Figure 24 we show the low frequency region of the wavelet power spectral density (which extends far beyond the data duration shown). While it is hard to discern structure in this region, one does expect coastal sea level measurements to exhibit cycles on the order of monthly or seasonal variation, which would barely be resolved by only 16 weeks of data.

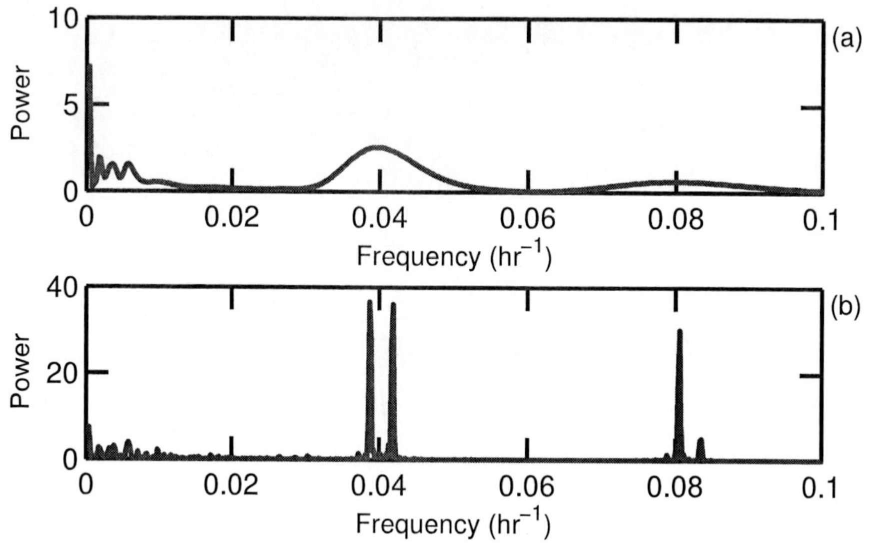

Figure 21. The time integrated wavelet power spectrum in (a) has a stronger low frequency response compared to the Fourier power spectrum in (b). The frequency resolution of the Fourier transform is superior at the cost of no temporal resolution.

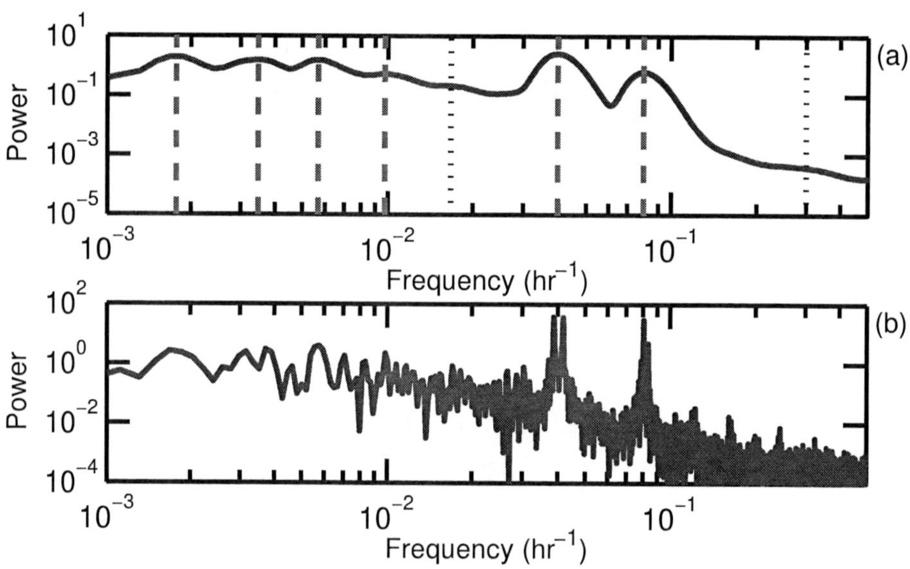

Figure 22. Comparison of the time integrated wavelet power spectrum (a) with the Fourier power spectrum (b) on logarithmic axes. The wavelet spectrum appears to resolve structure in the low frequency part of the spectrum. Peaks in the wavelet spectrum are indicated by the dashed lines, as are two bumps by dotted lines.

5. Conclusion

In this chapter we have investigated the theory of the continuous wavelet transform using the Morlet basis in the context of discrete signal sampling, paying particular regard to

Table 8. Locations of peaks and bumps in the wavelet power spectrum in units of both frequency and period

Units	Peaks			
hr^{-1}	0.0017725	0.0034698	0.0056831	0.0097703
hr	564.17	288.2	175.96	102.35
day	23.50729	12.00838	7.33168	4.26463
Units	Peaks		Bumps	
hr^{-1}	0.039696	0.079881	0.01674	0.3
hr	25.191	12.519	59.737	3.3333
day	1.04964	0.52161	2.48905	0.13889

maintaining a transparent relationship to the Fourier transform. The implementation of the wavelet basis functions as Gaussian windowed oscillations invites their comparison to the windowed Fourier transform using a fixed Gaussian window. We note that the wavelet coefficients for a given frequency correspond to a particular choice of fixed Gaussian window, so that each line in the wavelet transform essentially is taken from a different fixed window Fourier transform. It is remarkable that their union can be made to satisfy Parseval's theorem as well as nearly perfect reconstruction by the inclusion of a reconstruction factor in the wavelet normalization.

The characterization of the residual indicates that the reconstruction factor should have some slight frequency dependence. For signals bandwidth limited to frequencies well below the Nyquist limit, the residual is dominated by an apparent Gibbs effect at the locations of the jump discontinuities associated with the finite data duration. These residual oscillations appear to have a characteristic frequency equal to the Nyquist value and a decay width independent of the size of the jump. As the proportion of high frequency content increases, the quality of reconstruction decreases, until the Gibbs effect is no longer apparent.

An analysis of coastal sea level measurements has been performed, comparing the power spectral densities of the wavelet and Fourier transforms. Their spectral energies are in accord with the signal energy, and the expected cycles of lunar and solar influence on the tides are observed. The wavelet spectrum reveals the alternation in amplitude of the daily and semi-daily cycles and also supports the existence of low frequency cycles in the data with periods greater than one day.

By extending the domain of the wavelet coefficients to include every one which contributes to the reconstruction of the signal, and by incorporating an empirical factor into the wavelet normalization, a transform algorithm is developed with nearly perfect reconstruction and satisfaction of Parseval's theorem for discretely sampled data. The zero-mean admissibility condition is seen to alter the spectral leakage of the wavelet basis functions in an unnatural way and is not included here. There is still work to be done to understand fully the role of the reconstruction factor: why is it there, what does it do, and how can we get rid of it are questions remaining to be answered. Nonetheless, the wavelet transform is a useful tool for the spectral analysis of physical signals.

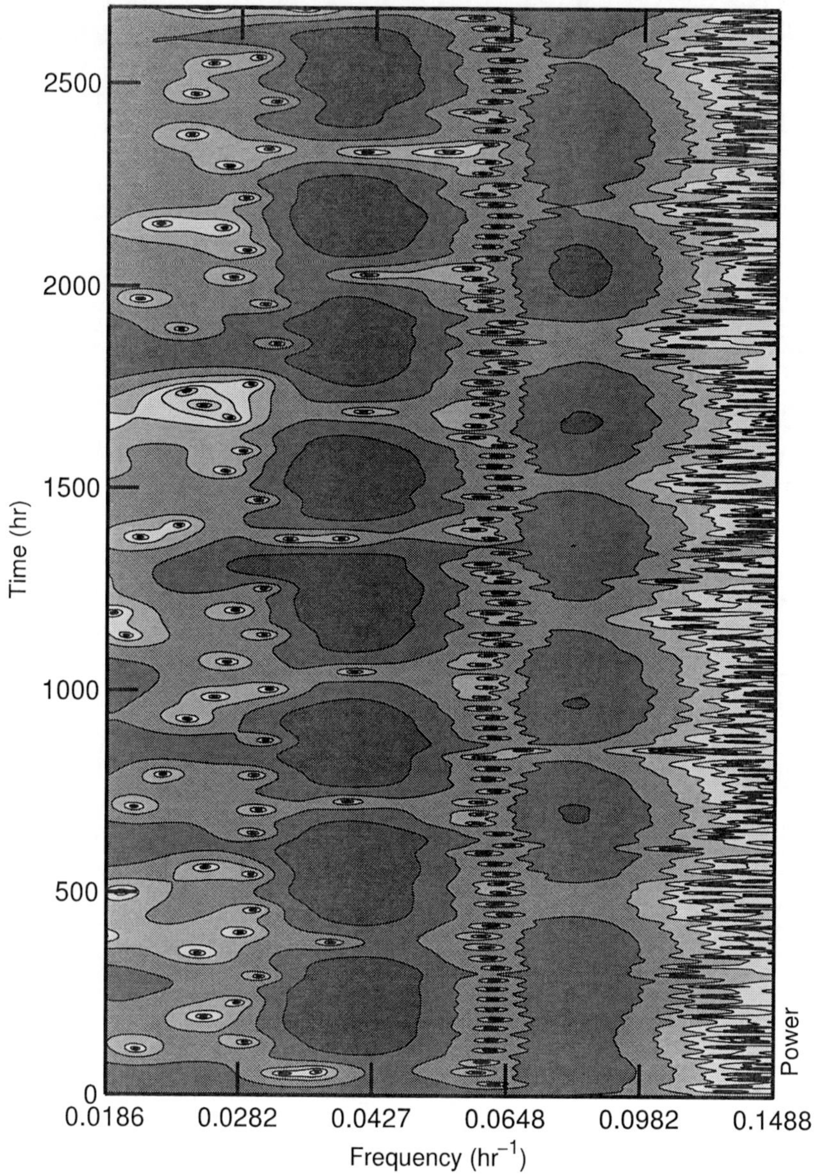

Figure 23. Wavelet power spectral density for the region of the higher frequency peaks. Clearly seen is the alternation in amplitude of the 24 and 12 hour cycles associated with the alignment of the moon and sun with earth.

Acknowledgments

Sea level measurement data provided by the Center for Operational Oceanographic Products and Services, National Ocean Service / National Oceanographic and Atmospheric Administration. The author thanks Yonggang Liu of the University of South Florida for help with collecting the data.

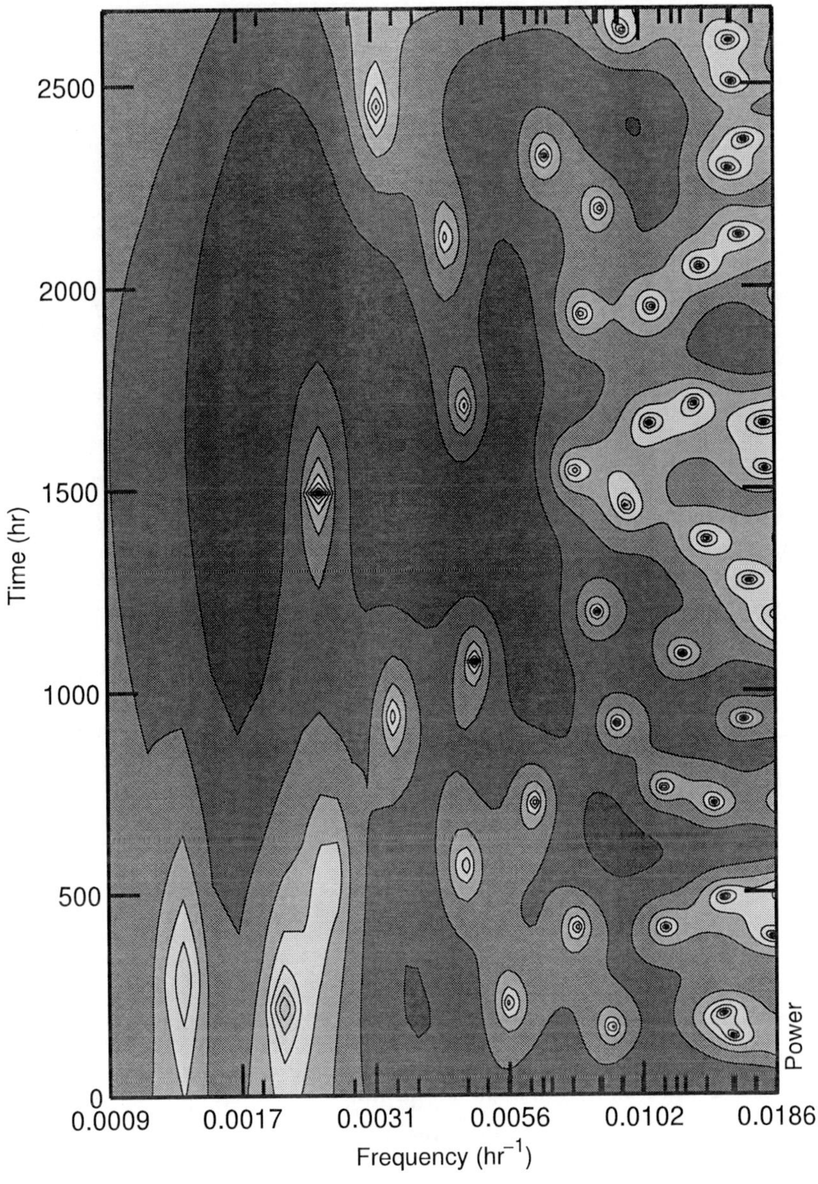

Figure 24. Wavelet power spectral density for the region of the low frequency peaks. While no structure is readily apparent, a significant amount of power is carried by components in this region.

References

[1] P. Goupillaud, A. Grossman, and J. Morlet, *Geoexploration* **23**, 85 (1984).

[2] S. Karakaz'yan, M. Skopina, and M. Tchobanou, *Int. J. Wavelets Multi.* **7**(3), 313 (2009).

[3] G. Kaiser, *A Friendly Guide to Wavelets.* Birkhauser, Cambridge, MA, (1994).

[4] W. Sweldens, *SIAM J. Math. Anal.* **29**(2), 511 (1998).

[5] O. Rioul and P. Flandrin, *IEEE T. Signal Proces.* **40**, 1746 (1992).

[6] C. H. Li, Z. F. Yang, G. H. Huang, and Y. P. Li, *Expert Syst. Appl.* **36**(2), 3309 (2009).

[7] J. Polygiannakis, P. Preka-Papadema, and X. Moussas, *Mon. Not. R. Astr. Soc.* **343**, 725 (2003).

[8] M. Fligge, S. K. Solanki, and J. Beer, *Astron. Astrophys.* **346**, 313 (1999).

[9] S. L. Weber, *Clim. Past* **1**, 9 (2005).

[10] M. P. S. Echer, E. Echer, D. J. Nordemann, N. R. Rigozo, and A. Prestes, *Climatic Change* **87**(3), 489 (2008).

[11] M. P. S. Echer, E. Echer, D. J. R. Nordemann, and N. R. Rigozo, *J. Atmos. Sol.-Terr. Phy.* **71**(1), 41 (2009).

[12] P. Piscaronoft, J. Kalvová, and R. Brázdil, *Int. J. Climatol.* **24**(13), 1661 (2004).

[13] G. M. Le, *Chin. J. Astron. Astrophys.* **4**(6), 578 (2004).

[14] G. M. Le and J. L. Wang, *Chin. J. Astron. Astrophys.* **3**(5), 391 (2003).

[15] S. Baliunas, P. Frick, D. Sokoloff, and W. Soon, *Geophys. Res. Lett.* **24**(11), 1351 (1997).

[16] P. Frick, D. Galyagin, D. V. Hoyt, E. Nesme-Ribes, K. H. Schatten, D. Sokoloff, and V. Zakharov, *Astron. Astrophys.* **328**, 670 (1997).

[17] R. W. Johnson, *J. Geophys. Res.-Space* **114**, A05105 (2009).

[18] Y. Liu, X. S. Liang, and R. H. Weisberg, *J. Atmos. Oceanic Technol.* **24**, 2093 (2007).

[19] W. H. Press, S. A. Teukolsky, W. T. Vetterling, and B. P. Flannery, *Numerical Recipes*. Cambridge University Press, Cambridge, England, (1992).

[20] G. V. Haines and A. G. Jones, *Geophys. J.* **92**(1), 171 (1988).

[21] N. G. de Bruijn In *Inequalities II,* Shisha, O., editor (Academic Press, New York, 1967).

[22] D. Swami, P. K. Mahanti, and M. Al-Fayoumi, *Eur. J. Sci. Res.* **18**(2), 332 (2007).

[23] C. Torrence and G. P. Compo, *B. Am. Meteorol. Soc.* **79**(1), 61 (1998).

[24] J. Lewalle, *Int. J. Wavelets Multi.* **7**(1), 131 (2009).

[25] E. B. Christopoulou, A. N. Skodras, and A. A. Georgakilas, *Digital Signal Processing, 2002. DSP 2002. 2002 14th International Conference on* **2**, 893 (2002).

[26] T. Kijewski and A. Kareem, *J. Sound Vib.* **256**(5), 980 (2002).

[27] P. Frick, S. L. Baliunas, D. Galyagin, D. Sokoloff, and W. Soon, *Astrophys. J.* **483**(1), 426 (1997).

[28] R. W. Johnson, *Astrophys. Space Sci.* **326**, 181 (2010).

[29] S. P. Nanavati and P. K. Panigrahi, *Resonance* **9**(3), 50 (2004).

[30] S. R. Jammalamadaka and A. SenGupta, *Topics in Circular Statistics.* World Scientific, Singapore, (2001).

[31] G. W. Forbes and M. A. Alonso, *Am. J. Phys.* **69**, 340 (2001).

[32] M. Farge, *Annu. Rev. Fluid Mech.* **24**(1), 395 (1992).

In: Wavelets: Classification, Theory and Applications
Editors: Manel del Valle et al.
ISBN: 978-1-62100-252-9

Chapter 7

CHARACTERIZATION AND CLASSIFICATION OF AGRI-FOOD PRODUCTS BY WAVELET SIGNAL ANALYSIS

Chandra B. Singh and Digvir S. Jayas*
Department of Biosystems Engineering, University of Manitoba,
Winnipeg, MB, Canada

ABSTRACT

Recent developments in sensing and optical instrumentation have shown tremendous potential for characterization and classification of agri-food material using non-destructive techniques such as gas sensors, near-infrared spectroscopy, ultrasound, X-rays, color machine vision, and hyperspectral imaging. Statistical methods are generally used to extract the features from the recorded signals of these instruments that describe the signal property. Wavelet transformation provides an alternative powerful tool to compress the signal and extract the features describing the signal. These extracted features can be used as input in pattern recognition, prediction, and supervised classification model development. In this review, various commonly used wavelet transforms are described and their applications in characterization and classification of agri-food products are discussed.

INTRODUCTION

Signals recorded from agricultural and food materials, using a variety of sensing and spectroscopic and optical imaging devices (e.g., spectroscopy, ultrasound, color imaging, hyperspectral imaging, X-ray imaging) contain vast amount of information about the product's morphology (shape, size), texture, color, and chemical composition. However, extracting the relevant information from the signal in quantitative form for objective

* E-mail address: Digvir_Jayas@Umanitoba.ca.

classification, product sorting, qualitative and compositional analysis is quite challenging due to complexity of signal caused by signal noise, interference by other chemical components, ambient conditions, and non-homogeneous nature of agri-food and other biological materials. Therefore, it is very important to choose an appropriate method to de-noise the signal, compress it, and extract the features that represent and objectively describe the product. Signals can be decomposed and represented by sequences (discrete) or functions (continuous). Fourier used sine and cosine functions to decompose a signal and to represent it by its coefficients. Fourier transform has been successfully used in sparse representation of signals; however, Fourier transform is only suitable for time-invariant or uniform signal analysis (constant frequency). Analysis of a transient signal at a particular time by Fourier transform is cumbersome and requires large computation power and time. Fast Fourier transform (FFT) also known as short time Fourier transform (STFT) provides an alternative to the traditional Fourier transform by breaking the total time interval into several small time intervals and then analysing each window by Fourier transform. This is also sometimes referred to as windowed Fourier transform. However, the accuracy of FFT is highly dependent on size of the window selected. Using several short time interval windows would give very good time resolution but compromise frequency resolution in low frequency signals. Similarly, use of large time window would yield better frequency resolution but might not give good time resolution in high frequency signals. Because of the transient nature of signals recorded from biological materials, Fourier transform or short-time Fourier transform can not process such signals effectively [1].

Similar to these functions, other functions called basis vectors (e.g., wavelets) with special characteristics can also be used for signal decomposition. Wavelets, which represent transient signals by well localized few coefficients, have fast computational power and can be implemented in real-time algorithms for signal analysis using variable size window (large time window for low frequency signals and small window for high frequency signals). Contrary to sinusoidal waves in Fourier transform that extend to infinity, wavelets have compact support thus localize their energy in a finite time and frequency space [2]. Wavelets consist of a family of basis functions also known as mother wavelets which are then shifted (translated) and scaled (stretched/dilated) to create a family of orthonormal (orthogonal and of unit norm) basis functions. By scaling and shifting the wavelet, it is matched with the signal frequency and time to extract frequency and time location information from the signal. A particular shape of wavelet may match the signal shape more closely than the other ones. The selection of a basis depends on class of function or sequence we intend to represent in the decomposition. Judgment on the quality of basis can be made on the basis of how closely it approximates (correlates) a given function with optimum number of coefficients. A number of wavelets are matched with the signal to find out the best correlation. A signal, transformed using a number of scaled (stretched) and translated (shifted) wavelets (analysis), can be rebuilt using the wavelet coefficients (synthesis). Highly stretched (high scale or low frequency) wavelets are suitable to find peaks and valleys in a long wave (low frequency) signals, whereas, short wavelets (low scale or high frequency) are suitable for detecting discontinuities in a signal. In wavelet transform, detail coefficients can be discarded or modified for a specific time interval. It is also possible to discard certain frequencies of Fourier transformed data but these frequencies would be removed from entire signal.

Wavelet transform is recognized as a valuable technique in chemometrics and digital signal and digital image processing [3]. Wavelet transform can be used in spectroscopy and hyperspectral imaging for data compression, separation of overlapping spectral bands, signal noise removal, smoothing, base line correction, and in removing multicollinearity effect of multidimensional spectra [4, 5]. Wavelet analysis of digital images can discriminate several spatial orientations and effectively analyze the information content of images and discriminate textures [6]. Wavelet signal analyses of optical sensors and imaging instruments has shown tremendous potential for applications in quality inspection of agricultural and food products [7-9]. This chapter covers the basic theory of wavelets and various application in signal analysis for quality evaluation of agricultural and food products. The details of wavelets theory can be found elsewhere [10-13].

WAVELET THEORY

A periodic function can be represented by a set of sinusoidal waves (sine and cosine functions) using Fourier transform. These sinusoidal waves are smooth, symmetric, stretched (can expand from $-\infty$ to $+\infty$), and sufficient to analyze constant frequency signals. Wavelets, which have a compact support are irregular and sometimes non-symmetrical, are used to analyze transient signals, anomalies, and pulses [12]. Wavelets, contrary to the Fourier transform, not only stretch but also shift the basis functions. The original basis function which is shifted and scaled is called the mother wavelet. Wavelet theory is generally categorized into discrete wavelet transform (DWT) and continuous wavelet transform (CWT). Discrete wavelet transform can remove the noise and compress the data (by separating and removing certain frequency signals) and reconstruct the original signal by inverse transform. Discontinuity in a signal can be detected using CWT. Discrete wavelet and continuous wavelet transforms are discussed in next sections. It is important to introduce filter banks, which are a key component of wavelet theory.

Filter Banks

Filter banks are the foundation of understanding the concept of wavelets for both discrete and continuous wavelet transforms discussed in later sections. The single level filter bank consisting of a set of four filters is shown in Figure 1. The two left side filters are used for decomposition or analysis and two right filters for reconstruction or synthesis of a signal. The two filters in the left side are highpass ($h'(n)$) and lowpass ($g'(n)$) decomposition or analysis filters. The right side filers are called highpass ($h(n)$) and lowpass ($g(n)$) reconstruction or synthesis filters. The system forms a two channel (lowpass and highpass) filer bank. A signal is filtered first by highpass and lowpass decomposition filters. The detailed (CD) and approximation (CA) coefficient are obtained from highpass and lowpass decomposition filters, respectively. These coefficients can be further decomposed to higher level of decomposition, depending upon the objective of analysis (multilevel decomposition not shown in figure). The detail and approximation coefficients are then transformed by highpass and lowpass synthesis filters, respectively, resulting into detailed and approximated signals. Highpass filtering gives details (up to some extent signal noise) and lowpass filtering gives smooth signal. By adding details and approximations, it is possible to perfectly reconstruct the original signal, provided none of the details or approximations was discarded (in noise removal and compression). However, reconstructing the perfect signal would most likely not be the goal of wavelet signal analysis for noise removal, feature extraction, or compression. A signal de-noising could be achieved by setting the details to zero, hence reconstructed signals contains only approximation part (smooth part). The combination of highpass decomposition and highpass synthesis filters forms a highpass halfband filter (Figure 2.). Similarly, combining the lowpass decomposition and synthesis filters form lowpass halfband filter (Figure 2).

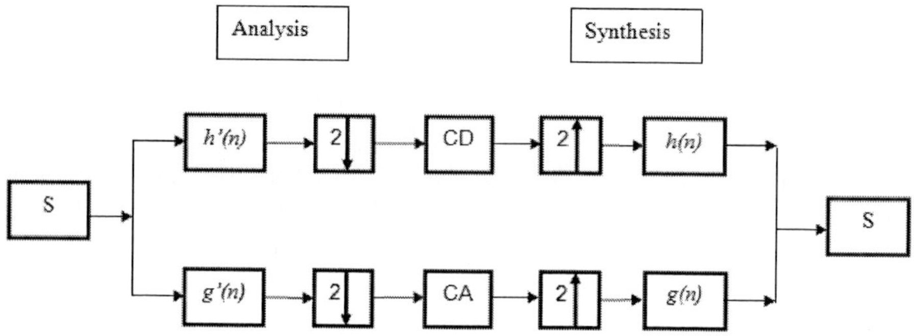

Figure 1. Filter Bank.

Two-channel and multi-channel filter banks are used to construct tree-structured filter banks and then used to implement DWT and wavelet packet transform (WPT). Tree-structured filter banks can have N-channel filter banks, however, two channel filter banks are commonly used as they are relatively simple and practically applicable in lowpass signal decomposition using j level octave band DWT decomposition. The DWT has strong polynomial approximation ability and can identify singularities in smooth signals.

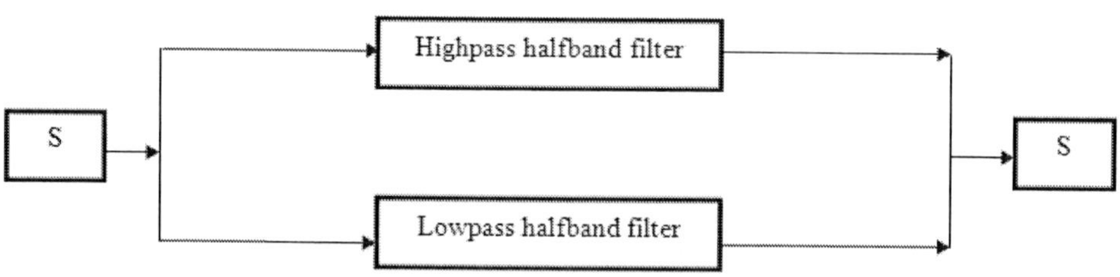

Figure 2. Halfband Filters.

Tree-Structure Filter Bank

The tree structure splits the lowpass signal further into highpass and lowpass signal recursively and the number of recursions (level of decomposition) depends on type of analysis. For example a signal from 0 to π is cut into 0 to $\pi/2$ (coarse) and $\pi/2$ to π (fine). The

0 to π/2 is further cut into 0 to π/4 (coarse) and π/4 to π/2 (fine). The 0 to π/4 is further split into 0 to π/8 (coarse) and π/8 to π/4 (fine). Thus the final 3rd level analysis signal will have one coarse and three detail (fine) components. This type of filer is also referred as an octave-band filter bank.

Discrete Wavelet Transform

The discrete wavelet transform is applied using the tree-structure octave band filter bank with a given level of decomposition. A third level, tree-structured DWT is shown in Figure 3. Multi-level decomposition analysis helps choosing the threshold frequency for setting it to zero. Due to down sampling by a factor of two at each decomposition level, it is also referred as dyadic DWT. The bases in DWT are orthogonal to each other. DWT is used for singularity detection in a signal. The delay in DWT is a serious problem for real-time signal processing with a total delay of $(2^j-1)D$, where D is delay of first block [10].

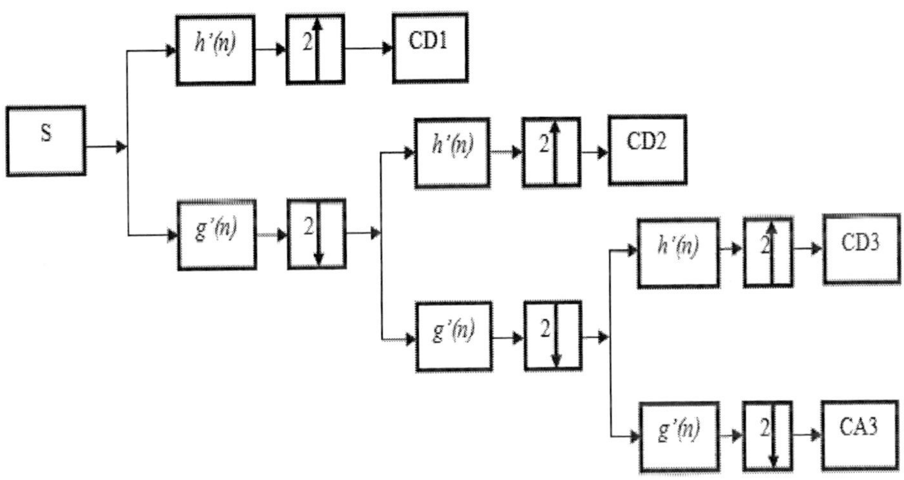

Figure 3. Tree-structured decomposition.

Wavelet Packet Filter Bank

In tree-structured decomposition in previous section, we saw that only coarse signal at the first level of decomposition is further decomposed to approximations and details. However, sometimes it might be beneficial to further decompose both coarse and fine details of first level of decomposition to desired j^{th} level.

This gives us flexibility of analyzing both coarse and fine details to a given decomposition level and select the best set of orthogonal bases using defined criteria. Tree-structured octave band filters are a subset of wavelet packets. In WPT, the detail of filter *h'(n)* is also decomposed in similar way as *g'(n)* (Figure 3).

Continuous Wavelet Transform

Wavelets are built using scaling function by scaling and shifting the original scale function by iterating a filter bank. In DWT and wavelet packet transform as discussed earlier, the iteration process is limited to a finite number with dyadic scaling. Continuous wavelet transformations are built by iterating the filter an infinite number of times. Suppose *g(n)* and *h(n)* are two filters (shown in Figure 1) given by equations:

$$g(n)=\frac{1}{\sqrt{2}}(\delta_n+\delta_{n-1}) \tag{1}$$

$$h(n)=\frac{1}{\sqrt{2}}(\delta_n-\delta_{n-1}) \tag{2}$$

The iteration of the *g(n)* and *h(n)* filters j times, results in the following scaling $\Phi^{(j)}(t)$ and wavelet $\psi^{(j)}(t)$ functions:

$$\Phi^{(j)}(t)=2^{j/2}g(n)^j \tag{3}$$

$$\psi^{(j)}(t)=2^{j/2}h(n)^j \tag{4}$$

These equations result in the formation of wavelet (Haar) series with basis functions of the following form [10]:

$$\Psi_{m,k}(t)=2^{-m/2}\psi(2^{-m}t-k)=\frac{1}{2^{m/2}}\Psi(\frac{t-2^m k}{2^m}) \tag{5}$$

$$\Phi_{m,k}(t)=2^{-m/2}\Phi(2^{-m}t-k)=\frac{1}{2^{m/2}}\Phi(\frac{t-2^m k}{2^m}) \tag{6}$$

These equations can be used to build frames having a redundant representation, which is some time beneficial. If we replace '2' by variable '*a*' and multiply '*k'* with variable 'b', the Haar equations (Eq. 5) can be written as:

$$\Psi_{m,k}(t)=a^{-m/2}\psi(a^{-m}t-bk)=\frac{1}{a^{m/2}}\Psi(\frac{t-a^m bk}{a^m}) \tag{7}$$

Frames give us flexibility to have different scales and shifts. The more redundant continuous wavelet transform can be developed from the above equations by reducing the constraints by replacing scale 'a^m' with real number '*s*' and shift '$a^m bk$' with real number 'τ':

$$\Psi_{s,\tau}(t)=\frac{1}{\sqrt{s}}\Psi(\frac{t-\tau}{s}) \tag{8}$$

The above Haar wavelet is centered at t=τ and scaled by a factor of *a*. The variables *s* and τ are the new dimensions of the signal after wavelet transformation. Scaling, which is the inverse of frequency, is also termed as dilation. The dilation process either stretches or squeezes the wavelet function along the time axis and translation or position shifts the wavelet function in time axis and tells its location. This flexibility in scaling and shifting provide the CWT capability to detect small variations (singularities) in a signal.

The original signal *f(t)* is obtained by taking the inverse wavelet transform using the following equation:

$$f(t) = \iint W(s,\tau)\psi_{s,\tau}(t)d\tau ds \tag{9}$$

where, *W*(*s*, τ) are the wavelet coefficients of the time signal *f(t)* at different scales (*s*) and translations (τ). Wavelet coefficients are determined using the following equation:

$$W(s,\tau) = \int f(t)\psi^{*}{}_{s,\tau}(t)dt \tag{10}$$

The properties of wavelet series, frames, and CWT such as orthogonality, linearity, shift in time, scale in time, smoothness, zero-moment property, Parseval's equality, time-frequency localization, and characterization of singularities and their proof is beyond the scope of this chapter and can be found elsewhere [10, 11, 12].

The CWT has some drawbacks that make it difficult to implement directly. The first drawback is the redundancy of the CWT, where the wavelet transform is calculated by continuously shifting a scalable function over a signal and calculating the correlation between the two. These scaled functions are continuous and therefore obtained wavelet coefficients are highly redundant. The second one is the infinite number of wavelets involved in the wavelet transform. The third issue is that for most of the functions the wavelet transform has no analytical solutions and they can be calculated only numerically. Therefore, discrete wavelet transform (DWT) algorithms have been developed which can overcome the drawbacks of the CWT. The DWT has the ability to solve a diverse set of problems [14].

Multi Resolution Analysis

Multi resolution analysis (MRA) is just opposite to building continuous time basis from iteration of discrete time filters used as building blocks in DWT. Wavelet series are built from multi resolution spaces. The MRA synthesizes the discrete signal data by adding the details to create a high resolution signal. In image processing, the low-resolution image can be analyzed first and then increase the resolution to the desired level.

Biorthogonality

Similar to orthogonal basis (filter *g* and *h*), biorthogonal basis are also linear and shift varying but they provide additional basis (*g'* and *h'*) from which we can select basis for analysis and synthesis. Hence, desired polynomial approximation properties and characterization of singularities is obtained with higher accuracy. Biorthogonal filter banks implement the expansion similar to orthogonal filer banks. Biorthogonal filters are less constrained than the orthogonal filters.

Families of Wavelets

The selected wavelet bases should efficiently approximate the signal (functions) of interest and produce only few non-zero wavelet coefficients. Therefore, wavelet must be optimized using a family of wavelets to produce highest number of wavelet coefficients with nearly zero values. The Haar wavelet, a square step function, is the earliest developed and simplest wavelet [15]. Haar filters have very high time resolution and are suitable for edge detection [12]. Other wavelet families include Daubechies, Biorthogonal, Coiflets, Symlets, Morlet, Mexican hat, linear spline of Battle-Lemarie (BLL), and Meyer [15, 16]. The selection of suitable mother wavelet depends on wavelet properties such as: (1) visual similarity between original signal and wavelet function, (2) vanishing wavelet moment (related to the order of function), (3) implementation and computational requirements, and (4) type of signal [17]. The wavelets which are continuous (smooth), have compact support (zero outside given interval), optimally performing, and least complex are often chosen for analysis. However, in real-time signal analysis, it is quite difficult to find a single ideal wavelet so a balance is made between performance and complexity. It is good to use more than one wavelet family for performance evaluation and select the wavelet with the best performance.

APPLICATIONS

Food Products

Yin et al. [18] analyzed gas sensor array data by wavelet packet transform to discriminate different types of Chinese vinegars. Each sensor output signal was decomposed to third level of wavelet packet decomposition using Daubechies basis resulting in eight frequency bands. The energy values were obtained at each frequency band using wavelet coefficients and normalized. The relative energy values were used as input in the principal component analysis (PCA) and radial basis function neural network (RBFNN) pattern recognition classifiers. Wavelet feature extraction method combined with pattern recognition classifiers successfully discriminated three types of vinegars. Lee et al. [19] characterized frozen orange juice by analysing ultrasound signal using wavelet transform. Wavelet analysis provided both time and frequency localization of signal. The variation in frequency as a function of time was used to calculate velocity dispersion. The higher the dispersion means larger time or lower velocity. Dispersive velocity obtained from wavelet analysis was used to interpret the juice characteristics. The dispersion was related to physical structure of food and its quality. Aktas et al. [20] predicted colorants in food products using ultraviolet-visible (UV-VIS) spectra and applying CWT wavelet transform. The Haar CWT transform coefficients obtained from CWT spectra of soft drink powders containing tartrazine (T) and allura red (A) colorants were used to develop linear regression models. The developed models were successfully tested using various synthetic ternary mixtures and real samples. Eriksson et al. [21] analysed fluorescence data of white sugar using wavelet transform after applying orthogonal signal correction (OSC) to raw data. They investigated two wavelet transform techniques namely WPT with best basis and DWT linked to MRA and used Daubechies-4, Symlet-8, and Coiflet-2 wavelet functions. They used the wavelets to compress the fluorescence data to 4% of original size. The regression model developed using wavelet reduced data set predicted ash content in white sugar with high accuracy. Data reduction by PCA is common method in near-infrared (NIR) spectroscopy. The wavelet analysis provided some advantages such as gain in computational speed and lowered demands for computer memory.

He et al. [22] analyzed NIR spectral data for variety discrimination of tea by Daubechies wavelet transform. The size of the spectral data was reduced by third level of wavelet decomposition and wavelet features were extracted. The size of extracted wavelet features was further reduced from 77 low frequency detail signals to 8 PC components by applying PCA. The PC components were used as input in training a back propagation neural network (BPNN) classifier. The BPNN classifier identified all eight tea varieties. The wavelets analysis was successfully used as a compression tool for NIR spectral data. Li et al. [23] pre-treated the NIR spectral data using wavelet transform for prediction of soluble solid content in tea. Spectral data was also pre-treated using conventional mathematical-pretreatment methods such as standard normal variate (SNV), multiple scatter correction (MSC), averaging, and derivative. The partial least square (PLS) calibration models were developed after applying wavelet transform and mathematical-pretreatment methods. When wavelet transform with SNV was used in developing PLS models, the highest prediction accuracy ($r = 0.98$) was achieved. These results also proved the capability of wavelets in de-noising the spectral signals. Chen et al. [7] improved the prediction accuracy of NIR spectroscopy by pre-treating the spectral data by wavelet transform and predicted oil content in the instant noodles. Wavelet transform was applied to NIR spectra with 5-8 level of decompositions. Spectral data were also pre-treated with first and second derivatives of spectra for comparison. Forward step linear regression model developed using 8th level (scale) wavelet transform coefficients gave the best prediction accuracy. Wavelet transform highlighted the major absorption peaks in the NIR spectra. Cocchi et al. [9] applied DWT on NIR spectra to classify wheat flour into two bread making quality classes. Three wavelet bases namely Daubechies, Symlets, and Coiflets were applied to extract wavelet features. The extracted wavelet coefficients were used as input in percentage of assignation (PA) and soft independent modeling of class analogy (SIMCA) algorithms to classify wheat flour. Principal component analysis followed by cluster analysis and SIMCA were also applied to wavelet transformed data. The SIMCA model with six wavelet coefficients gave the highest classification accuracy. The accuracy could be further improved by using large sample data set as this study used a relatively small sample size (training (36), monitoring (18), and test (18)). Fu et al. [24] predicted the reducing sugar content in vinegar using NIR spectroscopy and applying wavelet transform to NIR spectral data. Wavelet transformation separated the absorption peaks of reducing sugar from the composite NIR absorbance spectra (recorded spectra). Wavelet coefficients were used in developing forward-step linear regression model. The wavelet transformation coefficients with a scale of 7 gave the best prediction results and higher level of decomposition resulted in poor prediction accuracy. The prediction model developed using wavelet transformed data was compared with PLS model developed using NIR spectra. Wavelet feature based model gave better prediction results.

Tea is graded into different commercial grades according to the size of tea granules in the tea industry. Borah et al. [25] used Daubechies wavelet transform to describe surface roughness of tea granules of different sizes and grade them into different grades using textural information. Textural features of an image describe image patterns by roughness, smoothness or repetitiveness [26]. The bulk tea samples were imaged using a color machine vision system and acquired images were decomposed up to 4 levels by Daubechies wavelet transform. Three statistical features namely variance, energy, and entropy from each of the decomposed images were extracted. These extracted features were given as input to neural network classifiers. The classifier achieved accuracy of up to 80%. Some of the tea samples were misclassified in to other grades due to overlap of textural features.

Grains

Choudhary et al. [8] used a color machine vision system to classify cereal grains (Canada western red spring (CWRS) wheat, Canada western amber durum (CWAD) wheat, barley, oats, and rye). Wavelet, color, morphological, and textural features were extracted from the non-touching single kernels. Classification algorithms were developed using linear discriminant analysis (LDA) and quadratic discriminant analysis (QDA) and different combination of extracted features. Wavelet features were extracted from each color image by implementing two-dimensional (2D) DWT transform using fourth order Daubechies wavelet function for up to third level of decomposition. A 2D, DWT transform of an image gives four coefficients at each level of decomposition. First the rows of the input image (pixel values) are filtered through the lowpass and highpass filters then down sampled by a factor of 2. The down sampled images are further decomposed (by lowpass and highpass filters followed by down sampling by factor of 2) along the columns to obtain four wavelet coefficients namely approximation, horizontal, vertical, and diagonal details [26]. Three energy and three correlation features were extracted at each level of decomposition. Linear discriminant classifier gave the highest classification accuracy by using combined wavelet, color, morphological, and textural image features as input.

Near infrared spectroscopy has been widely used for analysis of agricultural and food products. The absorbance/reflectance spectra are used to develop prediction/classification models after applying appropriate pre-processing and data reduction methods. The NIR calibration models for wheat moisture content were successfully developed after treating the spectra with Haar wavelet transform [5]. Calibration models were developed using Haar mother and son wavelet coefficients and compared with genetic algorithms (GA). Spectral data must be treated for multi-colinearity problem before the prediction model development. The NIR spectral data treated with daughter Haar wavelet had least multi-colinearity effect, whereas, son transform gave simplest models with least errors. The prediction model developed using wavelet coefficients gave better performance than the GA calibration prediction model. Daughter Haar wavelets are scaled and translated version of mother wavelets for highpass filtering that give detail coefficients. Son wavelets are scaled and translated version of father wavelets also known as scaling function used as lowpass filters. Daughter wavelets generally have better data compression ability than son wavelets. Data compression by wavelet transform reduces the need of preprocessing methods. Highpass filtering of the signal (spectra) smoothes the signal (spectra) [4] and lowpass filter corrects baseline slope by removing low frequency signal [27]. Wang and Paliwal [28] used DWT, discrete Fourier transform (DFT), and PCA for NIR spectral data compression to classify wheat in different classes. The PCA achieved the highest compression ratio of 1:370 at 2.3 level of threshold and DFT achieved the lowest compression ratio of 1:18.5. Instead of varying compression ratios, all three compression methods retained above 99.75% of the original information. Without applying spectral pre-processing methods namely baseline correction, the DFT performed poorly with a compression ratio of only 1:1.8. The DWT was not affected by baseline correction indicating its ability to overcome/remove signal noise, hence, eliminating pre-processing steps in spectral data analysis. Spectral data compressed by DFT and DWT were further analyzed by PCA to reduce the dimensionality of DFT and DWT coefficients prior to classification.

Hyperspectral imaging is an advanced technique developed by combining spectroscopy and digital imaging. The samples are scanned at hundreds of wavelengths to form a large three dimensional spectral data set (hypercube). The dimensionality of the hyperspectral data should be reduced for development of accurate pattern recognition or prediction model. Principal component analysis is a common data reduction method, however, PCA transformation is time-consuming, uses memory hierarchy inefficiently, and causes large inter-processor communication overhead [29]. Choudhary et al. [30] analyzed NIR hyperspectral imaging data by wavelet transform to classify eight western Canada wheat classes. They used Daubechies-4 DWT decomposition for up to five levels. Two-dimensional wavelet transform was used to extract wavelet features from images (at each of 75 wavelengths) along horizontal, vertical and diagonal directions that captured the intensity variation in the respective directions within that image. An additional rotational invariant feature was calculated by adding these three features at each resolution level. The extracted wavelet features (energy and entropy at each level and each direction) were used in classifying cereal grains. The data size of extracted wavelet coefficients was further reduced by stepwise linear discriminant analysis before classification algorithm (LDA, QDA, BPNN) development. The LDA achieved highest classification accuracy of above 99% in classifying wheat. It was observed that wavelet energy features contributed more than the entropy features in classification. The rotation invariant features were more significant than other individual orientation features. The wavelet texture features at higher resolutions (finer) were more significant than those at the lower (coarser) resolutions.

Fruits

Nicolai et al. [31] developed kernel PLS algorithm for sugar content prediction in apples and compressed NIR spectral data using wavelet transform. Wavelet families of Daubechies, Symlets, and Coiflets were used for spectral data analysis and their performance was evaluated. Daubechies basis (4th order) gave the best results in terms of computational time and calibration performance. With large variables size data (uncompressed raw spectral data), kernel Gram matrix calculation in kernel PLS analysis required longer computation time. The computational time was significantly reduced by compressing the spectral data using wavelet transform before kernel PLS analysis. The wavelet analysis was advantageous over PCA in terms of computation time with large variables and observations in data set.

NIR imaging has shown potential for quality inspection of fruits and vegetables. Zhu et al. [32] analyzed NIR images of apples for quality inspection by applying wavelet transform. Gabor wavelet features were extracted and used as input to kernel PCA. Kernel PCA maps the non-linear features to linear space and then applies PCA to separate the image features (solves non-linearity problem). K-nearest neighbor classifier was developed using PCA reduced data set to discriminate healthy apples from blemished apples. The classification performance was compared with features extracted from other data reduction methods namely support vector machine (SVM), PCA, kernel PCA, and Gabor PCA. Gabor wavelet transform (up to 5 scales and 8 orientations) combined with kernel PCA gave the highest classification accuracy (90.5%).

Meats

Panigrahi et al. [33] used wavelet packet transforms (WPT) to reduce noise and compress signal from an electronic nose to detect *Salmonella* contamination in packaged meat. The detectors in the electronic nose recorded the signal produced from olfactory signature of the volatile organic compounds in the headspace of the meat packages. The recorded signals were analyzed by WPT and wavelet coefficients were used as input to develop bootstrapped linear and quadratic discriminant analysis classifiers. More than 82% *Salmonella* contaminated meat samples were correctly identified.

Wavelets have been successfully applied to determine quality of beef by extracting the wavelet features from beef color images that describe the beef texture [34]. Several wavelet transforms such as biorthogonal, reverse biorthogonal, discrete Meyer, Daubechies, symmetric modified Daubechies and Coifman modified Daubechies were investigated to extract the most useful beef textural features. Textural features were also extracted using classical run length and co-occurrence algorithms and compared with wavelet features for their ability to determine beef quality. Prediction models for beef acceptability, tenderness, juiciness, flavour and hardness were developed using features from each wavelet analysis and run length and co-occurrence algorithms. The developed models gave moderate to high accuracy (0.55-0.84 r^2) in predicting beef quality. The symmetric modified Daubechies basis gave the best overall performance. Zheng et al. [35] used color machine vision system to classify beef tenderness by extracting textural features using wavelet transform and developing discriminant model. Multiresolution technique was used to decompose the images to 4th level where image size (original image size was 256×256 pixels) reduced to 16×16 pixels (due to down sampling by two at each level). Three, 2D wavelet transforms were applied in horizontal, vertical and diagonal directions and energy of wavelets (12 energy values from three transforms at four levels) was calculated and used as textural feature. Textural features were also extracted by Gabor transformation using wavelet decomposed as well as original images. Gabor transformation function is a combination of Gaussian function which is used in spatial domain and Fourier function is used in frequency domain. The extracted wavelet and Gabor features were given as input to discriminant classifier and their performance was evaluated. Gabor features extracted from the original color images gave the poorest classification. Wavelet energy features gave better classification. Gabor features extracted from the wavelet decomposed images gave the highest classification accuracy. Du and Sun [36] evaluated tenderness of cooked pork ham by extracting textural image features namely first order gray level statistics (FGLS), run length matrix (RLM), grey level co-occurrence matrix (GLCM), fractal dimension (FD), and wavelet transform. Steerable wavelet transform was used to compress the color images of pork ham and energy at each compression level was calculated and used as textural feature. Reference tenderness of the pork was measured by shear force test. A partial least square (PLS) regression model was developed using measured shear force and extracted textural features to predict pork ham tenderness. The wavelet textural features gave the best prediction accuracy in predicting tenderness of the pork ham (correlation coefficient = 0.93 to 0.99). The wavelet textural features well described the texture of pork ham.

Ultrasound is another technique used for food quality analysis by generally characterizing tissue elasticity. In an elastrogram, external stress is applied and internal displacement of tissue elements is monitored by ultrasonic pulses and recorded and signal is displayed as image. Wavelet transform has been applied to extract the features that describe the structure and texture of tissue or food material. Huang et al. [37] evaluated meat quality by extracting the textural features from an elastrogram of meat using two-dimensional wavelet transform. A second order Daubechies wavelet filter (length 4) decomposed the 128 × 128 pixels size image. The decomposed image produced 29 sub-images after 7th level of decomposition ($128 = 2^7$) with last level image of only 1× 1 pixel size. Wavelet features were extracted from the decomposed images at each level. Linear regression model was developed using the wavelet features and measured stress value for prediction of meat tenderness. Haralick's textural features were also extracted and another regression model was developed using these features for comparison. The wavelet feature based model gave better prediction accuracy ($R^2 = 0.7$ to 0.9) than Haralick's features based regression model ($R^2 = 0.1$ to 0.7). In another study, Huang et al. [38] used artificial neural network instead of linear regression model to predict the meat tenderness. Artificial neural networks considered nonlinear relationship between wavelet features and meat tenderness. The neural network was found to be more effective in predicting meat tenderness ($R^2 = 0.98$). In these studies, a 128 × 128 size image at 7th level of decomposition ($128 = 2^7$) was reduced to just a single pixel that would not have enough details of the image. Wavelet coefficients at 3rd or 4th level of decomposition might have contained the most of the energy of the image data, which could be sufficient for classification. Chang and Kuo [39] suggested that the decomposed sub-image should not be less than 16 × 16 pixels size.

Kong et al. [40] applied wavelet transformation to analyze hyperspectral images for inspecting poultry skin tumor. Since hyperspectral data set is very large, authors first indentified significant NIR image bands (at significant wavelengths) by using PCA. The significant wavelength images were decomposed by 1-level Daubechies wavelet transform. Two features from approximation images were used as input in developing fuzzy inference classifier. The developed classifier achieved nearly 82% classification accuracy. Xu et al. [41] used hyperspectral reflectance imaging for poultry skin tumor detection and analyzed hyperspectral data using PCA, wavelet transform, and kernels discriminant analysis (KDA). Classifiers using PCA based features gave the lowest classification accuracy. KDA features based classifier gave maximum accuracy. Wavelet feature based classifier had slightly lower classification accuracy than KDA feature based classifier. The classification model developed by combining features from all the three methods (PCA, wavelet, KDA) further improved the classification accuracy.

X-ray imaging has been used for quality inspection of agricultural and food products. Chen and Tao [42] used an online x-ray imaging system to inspect deboned poultry. X-ray images were analyzed by applying wavelet transform. Wavelet decomposition of images detected the presence of objects such as bones, however, at higher level of decomposition, due to poor resolution (reduced pixels), the information about the shape and size of the small objects was lost, so it was difficult to detect the small bones. Therefore, upsampling was done to add more shape and size information (increased pixels). Though, authors did not develop any classification model, the features from these images at finer resolution can be used for developing pattern recognition techniques.

CONCLUSION

The large amount of data collected by sensors and spectroscopic and imaging devices require data compression to remove redundancy and computational complexity and improve classification/prediction accuracy of developed algorithms. Data compression by statistical methods such as principal component analysis (PCA) is based on linear transformation and may not produce the features that have high discrimination capability. Wavelet transformation has demonstrated its ability to successfully compress the data without any significant signal information loss and signals can be analyzed both in time and frequency domains. Along with data compression, wavelets have also demonstrated their capability in de-noising the spectra, separating the absorption peaks, and eliminating co-linearity, hence, eliminating the need of pre-processing methods such as standard normal variate (SNV), de-trending, orthogonal signal correction (OSC), smoothing, and derivatives. The discrete wavelet transform (DWT) is the common wavelet transform used due to its ability to compress the data, analyze signal, extract features, and can approximate functions very well. Wavelet packet transform (WPT) which decomposes both details and approximations (in DWT, only approximations are further decomposed) is considered suitable for textural analysis of images as details contain significant textural information. Wavelet texture analysis of digital images is a powerful tool for classification/pattern recognition. There are several wavelet basis functions namely Haar, Daubechies, Symlets, Coiflets, Meyer, Morlet, and Mallat. In the literature reviewed, the Daubechies wavelet family has shown the best performance in analyzing spectroscopic and imaging data obtained from agricultural and food products. Wavelet analysis has potential to effectively analyze the gas sensors and spectroscopic data as well as digital color imaging, hyperspectral imaging, X-ray imaging, and ultrasonic data for quality inspection of agriculture and food products.

REFERENCES

[1] B.P. Marchant, *Biosyst. Eng*. 85, 261 (2003).
[2] M.A. Colestock, *IEEE*, paper no. 0-7803- I.343-7/9, 54 (1993).
[3] K. Jetter, U. Depczynski, K. Molt and A. Niemöller, *Anal. Chim. Acta* 420, 169 (2000).
[4] V.J. Barclay, R.F. Bonner and I.P. Hamilton, *Anal. Chem.* 69, 78 (1997).
[5] C.E.W. Gributs and D.H. Burns, *Chemometr. Intell. Lab. Systems* 83, 44 (2006).

[6] S.G. Mallat, *IEEE Trans. Patt. Anal. Mach. Intel.* 11, 674 (1989).

[7] B. Chen, X. Fu and D. Lu, *J. Food Eng.* 53, 373 (2002).

[8] R. Choudhary, J. Paliwal, and D.S. Jayas, *Biosyst. Eng.* 99, 330 (2008).

[9] M. Cocchi, M. Corbellini, G. Foca, M. Lucisano, M.A. Pagani, L. Tassi and A. Ulrici, *Anal. Chim. Acta* 544, 100 (2005).

[10] M. Vetterli, J. Kovacevic and V.K Goyal, *Fourier and Wavelet Signal Processing*, Accessed online: http://www.fourierandwavelets.org/book.pdf (11-01-2011).

[11] S. Mallat, *A Wavelet Tour of Signal Processing: the Sparse Way*, Academic Press, Burlington (2009).

[12] D.L. Fungal, *Conceptual Wavelets in Digital Signal Processing*, Space and Signals Technical Publishing, San Diego (2006).

[13] C.B. Singh, R. Choudhary, D.S. Jayas and J. Paliwal, *Food Bio. Tech.* 3, 2 (2010).

[14] G. Jacques, J.L. Frymiare, J. Kounios, C. Clark and R. Polikar, *Proc. IEEE Int. Conf. on Acoustic, Speech, and Signal Proc.*, 5, 389 (2005).

[15] J.S. Walker, *A Primer on Wavelets and Their Scientific Applications*, CRC Press, Boca Raton, Florida (1998).

[16] M. Misti, Y. Misti, G. Oppenheim and J.M. Poggi, *Wavelet Toolbox User's Guide*, Maths Works Inc., Natick, Massachusetts (1996).

[17] L.S. Safavian, W. Kinsner and H. Turanli, *Canadian Conf. on Elect. Comp. Eng.*, Saskatoon, Canada, May 1-4 (2005).

[18] Y. Yin, H. Yu and H. Zhang, *Sens. Actuators* B134, 1005 (2008).

[19] S. Lee, L.J. Pyrak-Nolte, P. Cornillon and O. Campanella, *J. Sci. Food Agri.* 84, 405 (2004).

[20] H.A. Aktas, E. Dinc, G. Pekcan, U. Ozgur and A. Tas, , Wavelet transform for the simultaneous prediction of the colorants in food product, In: Tas et al. (eds.), *Mathematical Methods in Engineering*, Springer, Heidelberg (2007) pp. 257-264.

[21] L. Eriksson, J. Trygg, E. Johansson, R. Bro and W. Svante, *Anal. Chim. Acta,* 420, 181 (2000).

[22] Y. He, X. Li and X. Deng, J. Food Eng. 79, 1238 (2007).

[23] X. Li, Y. He, C. Wu and D.-W Sun, *J. Food Eng.* 82, 316 (2007).

[24] X. Fu, G. Yan, B. Chen and H. Li, *J. Food Eng.* 69, 461 (2005).

[25] S. Borah, E.L. Hines and M. Bhuyan, *J. Food Eng.* 79, 629 (2007).

[26] R.C. Gonzalez and R.E. Woods, *Digital Image Processing*, Pearson Education, Singapore (2002).

[27] H.-W. Tan and S.D. Brown, *J. Chemometr.* 16, 228 (2002).

[28] W. Wang and J. Paliwal, *Trans. ASABE* 49, 1607 (2006).

[29] E. El-Araby, T. El-Ghazawi, J.L. Moigne and K. Gaj, *Proc. IEEE Int. Conf. on Field-Prog. Tech.*, Brisbane, Australia, Dec 6-8 (2004).

[30] R. Choudhary, S. Mahesh, J. Paliwal and D.S. Jayas, *Biosyst. Eng.* 102, 115 (2009).

[31] B.M. Nicolai, K.I. Theron and J. Lammerty, *Chemometr. Intell. Lab. Systems* 85, 243-252(2007)..

[32] B. Zhu, L. Jiang, Y. Luo and Y. Tao, *J. Food Eng.* 81, 741 (2007).

[33] S. Panigrahi, Y. Chang, L.R. Khot, J. Glower and C.M. Logue, *IEEE Sens. Appl. Symp.*, Atlanta, GA, Feb 12-14 (2008).

[34] P. Jackman, D.-W. Sun and P. Allen, *Meat Sci.* 83, 82 (2009).

[35] C. Zheng, D.-W. Sun and L. Zheng, *Trans. ASAE*, 49, 1447 (2006).

[36] C.-J. Du and D.-W. Sun, *Trans. ASAE* 49, 441 (2006).
[37] Y. Huang, R.E. Lacey, L.L. Moore, R.K. Miller, A.D. Whittaker and J. Ophir, *Trans. ASAE*, 40,1741 (1997).
[38] Y. Huang, R.E. Lacey and A.D. Whittaker, *Trans. ASAE* 41, 1173 (1998).
[39] T. Chang and C.C.J. Kuo, *IEEE Trans. Image Proc.* 2, 429 (1993).
[40] S.G. Kong, Y.R. Chen, I. Kim and M.S. Kim, *Appl. Opt.* 43, 824 (2004).
[41] C. Xu, I. Kim and M.S. Kim, Poultry skin tumor detection in hyperspectral reflectance images by combining classifiers, In M. Kamel, and A. Campilho (eds.), *Proc. of the 4th International Conference on Image Analysis and Recognition,* Montreal, Canada, (2007), pp. 1289-1296.
[42] Z. Chen and Y. Tao, *Pattern Recog.* 34, 2331 (2001).

In: Wavelets: Classification, Theory and Applications
Editors: Manel del Valle et al.
ISBN: 978-1-62100-252-9

Chapter 8

WAVELET TRANSFORM IN BANACH SPACE AND QUANTUM TOMOGRAPHY

***M. Rezaei*[1,*], *M. A. Jafarizadeh*[1] *and M. Mirzaee*[2]**
[1]The Department of Theoretical and Astrophysics, The Faculty of Physics, Tabriz University, Tabriz,Iran
[2]Department of Physics, Faculty of Scince, University of Arak, Arak, Iran

Abstract

In this chapter of the book, first the wavelet transform and its corresponding inverse are defined in the Banach spaces in the three separate cases **(I)** Unitary representation of group **(II)** Homogeneous space representation for G/H and **(III)** Projective representation space. Here the intimate connection between the Banach space wavelet reconstruction method for each of the above three representations and some of the well-known quantum tomographies is established. For a unitary representation of a given group using wavelet transform and its corresponding inverse some of quantum tomographies such as: tomography of rotation group, spinor tomography and tomography of unitary group are presented. On the homogeneous spaces with both singular and nonsingular vacuum vectors some of the well-known quantum tomographies such as: Moyal-representation for a spin, discrete phase space tomography, tomography of a free particle, Homodyne tomography, phase space tomography and $SU(1,1)$ tomography are presented. The connection between the wavelet formalism on Banach space and Q-function is discussed. Wavelet reconstruction based on the projective representation with both singular and nonsingular vacuum vectors, and some of the quantum tomographies as Discrete phase space tomography, $SL(2,z)$ tomography, Permutation group tomography, Homodyne and $SU(1,1)$ tomography is revealed. However, both the atomic decomposition and the Banach frame nature of these quantum tomographic examples are also revealed in details.

Introduction

The mathematical theory of wavelet Transform finds nowadays an enormous success in various fields of science and technology, including treatment of large databases, data

*E-mail address: karamaty@tabrizu.ac.ir

and image compression, signal processing, telecommunication and many other applications [1–3]. After the empirical discovery by Morlet [4], it was recognized from the very beginning by Grossmann, Morlet, Paul [5,6] and daubechies [7] that wavelets are simply coherent states associated to affine group of the line (dilations and translations). Thus, immediately the stage was set for a far reaching generalization [7,8]. Unlike function which form orthogonal bases for space, Morlet wavelets are not orthogonal and form frames. Frames are the set of functions which are not necessarily orthogonal and which are not linearly independent. Actually, frames are a repeatable set of vectors in Hilbert space which produces each vectors in space with a natural representation.

Recently another concept called atomic decomposition have played a key role in further mathematical development of wavelet theory. Indeed atomic decomposition for any space of function or distribution aims at representing any element in the form of a set of simple function which are called atoms [9]. As far as the Banach space is concerned, Feichtinger-Grocheing [10] provided a general and very flexible way to construct coherent atomic decompositions and Banach frames for Banach spaces.

The concept of a quantum state represents one of the most fundamental pillars of the paradigm of quantum theory. Usually the quantum state is described either by state vector in Hilbert space, or density operator or a phase space probability density distribution (quasi-distributions). The quantum states can be determined completely from the appropriated experimentally data by using the well known technic of quantum tomography or better to say tomographic transformation.

A general framework is already presented for the unification of the Hilbert space wavelets transformation on the one hand, and quasi-distributions and tomographic transformation associated with a given pure quantum states on the other hand [11]. Here in this manuscript we are trying to establish the intimate connection between the Banach space wavelet reconstruction method developed by Feichtinger-Grocheing [10] kisil [12] and some of well known quantum tomographies associated with mixed states, such as: tomography of rotation group [13–16], Spinor tomography [17, 18], discrete spin tomography [19] and tomography of Unitary group [20, 21], all which can be represented by density matrices. Since the density matrix can be presented through Banach space in quantum physics [22]. Therefore, it is natural to do quantum tomography of any density matrix by using the wavelet transform and its inverse in Banach space connected with the corresponding group representation associated with that density matrix. This obtained quantum tomography by Banach space wavelet method for density states is completely consistent with the Quantum tomography obtained by other methods. Also both the atomic decomposition and Banach frame nature of these quantum tomographic examples is revealed in details.

This chapter is organized as follows:
In section-2 the wavelet transform and its inverse for each unitary representation of a given group in Banach space in the three separate cases **(I)** Unitary representation of group **(II)** Homogeneous space representation for G/H and **(III)** Projective representation space is defined and then their atomic decomposition and Banach frame in Banach space are obtained. In section -3 first the quantum tomography associated with each one of these three representations in Banach space are obtained and then some typical quantum tomographic examples are introduced.

1. Wavelet Transform on Banach Space

In this chapter of the book, first the wavelet transform and its corresponding inverse are defined in the Banach spaces in the three separate cases **(I)** Unitary representation of group **(II)** Homogeneous space representation for G/H and **(III)** Projective representation space.

1.1. Wavelet Transform on Banach Spaces for Unitary Groups

The following is a brief recapitulation of some aspects of the theory of wavelets, atomic decomposition, and Banach frame. We only mention those concepts that will be needed in the sequel, a more detailed treatment may be found in, for example, [9, 10, 23, 24].
Let G be locally compact group with left Haar measure $d\mu$ and let π be a continuous representation of a group G in a (complex) Banach space $\mathcal{B}$.
A representation for group $G\times G$ in the space $\mathcal{L}(\mathcal{B})$ of bounded linear operators acting on Banach space $\mathcal{B}$: is defined as:

$$\hat{\pi}: G\times G \rightarrow \mathcal{L}(\mathcal{L}(\mathcal{B})): \hat{O}\rightarrow U(g_1)\hat{O}U(g_2^{-1}), \tag{1}$$

where if, g_1 is equal to g_2, the representation is called adjoint representation, and if g_2 is equal to identity operator, the representation is called left representation of group. We will say that set of vectors $b_{g_1,g_2}=\hat{\pi}(g_1,g_2)b_0$ form a family of coherent states if there exists continuous non-zero linear functionals $l_0, l_0^{'}\in\mathcal{B}^*$ (called test functional) and vectors $b_0^{'}, b_0\in\mathcal{B}$ (called vacuum vector) such that

$$C(b_0,b_0^{'}) = \int_G <\hat{\pi}(g_1^{-1},g_2^{-1})b_0, l_0><\hat{\pi}(g_1,g_2)b_0^{'}, l_0^{'}> d\mu(g_1,g_2), \tag{2}$$

is non-zero and finite, which is known as the admissibility relation. For unitary representation in Hilbert spaces, the condition (2) is known as square integrability. Thus our definition describes an analog of square integrable representation for Banach space.
The wavelet transform $\mathcal{W}$ from Banach space $\mathcal{B}$ to a space of function $F(G\times G)$ is defined by a representation $\hat{\pi}$ of $G\times G$ on $\mathcal{B}$ and a test functional l_0, is given by the following formula:

$$\begin{aligned}\mathcal{W}: \mathcal{B}\rightarrow F(G\times G): \hat{O}\rightarrow \hat{O}(g_1,g_2) = [\mathcal{W}\hat{O}](g_1,g_2)\\ =<\hat{\pi}(g_1^{-1},g_2^{-1})\hat{O}, l_0> = <\hat{O},\hat{\pi}^*(g_1,g_2)l_0> \quad \forall g_1,g_2\in G\end{aligned} \tag{3}$$

where $\hat{\pi}^*$ is dual of $\hat{\pi}$. The inverse wavelet transform $\mathcal{M}$ with a vacuum vector b_0 from $F(G\times G)$ to $\mathcal{B}$ is given by the formula:

$$\begin{aligned}\mathcal{M}: F(G\times G)\rightarrow \mathcal{B}: \hat{O}(g_1,g_2)\rightarrow \mathcal{M}[\mathcal{W}\hat{O}]\\ = \int_G \hat{O}(g_1,g_2)b_{g_1,g_2}d\mu(g_1,g_2) = \int_G \hat{O}(g_1,g_2)\hat{\pi}(g_1,g_2)b_0 d\mu(g_1,g_2)\end{aligned} \tag{4}$$

The operator $PI=\mathcal{M}\mathcal{W}:\mathcal{B}\longmapsto\mathcal{B}$ is a projection of $\mathcal{B}$ into its linear subspace in which b_0 is cyclic (i.e. the set $\{T(g)b_0|g\in G\}$ span Banach space $\mathcal{B}$), and $\mathcal{M}\mathcal{W}(\hat{O}) = P(\hat{O})$ where the constant P is equal to $\frac{c(b_0,b_0^{'})}{<b_0,l_0^{'}>}$. Especially, if in cases, left representation U is an irreducible representation; then, the inverse wavelet transform $\mathcal{M}$ is a left inverse operator

of wavelet transform $\mathcal{W}$ on $\mathcal{B}$, i.e, MW$=I$.
Frames can be seen as a generalization of basis in Hilbert or Banach space [9]. Banach frames and atomic decomposition are sequences that have basis-like properties but which need not to be bases. Atomic decomposition has played a key role in the recent development of wavelet theory. Now we define a decomposition of a Banach space as follows:

Definition of atomic decomposition: let $\mathcal{B}$ be a Banach space , and $\mathcal{B}_2$ be an associated Banach space of scalar-valued sequences indexed by $N=\{1,2,3,...\}$, and let $\{y_{i,j}=\hat{\pi}^*(g_i,g_j)l_0\}_{i,j\in N}\subset\mathcal{B}^*$ where $\mathcal{B}^*$ is dual Banach space and $\{b_{i,j}=\hat{\pi}(g_i,g_j)b_0\}_{i,j\in N}\subset\mathcal{B}$ be given. If

a) $\{<\hat{O},y_{i,j}>\}\in\mathcal{B}_2$ for each $\hat{O}\in\mathcal{B}$,
b) The norms $\|\hat{O}\|_{\mathcal{B}}$ and $\|\{<\hat{O},y_{i,j}>\}\|_{\mathcal{B}_2}$ are equivalent,
c) $\hat{O}=\sum_{i,j=1}^{\infty}<\hat{O},y_{i,j}>b_{i,j}$ for each $\hat{O}\in\mathcal{B}$, then $(\{y_{i,j}\},\{b_{i,j}\})$ is an atomic decomposition of $\mathcal{B}$ with respect to $\mathcal{B}_2$ and, if the norm equivalence is given by

$$A\|\hat{O}\|_{\mathcal{B}}\leq\|\{<\hat{O},y_{i,j}>\}\|_{\mathcal{B}_2}\leq\mathcal{B}\|\hat{O}\|_{\mathcal{B}}, \tag{5}$$

then A, B are a choice of atomic bounds for $(\{y_{i,j}\},\{b_{i,j}\})$.

Definition of Banach frame: let $\mathcal{B}$ be a Banach space, and $\mathcal{B}_2$ be an associated Banach space of scalar-valued sequences indexed by $\mathcal{N}$ and let $\{y_{i,j}=\hat{\pi}^*(g_i,g_j)l_0\}_{i\in N}\subset\mathcal{B}^*$ and $T:\mathcal{B}_2\longrightarrow\mathcal{B}$ be given. If

a) $\{<\hat{O},y_{i,j}>\}\in\mathcal{B}_2$ for each $O\in\mathcal{B}$,
b) The norms $\|x\|_{\mathcal{B}}$ and $\|\{<\hat{O},y_{i,j}>\}\|_{\mathcal{B}_2}$ are equivalent, so that

$$A\|\hat{O}\|_{\mathcal{B}}\leq\|\{<\hat{O},y_{i,j}>\}\|_{\mathcal{B}_2}\leq\mathcal{B}\|\hat{O}\|_{\mathcal{B}}, \tag{6}$$

c) T is bounded and linear, and $T\{<\hat{O},y_{i,j}>\}=\hat{O}$ for each $O\in\mathcal{B}$,
then $(\{y_{i,j}\},T)$ is a Banach frame for $\mathcal{B}$ with respect to $\mathcal{B}_2$. The mapping T is called the reconstruction operator (inverse wavelet transform). If the norm equivalence is given by $A\|\hat{O}\|_{\mathcal{B}}\leq\|\{<\hat{O},y_{i,j}>\}\|_{\mathcal{B}_2}\leq B\|\hat{O}\|_{\mathcal{B}}$, then A, B will be a choice of frame bounds for $(\{y_{i,j}\},T)$.
Obviously one can show that the admissibility condition is the same as frame condition.

From definition of banach frame all states that a separated set $\{y_{i,j},i,j\in N\}$ is a set of sampling for $\mathcal{B}^*$ if and only if the set of $\{y_{i,j},i,j\in N\}$ is a frame for $\mathcal{B}^*$. It is well known from frame theory that there exists a dual frame $\{b_{i,j},\ i,j\in N\}\subset\mathcal{B}$ that allows us to reconstruct functions $\{\hat{O}_{i,j}=<\hat{O},y_{i,j}>\}\subset\mathcal{B}$ explicitly as

$$\mathcal{M}\mathcal{W}\hat{O}=\sum_{i,j}<\hat{O},y_{i,j}>b_{i,j}, \tag{7}$$

where $\mathcal{M}\mathcal{W}$ is a map from $\mathcal{B}$ to $\mathcal{B}$ then $S=\mathcal{M}\mathcal{W}$ is a frame operator similar to definition of frame operator in Hilbert space [9]. The frame condition can be expressed in term of the frame operator as:

$$AI\leq S\leq BI. \tag{8}$$

Now according to the Schure's lemma a representation that commutes with all irreducible representation must be a constant times the identity. This lemma reduced to orthogonality relation and resolution of identity for any coherent state. In the other cases if b_0 be cyclic it satisfy the resolution of identity. Then if the representation be irreducible or b_0 be cyclic, $S = \mathcal{M}\mathcal{W}$ is proportional to identity. Therefore, frame bounds A, B is equal to a constant(tight frame) [8].

1.2. Wavelet Transform in Banach Spaces on Homogeneous Space

The following is a brief recapitulation of some aspects of the theory of wavelets, the atomic decomposition and the banach frame on homogeneous space. We have only mentioned the concepts which are needed in the sequel, a more detailed treatment can be found in [10, 12]. Let G be a locally compact group and H be its closed normal subgroup. Let $X = G/H$ be the corresponding homogeneous space with an invariant measure $d\mu$ and since in general the representation is not directly defined on X, $x \in G/H$ can be defined only after an embedding $\sigma : G/H \to G$ which is fixed in the principal bundle $G \to G/H$.

In this part of our discussions for simplicity, we write $U(x)$ instead of $U(\sigma(x))$, for a fixed section $\sigma : G/H \to G$. Given a banach space $\mathcal{B}$, we could define a new representation $\hat{\pi}$ for embedding $X \times X$ in the space $\mathcal{L}(\mathcal{B})$ of bounded linear operators $\mathcal{B} \to \mathcal{B}$, let U be a continuous representation of a group:

$$\hat{\pi} : X \times X \to \mathcal{L}(\mathcal{L}(\mathcal{B})) : \hat{O} \to U(x_1)\hat{O}U(x_2^{-1}), \tag{9}$$

where if x_1 is equal to x_2, the representation is called an adjoint representation, and, if x_2 is equal to the identity operator, the representation is called a left representation of the homogeneous space.

Let $\mathcal{L}(\mathcal{B})$ be the space of the bounded linear operators $\mathcal{B} \to \mathcal{B}$ in the banach space, we will say that $b_0 \in \mathcal{B}$ is a vacuum vector if for all $h \in H_1 \times H_2$ with normal subgroups H_1 and H_2 we have $\hat{\pi}(h)b_0 = \chi(h)b_0$ where χ is a character of the subgroup H and also the set of vectors $b_{x_1,x_2} = \hat{\pi}(x_1,x_2)b_0$ forms a family of coherent states, if there exists a continuous non-zero linear functional $l_0 \in \mathcal{B}^*$ (called test functional where $\mathcal{B}^*$ is dual of $\mathcal{B}$) such that

$$C(b_0, b_0') = \int_X \int_X < \hat{\pi}(x_1^{-1}, x_2^{-1})b_0, l_0 >< \hat{\pi}(x_1,x_2)b_0', l_0' > d\mu(x_1,x_2), \tag{10}$$

is non-zero and finite, which is known as the admissibility relation.
If the subgroup H is non-trivial, one does not need to know the wavelet transform on the whole group G, but it should be defined on only the homogeneous space G/H, then the reduced wavelet transform $\mathcal{W}$ to a homogeneous space of the function L_2 is defined by a representation $\hat{\pi}$ of $X \times X$ on $\mathcal{B}$ and a test functional $l_0 \in \mathcal{B}^*$ such that [12]

$$\mathcal{W} : \mathcal{B} \to L_2(X \times X) : \hat{O} \to \hat{O}(x_1,x_2) = [\mathcal{W}\hat{O}](x_1,x_2)$$

$$=< \hat{\pi}(x_1^{-1}, x_2^{-1})\hat{O}, l_0 >=< \hat{O}, \hat{\pi}^*(x_1,x_2)l_0 > \quad \forall x_1, x_2 \in X, \tag{11}$$

where $\hat{\pi}^*$ is dual of $\hat{\pi}$. The inverse wavelet transform $\mathcal{M}$ with a vacuum vector $b_0 \in \mathcal{B}$ from $L_2(X \times X)$ to $\mathcal{B}$ is given by the formula:

$$\mathcal{M} : L_2(X \times X) \to \mathcal{B} : \hat{O}(x_1,x_2) \to \mathcal{M}[\hat{O}]$$

$$\mathcal{M}\mathcal{W}(\hat{O}) = \int_X \int_X \hat{O}(x_1,x_2) b_{x_1,x_2} d\mu(x_1,x_2) = \int_X \int_X \hat{O}(x_1,x_2)\hat{\pi}(x_1,x_2)b_0 d\mu(x_1,x_2). \tag{12}$$

The operator $PI = \mathcal{M}\mathcal{W} : \mathcal{B} \longmapsto \mathcal{B}$ is a projection of $\mathcal{B}$ into the linear subspace in which b_0 is cyclic (i.e., the set $\{\hat{\pi}(x_1,x_2)b_0 | x_1,x_2 \in X\}$ spans the banach space $\mathcal{B}$), and $\mathcal{M}\mathcal{W}(\hat{O}) = P(\hat{O})$ in which the constant P is equal to $\frac{C(b_0,b_0')}{<b_0,l_0'>}$. There are two different cases which correspond to different choices of the vacuum vector:

a) **Non-singular cases:**

In this case, $\hat{\pi}$ is an irreducible representation, then the inverse wavelet transform $\mathcal{M}$ is a left inverse operator on $\mathcal{B}$ for the wavelet transform $\mathcal{W}$ i.e., $\mathcal{M}\mathcal{W}$=I for which admissibility relation (10) holds and the vector b_0 can be found in $\mathcal{B}$.

b) **Singular cases:**

In this case the representation $\hat{\pi}$ of G is neither a square-integrable nor a square-integrable modulo of a subgroup H. Therefore, the vacuum vector b_0 could not be selected within the original Banach space $\mathcal{B}$ (the representation space of U). Then, in the singular case, we assume that there is a probe vector $p_0 \in \mathcal{B}$ where the following integral converges:

$$C(b_0,p_0) = < \int_X \int_X < \hat{\pi}(x_1^{-1},x_2^{-1})p_0, l_0 > \hat{\pi}(x_1,x_2)b_0 d\mu(x_1,x_2), l_0 > . \qquad (13)$$

The choice of a probe vector is similar to regularization [41], which has been used in our calculations. According to the theory of distributions, the smoothness, regularity, and localization of a tempered distribution can be improved by a function of the Schwartz class. Various regulators can be used for numerical computations.

The frames can be seen as a generalization of bases in the Hilbert or the banach space [9]. The banach frames and the atomic decomposition are sequences that have basis-like properties but which need not to be bases. The atomic decomposition has played a key role in the recent development of the wavelet theory.

Now we define a decomposition of a banach space on homogeneous space as follow:

The definition of the coorbit space: Let $\mathcal{B}$ be a banach space. The coorbit space is the collection of all functions for which the wavelet transform is contained in L_2. Similar to the definition of the coorbit space in the group, we can define the coorbit spaces for $X = G/H$ as [40]:

$$M_2 = \{\hat{O} \in \mathcal{B} : \mathcal{W}\hat{O} \in L_2\} \quad with \qquad norm \ \ ||\hat{O}||_{M_2} = ||\mathcal{W}\hat{O}||_{L_2}. \qquad (14)$$

The definition of the atomic decomposition: Let M_2 be a coorbit space and let L_2 be an associated banach space. Let $\{y_{i,j} = \hat{\pi}^*(x_i,x_j)l_0\}$ and $\{\hat{b}_{i,j} = \hat{\pi}(x_i,x_j)b_0\}$ be given. The set [40] $(\{y_{i,j}\},\{b_{i,j}\})$ is an atomic decomposition of $\mathcal{B}$ with respect to L_2 and, the norm equivalence is given by:

$$A\|\hat{O}\|_{M_2} \leq \|\{<\hat{O},y_{i,j}>\}\|_{L_2} \leq B\|\hat{O}\|_{M_2}, \qquad (15)$$

then A, B are a choice of the atomic bounds for $(\{y_{i,j}\},\{b_{i,j}\})$. And $\hat{O} = \sum_{i,j=1}^{\infty} < O, y_{ij} > b_{ij}$ for each $\hat{O}$.

The definition of the banach frame: Let T is a bounded linear operator such that $T\{<\hat{O},y_{i,j}>\} = \hat{O}$ for each $\hat{O} \in M_2$.

Then $(\{y_{i,j}\},T)$ is a banach frame for M_2 with respect to L_2. The mapping T is a reconstruction operator (inverse wavelet transform). If the norm equivalence is given by

$A\|\hat{O}\|_{M_2} \leq \|\{< \hat{O}, y_{i,j} >\}\|_{L_2} \leq B\|\hat{O}\|_{M_2}$, then A, B are a choice of the frame bounds for $(\{y_{i,j}\}, T)$.
It is a remarkable fact that the admissibility condition is an analogous relation to the existence of frame bounds.

From the definition of the banach frame all of the states that define a separated set $\{y_{i,j}\}$ are a set of sampling for L_2 if and only if the set of $\{y_{i,j}\}$ is a frame for L_2. It is inferred from frame theory that there exists a dual frame $\{b_{i,j}\} \subset L_2$ that allows us to reconstruct functions $\{\hat{O}_{i,j} =< \hat{O}, y_{i,j} >\} \subset L_2$ explicitly as

$$\mathcal{M}\mathcal{W}\hat{O} = \sum_{i,j} < \hat{O}, y_{i,j} > b_{i,j}, \tag{16}$$

where $\mathcal{M}\mathcal{W}$ is a map from M_2 to M_2 then $S = \mathcal{M}\mathcal{W}$ is a frame operator similar to the definition of the frame operator in the Hilbert space [9]. The frame condition can be expressed in terms of the frame operator as:

$$AI \leq S \leq BI. \tag{17}$$

Now according to the Schur's lemma a representation that commutes with all irreducible representations must be a constant multiple of the identity. This lemma reduces to an orthogonality relation and the resolution of the identity for any coherent state. In other cases if b_0 is cyclic it satisfies the resolution of the identity. Then if the representation be irreducible or b_0 is cyclic, $S = \mathcal{M}\mathcal{W}$ is proportional to the identity. Therefore, frame bounds A, B are constant(tight frame) [8].

1.3. Wavelet Transform in Banach Spaces Based on Projective Representation of Group

The following is a brief recapitulation of some aspects of the theory of wavelets, atomic decomposition and Banach frame connected with projective representation. We only mention those concepts that will be needed in the sequel, a more detailed treatment may be found for example in [10, 12, 47].

Projective representation of a group:
We present some basic background material on the projective representations of groups which is required later. Let G be a group with identity 1 of order $|G|, C$ the field of complex numbers, $C^* = C \setminus \{0\}$ and GL(n,C) the group of invertible $n \times n$ matrices over C. A projective representation of degree n of G is a map $\hat{\pi} : G \longrightarrow GL(n,C)$ such that for $g, h \in G$

$$\hat{\pi}(g)\hat{\pi}(h) = \chi(g,h)\hat{\pi}(gh) \tag{18}$$

and $\hat{\pi}(1) = I_n$, where I_n is the identity $n \times n$ matrix and $\chi(g,h) \in C^*$. Since multiplication in G and GL(n,C) is associative, it follows that

$$\chi(g,h)\chi(gh,k) = \chi(g,hk)\chi(h,k) \tag{19}$$

for all $g, h, k \in G$. A map $\chi : G \times G \rightarrow C^*$ which satisfies (19) is called a 2- cocycle(factor set) of G in C and we shall say that $\hat{\pi}$ is a projective representation with 2-cocycle χ. Projective representations $\hat{\pi}$ and $\hat{\zeta}$ of degree n with 2-cocycles χ and η respectively are said to be projectively equivalent if there exists a map $\mu : G \to C^*$ and a matrix $S \in GL(n,C)$ such that

$$\hat{\zeta}(g) = \mu(g)S^{-1}\hat{\pi}(g)S \tag{20}$$

for all $g \in G$. If $\hat{\pi}$ and $\hat{\zeta}$ are projectively equivalent, it follows that

$$\eta(g,h) = \frac{\mu(g)\mu(h)}{\mu(gh)}\chi(g,h) \tag{21}$$

for all $g,h \in G$. The corresponding 2-cocycles χ and η are then said to be equivalent. Let $H^2(G,C^*)$ denote the set of equivalence classes of 2-cocycles; then $H^2(G,C^*)$ is an abelian group which is called the Schur multiplier of G. The Schur multiplier gives a measure of the number of different classes of projectively inequivalent representations which a group G possesses. If G is a finite group, then $H^2(G,C^*)$ is a finite abelian group. All projective representations of G may be obtained from ordinary representations of a larger group; thus the problem of determining all the projective representations of a group G is essentially reduced to that of determining ordinary representations of a larger finite group. **A central extension** (H,ϕ) of a group G is a group H together with a homomorphism $\phi : H \to G$ such that $ker\phi \subset Z(H)$, where Z(H) is the centre of H, that is,

$$1 \to ker\phi \to H \xrightarrow{\phi} G \to 1 \tag{22}$$

is exact. Let $A = ker\phi$, and let $\{\gamma(g)|g \in G\}$ be a set of coset representatives of A in H which are in 1 - 1 correspondence with the elements of G; thus

$$H = \bigcup_{g \in G} A\gamma(g). \tag{23}$$

Then, for all $g,h \in G$, let a(g, h) be the unique element in A such that

$$\gamma(g)\gamma(h) = a(g,h)\gamma(gh). \tag{24}$$

The associative law in H and G implies that

$$a(g,h)a(gh,k) = a(g,hk)a(h,k) \tag{25}$$

for all $g,h,k \in G$.

Now, let γ be a linear character of the abelian group A and put

$$\chi(g,h) = \gamma(a(g,h)) \tag{26}$$

for all $g,h \in G$, then (25) implies that χ is a 2-cocycle of G.

Now, let U be an ordinary irreducible representation of H of degree n put $\hat{\pi}(g) = U(\gamma(g))$ for all $g \in G$, then $\hat{\pi}$ is a projective representation of G with 2-cocycle χ. A projective representation $\hat{\pi}$ of G arising from an irreducible ordinary representation U of H in this way is said to be linearized by the ordinary representation U.

If G is a finite group, then there exists a central extension H of G with kernel $H^2(G,C^*)$

which linearizes every projective representation of G. Such a group H is called a representation group of G; this implies that every finite group has at least one representation group. Thus, the problem of determining all the irreducible projective representations of G for all possible 2-cocycles is reduced to determining all the ordinary irreducible representations of a representation group H.

In practice, we shall see that it will be sufficient to determine a complete set of irreducible projective representations of a group G for a fixed 2-cocycle χ whose values are roots of unity. In that case, we can calculate in terms of a subgroup of the representation group of G which will be called a χ-covering group of G. Let χ be a 2-cocycle such that $\{\chi\}$ has order n and let ω be a primitive n-th root of unity, then $\chi(g,h) = \omega^{\theta(g,h)}$ for some $0 \leq \theta(g,h) < n$. Suppose that $\{\nu(g)|g \in G\}$ is a set of distinct symbols in one-one correspondence with the elements of G. Let $G(\chi) = \{(\chi^j, \nu(g))|0 \leq j < n, g \in G\}$, then it is easily verified that $G(\chi)$ is a group with composition defined by

$$(\chi^j, \nu(g))(\chi^k, \nu(h)) = (\chi^{j+k+\theta(g,h)}, \nu(gh)) \tag{27}$$

for all $g, h \in G$, $0 \leq j, k < n$.

If now $\hat{\pi}$ is a projective representation of G of degree n with 2-cocycle χ, then define U : $G(\chi) \to GL(n,C)$ by

$$U(\chi^j, \nu(g)) = \omega^j \hat{\pi}(g), \tag{28}$$

then U is an ordinary representation of $G(\chi)$ [55]. That is, $\hat{\pi}$ has been lifted to an ordinary representation of $G(\chi)$. Such a group $G(\chi)$ is called an χ-covering group of the group G.

Wavelet transform in Banach spaces on Projective Representation space:

Let G be a locally compact group with left Haar measure $d\mu$ and Z(H) be a center of group H. Let U be a continuous representation of the group G and $X = G/Z(H)$ be a central extension. In the last subsection we saw that $\hat{\pi}$ has been lifted to an ordinary representation of $G(\chi)$.

Let $\mathcal{L}(\mathcal{B})$ be the space of bounded linear operator $\mathcal{B} \to \mathcal{B}$ in Banach space. We will say that $b_0 \in \mathcal{B}$ is a vacuum vector if for all $h \in Z(H)$ we have $\hat{\pi}(h)b_0 = \chi(h)b_0$ and also the set of vectors $b_x = \hat{\pi}(x)b_0$ forms a family of coherent states, if there exists a continuous non-zero linear functional $l_0 \in \mathcal{B}^*$ (called test functional) and a vector $b_0 \in \mathcal{B}$ (called vacuum vector) such that

$$C(b_0, b_0') = \int_X < \hat{\pi}(x^{-1})b_0, l_0 >< \hat{\pi}(x)b_0', l_0' > d\mu(x), \tag{29}$$

is non-zero and finite, which is known as the admissibility relation.

If the center $Z(H)$ is non-trivial, one does not need to know wavelet transform on the whole group G, but it should be defined on only the central extension $G/Z(H)$, then the reduced wavelet transform $\mathcal{W}$ to a central extension space of function L_2 is defined by a projective representation $\hat{\pi}$ of $G \times G$ on $\mathcal{B}$ and a test functional $l_0 \in \mathcal{B}^*$ such that [12]

$$\mathcal{W} : \mathcal{B} \to L_2(X) : \hat{O} \to \hat{O}(x) = [\mathcal{W}\hat{O}](x)$$

$$= < \hat{\pi}(x^{-1})\hat{O}, l_0 > = < \hat{O}, \hat{\pi}^*(x)l_0 > \quad \forall x \in X, \tag{30}$$

where $\hat{\pi}^*$ is dual of $\hat{\pi}$. The inverse wavelet transform $\mathcal{M}$ with a vacuum vector $b_0 \in \mathcal{B}$ from $L_2(X)$ to $\mathcal{B}$ is given by the formula:

$$\mathcal{M} : L_2(X) \to \mathcal{B} : \hat{O}(x) \to \mathcal{M}[\hat{O}]$$

$$\mathcal{M}\mathcal{W}(\hat{O}) = \int_X \hat{O}(x) b_x d\mu(x) = \int_X \hat{O}(x)\hat{\pi}(x) b_0 d\mu(x). \tag{31}$$

The operator $PI = \mathcal{M}\mathcal{W} : \mathcal{B} \longmapsto \mathcal{B}$ is a projection of B into its linear subspace in which b_0 is cyclic (i.e., the set $\{\hat{\pi}(x)b_0 | x \in X\}$ span Banach space $\mathcal{B}$), and $\mathcal{M}\mathcal{W}(\hat{O}) = P(\hat{O})$ in which the constant P is equal to $\frac{c(b_0,b_0^{'})}{<b_0,l_0^{'}>}$. There are two different cases which correspond to different choices of vacuum vector:

a) **Non-singular cases:**

In this case, $\hat{\pi}$ is an irreducible representation, then the inverse wavelet transform $\mathcal{M}$ is a left inverse operator on $\mathcal{B}$ for the wavelet transform $\mathcal{W}$ i.e., $\mathcal{M}\mathcal{W}$=I for which admissibility relation (29) holds.

b) **Singular cases:**

In this case the representation $\hat{\pi}$ of G is neither square-integrable nor square-integrable modulo a subgroup H. Therefore, the vacuum vector b_0 could not be selected within the original Banach space $\mathcal{B}$ (representation space of U). Then, in the singular case, we assume that there is a topological linear space $\hat{\mathcal{B}}$ with $\mathcal{B}$ as its subset such that:

1- $\mathcal{B}$ is dense in $\hat{\mathcal{B}}$ and representation U could be uniquely extended to the projective continuous representation $\hat{\pi}$ on $\hat{\mathcal{B}}$.

2- There exists $b_0 \in \hat{\mathcal{B}}$ such that the following relation holds for all $h \in H$

$$\hat{\pi}(h) b_0 = \chi(h) b_0, \qquad \chi(h) \in C.$$

3- There exists a continuous non-zero linear functional $l_0 \in \mathcal{B}^*$ such that $\hat{\pi}(h)^* l_0 = \chi(\bar{h}) l_0$.

4- The following relation holds for a probe vector $p_0 \in \mathcal{B}$

$$C(b_0, p_0) =< \int_X < \hat{\pi}(x^{-1}) p_0, l_0 > \hat{\pi}(x) b_0 d\mu(x), l_0 >, \tag{32}$$

where the integral converges in the weak topology of $\hat{\mathcal{B}}$.

5- The composition $\mathcal{M}\mathcal{W} : \mathcal{B} \to \hat{\mathcal{B}}$ of the wavelet transform and the inverse wavelet transform map $\mathcal{B}$ to $\mathcal{B}$.

The choice of probe vector is similar to regularization [41], which have been used in our calculations. According to the theory of distribution, the smoothness, regularity, and localization of a temper distributions can be improved by a function of the Schwartz class. Various regularizers can be used for numerical computations.

A good example is the Gaussian distribution :

$$R_\delta(x) = exp(-\frac{x^2}{2\delta^2}),$$

where R_δ is a regularizer which has the properties [41]

$$Lim_{\delta \longrightarrow \infty} R_\delta(x) = 1, \quad R_\delta(0) = 1.$$

Frames can be seen as a generalization of basis in Hilbert or Banach space [9]. Banach frames and atomic decomposition are sequences that have basis-like properties but which need not to be bases. Atomic decomposition has played a key role in the recent development of wavelet theory.

Now we define a decomposition of a Banach space on central extension space as follow:

Definition of Coorbit space: let $\mathcal{B}$ be a Banach space. The coorbit space is the collection of all functions for which wavelet transform is contained in L_2. Similar to the definition of coorbit space in group, we can define coorbit spaces for X=G/Z(H) as [40]:

$$M_2 = \{\hat{O} \in \mathcal{B} : \mathcal{W}\hat{O} \in L_2\} \quad with \qquad norm \ ||\hat{O}||_{M_2} = ||\mathcal{W}\hat{O}||_{L_2}. \tag{33}$$

Definition of atomic decomposition: let M_2 be a coorbit space and let L_2 be an associated Banach space of scalar-valued sequences indexed by $N = \{1,2,3,...\}$. Let $\{y_i = \hat{\pi}^*(x_i)l_0\}_{i \in N} \subset \mathcal{B}^*$ and $\{\hat{b}_i = \hat{\pi}(x_i)b_0\}_{i \in N} \subset M_2$ be given. If [40]:

a) $\{< \hat{O}, y_i >\} \in L_2$ for each $\hat{O} \in M_2$,

b) The norms $\|\hat{O}\|_{M_2}$ and $\|\{< \hat{O}, y_i >\}\|_{L_2}$ are equivalent,

c) $\hat{O} = \sum_{i=1}^{\infty} < O, y_i > b_i$ for each $\hat{O} \in M_2$,

then $(\{y_i\}, \{b_i\})$ is an atomic decomposition of $\mathcal{B}$ with respect to L_2 and, if the norm equivalence is given by:

$$A\|\hat{O}\|_{M_2} \le \|\{< \hat{O}, y_i >\}\|_{L_2} \le B\|\hat{O}\|_{M_2}, \tag{34}$$

then A, B are a choice of atomic bounds for $(\{y_i\}, \{b_i\})$. If **i** is a continuous index then $\sum_i \to \int d\mu(X)$.

Definition of Banach frame: let M_2 be a coorbit space and let L_2 be an associated Banach space of scalar-valued sequences indexed by $N = \{1,2,3,...\}$. Let $\{y_i = \hat{\pi}^*(x_i)l_0\}_{i \in N} \subset \mathcal{B}^*$ and $\{b_i = \hat{\pi}(x_i)b_0\}_{i \in N} \subset M_2$ and $T : L_2 \longrightarrow M_2$ be given. If [40]

a) $\{< \hat{O}, y_i >\} \in L_2$ for each $\hat{O} \in M_2$,

b) The norms $\|\hat{O}\|_{M_2}$ and $\|\{< \hat{O}, y_i >\}\|_{L_2}$ are equivalent. so that,

$$A\|\hat{O}\|_{M_2} \le \|\{< \hat{O}, y_i >\}\|_{L_2} \le B\|\hat{O}\|_{M_2}, \tag{35}$$

c) T is bounded and linear, and $T\{< \hat{O}, y_i >\} = \hat{O}$ for each $\hat{O} \in M_2$.

Then $(\{y_i\}, T)$ is a Banach frame for M_2 with respect to L_2. The mapping T is a reconstruction operator (inverse wavelet transform). If the norm equivalence is given by $A\|\hat{O}\|_{M_2} \le \|\{< \hat{O}, y_i >\}\|_{L_2} \le B\|\hat{O}\|_{M_2}$, then A, B are a choice of frame bounds for $(\{y_i\}, T)$. It is a remarkable fact that the admissibility condition is a relation analogous to the existence of frame Bounds. Again if **i** is a continuous index then $\sum_i \to \int d\mu(X)$.

From definition of banach frame all states that a separated set $\{y_i, i \in N\}$ is a set of sampling for L_2 if and only if the set of $\{y_i, i \in N\}$ is a frame for L_2. It is well known from frame theory that there exists a dual frame $\{b_i,\ i \in N\} \subset L_2$ that allows us to reconstruct functions $\{\hat{O}_i =< \hat{O}, y_i >\} \subset L_2$ explicitly as

$$\mathcal{M}\mathcal{W}\hat{O} = \sum_i < \hat{O}, y_i > b_i, \tag{36}$$

where $\mathcal{MW}$ is a map from M_2 to M_2 then $S = \mathcal{MW}$ is a frame operator similar to definition of frame operator in Hilbert space [9]. The frame condition can be expressed in term of the frame operator as:

$$AI \leq S \leq BI. \tag{37}$$

Now according to the Schure's lemma a representation that commutes with all irreducible representation must be a constant times the identity. This lemma reduced to orthogonality relation and resolution of identity for any coherent state. In the other cases if b_0 be cyclic it satisfy the resolution of identity. Then if the representation be irreducible or b_0 be cyclic, $S = \mathcal{MW}$ is proportional to identity. Therefore, frame bounds A, B is equal to a constant(tight frame) [8].

2. Quantum Tomography via Group Theory with Wavelet Transform on Banach Space

Group tomography of a compact group G, with an irreducible unitary representation U acting on separable Hilbert space $\mathcal{H}$, means that, every element of $\mathcal{B}(\mathcal{H})$, the Banach algebra of bounded linear operators acting on $\mathcal{H}$, can be constructed by the set $\{U(g), g \in G\}$ according to formula (39), where the set $\{U(g), g \in G\}$ is known as tomographic set and $Tr[U^{\dagger}(g)\hat{O}]$ is sampling set or tomogram set of a given operator $\hat{O}$ [25]. When $\mathcal{H}$ is finite-dimensional, the hypothesis that $\{U(g)\}$ is a tomographic set is sufficient to reconstruct any given operator from the tomographic set by using (39), but the case of $\dim(\mathcal{H}) = \infty$ needs a further condition to make sure that every expression converges and that it can be attributed to a precise mathematical meaning. More explicitly, U needs to fulfill the following equality:

$$\int d\mu(g) < f_1, U(g) f_2 >< f_3 U(g), f_4 >=< f_1, f_4 >< f_3, f_2 > \qquad \forall |f_1>, |f_2>, |f_3>, |f_4> \in \mathcal{H}, \tag{38}$$

which is known as the square integrability of the representation U(g). If O is a trace-class operator on $\mathcal{H}$ and $\{U(g)\}$ is a tomographic set and satisfies (38) then we have

$$\hat{O} = \int d\mu(g) Tr[U^{\dagger}(g)\hat{O}] U(g). \tag{39}$$

Now we try to obtain the above explained tomography via wavelets transform in Banach space.

In order to do so, we need choose the tomographic set U(g) as a continuous representation of the wavelet transformation and the identity operator as a vacuum vector. Therefore, the corresponding wavelet transformation takes the following form:

$$\begin{aligned} \mathcal{W} : \mathcal{B} \mapsto F(g) : \hat{O} \mapsto \hat{O}(g) &=< \hat{O}, l_g > \\ &=< \hat{O}, U(g) l_0 >=< \hat{O} U(g)^{\dagger}, l_0 >= tr(\hat{O} U(g)^{\dagger}). \end{aligned} \tag{40}$$

With those conditions, the inverse wavelet transform $\mathcal{M}$ becomes a left inverse operator of the wavelet transform $\mathcal{W}$:

$$\mathcal{MW} = I \Rightarrow \mathcal{M} : F(g) \mapsto \mathcal{B} : \hat{O}(g) \mapsto \mathcal{M}[\hat{O}] = \mathcal{MW}(\hat{O}) = \hat{O} = \int d\mu(g) < \hat{O}, l_g > b_g. \tag{41}$$

Therefore, with the choice of $b_0 = I$ (identity operator), the tomography relation can be written as:

$$\hat{O} = \int d\mu(g) Tr(\hat{O}U(g)^\dagger)U(g). \tag{42}$$

By choosing $b_0 = I$, $b'_0 = |f_2 >< f_1|$, $l_0(\hat{O}) = Tr(\hat{O}|f_4 >< f_3|)$ and $l'_0(\hat{O}) = Tr(\hat{O})$(with f_1, f_2, f_3, f_4 as arbitrary vectors in $\mathcal{H}$ and $\hat{O}$ as an arbitrary bounded operator), the admissibility condition for wavelet transform on Banach space (2) can be reduced to the square integrability for group theory tomography (38).

Similarly, by the same choice as above for vacuum vectors and test functions, we can get the atomic decomposition and Banach frame for this example. To do so we need further to choose set $\{U(g)l_0\} \subset \mathcal{B}^*$ as the index sequence of functionals (with index set G) which belong to dual Banach space, then we can show that:

a) $\{< \hat{O}, U(g)l_0 >\} = \{Tr(\hat{O}U^\dagger(g))\} \in \mathcal{B}_2$ for each $\hat{O} \in \mathcal{B}$,

b) The norms $||\hat{O}||_{\mathcal{B}}$ and $||\{Tr(\hat{O}U^\dagger(g))\}|| = [\int Tr(\hat{O}U^\dagger(g))\overline{Tr(\hat{O}U^\dagger(g))}d\mu(g)]^{\frac{1}{2}}$ are equivalent since

$$||\{Tr(\hat{O}U^\dagger(g))\}|| = [\int Tr(\hat{O}U^\dagger(g))\overline{Tr(\hat{O}U^\dagger(g))}d\mu(g)]^{\frac{1}{2}}$$

$$= |\int tr(U^\dagger(g)\hat{O})tr(U(g)\hat{O}^\dagger)d\mu(g)|^{1/2} = |tr(\int tr(U^\dagger(g)\hat{O})U(g)d\mu(g)\hat{O}^\dagger)|^{1/2},$$

if we use the tomography relation (42) we obtain

$$||\{Tr(\hat{U}^\dagger(g)\hat{O})\}|| = |tr(\hat{O}\hat{O}^\dagger)|^{1/2} = ||\hat{O}||,$$

such that they can saturate the inequality (5) with the atomic bounds A=B=1, provided that we use the the Hilbert-Schmidt norm for the operator $\hat{O}$ and if we use the relation (39) we have:

c) $\hat{O} = \int Tr(\hat{O}U^\dagger(g))U(g)d\mu(g)$.

Therefore, $\{U(g)b_0, U(g)l_0\}$ is an atomic decomposition of Banach space of bounded operators acting on representation space with respect to $\mathcal{B}_2$ with atomic bounds A=B=1.

At the end by the same choice of vacuum vector, test functional and index sequence of functional as in the atomic decomposition case, we can show that the required conditions (a),(b) for the existence of Banach frame is the same as the atomic decomposition one, and in order to have the last condition for the existence of atomic decomposition, we can define the reconstruction operator S as follows:

$T\{Tr(\hat{O}U^\dagger(g))\} = \int Tr(\hat{O}U^\dagger(g))U(g)d\mu(g) = \hat{O}$ for each $\hat{O} \in \mathcal{B}$.

It is straightforward to show that the operator T as defined above is a linear bounded operator.

Therefore $\{U(g)l_0, T\}$ is a Banach frame for $\mathcal{B}$ with respect to $\mathcal{B}_2$ with frame bounds A=B=1.

2.1. Tomography for Rotation Group

The most general(unnormalized) density matrix $\hat{\rho}$ is a $(2s+1) \times (2s+1)$ Hermitian matrix with $(2s+1)^2$ real parameters. Various methods have been proposed to determine

$\hat{\rho}$. The expectations of $(2s+1)^2-1$ linearly independent spin multipoles do fix a unique (normalized) density operator [26]. In order to establish a down-to-earth approach, it is natural to restrict the measurements to those performed with a standard Stern-Gerlach apparatus, the quantization axis of which can be arbitrarily oriented in space. Therefore, we are dealing here with Hilbert space spin tomography i.e., angular momentum tomography [13, 14, 16, 19] with Hilbert space $H = C^{2s+1}$, **s** being the spin of the particle, and the corresponding group being $SU(2)$.
Obviously, the diagonal elements of the density matrix are non-negative and their sum is equal to unity, their physical meaning being the probabilities of observing the value of spin projection on fixed axis in space. Obviously the relevant representation of SU(2) group is irreducible unitary representation $U(\Omega) = U(\alpha,\beta,\gamma)$, where α,β,γ are Euler angles which are also our tomographic set at the same time. The haar invariant measure for $SU(2)$ is given by [27]:

$$dg(\alpha,\beta,\gamma) = \frac{2s+1}{4\pi^2} sin(\beta) d\alpha d\beta d\gamma, \tag{43}$$

and finally spin tomography can be written as:

$$\rho = \frac{2s+1}{4\pi^2}\int_0^{2\pi} sin(\beta) d\alpha d\beta d\gamma Tr[\rho U^{\dagger}(\alpha,\beta,\gamma)]U(\alpha,\beta,\gamma). \tag{44}$$

Now we try to obtain the above explained tomography via wavelets transform in Banach space. In order to do so, we need choose the tomographic set $U(\alpha,\beta,\gamma)$ as a continuous representation of the wavelet transformation and the identity operator as a vacuum vector. Therefore, the corresponding wavelet transformation takes the following form:

$$\mathcal{W} : \mathcal{B} \mapsto F(G) : \hat{\rho} \mapsto \hat{\rho}(\alpha,\beta,\gamma) =$$

$$<\hat{\rho}, l_{(\alpha,\beta,\gamma)}> = <\hat{\rho}, U(\alpha,\beta,\gamma) l_0> = <\hat{\rho} U^{\dagger}(\alpha,\beta,\gamma), l_0> = Tr(\hat{\rho} U^{\dagger}(\alpha,\beta,\gamma)). \tag{45}$$

With this condition, inverse wavelet transform in $\mathcal{M}$ is a left inverse operator on B for the wavelet transform $\mathcal{W}$:

$$\mathcal{M} : F(G) \mapsto \mathcal{B} : \hat{\rho}(\alpha,\beta,\gamma) \mapsto \mathcal{M}[\hat{\rho}] = \mathcal{M}(\hat{\rho})$$

$$\mathcal{M}\mathcal{W}(\hat{\rho}) = \int d\mu(\alpha,\beta,\gamma) <\hat{\rho}, l_{(\alpha,\beta,\gamma)}> b_{(\alpha,\beta,\gamma)} \tag{46}$$

Also the constant on left hand side of (29) becomes proportional to the dimension of unitary representation, that is, $C(b_0, b_0') = 2s+1 = d$, where **d** is dimensional of representation. Finally the constant P becomes equal to one, i.e., $P = \frac{c(b_0,b_0')}{<b_0,l_0'>} = 1$, and the reconstruction procedure of wavelet transform (operation of the combination of wavelet transform and its inverse one, $\mathcal{M}\mathcal{W}$ on the density operator $\hat{\rho}$) leads to the tomography relation(44).
Again, by the same choice as above for vacuum vectors and test functions, we can get the atomic decomposition and Banach frame for this example. To do it, we need just choose the set $\{U(\alpha,\beta,\gamma)l_0\} \subset \mathcal{B}^*$ as the index sequence of functionals (with index set G) which belong to dual Banach space, then we can show that:

a) $\{<\hat{\rho}, U(\alpha,\beta,\gamma)l_0>\} = \{Tr(\hat{\rho}U^{\dagger}(\alpha,\beta,\gamma))\} \in \mathcal{B}_2$ for each $\hat{\rho} \in \mathcal{B}$, b) The norms $||\hat{\rho}||_{\mathcal{B}}$ and $||\{Tr(\hat{\rho}U^{\dagger}(\alpha,\beta,\gamma))\}||$ are equivalent in the sense that they satisfy the inequality (5) with the atomic bounds A=B=1, provided that we use the Hilbert-Schmidt norm for the operator $\hat{\rho}$:

$$\int ||Tr(\hat{\rho}U^{\dagger}(\alpha,\beta,\gamma))||d\mu(\alpha,\beta,\gamma) = \int Tr(\hat{\rho}U^{\dagger}(\alpha,\beta,\gamma))\overline{Tr(\hat{\rho}U^{\dagger}(\alpha,\beta,\gamma))}d\mu(\alpha,\beta,\gamma) \quad (47)$$

$$= \sum_{m,m',n,n'} \int \hat{\rho}^{J}_{mm'}\hat{\rho}^{*J}_{nn'}U^{J}_{nn'}(\alpha,\beta,\gamma)U^{*J}_{mm'}(\alpha,\beta,\gamma)d\mu(\alpha,\beta,\gamma) = ||\ \hat{\rho}\ ||^2.$$

and if we use the relation (44), we have:

c) $\hat{\rho} = \int Tr(\hat{\rho}U^{\dagger}(\alpha,\beta,\gamma))U(\alpha,\beta,\gamma)d\mu(\alpha,\beta,\gamma)$,

Therefore, $\{U(\alpha,\beta,\gamma)b_0, U(\alpha,\beta,\gamma)l_0\}$ is an atomic decomposition of Banach space of bounded operators acting on spin **s** representation space with respect to $\mathcal{B}_2$ with atomic bounds A=B=1.

Finally, the same choice of vacuum vector, test functional and index sequence of functional as in the atomic decomposition case yield the required conditions (a) and (b) for the existence of Banach frame, which is the same as the atomic decomposition one, and in order to have the last condition for the existence of atomic decomposition, we define the reconstruction operator T as follows:

$T\{Tr(\hat{\rho}U^{\dagger}(\alpha,\beta,\gamma))\} = \int Tr(\hat{\rho}U^{\dagger}(\alpha,\beta,\gamma))U(\alpha,\beta,\gamma)d\mu(\alpha,\beta,\gamma) = \hat{\rho}$ for each $\hat{\rho} \in \mathcal{B}$.

Also it is straightforward to show that the operator S as defined above is a linear bounded operator.

Therefore, $\{U(\alpha,\beta,\gamma)l_0, T\}$ is a Banach frame for Banach space of operators acting on spin representation space with respect to $\mathcal{B}_2$ with frame bounds A=B=1.

Let us first introduce $(2s+1)^2$ irreducible multipole tensor operators T_{LM} [28]:

$$\widehat{T}^{s}_{LM} = \sqrt{\frac{2L+1}{2S+1}} \sum_{m,m'=-s}^{s} |sm><sm'|C^{sm}_{sm'LM} \quad (48)$$

By substituting the following identity :

$$\widehat{U}^{s}(\alpha,\beta,\gamma) = \sum_{L,M,m,m'} \sqrt{\frac{2l+1}{2s+1}} C^{sm}_{sm'LM} U^{s}{}_{mm'}(\Omega)\widehat{T}^{s}_{LM}. \quad (49)$$

in Equation (44) where $C^{sm}_{sm'LM}$ are Clebsch-Gordan coefficients, and by using the sums involving product of two Clebsch-Gordan coefficients, we get:

$$\hat{\rho} = \sum_{LM} Tr[\hat{\rho}\widehat{T}^{s\dagger}_{LM}]\widehat{T}^{s}_{LM}. \quad (50)$$

On the other hand, we know that these operators are the most convenient basis for $\mathcal{B}$, Banach space of operators acting on $(2s+1)$-dimensional Hilbert space associated with the representation of rotation group, in sense that any linear bounded operator acting in the (2s+1)-dimensional spin state space can be expanded in terms of these multipole tensor operators [15].

2.2. Tomography of Quantum Spinor States

The family of the probability distribution functions of the 1/2-spin projection is parameterized by the point's coordinates θ and φ on the sphere of unity radius. This parameterization coincides with the physical meaning of the marginal distribution in the sense that the distribution function $w(m,\Omega)$ is the probability to observe the spin projection **m** if we measure this spin projection on the quantization axis which is parallel to the vector normal to the surface of the sphere of the unit radius at the point with the coordinates θ and φ. If we know the positive and normalized marginal distribution $w(m,\Omega)$, then, as it was shown in [17, 18, 29], the matrix elements $\hat{\rho}_{ik}^{(j)}$ can be calculated with the help of the measurable marginal distribution $w(m,\Omega)$ of the particle with an arbitrary spin j and the values of indices $m=-j,-j+1,\ldots,j$ by means of the relation:

$$\begin{aligned}(-1)^k\hat{\rho}_{ik}^{(j)} &= \sum_{J'=0}^{2j}\sum_{M=-J'}^{J'}(2J'+1)^2\sum_{m=-j}^{j}(-1)^m \\ &\otimes\int w(m,\Omega)D_{0M}^{J'}(\Omega)\frac{d\Omega}{8\pi^2}\begin{pmatrix} j & j & J' \\ m & -m & 0\end{pmatrix}\begin{pmatrix} j & j & J' \\ i & -k & M\end{pmatrix},\end{aligned} \tag{51}$$

where $i,k=-j,-j+1,\ldots,j$ and the integration has been operationalized according to the rotation angles φ,θ,ψ

$$\int d\Omega = \int_0^{2\pi} d\varphi\int_0^{2\pi} d\psi\int_0^{\pi}\sin\theta d\theta. \tag{52}$$

Since in this example adjoint representation is used, the ψ parameter is removed and the representation is limited to homogeneous space, but the representation which we used was group representation. Therefore, the spin tomography can be obtained via wavelet transform in Banach space on Homogeneous space $S^2=SU(2)/U(1)$, for more details see [30] Then definition of the wavelet transforms for adjoint representation is given by:

$$\hat{\rho}(\Omega)=<\hat{\rho},l_\Omega>=<\pi(\Omega)\hat{\rho},l_0>=<U(\Omega)\hat{\rho}(\Omega)^\dagger,l_0>, \tag{53}$$

where U is irreducible representation of $SU(2)$ group for spin J. In this case, by choosing the test functional as $l_0(\hat{\rho})=Tr[\hat{\rho}\mid j,m_1><j,m_2\mid]$, the corresponding wavelet transform becomes:

$$\begin{aligned}\hat{\rho}(\Omega)&=<U(\Omega)\mid j,m_1><j,m_2\mid U(\Omega)^\dagger,\hat{\rho}> \\ &=<j,m_1\mid U(\Omega)\hat{\rho}U(\Omega)^\dagger\mid j,m_2>=\omega(\Omega,m_1,m_2)\end{aligned} \tag{54}$$

The inverse wavelet transform is:

$$\mathcal{M}(\hat{\rho})=\int d\Omega\,\omega(\Omega,m_1,m_2)U^\dagger(\Omega)b_0U(\Omega), \tag{55}$$

if we choose $b_0=\mid j,m_1><j,m_2\mid$, the inverse wavelet transform becomes:

$$\mathcal{M}(\hat{\rho})=\int d\Omega\,\omega(\Omega,m_1,m_2)U^\dagger(\Omega)\mid j,m_1><j,m_2\mid U(\Omega). \tag{56}$$

Therefore, the matrix elements $\mathcal{M}(\hat{\rho})_{ik}$ take the following form:

$$\mathcal{M}(\hat{\rho})_{ik} = \int d\Omega\, \omega(\Omega, m_1, m_2) < j, i \mid U^{\dagger}(\Omega) \mid j, m_1 >< j, m_2 \mid U(\Omega) \mid j, k > . \quad (57)$$

By putting $m_1 = m_2 = m$, and taking the average sum over m, we get:

$$\mathcal{M}(\hat{\rho})_{ik} = (\frac{1}{2j+1})\sum_m \int d\Omega\, \omega(\Omega, m)(-1)^{m-j} D^j_{im}(\Omega) D^j_{-k-m}(\Omega), \quad (58)$$

now, if we expand the product of D functions in the terms of D function (by using addition rule of D functions), after some algebra we get:

$$\begin{aligned} \mathcal{M}\mathcal{W}\hat{\rho}^{(j)}_{ik} &= (\frac{1}{2j+1})\sum_{J'=0}^{2j}\sum_{M=-J'}^{J'}(2J'+1)^2\sum_{m=-j}^{j}(-1)^{m-k} \\ &\otimes \int w(m,\Omega) D^{J'}_{0M}(\Omega)\frac{d\Omega}{8\pi^2}\begin{pmatrix} j & j & J' \\ m & -m & 0 \end{pmatrix}\begin{pmatrix} j & j & J' \\ i & -k & M \end{pmatrix}, \end{aligned} \quad (59)$$

also after some routine calculation, we can show that the constant on left hand side of (10) is $C(b_0, b_0^{'}) = 2j+1$, the constant $P = \frac{c(b_0, b_0^{'})}{<b_0, l_0^{'}>}$ is equal to $\frac{1}{2j+1}$, and finally the reconstruction procedure of wavelet transform (operation of the combination of wavelet transform and its inverse one, $\mathcal{M}\mathcal{W}$ on the density operator $\hat{\rho}$) leads to the tomography relation(51).
Again, by the above choice of vacuum vectors and test functions, we can get the atomic decomposition and Banach frame for this example. To do it, we need just choose the set $\{\pi(\Omega)l_0\} \subset \mathcal{B}^*$ as the index sequence of functionals (with index set G) which belong to dual Banach space, then we can show the following conditions:

a) $\{< \hat{\rho}, \pi(\Omega)l_0 >\} = \{Tr(\pi^{\dagger}(\Omega)\hat{\rho})\} \in \mathcal{B}_2$ for each $\hat{\rho} \in \mathcal{B}$,

b) The norms $||\hat{\rho}||_{\mathcal{B}}$ and $||\{Tr(U(\Omega)\hat{\rho}U^{\dagger}(\Omega))\}||$ are equivalent in the sense that they satisfy the inequality (5) with the atomic bounds $A = B = \frac{1}{2j+1}$, providing that we use the Hilbert-Schmidt norm for the density operator $\hat{\rho}$:

$$\int ||Tr(U^{\dagger}(\Omega)\hat{\rho}U(\Omega))|| d\mu(\Omega) = \int Tr(U(\Omega^{-1})\hat{\rho}U(\Omega))\overline{Tr(U(\Omega^{-1})\hat{\rho}U(\Omega))} d\mu(\Omega), \quad (60)$$

if we use the relation (59), we have:

c) $\hat{\rho} = \int Tr(U(\Omega^{-1})\hat{\rho}U(\Omega))U(\Omega)b_0U(\Omega^{-1})d\mu(\Omega)$.

Therefore, $\{\pi(\Omega)b_0, \pi(\Omega)l_0\}$ is an atomic decomposition of Banach space of bounded operators acting on spin **s** representation space with respect to $\mathcal{B}_2$ with atomic bounds $A = B = \frac{1}{2j+1}$.
Finally, the same choice of vacuum vector, test functional and index sequence of functional as in the atomic decomposition case yield the required conditions (a) and (b) for the existence of Banach frame which is the same as the atomic decomposition one, and in order to have the last condition for the existence of atomic decomposition, we define the reconstruction operator T as follows:

$T\{Tr(U(\Omega^{-1})\hat{\rho}U(\Omega))\} = \int Tr(U(\Omega^{-1})\hat{\rho}U(\Omega))U(\Omega)b_0U(\Omega^{-1})d\mu(\Omega) = \hat{\rho}$ for each $A \in \mathcal{B}$.

Also it is straightforward to show that the operator T as defined above is a linear bounded operator.
Therefore, $\{\pi(\Omega)l_0, T\}$ is a Banach frame for Banach space of operators acting on spin representation space with respect to $\mathcal{B}_2$ with frame bounds $A = B = \frac{1}{2j+1}$.

2.3. Discrete Spin Tomography

It is easy to characterize a class of discrete representations whose reconstruction is similar to the continuous representation.The most general form of a density matrix of a single qubit can be written as:

$$\rho(\vec{n}) = \frac{1}{2}(1 + a\vec{n}.\vec{\sigma}),$$

where $0 \leq a \leq 1$ and the unit vector in Bloch sphere $\vec{n}$ is the direction along which the spin is pointing up, and it satisfies the following conditions [31]:

$$\Sigma_{\alpha=1}^{K}\vec{n}_{\alpha} = 0, \tag{61}$$

$$\frac{1}{3}\delta_{jk} = \frac{1}{K}\Sigma_{\alpha}(n_{\alpha})_j(n_{\alpha})_k.$$

In the dihedral and tetrahedral subgroups of $SU(2)$, these condition are satisfied by unit vectors that point to the vertices of any regular polyhedron inscribed within the Bloch sphere i.e., a tetrahedron, octahedron, cube, icosahedron, or dodecahedron. The four projectors for a tetrahedron are linearly independent. An octahedron gives rise to the six cardinal-direction representations on the Bloch sphere. The regular polyhedral by no means exhausts the possibility of representations of this sort. we can use simultaneously the vertices forms of any number of polyhedral.

For spin s=1 it is possible to find a finite group instead of SU(2). In fact, consider the 12 element Tetrahedric group composed of the $\pm\frac{2\pi}{3}$ rotations around the versos:

$$\{\vec{n}_1 = \frac{1}{\sqrt{3}}(1,1,1), \vec{n}_2 = \frac{1}{\sqrt{3}}(1,-1,-1), \vec{n}_3 = \frac{1}{\sqrt{3}}(-1,1,-1), \vec{n}_4 = \frac{1}{\sqrt{3}}(-1,-1,1)\},$$

of the π rotations around

$$\{\vec{n}_5 = (1,0,0), \vec{n}_6 = (0,1,0), \vec{n}_7 = (0,0,1)\},$$

and of the identity. It induces a unitary irreducible representation on the space C^3, given by 3×3 rotation matrices. Hence, the tomography for s=1 can be written as [16,19]:

$$\hat{\rho} = \frac{1}{4}\sum_{m=-1}^{1}\sum_{j=1}^{7} P(\vec{n}_j, m)K_j(m - \vec{s}.\vec{n}_j) + \frac{1}{4}I, \tag{62}$$

where $P(\vec{n}_j, m)$ is the probability of having outcome **m** which is the result of measuring the operator $\vec{s}.\vec{n}$ and $K_j(m - \vec{s}.\vec{n}_j)$ is a kernel function and $U_n(\Psi) = e^{i\Psi(\vec{s}.\vec{n})}$ is a unitary representation. By choosing the identity operator as the vacuum vector, the corresponding wavelet transform becomes:

$$\mathcal{W} : \mathcal{B} \mapsto F(n,\Psi) : \hat{\rho} \mapsto \hat{\rho}(n,\Psi) =$$

$$< \hat{\rho}, l_{(n,\Psi)} >=< \hat{\rho}, U_n(\Psi)l_0 >=< U_n^{\dagger}(\Psi)\hat{\rho}, l_0 >= tr(U_n^{\dagger}(\Psi)\hat{\rho}). \tag{63}$$

For the choice of the identity operator as a vacuum vector and the test functional $l_0(\hat{\rho}) = Tr[\hat{\rho}]$, the inverse wavelet transform $\mathcal{M}$ (4) becomes left inverse operator of the wavelet transform $\mathcal{W}$:

$$\mathcal{M}\mathcal{W} = PI \Rightarrow \mathcal{M} : F(n,\Psi) \mapsto \mathcal{B} : \hat{\rho}(n,\Psi) \mapsto \mathcal{M}[\hat{\rho}] = \mathcal{M}\mathcal{W}(\hat{\rho}) = \sum_{\Psi,n} \frac{1}{P} < \hat{\rho}, l_{(n,\Psi)} > b_{(n,\Psi)} \tag{64}$$

Evaluating the trace over the complete set of vectors $|\overrightarrow{n},m>$, which are the eigenstate of $\overrightarrow{s}.\overrightarrow{n}$, with eigenvalues m, and by taking into account $< \overrightarrow{n},m|\hat{\rho}|\overrightarrow{n},m >= p(\overrightarrow{n},m)$, we get:

$$\mathcal{M}\mathcal{W}(\hat{\rho}) = \frac{1}{P} \sum_{m=-1}^{1} \sum_{j=1}^{7} P(\vec{n}_j,m) K_j(m - \vec{s}.\vec{n}_j) + \frac{1}{P} I. \tag{65}$$

we can show that the constant, on left hand side of (29) $C(b_0,b_0^{'}) = 12$, the constant $P = \frac{c(b_0,b_0^{'})}{<b_0,l_0^{'}>} = 4$, and finally the reconstruction procedure of wavelet transform (operating the combination of wavelet transform and its inverse one, $\mathcal{M}\mathcal{W}$ on the operator $\hat{\rho}$) leads to the tomography relation(62).

Again, by the above choice for vacuum vectors and test functions, we can get the atomic decomposition and Banach frame for this example similar to rotation group with bound A=B=4.

2.4. Unitary Group Tomography

Here in this case we consider a state with d levels. Firstly we prepare the generators for SU(d) systems and thereby construct the density matrices for a qudit system. By choosing an irreducible square integrable representation of $SU(d)$ group as $U(\Omega) = e^{i\hat{J}.\hat{n}\psi}$, the tomography relation is given by:

$$\rho = \int d\mu(\Omega) Tr[U^{\dagger}(\Omega)\rho] U(\Omega), \tag{66}$$

Now we try to obtain the above explained tomography via wavelets transform in Banach space. The wavelet transform $\mathcal{W}$ from Banach space $\mathcal{B}$ to a space of function F(G) is defined by a representation $U(\Omega) = e^{i\hat{J}.\hat{n}\psi}$ of G on $\mathcal{B}$, with the selection of a vacuum vector b_0 which is equal to identity, and a test functional $l_0(\hat{\rho}) = Tr(\hat{\rho})$ is given by the following formula:

$$\mathcal{W} : \mathcal{B} \longmapsto F(\Omega) : \rho \longmapsto \widehat{\rho}(\Omega) =< \rho, l_{\Omega} >=< U^{\dagger}(\Omega)\rho, l_0 >= Tr[U^{\dagger}(\Omega)\rho], \tag{67}$$

Also the constant on left hand side of (10) becomes proportional to the dimension of the unitary representation, that is, $C(b_0,b_0^{'}) = d$, where d is dimensional of representation. Finally the constant P becomes equal to one, i.e., $P = \frac{c(b_0,b_0^{'})}{<b_0,l_0^{'}>} = 1$, and the reconstruction procedure of wavelet transform (operating the combination of wavelet transform and its inverse one, $\mathcal{M}\mathcal{W}$ on the density operator $\hat{\rho}$) leads to the tomography relation (66).

We know that we can expand the $SU(d)$ representation in terms of the generators J as:

$$U(\Omega) = \sum_{i=0}^{d^2-1} a_i(\Omega) J_i \quad , \quad a_i(\Omega) = \frac{1}{d} Tr[U(\Omega)J_i], \tag{68}$$

and using the relation $Tr[J_iJ_j] = d\delta_{ij}$, the tomography relation can be written as:

$$\rho = \int d\mu(\Omega) \sum_{i,j} Tr[J_i\rho]J_j a_i^*(\Omega)a_j(\Omega). \tag{69}$$

Now using the identity:

$$\int d\mu(\Omega) a_i^*(\Omega)a_j(\Omega) = \frac{1}{d}\delta_{ij}, \tag{70}$$

the tomography relation reduces to:

$$\rho = \frac{1}{d}\sum_j Tr[J_i\rho]J_i. \tag{71}$$

The generators of $SU(d)$ group may be conveniently constructed by the elementary matrices of d-dimensions, $\{e_j^k | k, j = 1 \ldots d\}$.

$$(e_j^k)_{\mu\nu} = \delta_{\nu j}\delta_{\mu k}, \qquad 1 \leq \nu, \mu \leq d, \tag{72}$$

There are $d(d-1)$ traceless matrices,

$$\Theta_j^k = e_j^k + e_k^j \tag{73}$$

$$\beta_j^k = -i(e_j^k - e_k^j), \quad 1 \leq k < j \leq d, \tag{74}$$

which are the off-diagonal generators of the $SU(d)$ group. We add the $d-1$ traceless matrices

$$\eta_r^r = \sqrt{\frac{2}{r(r+1)}}\left[\sum_{j=1}^{r} e_j^j - r e_{r+1}^{r+1}\right], \tag{75}$$

as the diagonal generators and obtain a total of $d^2 - 1$ generators. $SU(2)$ generators are, for instance, given as $\{X = \Theta_2^1 = e_2^1 + e_1^2, Y = \beta_2^1 = -i(e_2^1 - e_1^2), Z = \eta_1^1 = e_1^1 - e_2^2\}$.

We now define the λ-matrices, which are similar to Pauli matrices in SU(2) case:

$$\lambda_{(j-1)^2+2(k-1)} = \Theta_j^k \tag{76}$$

$$\lambda_{(j-1)^2+2k-1} = \beta_j^k \tag{77}$$

$$\lambda_{j^2-1} = \eta_{j-1}^{j-1}, \tag{78}$$

In conjunction with a scaled d-dimensional identity operator these form a complete hermitian operator basis [20]. If we replace J_i with $\lambda_i/2$, unitary group tomography relation (71), reduced to

$$\hat{\rho}_d = \frac{1}{d}\sum_{j=0}^{d^2-1} r_j\hat{\lambda}_j. \tag{79}$$

In which $\widehat{\rho}_d$ is a density matrix of dimension d, a qudit, and $Tr[\widehat{\rho}_d] = 1$ implies that the coefficient r_0 is one. The condition $\mathrm{Tr}\left[\rho_d^2\right] \leq 1$ requires $\sum_{j=1}^{d^2-1} r_j^2 \leq d(d-1)/2$.

At the end we can show that $\{U(\Omega)b_0, U(\Omega)l_0\}$ is atomic decomposition and $(\{U(\Omega)l_0\}, T)$ is Banach frame with atomic bounds A=B=1.

we can extend these results to n-qudits. It is shown that for multiple qudits one needs only to consider a space of operators defined by the tensor product of the generators, $SU(d) \otimes SU(d) \otimes \ldots \otimes SU(d)$, and the representation is the tensor product of the representations, where this representation is irreducible. If we choose the vacuum vector as $b_0 = I_d \otimes I_d \otimes \ldots \otimes I_d$ for admissibility condition we will have $C(b_0, b_0') = d^n$. Then n-qudit tomography can be written as:

$$\hat{\rho}_{nd} = \frac{1}{d^n} \sum_{j1\ldots jn=0}^{d^2-1} r_{j1\ldots jn} \hat{\lambda}_{j1} \otimes \ldots \otimes \hat{\lambda}_{jn}., \tag{80}$$

where, $r_{00\ldots0} = 1$ and $r_{j1,j2,\ldots,jn} = \frac{d^n}{2^n} Tr[\lambda_{j1} \otimes \lambda_{j2} \ldots \otimes \lambda_{jn} \rho] \quad (j1, j2, \ldots, jn = 1, 2, \ldots, d)$.

3. Quantum Tomography with Wavelet Transform on Homogeneous Space (Non-singular Case)

3.1. Moyal-Type Representations for a Spin

For a spin s, in [42] a 'Stratonovich-Weyl' correspondence as a rule which maps each operator $\widehat{O}$ on the $(2s+1)$-dimensional Hilbert space $\mathcal{H}_s$ to a function on the phase space of the classical spin, S^2 is defined. A *discrete* Moyal formalism is defined as [32].

$$\hat{\Delta}_{\mathbf{ns}} = \hat{U}_{\mathbf{n}} \hat{\Delta}_{\mathbf{n_z s}} \hat{U}_{\mathbf{n}}^{\dagger}, \tag{1}$$

where $\hat{U}_{\mathbf{n}}$ represents a rotation which maps the vector $\mathbf{n}_z$ to $\mathbf{n}$.
The associated kernels are defined as

$$\hat{\Delta}_{\mathbf{ns}} = |s, \mathbf{n}\rangle\langle s, \mathbf{n}| \equiv |\mathbf{n}\rangle\langle\mathbf{n}|, \tag{2}$$

$$\hat{\Delta}^{\mathbf{ns}} = \sum_{m=-s}^{s} \Delta^m |m, \mathbf{n}\rangle\langle m, \mathbf{n}|. \tag{3}$$

In the wavelet notation, the banach space is $(2s+1)^2$-dimensional and the group is $SU(2)$, the subgroup is U(1) and the measure is $d\mu(n) = \frac{2s+1}{4\pi} d(\mathbf{n})$ and the unitary irreducible representation of the group is U_n which is the result of the adjoint representation on any operators in the banach space:

$$\hat{\pi}(n)\hat{O} = \hat{U}_n \hat{O} \hat{U}_n^{\dagger}. \tag{4}$$

Then the wavelet transform in this banach space with the test functional,

$$l_0(\hat{O}) = Tr(\hat{O} \sum_m \Delta^m |m, n_z\rangle\langle m, n_z|), \tag{5}$$

is given by:

$$\mathcal{W}\hat{\rho} = \hat{\rho}(n) = < \hat{\pi}(n)^{\dagger}\hat{O}, l_0 > = Tr(\hat{U}_n^{\dagger} \hat{O} \hat{U}_n \sum_m \Delta^m \mid m, n_z >< m, n_z \mid), \tag{6}$$

then we have:

$$\hat{O}(n) = Tr(\hat{O} \hat{U}_n \sum_m \Delta^m \mid m, n_z >< m, n_z \mid \hat{U}_n^{\dagger}) = Tr(\hat{O} \hat{\Delta}^{ns}).$$

If we choose the vacuum vector $b_0 = \mid s, n_z >< s, n_z \mid$, the inverse wavelet transform $\mathcal{M}$ becomes the left inverse operator of the wavelet transform $\mathcal{W}$:

$$\mathcal{M}\mathcal{W}\hat{\rho} = \int < \hat{\pi}^{\dagger}(n)\hat{O}, l_0 > \pi\hat{(n)} b_0 \tag{7}$$

$$= \int d\mu(n) Tr(\hat{O}\hat{\Delta}^{ns})\hat{U}_n \mid s, n_z >< s, n_z \mid \hat{U}_n^{\dagger} \Rightarrow \hat{O} = \frac{1}{P}(\frac{2s+1}{4\pi}\int dn Tr(\hat{O}\hat{\Delta^{ns}})\hat{\Delta_{ns}}).$$

By using the relations:

$$\frac{2s+1}{4\pi}\int_{S^2} d\mathbf{n} \text{ Tr } \left[\hat{\Delta}_{\mathbf{ms}}\hat{\Delta}^{\mathbf{ns}}\right]\hat{\Delta}_{\mathbf{ns}} = \hat{\Delta}_{\mathbf{ms}},$$

and

$$\text{Tr } \left[\hat{\Delta}_{\mathbf{n_z s}}\hat{\Delta}^{\mathbf{ns}}\right] = \sum_{l=0}^{2s}\frac{2l+1}{2s+1}P_l(\cos\theta),$$

one can show that the constant on the left hand side of (10) is

$$C(b_0, b_0') = \int_X < \hat{\pi}^{\dagger}|s, n_z\rangle\langle s, n_z|, l_0 >< \hat{\pi}|s, n_z\rangle\langle s, n_z|, l_0 > d\mu(n),$$

now using definitions (4) and (5) we have

$$C(b_0, b_0') = \int_X < \hat{U}_n^{\dagger}|s, n_z\rangle\langle s, n_z|\hat{U}_n, l_0 >< \hat{U}_n|s, n_z\rangle\langle s, n_z|\hat{U}_n^{\dagger}, l_0 > d\mu(n)$$

$$= \int tr(\hat{\Delta}_{n_z s}\hat{\Delta}^{ns}) tr(\hat{\Delta}^{n_z s}\hat{\Delta}_{ns}) d\mu(n) = tr(\hat{\Delta}^{n_z s}\hat{\Delta}_{n_z s}) = 2s+1,$$

then the constant $P = \frac{C(b_0, b_0')}{<b_0, l_0>} = 1$, and finally the reconstruction procedure of the wavelet transform (operating the combination of the wavelet transform and its inverse one, $\mathcal{M}\mathcal{W}$ on the operator $\hat{O}$) leads to the following tomography relation:

$$\hat{O} = \frac{(2s+1)}{4\pi}\int_{S^2} dn Tr[\hat{O}\hat{\Delta}_{\mathbf{ns}}]\hat{\Delta}^{\mathbf{ns}}. \tag{8}$$

By the same choice as above for the vacuum vectors and test functions, we can get the atomic decomposition and the banach frame for this example. Then we can show the following conditions:

The norms $||\hat{O}||_{M_2}$ and $||\{Tr(\hat{\pi}^{\dagger}(n)\hat{O})\}|| = [\int Tr(\hat{\pi}^{\dagger}(n)\hat{O})\overline{Tr(\hat{\pi}^{\dagger}(n)\hat{O})} d\mu(n)]^{\frac{1}{2}}$ are equivalent since

$$||\{Tr(\hat{\pi}^{\dagger}(n)\hat{O})\}|| = [\int Tr(\hat{\pi}^{\dagger}(n)\hat{O})\overline{Tr(\hat{\pi}^{\dagger}(n)\hat{O})} d\mu(n)]^{\frac{1}{2}}$$

$$= |\int tr(\hat{\pi}^{\dagger}(n)\hat{O}) tr(\hat{\pi}(n)\hat{O}^{\dagger}) d\mu(n)|^{1/2} = |tr(\int tr(\hat{\pi}^{\dagger}(n)\hat{O})\hat{\pi}(n) d\mu(n)\hat{O}^{\dagger})|^{1/2},$$

if we use the tomography relation (8) we obtain

$$||\{Tr(\hat{\pi}^{\dagger}(n)\hat{O})\}|| = |tr(\hat{O}\hat{O}^{\dagger})|^{1/2} = ||\hat{O}||,$$

such that they can saturate the inequality (34) with the atomic bounds A=B=1, provided that we use the the Hilbert-Schmidt norm for the operator $\hat{O}$ and if we use the relation (8) we have:

$$\hat{O} = \int Tr(\hat{\pi}^{\dagger}(n)\hat{O})\hat{\pi}(n)b_0 d\mu(n)$$

Therefore, $\{\hat{\pi}(n)b_0, \hat{\pi}(n)l_0\}$ is an atomic decomposition of M_2 of bounded operators acting on representation space with respect to L_2 with atomic bounds A=B=1.

Finally, for the banach frame we use the same choice of the vacuum vector and the test functional as in the atomic decomposition case. In order to have the banach frame condition for the existence of the atomic decomposition, we can define the reconstruction operator T as follows:

$$T\{Tr(\hat{\pi}^{\dagger}(n)\hat{O})\} = \int Tr(\hat{\pi}^{\dagger}(n)\hat{O})\hat{\pi}(n)d\mu(n) = \hat{O} \quad \text{for each} \quad \hat{O} \in M_2,$$

It is straightforward to show that the operator T as defined above is a linear bounded operator. Therefore, $\{\hat{\pi}(n)l_0, T\}$ is a banach frame for M_2 with respect to L_2 with frame bounds A=B=1.

3.2. Discrete Phase Space Tomography

In [33] the formalism was applied to represent the states and the evolution of a quantum system in the phase space in the finite dimensional Hilbert space. For discrete systems we can define the finite translation operators $\hat{Q}$ and $\hat{V}$, which respectively generate the finite translation in position and momentum. Now by identifying the corresponding displacement operators, the discrete analogue of the phase space translation operator is given by:

$$\hat{U}(q,p) = \hat{Q}^q \hat{V}^p exp(i\pi pq/N). \tag{9}$$

Here we can define the point operator as:

$$\hat{A}(q,p) = \frac{1}{(2N)^2} \sum_{n,m=0}^{2N-1} \hat{U}(m,k) exp(-2\pi i \frac{(kq-mp)}{2N}), \tag{10}$$

or:

$$\hat{A}(\alpha) = \frac{1}{2N} \hat{Q}^q \hat{R} \hat{V}^{-p} exp(i\pi pq/N), \tag{11}$$

where $\hat{R}$ is a parity operator and it is worth noting that the phase space point operators have been defined on a lattice with $2N \times 2N$ points, but it has been shown that there are only N^2 independent phase space point operators on the set $G_N = \{\alpha = (q,p); 0 \le q,p \le N-1\}$.

Now in order to obtain the tomography equation via the wavelets transform in banach space we can define: the group, the subgroup and the representation are respectively the finite Heisenberg group, its center and $U(\alpha)$. Then the wavelet transform with the test functional

$$l_0(O) = Tr(O) \text{ for any operator } \quad O,$$

is given by

$$\mathcal{W}\hat{\rho} = \hat{\rho}(\alpha) = <\hat{\rho}, l_\alpha> = Tr(\hat{U}^\dagger(\alpha)\hat{\rho}). \tag{12}$$

Since the representation is an irreducible representation, the inverse wavelet transform $\mathcal{M}$ will be the left inverse operator of wavelet transform $\mathcal{W}$:

$$\mathcal{M}\mathcal{W}\hat{\rho} = \sum_{\alpha\in G_N} <\hat{\rho}, l_\alpha> b_\alpha = \sum_{\alpha\in G_N} <\hat{U}^\dagger(\alpha)\hat{\rho}, l_0> \hat{U}(\alpha)b_0. \tag{13}$$

We can obtain the tomography relation as follows:

$$\hat{\rho} = 1/N \sum_{\alpha\in G_N} Tr(\hat{\rho}\hat{U}^\dagger(\alpha))\hat{U}(\alpha) = 4N \sum_{\alpha\in G_N} Tr(\hat{\rho}\hat{A}(\alpha))\hat{A}(\alpha), \tag{14}$$

where $W(\alpha) = Tr(\hat{A}(\alpha)\hat{\rho})$ is a Wigner function, for the admissible $b_0 = I/N$.

Through the same choice as above for the vacuum vector and the test functions, we can get the atomic decomposition and the banach frame for this example. Then we can show that:

The norms $||\hat{\rho}||_{M_2}$ and $||\{Tr(\hat{\rho}\hat{U}^\dagger(\alpha))\}||$ are equivalent since

$$||\{Tr(\hat{U}^\dagger(\alpha)\hat{O})\}|| = [\int Tr(\hat{U}^\dagger(\alpha)\hat{O})\overline{Tr(\hat{U}^\dagger(\alpha)\hat{O})}d\mu(\alpha)]^{\frac{1}{2}}$$

$$= |\int tr(\hat{U}^\dagger(\alpha)\hat{O})tr(\hat{U}(\alpha)\hat{O}^\dagger)d\mu(\alpha)|^{1/2} = |tr(\int tr(\hat{U}^\dagger(\alpha)\hat{O})\hat{U}(\alpha)d\mu(\alpha)\hat{O}^\dagger)|^{1/2},$$

if we use the tomography relation (18) we obtain

$$||\{Tr(\hat{U}^\dagger(\alpha)\hat{O})\}|| = |tr(N\hat{O}\hat{O}^\dagger)|^{1/2} = \sqrt{N}||\hat{O}||.$$

Therefore they saturate the inequality (34) with the atomic bounds A=B=$\sqrt{N}$, provided that we use the Hilbert-Schmidt norm for the operator $\hat{O}$. Also if we use the relation (17), we have,

$$\hat{\rho} = \sum_\alpha Tr(\hat{\rho}\hat{U}^\dagger(\alpha))\hat{U}(\alpha)b_0,$$

hence $\{\hat{U}(\alpha)b_0, \hat{U}(\alpha)l_0\}$ is a linear atomic decomposition of M_2 with respect to L_2.

Finally, we can define the reconstruction operator T as follows:

$$T\{Tr(\hat{\rho}\hat{U}^\dagger(\alpha))\} = \sum_\alpha Tr(\hat{\rho}\hat{U}^\dagger(\alpha))\hat{U}(\alpha) = \hat{\rho}$$

for each $\hat{\rho} \in M_2$ therefore, $\{\hat{U}(\alpha)l_0, T\}$ is a banach frame for M_2 with respect to L_2 with the frame bounds A=B=$\sqrt{N}$.

4. Quantum Tomography with Wavelet Transform on Homogeneous Space (Singular Case)

4.1. Homodyne Tomography

The problem of measuring the density matrix $\hat{\rho}$ of radiation has been extensively considered both experimentally and theoretically [38]. The Homodyne tomography is rather the only method that can be used to achieve such measurement [35]. This method is based on the idea that the density matrix for the radiation states can be evaluated in the optical Homodyne experiments by using a collection of the quadrature probability distribution .

Now we wish to obtain the tomography equation via the wavelets transform in the banach space. Obviously the group is the Heisenberg group H^R. Since the representation of H^R fails to be square-integrable, according to Stone-Von Neumann [39], we can factor out the center H^R and consider only the factor space.
For the vacuum vector and the test functional, we need to choose the identity operator and $l_0(O) = Tr[O]$ for any operator O, respectively. Then the wavelet formula for the left translation is given by:

$$\mathcal{W}\hat{\rho} = \hat{\rho}(\alpha) = <\hat{\rho}, l_\alpha> = Tr(\hat{\rho}\hat{U}^\dagger(\alpha)), \tag{1}$$

where $\hat{U}(\alpha) = exp(\alpha a^\dagger - \alpha^* a)$ is a displacement operator. But the above reference state is not admissible. Thus according to the prescription of singular cases, we have to select a probe vector $p_0 \in \mathcal{B}$ in which the equation (32) is non-zero and finite. In this case, the probe vector can be selected as:

$$p_0 = \int |\alpha><\alpha| e^{(\frac{-|\alpha|^2}{\Delta})} \frac{d^2\alpha}{\pi}, \tag{2}$$

where Δ is non-zero and finite and $b_0 \in \mathcal{B}$ is an identity operator. Since the representation is irreducible and

$$C(b_0, p_0) = < \int_\alpha < \hat{U}^\dagger(\alpha) \int_{\alpha'} |\alpha'><\alpha'| e^{(\frac{-|\alpha'|^2}{\Delta})} \frac{d^2\alpha'}{\pi}, l_0 > \hat{U}(\alpha) \frac{d^2\alpha}{\pi}, l_0 >$$

$$= \int_\alpha \int_{\alpha'} tr(\hat{U}^\dagger(\alpha) |\alpha'><\alpha'| e^{(\frac{-|\alpha'|^2}{\Delta})} \frac{d^2\alpha'}{\pi}) tr(\hat{U}(\alpha)) \frac{d^2(\alpha)}{\pi} = \Delta,$$

where $tr(\hat{U}(\alpha)) = \pi\delta^2(\alpha)$ then the inverse wavelet transform in $\mathcal{M}$ is a left inverse operator on $\mathcal{B}$ for the wavelet transform $\mathcal{W}$:

$$\mathcal{M}\mathcal{W}\hat{\rho} = \int d\mu(\alpha) < \hat{\rho}, l_\alpha > b_\alpha = \int d\mu(\alpha) Tr(\hat{\rho}\hat{U}^\dagger(\alpha))\hat{U}(\alpha) b_0, \tag{3}$$

where $d\mu(\alpha) = \frac{d^2\alpha}{\pi}$ is an invariant measure. For $b_0 = I$, the reconstruction procedure of wavelet transform (9) leads to the tomography relation:

$$\hat{\rho} = \int_C \frac{d^2\alpha}{\pi} Tr[\hat{\rho}\hat{U}^\dagger(\alpha)]\hat{U}(\alpha). \tag{4}$$

In tomography relation (1), the expression $Tr(\hat{\rho}\hat{U}^\dagger(\alpha))$ is a Wigner characteristic function. We can also obtain other quasi-distribution characteristic functions by choosing different representations according to the following table.

Table 1. The correspondence between different characteristic functions and representations

charasteristic function	representation	quasi distribution function
$Tr(\hat{\rho}\hat{U}_{an}(\alpha))$	$\hat{U}_{an}(\alpha) = e^{\alpha\hat{a}^{\dagger}}e^{-\alpha^{*}\hat{a}}$	**P-function**
$Tr(\hat{\rho}\hat{U}_{n}(\alpha))$	$\hat{U}_{n}(\alpha) = e^{-\alpha^{*}a}e^{\alpha a^{\dagger}}$	**Q-function**
$Tr(\hat{\rho}\hat{U}_{h}(\alpha))$	$\hat{U}_{h}(\nu) = e^{-\nu^{*}b}e^{\nu b^{\dagger}}$ ($\hat{b} = \mu\hat{a} + \nu\hat{a}^{\dagger}$ **and** $\mu^2 - \nu^2 = 1$)	**Husimi function**
$Tr(\hat{\rho}\hat{U}_{s}(\xi.\eta))$	$\hat{U}_{s}(\xi.\eta) = e^{i\xi\hat{q}}e^{i\eta\hat{p}}$	**Standard-ordered function**
$Tr(\hat{\rho}\hat{U}_{as}(\xi.\eta))$	$\hat{U}_{as}(\xi.\eta) = e^{i\eta\hat{p}}e^{i\xi\hat{q}}$	**Antistandard-ordered function**

For the complex Fourier transform of the displacement operator $\hat{U}$ [34]

$$\hat{\mathcal{U}}(\alpha) = \int \frac{d^2\xi}{\pi}\hat{U}(\xi)exp(\alpha\xi^{*} - \alpha^{*}\xi), \tag{5}$$

the expansion of the operator in terms of the operator $\hat{\mathcal{U}}(\alpha)$ is given by

$$\hat{\rho} = \int \frac{d^2\alpha}{\pi}W(\alpha)\hat{\mathcal{U}}(\alpha), \tag{6}$$

where $W(\alpha)$ is a Wigner function. Also by defining the complex Fourier transform for each of the above representations, we can get the tomography relation for each quasi-distribution. Now we try to obtain the atomic decomposition and the banach frame for this example. Applying the same choice of the vacuum vector and the test functional we can show that the required atomic decomposition conditions are satisfied by the atomic bounds A=B=1, since we have

$$||\{Tr(\hat{U}^{\dagger}(\alpha)\hat{O})\}|| = [\int Tr(\hat{U}^{\dagger}(\alpha)\hat{O})\overline{Tr(\hat{U}^{\dagger}(\alpha)\hat{O})}d\mu(\alpha)]^{\frac{1}{2}}$$

$$= |\int tr(\hat{U}^{\dagger}(\alpha)\hat{O})tr(\hat{U}(\alpha)\hat{O}^{\dagger})d\mu(\alpha)|^{1/2} = |tr(\int tr(\hat{U}^{\dagger}(\alpha)\hat{O})\hat{U}(\alpha)d\mu(\alpha)\hat{O}^{\dagger})|^{1/2},$$

if we use the tomography relation (1) we obtain

$$||\{Tr(\hat{U}^{\dagger}(\alpha)\hat{O})\}|| = |tr(\hat{O}\hat{O}^{\dagger})|^{1/2} = ||\hat{O}||.$$

Therefore, $\{\hat{U}(\alpha)b_0, \hat{U}(\alpha)l_0\}$ is a linear atomic decomposition of M_2 with respect to L_2. Similarly, by using the relation (1) and definition T, $\{\hat{U}(\alpha)b_0, T\}$ is a banach frame for M_2 with respect to L_2 with frame bounds A=B=1. We can also generalize the single mode Homodyne tomography to a multimode state. In the wavelet notation, the irreducible representation is $\hat{U} = \hat{U}_0 \otimes \hat{U}_1 \otimes \ldots \otimes \hat{U}_m$, with $\hat{U}_j = exp(z_j\hat{a}_j{}^{\dagger} - z_j^{*}\hat{a}_j)$, and the tomography formula with the vacuum vector $b_0 = \hat{I} \otimes \hat{I} \otimes \ldots \otimes \hat{I}$ is given by

$$\hat{\rho} = \int_C \frac{d^2z_0}{\pi}\int_C \frac{d^2z_1}{\pi}\cdots\int_C \frac{d^2z_m}{\pi}Tr[\hat{\rho}\hat{U}^{\dagger}(z_0, z_1, \ldots, z_m)]\hat{U}(z_0, z_1, \ldots, z_m). \tag{7}$$

The atomic decomposition and the banach frame is similar to a one mode Homodyne, and the frame bounds A , B are equal to the identity.

4.2. Phase Space Tomography

Any marginal distribution is defined as the Fourier transform of the characteristic function $\mathcal{W}(X,\mu,\nu)=\int dke^{-ikX} < e^{ik(\mu\hat{q}+\nu\hat{p})} >$. This marginal distribution is related to the state of the quantum system which is expressed in terms of its Wigner function $W(q,p)$ [36], as follows

$$\mathcal{W}(X,\mu,\nu)=\int dke^{-ik(X-\mu\hat{q}-\nu\hat{p})}W(q,p)\frac{dkdqdp}{(2\pi)^2}. \tag{8}$$

Now we can obtain the tomography equation via wavelets transform in the banach space. Obviously the group is the Heisenberg group in phase space. For the vacuum vector and the test functional we need to choose the identity operator and $l_0(O)=Tr[O]$ for any operator O, respectively. If we apply the reduced wavelet transform for the left translation $\hat{U}(\mu,\nu)=e^{-i(\mu\hat{q}+\nu\hat{p})}$, we have:

$$\mathcal{W}\hat{\rho}=\hat{\rho}(\mu,\nu)=<\hat{\rho},l_{(\mu,\nu)}>=Tr(\hat{\rho}\hat{U}^{\dagger}(\mu,\nu)). \tag{9}$$

The vacuum vector $b_0=\hat{I}$ is not admissible, then we choose a probe vector with coherent states in the phase space [8] which is a shifted Gaussian wave packet:

$$\eta_{\sigma(q,p)}(x)=(\pi^{-1/4})exp[-i(\frac{q}{2}-x)p]exp[-\frac{(x-q)^2}{2}] \tag{10}$$

$$p_0=\int|\eta_{\sigma(q,p)}><\eta_{\sigma(q,p)}|\,exp[\frac{-(q^2+p^2)}{\Delta}]dqdp, \tag{11}$$

and the singularity condition gives $C(b_0,p_0)=\Delta$.
Since the representation is irreducible, the inverse wavelet transform $\mathcal{M}$ is a left inverse operator on $\mathcal{B}$ for the wavelet transform $\mathcal{W}$:

$$\mathcal{M}\mathcal{W}\hat{\rho}=\int d\mu(\mu,\nu)<\hat{\rho},l_{(\mu,\nu)}>b_{(\mu,\nu)}=\int d\mu d\nu Tr[\hat{\rho}\hat{U}^{\dagger}(\mu,\nu)]\hat{U}(\mu,\nu)b_0, \tag{12}$$

Then for $b_0=\hat{I}$, we have:

$$\hat{\rho}=\int d\mu d\nu Tr[\hat{\rho}\hat{R}^{\dagger}(\mu,\nu)]\hat{R}(\mu,\nu)=\int d\mu d\nu Tr[\hat{\rho}e^{i(\mu\hat{q}+\nu\hat{p})}]e^{-i(\mu\hat{q}+\nu\hat{p})}. \tag{13}$$

With a simple calculation, we can obtain the tomography relation as follows:

$$\hat{\rho}=\int dXd\mu d\nu\mathcal{W}(X,\mu,\nu)\hat{K}_{\mu\nu}, \tag{14}$$

where the kernel operator has the form:

$$\hat{K}_{\mu\nu}=\frac{1}{2\pi}e^{iX}e^{i\mu\nu}e^{-i\nu\hat{p}}e^{-i\mu\hat{q}}. \tag{15}$$

The atomic decomposition and the banach frame are similar to a one mode Homodyne, and the frame bounds A,B are equal to the identity.

4.3. $SU(1,1)$ Tomography

The Lie algebra $su(1,1)$ of the $SU(1,1)$ group is spanned by the operators $\hat{K}_+, \hat{K}_-, \hat{K}_z$. The Casimir invariant operator that labels all the unitary irreducible representations of the group is given by $(\hat{K}_z)^2 - 1/2(\hat{K}_+\hat{K}_- + \hat{K}_-\hat{K}_+) = k(k+1)\hat{I}$, where the eigenvalue k is also called the Bargmann index [37].
Now we obtain the tomography equation via the wavelets transform in the banach space. Obviously the group is $SU(1,1)$, and the subgroup is U(1) with the reference state $b_o = I$. By choosing

$$\hat{\pi}^*(x) = \hat{u}^\dagger(x)\hat{K}_z\hat{u}(x), \tag{16}$$

$$\hat{U}(x) = \{(-1)^{\hat{K}_z} e^{\theta(e^{i\phi}\hat{K}_+ - e^{-i\phi}\hat{K}_-)}, \hat{K}_z\}_+, \tag{17}$$

where $\hat{u}(\theta,\phi) \equiv e^{-i\theta/2(e^{-i\phi}\hat{K}_+ + e^{i\phi}\hat{K}_-)}$ [37], the wavelet transform is :

$$\mathcal{W}\hat{\rho} = \hat{\rho}(x) =< \hat{U}(x^{-1})\hat{\rho}, l_0 >=< \hat{\rho}, \hat{\pi}^*(x) l_0 >= Tr[\hat{U}^\dagger(x)\rho], \tag{18}$$

and the inverse wavelet transform is shown by

$$\mathcal{M}\mathcal{W}[\hat{\rho}] = \int_X \hat{\rho}(x) b_x d\mu(x) = \int_X \hat{\rho}(x)\hat{\pi}(x) b_0 d\mu(x), \tag{19}$$

where $\hat{\pi}^*(x)$ is dual of $\hat{U}(x)$. The reference state is $b_0 = I$ but this reference state is not admissible, therefore, again according to the prescription of singular cases, we have to select a probe vector $p_0 \in \mathcal{B}$ in which equation (32) is non-zero and finite. In this case, the probe vector can be selected as

$$p_0 = \sum_r b^r \mid r >< r \mid, \tag{20}$$

where this probe vector is similar to the thermal states described by the density operator ρ_T

$$\rho_T = \frac{1}{1+\widetilde{N}} \sum_r (\frac{\widetilde{N}}{1+\widetilde{N}})^r \mid r >< r \mid, \tag{21}$$

with $\widetilde{N} \equiv < \rho_T N >= \frac{1}{exp(\hbar\omega/KT)-1}$, and $N = a^\dagger a$. At high temperatures the thermal state is proportional with the identity operator. Since the representation is irreducible and $C(b_0, p_0) = \frac{1}{1-b}$, the inverse wavelet transform in $\mathcal{M}$ is a left inverse operator on $\mathcal{B}$. Then the tomography formula for $SU(1,1)$ group is given by

$$\hat{\rho} = \frac{1}{\pi}\int_0^{2\pi} d\phi \int_0^{\pi} d\theta \tanh(\theta) Tr[\{(-\mathbf{1})^{\hat{\mathbf{K}}_z} \mathbf{e}^{\theta(\mathbf{e}^{-i\phi}\hat{\mathbf{K}}_- - \mathbf{e}^{i\phi}\hat{\mathbf{K}}_+)}, \hat{\mathbf{K}}_z\}_+ \hat{\rho}] \times \tag{22}$$
$$e^{i\theta/2(e^{-i\phi}\hat{K}_+ + e^{i\phi}\hat{K}_-)} \hat{K}_z e^{-i\theta/2(e^{-i\phi}\hat{K}_+ + e^{i\phi}\hat{K}_-)}.$$

Here we obtain the atomic decomposition and the banach frame for this example. The norms $||\hat{\rho}||_{M_2}$ and $||\{Tr(\hat{\rho}\hat{B}^\dagger(x))\}||$ are equivalent in the sense that they satisfy the inequality (34) with the atomic bounds A=B=1, provided that we use the Hilbert-Schmidt norm for the operator $\hat{\rho}$

$$||Tr(\hat{\rho}\hat{U}^\dagger(x))||^2 = \int d\mu(x) Tr(\hat{\rho}\hat{U}^\dagger(x))\overline{Tr(\hat{\rho}\hat{\pi}^*(x))}, \tag{23}$$

Since the dual couple $\hat{U}(x)$ and $\hat{\pi}^*(x)$ satisfy the orthogonality relation [37]:

$$\delta_{mk}\delta_{nl} = \int d\mu(x) < m|B^\dagger(x)|n >< l|C(x)|k >,$$

then;

$$||Tr(\hat{\rho}\hat{U}^\dagger(x))||^2 = \int d\mu(x)\rho_{mn}U^*_{mn}(x)\rho^*_{kl}\pi_{kl}(x) = ||\hat{\rho}||^2,$$

and if we use the relation (24), we have:

$$\hat{\rho} = \int d\mu(x)Tr(\hat{\rho}\hat{U}^\dagger(x))\hat{\pi}^*(x),$$

Therefore, $\{\hat{\pi}^*(x)b_0, \hat{\pi}^*(x)l_0\}$ is an atomic decomposition of M_2 with respect to L_2 with the atomic bounds A=B=1. Similar to the atomic decomposition, $\{\hat{\pi}^*(x)l_0, T\}$ is a banach frame for the coorbit space of the operators with respect to L_2 with frame bounds $A = B = 1$

.

4.4. Tomography of a Free Particle

In this section we intend to consider the tomography of a free particle. For simplicity we suppose a particle with a unit mass and use the normalized unit $\hbar/2 = 1$, so the free Hamiltonian is given by $\hat{H}_F = \hat{p}^2$. The base is constituted by the set of operators $\hat{R}(x,\tau) = e^{-i\hat{p}^2\tau}|x\rangle\langle x|e^{i\hat{p}^2\tau}$ [34].

Again we try to obtain the tomography equation via the wavelets transform in the banach space. Obviously the group is $\{\hat{P}, \hat{X}, \hat{P}^2, I\}$ and the subgroup is $\{\hat{X}, I\}$. The relevant representation for this example is an adjoint representation:

$$\hat{\pi}(x,\tau)\hat{\rho} = \hat{U}(x,\tau)\hat{\rho}\hat{U}^{-1}(x,\tau) \qquad with\ \hat{U}(x,\tau) = e^{-i\hat{P}^2\tau}\hat{D}(x).$$

In this representation, $\hat{D}(x)$ is a translation operator, such that $\hat{D}(x)|0\rangle = |x\rangle$, where $|x\rangle$ is eigenstate of the position operator and $\hat{P}$ is the momentum operator. On the other hand if we define:

$$< \hat{\rho}, l_0 >= l_0(\hat{\rho}) = Tr(\hat{\rho}\,|\,0 >< 0\,|),$$

the wavelet transform formula is given by:

$$\mathcal{W}\hat{\rho} = \hat{\rho}(x,\tau) =< \hat{\rho}, l_{(x,\tau)} >= Tr(\hat{\rho}e^{-i\hat{P}^2\tau}\,|\,x >< x\,|\,e^{i\hat{P}^2\tau}). \tag{24}$$

Also the inverse wavelet transform $\mathcal{M}$ associated with the wavelet transform $\mathcal{W}$ is:

$$\mathcal{M}\mathcal{W}[\hat{\rho}] = \int d\mu(x,\tau) < \hat{\rho}, l_{(x,\tau)} > b_{(x,\tau)} = \int dxd\tau Tr[\hat{\rho}e^{-i\hat{P}^2\tau}\,|\,x >< x\,|\,e^{i\hat{P}^2\tau}]\hat{\pi}(x,\tau)b_0. \tag{25}$$

The vacuum vector is defined as $b_0 = |0 >< 0|$, but it is not admissible. Thus according to the prescription of the singular cases, we have to select a probe vector $p_0 \in \mathcal{B}$ in which equation (32) is non-zero and finite. In this case, the probe vector can be selected as

$$p_0 =|\,D >< D\,|, \tag{26}$$

where $< D \mid p >= e^{-\frac{p^2}{\delta}}$ such that δ is a real and nonzero parameter. Using the bi-orthogonality and the following relations [34] (for $|j\rangle$, $j = p_1, p_2, p_3, p_4$)

$$\begin{aligned}
\int_R \int_R \quad dx\,d\tau \quad & \langle p_1|\hat{R}(x,\tau)|p_2\rangle \langle p_3|\hat{R}(x,\tau)|p_4\rangle \\
= \quad & \int_R \int_R dx\,d\tau\, e^{-i\tau(p_2^2-p_1^2+p_3^2-p_4^2)} \langle p_1|x\rangle \langle x|p_2\rangle \langle p_3|x\rangle \langle x|p_4\rangle \\
= \quad & \int_R \int_R dx\,d\tau\, e^{-i\tau(p_2^2-p_1^2+p_3^2-p_4^2)}\, e^{ix(p_1-p_2+p_3-p_4)} \\
= \quad & \delta(p_1-p_3)\,\delta(p_2-p_4)\,, \qquad (27)
\end{aligned}$$

we can show that the constant coefficient on the left of (32) is $C(b_0, p_0) = \delta/2\sqrt{\pi}$ and finally the reconstruction procedure of the wavelet transform leads to the tomography relation as follow:

$$\hat{\rho} = \int_R \int_R dx\,d\tau\, p(x,\tau)\, \hat{R}(x,\tau), \qquad (28)$$

where $p(x,\tau) = \mathrm{Tr}[\hat{\rho}\,\hat{R}(x,\tau)]$ is the probability density of the particle to be at position x at time τ. In order to obtain the atomic decomposition and the banach frame for this example by the same choice of the vacuum vector and the test functional, we can show that the required atomic decomposition conditions are satisfied by atomic bounds $A = B = 1$. Therefore, $\{\hat{\pi}(x,\tau)b_0, \hat{\pi}(x,\tau)l_0\}$ is a linear atomic decomposition of M_2 with respect to L_2. Similarly, by using the relation (28) and the definition T, $\{\hat{\pi}(x,\tau)b_0, T\}$ is a banach frame for M_2 with respect to L_2 with frame bounds $A = B = 1$.

4.5. Wavelet Transform and Q-function

Let $g \in \mathcal{L}^2(R)$ with $\| g \| = 1$ and the time-frequency translation of g be:

$$g^{[x_1,x_2]}(t) = e^{2\pi i t x_1} g(t + x_2) = U[x_1, x_2, 0]g(t), \qquad (29)$$

where U is the unitary irreducible representation of the Heisenberg group H^R. For an arbitrary function $f \in L^2(R)$, the pure state sampling [23] associated with it, can be given in terms of inner product as:

$$F(x_1, x_2) = < f, g^{[x_1,x_2]} >, \qquad (30)$$

where $g^{[x_1,x_2]} = U[x_1,x_2]g(t)$ is a coherent state. In pure states, the square of sampling is Q-function.

Now we try to obtain Q-function via wavelet and show that the wavelet transform in the Banach space is a Q-function. The group is the Heisenberg group and the subgroup is the identity and the representation is an adjoint. Then the wavelet transform is given by:

$$\mathcal{W}\hat{\rho} = \hat{\rho}(\alpha) = < \hat{\rho}, l_\alpha > = < \hat{\pi}(\alpha)^\dagger \hat{\rho}, l_0 > . \qquad (31)$$

On the other hand if we choose:

$$< \hat{\rho}, l_0 > = l_0(\hat{\rho}) = Tr[\hat{\rho}|0\rangle\langle 0|], \qquad (32)$$

then the wavelet transform for the adjoint representation is given by:

$$\mathcal{W}\hat{\rho} = \hat{\rho}(\alpha) = Tr\{\hat{\pi}(g)^\dagger \hat{\rho}|0\rangle\langle 0|\} = \langle 0|\hat{U}(\alpha)^\dagger(\hat{\rho})\hat{U}(\alpha)|0\rangle = \langle \alpha|\hat{\rho}|\alpha\rangle = Q(\alpha). \qquad (33)$$

5. Quantum Tomography with Wavelet Transform for Projective Representation of Group(Non-singular Case)

5.1. Tomography of Dihedral Group

Here we consider the tomography of dihedral group by using irreducible projective representation of this group.

The dihedral group D_n of order $2n$ defined by [51, 53]

$$D_n = < a, b | a^n = 1, b^2 = 1, bab^{-1} = a^{-1} > . \tag{1}$$

Let ε be a primitive nth root of 1 and let

$$\chi : D_n \times D_n \to c^*, \tag{2}$$

be defined by [52]

$$\chi(a^i, a^j b^k) = 1 \quad and \quad \chi(a^i b, a^j b^k) = \varepsilon^j, \tag{3}$$

for all $i, j \in \{0, 1, 2, ..., n-1\}$ and $k \in \{0, 1\}$.

For n=2m is even, for each $r \in \{1, ..., m-1\}$ put

$$A_r = \begin{pmatrix} \varepsilon^r & 0 \\ 0 & \varepsilon^{-r} \end{pmatrix} \quad A_m = \begin{pmatrix} \varepsilon^m & 0 \\ 0 & \varepsilon^m \end{pmatrix}, \quad B_r = \begin{pmatrix} 0 & 1 \\ 1 & 0 \end{pmatrix}, \tag{4}$$

and for n=2m+1 is odd, for each $r \in \{1, ..., m\}$ put [48] is

$$A_r = \begin{pmatrix} \varepsilon^r & 0 \\ 0 & \varepsilon^{-r} \end{pmatrix}, \quad B_r = \begin{pmatrix} 0 & 1 \\ 1 & 0 \end{pmatrix}, \tag{5}$$

and let $\hat{\pi}_r : D_n \to GL(2, c)$ be defined by

$$\hat{\pi}(i, j) = \hat{\pi}_r(a^i b^j) = A_r^i B_r^j \ , \quad \{i \in \{0, 1, ..., n-1\}, j = \{0, 1\}\}. \tag{6}$$

Now we try to obtain the above explained tomography via wavelets transform in Banach space based on projective representation. In order to do so, we need to choose the tomographic set $\hat{\pi}(i, j)$ as a projective representation of the wavelet transformation and the identity operator as a vacuum vector. Therefore, the corresponding wavelet transformation takes the following form:

$$\mathcal{W} : \mathcal{B} \mapsto L_2(G) : \hat{\rho} \mapsto \hat{\rho}(i, j) =$$

$$< \hat{\rho}, l_{(i,j)} > = < \hat{\rho}, \hat{\pi}(i, j) l_0 > = < \hat{\rho}\hat{\pi}^\dagger(i, j), l_0 > = Tr(\hat{\rho}\hat{\pi}^\dagger(i, j)). \tag{7}$$

With this condition, inverse wavelet transform in $\mathcal{M}$ is a left inverse operator on B for the wavelet transform $\mathcal{W}$:

$$\mathcal{M} : L_2(G) \mapsto \mathcal{B} : \hat{\rho}(i, j) \mapsto \mathcal{M}[\hat{\rho}] = \mathcal{M}(\hat{\rho})$$

$$\mathcal{M}\mathcal{W}(\hat{\rho}) = \frac{1}{p} \sum_{i,j} < \hat{\rho}, l_{(i,j)} > b_{(i,j)} \tag{8}$$

Also the constant on left hand side of (29) becomes $C(b_0, b_0^{'}) = 4$, finally the constant P becomes equal to one, i.e., $P = \frac{c(b_0,b_0^{'})}{<b_0,l_0^{'}>} = 2$, and the reconstruction procedure of wavelet transform (operation of the combination of wavelet transform and its inverse one, $\mathcal{MW}$ on the density operator $\hat{\rho}$) leads to the tomography relation of the following form

$$\rho = \frac{1}{2}\sum_{i,j} Tr[\hat{\rho}\hat{\pi}^{\dagger}(i,j)]\hat{\pi}(i,j) \tag{9}$$

For an example for $n = 2$, relative dihedral group is D_2 and projective irreducible representation of this group is same as Pauli's matrices and finally reconstruction procedure of wavelet transform is given by

$$\rho = \frac{1}{2}\sum_{j=0}^{4} Tr[\hat{\rho}\sigma_j]\sigma_j, \tag{10}$$

that is, similar to spin tomography relation [13, 14, 16, 19].
By the same choice as above for vacuum vectors and test functions, we can get the atomic decomposition and Banach frame for this example. To do it, we need further to choose the set $\{\hat{\pi}(i,j)l_0\} \subset \mathcal{B}^*$ as the index sequence of functional which belongs to dual Banach space, then we can show the following conditions:

a) $\{<\hat{O}, \hat{\pi}(i,j)l_0>\} = \{Tr(\hat{\pi}^{\dagger}(i,j)\hat{O})\} \in L_2$ for each $\hat{O} \in M_2$,

b) The norms $||\hat{O}||_{M_2}$ and $||\{Tr(\hat{\pi}^{\dagger}(i,j)\hat{O})\}|| = [\sum_{i,j} Tr(\hat{\pi}^{\dagger}(i,j)\hat{O})\overline{Tr(\hat{\pi}^{\dagger}(i,j)\hat{O})}]^{\frac{1}{2}}$ are equivalent since

$$||\{Tr(\hat{\pi}^{\dagger}(i,j)\hat{O})\}|| = [\sum_{i,j} Tr(\hat{\pi}^{\dagger}(i,j)\hat{O})\overline{Tr(\hat{\pi}^{\dagger}(i,j)\hat{O})}]^{\frac{1}{2}}$$

$$= |\sum_{i,j} Tr(\hat{\pi}^{\dagger}(i,j)\hat{O})Tr(\hat{\pi}(i,j)\hat{O}^{\dagger})|^{1/2} = |Tr(\sum_{i,j} Tr(\hat{\pi}^{\dagger}(i,j)\hat{O})\hat{\pi}(i,j)\hat{O}^{\dagger})|^{1/2},$$

if we use the tomography relation (9) we obtain

$$||\{Tr(\hat{\pi}^{\dagger}(i,j)\hat{O})\}|| = |2Tr(\hat{O}\hat{O}^{\dagger})|^{1/2} = \sqrt{2}||\hat{O}||,$$

such that they can saturate the inequality (9) with the atomic bounds $A = B = \sqrt{2}$, provided that we use the the Hilbert-Schmidt norm for the operator $\hat{O}$ and if we use the relation (9) we have:

c) $\hat{O} = \frac{1}{2}\sum_{i,j} Tr(\hat{\pi}^{\dagger}(i,j)\hat{O})\hat{\pi}(i,j)b_0$,

Therefore, $\{\hat{\pi}(i,j)b_0, \hat{\pi}(i,j)l_0\}$ is an atomic decomposition of M_2 of bounded operators acting on representation space with respect to L_2 with atomic bounds $A = B = \sqrt{2}$.

Finally, the same choice of vacuum vector, test functional and index sequence of functional as in the atomic decomposition case, yield the required conditions (a) and (b) for the existence of Banach frame as the atomic decomposition one, and in order to have the last condition for the existence of atomic decomposition, we can define the reconstruction operator T as follows:

$T\{Tr(\hat{\pi}^{\dagger}(i,j)\hat{O})\} = \frac{1}{2}\sum_{i,j} Tr(\hat{\pi}^{\dagger}(i,j)\hat{O})\hat{\pi}(i,j) = \hat{O}$ for each $\hat{O} \in M_2$,

It is straightforward to show that the operator T as defined above is a linear bounded operator. Therefore, $\{\hat{\pi}(i,j)l_0, T\}$ is Banach frame for M_2 with respect to L_2 with frame bounds $A = B = \sqrt{2}$.

5.2. Discrete Phase Space Tomography

In ref [33] formalism was applied to represent the states and the evolution of a quantum system in phase space in finite dimensional Hilbert space and, finally, it was discussed how to perform direct measurement to determine the wigner function. This approach was based on the use of phase space point operator to define Wigner function. For discrete systems we can define finite translation operators $\hat{Q}$ and $\hat{V}$, which generate finite translation in position and momentum, respectively. The translation operator $\hat{Q}$ generates cyclic shifts in the position basis and is diagonal in momentum basis:

$$\hat{Q}^m \mid n >=\mid n+m >, \qquad \hat{Q}^m \mid k >= exp(-2\pi imk/N) \mid k > . \tag{11}$$

Similarly, the operator $\hat{V}$ is a shift in the momentum basis and is diagonal in position basis:

$$\hat{V}^m \mid k >=\mid k+m >, \qquad \hat{V}^m \mid n >= exp(2\pi imn/N) \mid n > . \tag{12}$$

Now by identifying the corresponding displacement operators, the discrete analogue of the phase space translation operator is given by:

$$\hat{\pi}(q,p) = \hat{Q}^q \hat{V}^p exp[(i\pi pq/N)]. \tag{13}$$

These operators obey the simple composition rule

$$\hat{\pi}(q_1,p_1)\hat{\pi}(q_2,p_2) = \hat{\pi}(q_1+q_2,p_1+p_2)\exp[i\pi(p_1q_2-q_1p_2)/N]. \tag{14}$$

In comparison with projective representation (18) factor set $\chi(q,p)$ is defined by

$$\chi((q_1,p_1),(q_2,p_2)) = \exp[i\pi(p_1q_2-q_1p_2)/N]. \tag{15}$$

Here we can define the point operator as:

$$\hat{A}(q,p) = \frac{1}{(2N)^2}\sum_{n,m=0}^{2N-1}\hat{\pi}(m,k)exp(-2\pi i\frac{(kq-mp)}{2N}), \tag{16}$$

or as:

$$\hat{A}(\alpha) = \frac{1}{2N}\hat{Q}^q\hat{R}\hat{V}^{-p}exp(i\pi pq/N). \tag{17}$$

That $\hat{R}$ is parity operator and it is worth noting that the phase space point operators have been defined on a lattice with $2N\times 2N$ points, but it has be shown that there are only N^2 independent phase space point operators on the set $G_N = \{\alpha = (q,p); 0 \leq q,p \leq N-1\}$. The tomography relation is given by:

$$\hat{\rho} = 1/N\sum_{\alpha\in G_N} Tr(\hat{\rho}\hat{\pi}^{\dagger}(\alpha))\hat{\pi}(\alpha) = 4N\sum_{\alpha\in G_N} Tr(\hat{\rho}\hat{A}(\alpha))\hat{A}(\alpha). \tag{18}$$

where $W(\alpha) = Tr(\hat{A}(\alpha)\hat{\rho})$ is Wigner function.
Now we try to obtain the tomography equation (18) via wavelets transform in Banach space. Obviously the group, subgroup and representation are finite Heisenberg, its center and projective representation $\hat{\pi}(\alpha)$, respectively. Then the wavelet transform with the test functional

$$l_0(O) = Tr(O) \text{ for any operator O}$$

is given by

$$\mathcal{W} : \mathcal{B} \longmapsto L_2(\alpha) : \hat{\rho} \longrightarrow \hat{\rho}(\alpha) = < \hat{\rho}, l_\alpha > = < \hat{\pi}^\dagger(\alpha)\hat{\rho}, l_0 > = Tr(\hat{\pi}^\dagger(\alpha)\hat{\rho}). \quad (19)$$

Since the representation is an irreducible representation, the inverse wavelet transform $\mathcal{M}$ will be the left inverse operator of wavelet transform $\mathcal{W}$:

$$\mathcal{M} : L_2(\alpha) \longmapsto \mathcal{B} : \hat{\rho}(\alpha) \longrightarrow \mathcal{M}[\hat{\rho}] = \hat{\rho} = \sum_{\alpha \in G_N} < \hat{\rho}, l_\alpha > b_\alpha = \sum_{\alpha \in G_N} < \hat{\pi}^\dagger(\alpha)\hat{\rho}, l_0 > \hat{\pi}(\alpha) b_0, \quad (20)$$

We can obtain tomography relation (18), for the admissible $b_0 = I/N$. By the same choice as above for vacuum vector and test functions, we can get the atomic decomposition and Banach frame for this example. To do it, we need just to choose the $\{\pi(\hat{\alpha}) l_0\} \subset \mathcal{B}^*\}$, then we can show that:

a) $\{< \hat{\rho}, \pi(\hat{\alpha}) l_0 >\} = \{Tr(\hat{\rho}\hat{\pi}^\dagger(\alpha))\} \in L_2$ for each $\hat{\rho} \in M_2$,

b) The norms $||\hat{\rho}||_{M_2}$ and $||\{Tr(\hat{\rho}\hat{\pi}^\dagger(\alpha))\}||$ are equivalent and in the sense that they satisfy the inequality (34) with the atomic bounds $A = B = \sqrt{2}$, provided that we use the Hilbert-Schmidt norm for the operator $\hat{O}$ and if we use the relation (17), we have,

c) $\hat{\rho} = \frac{1}{N} \sum_\alpha Tr(\hat{\rho}\hat{\pi}^\dagger(\alpha)) \hat{\pi}(\alpha) b_0$,

then $\{\hat{\pi}(\alpha) b_0, \hat{\pi}(\alpha) l_0\}$ is a linear atomic decomposition of M_2 with respect to L_2 with frame bounds $A = B = \sqrt{2}$.

Finally by the same choice of vacuum vector, test functional and index sequence of functional as in the atomic decomposition case, we can show that the required conditions (a) and (b) for the existence of Banach frame as the atomic decomposition one, and in order to have the last condition for the existence of atomic decomposition, we can define the reconstruction operator S as follows

$S\{Tr(\hat{\rho}\hat{\pi}^\dagger(\alpha))\} = \sum_\alpha Tr(\hat{\rho}\hat{\pi}^\dagger(\alpha)) \hat{\pi}(\alpha) = \hat{\rho}$ for each $\hat{\rho} \in M_2$,

then $\{\hat{\pi}(\alpha) l_0, S\}$ is a Banach frame for M_2 with respect to L_2 with frame bounds A=B=1.

Now we can generalize single particle discrete phase space to multi-particle discrete phase space. For $i = 1, 2, \ldots, m$ let $\hat{\pi}_i$ be a projective unitary representation of G on a Hilbert space $\mathcal{H}_i$. Then the naturally defined tensor product representation $\hat{\pi}_1 \otimes \hat{\pi}_2 \otimes \ldots \otimes \hat{\pi}_m$ is easily seen to be a projective unitary representation of G on the Hilbert space $\mathcal{H}_1 \otimes \mathcal{H}_2 \otimes \ldots \otimes \mathcal{H}_m$ associated with the product cocycle $\chi_1 \chi_2 \cdots \chi_m$. Therefore, the irreducible projective representation is $\hat{\pi}(\alpha_1, \ldots, \alpha_m) = \hat{\pi}(\alpha_1) \otimes \ldots \otimes \hat{\pi}(\alpha_m)$ and reduced wavelet formula with the choice of $b_0 = \hat{I}/N \otimes \ldots \otimes \hat{I}/N$ is given by:

$$\mathcal{W} : \mathcal{B} \longmapsto L_2(\alpha_1, \ldots, \alpha_m) : \hat{\rho} \longrightarrow \hat{\rho}(\alpha_1, \ldots, \alpha_m) = < \hat{\rho}, l_{\alpha_1, \ldots, \alpha_m} > =$$
$$< \hat{\pi}^\dagger(\alpha_1, \ldots, \alpha_m)\hat{\rho}, l_0 > = Tr(\hat{\pi}^\dagger(\alpha_1, \ldots, \alpha_m)\hat{\rho}). \quad (21)$$

Since the representation is an irreducible representation, the inverse wavelet transform $\mathcal{M}$ will be the left inverse operator of wavelet transform $\mathcal{W}$:

$$\mathcal{M} : L_2(\alpha_1, \ldots, \alpha_m) \longmapsto \mathcal{B} : \hat{\rho}(\alpha_1, \ldots, \alpha_m) \longrightarrow \mathcal{M}[\hat{\rho}] = \hat{\rho} = \sum_{\alpha_1, \ldots, \alpha_m \in G_{N \times m}} < \hat{\rho}, l_{\alpha_1, \ldots, \alpha_m} > b_{\alpha_1, \ldots, \alpha_m}$$
$$= \sum_{\alpha_1, \ldots, \alpha_m \in G_{N \times m}} < \hat{\pi}^\dagger(\alpha_1, \ldots, \alpha_m)\hat{\rho}, l_0 > \hat{\pi}(\alpha_1, \ldots, \alpha_m) b_0, \quad (22)$$

where $G_{N\times m} = \{\alpha_i = (q_i, p_i); 0 \leq q_i, p_i \leq N-1; i = 1,...,m\}$. Then the tomography relation is given by:

$$\hat{\rho} = 1/N^m \sum Tr(\hat{\rho}\hat{\pi}^{\dagger}(\alpha_1,...,\alpha_m))\hat{\pi}(\alpha_1,...,\alpha_m) \tag{23}$$

Similar to discrete phase space tomography, we can introduce the tomography relation for direct products of cyclic groups of the same order. Suppose first that G is a cyclic group of order d. For odd d, let P and Q be the $d \times d$ matrices defined by [52]

$$P = \begin{pmatrix} 0 & 1 & 0 & \ldots & 0 \\ 0 & 0 & 1 & \ldots & 0 \\ : & : & : & : & : \\ 0 & 0 & 0 & \ldots & 1 \\ 1 & 0 & 0 & \ldots & 0 \end{pmatrix}, \quad Q = \begin{pmatrix} 0 & \varepsilon & 0 & \ldots & 0 \\ 0 & 0 & \varepsilon^2 & \ldots & 0 \\ : & : & : & : & : \\ 0 & 0 & 0 & \ldots & \varepsilon^{d-1} \\ 1 & 0 & 0 & \ldots & 0 \end{pmatrix}, \tag{24}$$

where P is shift operator and Q is phase shift modulation operator defined relative to phase operator and shift operator by following relation

$$Q = PD \ , \ D = \begin{pmatrix} 1 & 0 & 0 & \ldots & 0 \\ 0 & \varepsilon & 0 & \ldots & 0 \\ 0 & 0 & \varepsilon^2 & \ldots & 0 \\ : & : & : & : & : \\ 0 & 0 & 0 & \ldots & \varepsilon^{d-1} \end{pmatrix} \tag{25}$$

and ε is a primitive dth root of 1. Putting

$$R = P^{d-1}Q, \tag{26}$$

if G denotes the m direct product of cyclic group, we have two cases:
I)if $m = 2k$ is even, then

$$\begin{cases} E_i = R \otimes R \otimes \ldots \otimes P \otimes I \otimes \ldots \otimes I \\ E_{i+k} = R \otimes R \otimes \ldots \otimes Q \otimes I \otimes \ldots \otimes I, \end{cases} \tag{27}$$

where P and Q appear in the ith position and each tensor product has k factors (here I denotes the identity matrix).
Therefore, the map $\hat{\pi} : G \rightarrow GL(n^k, c)$ defined by [52]

$$\hat{\pi}_{i_1,i_2,\ldots,i_{2k}} = E_1^{i_1} E_2^{i_2} \ldots E_{2k}^{i_{2k}} \ , \ i_1, i_2, \ldots, i_{2k} = \{0, 1, \ldots, d-1\}, \tag{28}$$

is the only up to linear equivalency irreducible χ representation of G.
II) if $m = 2k+1$ is odd, then

$$\begin{cases} E_i = R \otimes R \otimes \ldots \otimes P \otimes I \otimes \ldots \otimes I \\ E_{i+k} = R \otimes R \otimes \ldots \otimes Q \otimes I \otimes \ldots \otimes I \\ E_{2k+1} = R \otimes R \otimes \ldots \otimes R \ , \ k factors, \end{cases} \tag{29}$$

and G has precisely n irreducible linearly nonequivalent χ representations, namely [52]

$$\hat{\pi}_{i_1,i_2,\ldots,i_{2k+1}} = (\varepsilon^{j_1} E_1)^{i_1} (\varepsilon^{j_2} E_2)^{i_2} \ldots (\varepsilon^{j_{2k+1}} E_{2k+1})^{i_{2k+1}} \tag{30}$$

$$i_1, i_2, ..., i_{2k+1} = \{0, 1, ..., d-1\}, j_1, j_2, ..., j_{2k+1} \in \{0, 1, ..., d-1\}.$$

These cases for representation of a group G can play a very important role for stabilizer cods in quantum computing, too [54].
If d is even, let P and Q be the $d \times d$ matrices defined by

$$P = \begin{pmatrix} 0 & 1 & 0 & \dots & 0 \\ 0 & 0 & 1 & \dots & 0 \\ \vdots & \vdots & \vdots & \vdots & \vdots \\ 0 & 0 & 0 & \dots & 1 \\ 1 & 0 & 0 & \dots & 0 \end{pmatrix}, \quad Q = \begin{pmatrix} 0 & \delta & 0 & \dots & 0 \\ 0 & 0 & \delta^2 & \dots & 0 \\ \vdots & \vdots & \vdots & \vdots & \vdots \\ 0 & 0 & 0 & \dots & \delta^{d-1} \\ 1 & 0 & 0 & \dots & 0 \end{pmatrix}, \tag{31}$$

where δ is a primitive 2dth root of 1 such that $\delta^2 = \varepsilon$. Putting

$$R = \delta P^{d-1} Q, \tag{32}$$

two cases $m = 2k$ (even) and $m = 2k+1$ (odd) are similar to cases (I) and (II) in above by using relation (32).
Now we try to obtain the above explained tomography via wavelets transform in Banach space. The wavelet transform $\mathcal{W}$ from Banach space $\mathcal{B}$ to a space of function F(G) is defined by a representation $\hat{\pi}(i_1, i_2, ..., i_m)$ of G on $\mathcal{B}$, with the selection of a vacuum vector b_0 which is equal to identity, and a test functional $l_0(\hat{\rho}) = Tr(\hat{\rho})$ is given by the following formula:

$$\mathcal{W} : \mathcal{B} \longmapsto L_2(\Omega) : \rho \longmapsto \widehat{\rho}(i_1, i_2, ..., i_m) = <\rho, l_{i_1, i_2, ..., i_m}> = <\hat{\pi}^\dagger(i_1, i_2, ..., i_m)\rho, l_0>$$
$$= Tr[\hat{\pi}^\dagger(i_1, i_2, ..., i_m)\rho], \tag{33}$$

Also the constant on left hand side of (29) becomes proportional to the dimension of the unitary representation, that is, $C(b_0, b_0^{'}) = d^{2k}$, where d is dimensional of representation. Finally the constant P becomes equal to d^k, i.e., $P = \frac{c(b_0, b_0^{'})}{<b_0, l_0^{'}>} = d^k$, and the reconstruction procedure of wavelet transform (operating the combination of wavelet transform and its inverse one, $\mathcal{M}\,\mathcal{W}$ on the density operator $\hat{\rho}$) leads to the following tomography relation

$$\rho = \frac{1}{d^k} \sum_{i_1, i_2, ..., i_m} Tr[\hat{\pi}^\dagger(i_1, i_2, ..., i_m)\rho] \hat{\pi}(i_1, i_2, ..., i_m). \tag{34}$$

For example for m=2, one part qudit system tomography leads to the tomography relation [30]

$$\rho = \frac{1}{d} \sum_i Tr[\rho U_i^\dagger] U_i \tag{35}$$

where U_i is irreducible representation of SU(d) group.
At the end we can show that $\{\hat{\pi}(i_1, i_2, ..., i_m)b_0, \hat{\pi}(i_1, i_2, ..., i_m)l_0\}$ is atomic decomposition and $(\{\hat{\pi}(i_1, i_2, ..., i_m)l_0\}, S)$ is Banach frame with atomic bounds $A = B = d^{\frac{k}{2}}$.

5.3. Tomography Based on Projective Representation of the Subgroup of $SL(2,Z)$

The quantization of the linear automorphisms of the two-torus was first developed by Hannay and Berry [58], who were able to determine the quantum propagators for the subgroup of $SL(2,Z)$ defined by

$$\Gamma_\theta = \left\{ A = \begin{pmatrix} a & b \\ c & d \end{pmatrix} \in SL(2,Z) \mid ab \equiv cd \equiv 0 \ mod \ 2 \right\}. \tag{36}$$

This set of matrices is often known as the theta group. It turns out that it is possible to associate to every element of $SL(2,Z)$ a propagator [60, 61], or quantum map, which is a unitary operator acting on a finite dimensional Hilbert space.

An exact form of Egorov theorem characterizes the quantum cat maps, which, as a consequence, are multiplicative up to a phase factor, i.e.

$$\hat{\pi}_N(AB) = e^{2\pi i \zeta(A,B)} \hat{\pi}_N(A)\hat{\pi}_N(B), \ A,B \in \Gamma \subset SL(2,Z). \tag{37}$$

where $\hat{\pi}_N(A)$ denotes the propagator, Γ is a subgroup of $SL(2,Z)$ and N is the dimension of the Hilbert space in which $\hat{\pi}_N(A)$ acts. In other words, the quantization determines a projective representation of Γ , which is sometimes referred to as Weil representation. The theta group has index three in SL(2,Z) and is the largest subgroup of SL(2,Z). It is shown that EgorovŠs theorem determines a projective representation of Γ_θ [57].

Theorem: For any positive integer N, there exists a choice of phases in the definition of the propagators $\hat{\pi}_N(A)$ such that

$$\hat{\pi}_N(AB) = \hat{\pi}_N(A)\hat{\pi}_N(B), \ A,B \in \Gamma. \tag{38}$$

A choice of propagators that realizes (38) is the following. If

$$S^{\pm} = \begin{pmatrix} 0 & \mp 1 \\ \pm 1 & 0 \end{pmatrix}, \tag{39}$$

then the propagator is the discrete Fourier transform:

$$[\hat{\pi}_N(S^{\pm})\Phi](Q) := \frac{1}{\sqrt{N}} \sum_{Q' \ mod \ N} e^{(\pm \frac{QQ'}{N})} \Phi(Q'), \quad \Phi \in \mathcal{H}_N. \tag{40}$$

The parity matrix

$$P = \begin{pmatrix} -1 & 0 \\ 0 & -1 \end{pmatrix}, \tag{41}$$

has the obvious quantization

$$[\hat{\pi}_N(P)\Phi](Q) := \Phi(-Q) \ \Phi \in \mathcal{H}_N, \tag{42}$$

If b = 0, i.e.

$$T_m^{\pm} = \begin{pmatrix} \pm 1 & 0 \\ m & \pm 1 \end{pmatrix}, \quad m \equiv 0 \ mod \ 2, \tag{43}$$

then we have

$$[\hat{\pi}_N(T_m^{\pm})\Phi](Q) := e^{(\pm\frac{mQ^2}{2N})}\Phi(\pm Q), \quad \Phi \in \mathcal{H}_N. \tag{44}$$

Similarly, if $a = 0$, i.e.

$$W_{\omega}^{\pm} = \begin{pmatrix} 0 & \pm 1 \\ \pm 1 & \omega \end{pmatrix}, \omega \equiv 0 \ mod \ 2, \tag{45}$$

then we set

$$[\hat{\pi}_N(W_{\omega}^{\pm})\Phi](Q) := \frac{1}{\sqrt{N}} \sum_{Q' \ mod \ N} e^{\pm\frac{1}{2N}(\omega Q^2 - 2QQ')}\Phi(Q'), \quad \Phi \in \mathcal{H}_N. \tag{46}$$

In all the other cases we define

$$[\hat{\pi}_N(A)\Phi](Q) := \frac{h(a,b)}{\sqrt{N_b}} \sum_{Q' \ mod \ N} G(\alpha,\beta,\gamma(Q,Q'))e^{\frac{1}{2N_b}(dQ^2 - 2QQ' + aQ'^2)}\Phi(Q'), \quad \Phi \in \mathcal{H}_N. \tag{47}$$

where $h(a,b)$ and $G(\alpha,\beta,\gamma(Q,Q'))$ are defined in equations (48) and (50) respectively. Formula (47) needs a few words of explanation. Firstly, we have

$$h(a,b) = \begin{cases} (\frac{|a|}{|b|})e^{(sign(ab))(\frac{|b|-1}{8})} & \text{if a is even,} \\ (\frac{|b|}{|a|})e^{(-sign(ab))\frac{|a|}{8}} & \text{if a is odd} \end{cases}, \tag{48}$$

where $(\frac{p}{q})$ is the Jacobi symbol and sign(x) is the sign function

$$sign(x) = \begin{cases} 1 & if \quad x > 0, \\ -1 & if \quad x < 0, \\ 0 & if \quad x = 0 \end{cases}. \tag{49}$$

The function $h(a,b)$ is a phase factor which distinguishes definition (47) from the quantization that Hannay and Berry introduced in their original article [58]. More precisely, Hannay and BerryŠs propagator is $\hat{\pi}_N^{HB}(A) = \sqrt{i}\hat{\pi}_N(A)/h(a,b)$. The term $G(\alpha,\beta,\gamma(Q,Q'))$ is a normalized Gauss sum:

$$G(\alpha,\beta,\gamma(Q,Q')) = \frac{1}{\sqrt{|\beta|}} \sum_{k \ mod \ |\beta|} e^{\frac{1}{2\underline{e}_{ta}}(\alpha k^2 + \gamma(Q,Q'k))}, \tag{50}$$

where $\alpha = N_b a, \ \beta = b', \ \gamma = 2(aQ' - Q)/(b,N)$ and (b,N) is the greatest common divisor of b and N, which we shall always take to be positive. In the previous definitions, in (47) and in what follows we use the notation

$$N_b = N/(b,N) \quad b' = b/(b,N). \tag{51}$$

Now we try to obtain the above explained tomography via wavelets transform in Banach space based on projective representation of theta group. In order to do so, we need to choose the tomographic set $\hat{\pi}_m$ as a projective representation of the wavelet transformation and the identity operator as a vacuum vector. Therefore, the corresponding wavelet transformation takes the following form:

$$\mathcal{W}:\mathcal{B}\mapsto L_2(G):\hat{\rho}\mapsto\hat{\rho}(m)=$$

$$<\hat{\rho},l_{(m)}>=<\hat{\rho},\hat{\pi}_m l_0>=<\hat{\rho}\hat{\pi}_m^{\dagger},l_0>=Tr(\hat{\rho}\hat{\pi}_m^{\dagger}). \tag{52}$$

With this condition, inverse wavelet transform in $\mathcal{M}$ is a left inverse operator on B for the wavelet transform $\mathcal{W}$:

$$\mathcal{M}:L_2(G)\mapsto\mathcal{B}:\hat{\rho}(m)\mapsto\mathcal{M}[\hat{\rho}]=\mathcal{M}(\hat{\rho})$$

$$\mathcal{M}\mathcal{W}(\hat{\rho})=\frac{1}{p}\sum_m<\hat{\rho},l_{(m)}>b_{(m)} \tag{53}$$

According to relation (29), $C(b_0,b_0^{'})$ and $P=\frac{c(b_0,b_0^{'})}{<b_0,l_0^{'}>}$ are finite and the reconstruction procedure of wavelet transform (operation of the combination of wavelet transform and its inverse one, $\mathcal{M}\mathcal{W}$ on the density operator $\hat{\rho}$) leads to the tomography relation of the following form

$$\rho=\frac{1}{P}\sum_{m=0}^{N-1}Tr[\hat{\rho}\hat{\pi}_m]\hat{\pi}_m \tag{54}$$

By the same choice as above for vacuum vectors and test functions, we can get the atomic decomposition and Banach frame for this example with finite bound A,B.

5.4. Tomography Based on Projective Representation of Permutation Group

Let us consider projective representations of the symmetric (permutation) groups that have long been known to mathematicians, but received little attention from physicists. Such representations were overlooked in physics much like projective representations of the rotation groups were overlooked in the early days of quantum mechanics. One especially useful presentation of the symmetric group S_n on n elements is given by

$$S_n=\{t_1,\ldots,t_{n-1}:t_i^2=1,(t_jt_{j+1})^3=1,t_mt_l=t_lt_m\}, \tag{55}$$

where $1\leq i\leq n-1, 1\leq j\leq n-2, m\leq l-2$. Here t_i are transpositions,

$$t_1=(12),t_2=(23),\ldots,t_{n-1}=(n-1\ n). \tag{56}$$

Closely related to S_n is the group $\tilde{S}_n$,

$$\tilde{S}_n=\{z,t_{1;2}^{'},\ldots,t_{n-1;n}^{'}\,|\,z^2=1,zt_{i;i+1}^{'}=t_{i;i+1}^{'}z,t_{1;2}^{'\,2}=z,(t_{j;j+1}^{'}t_{j+1;j+2}^{'})^3=z,$$

$$t_{m;m+1}^{'}t_{l;l+1}^{'}=zt_{l;l+1}^{'}t_{m;m+1}^{'}\}, \tag{57}$$

where $1\leq i\leq n-1, 1\leq j\leq n-2, m\leq l-2$.

A celebrated theorem of Schur (Schur, 1911 [62]) states the following:

(i) The group $\tilde{S}_n$ has order $2(n!)$.

(ii) The subgroup $\{1,z\}$ is central, and is contained in the commutator subgroup of $\tilde{S}_n$, provided n = 4.

(iii) $\tilde{S}_n/\{1,z\} \simeq S_n$.

(iv) If $n < 4$, then every projective representation of S_n is projectively equivalent to a linear representation.

(v) If $n \leq 4$, then every projective representation of S_n is projectively equivalent to a representation $\hat{\pi}$,

$$\rho(S_n) = \{\hat{\pi}(t_1),...,\hat{\pi}(t_{n-1}) : \hat{\pi}(t_i)^2 = z, (\hat{\pi}(t_j)\hat{\pi}(t_{j+1}))^3 = z, \hat{\pi}(t_m)\hat{\pi}(t_l) = z\hat{\pi}(t_l)\hat{\pi}(t_m)\}, \quad (58)$$

where $1 \leq i \leq n-1, 1 \leq j \leq n-2, m \leq l-2$ and $z = \pm 1$.

In the case $z = +1$, $\hat{\pi}$ is a linear representation of S_n.

The group $\tilde{S}_n$ is called the representation group for S_n.

The most elegant way to construct a projective representation $\hat{\pi}(S_n)$ of S_n is by using the complex Clifford algebra $\mathrm{Cliff}_C(V,g) \equiv C_n$ associated with the real vector space $V = n\mathcal{R}$,

$$\{\gamma_i, \gamma_j\} = 2g(\gamma_i, \gamma_j) \quad (59)$$

Here $\{i\}_{i=1}^n$ is an orthonormal basis of V with respect to the symmetric bilinear form

$$g(\gamma_i, \gamma_j) = +\delta_{ij}. \quad (60)$$

Clearly, any subspace $\bar{V}$ of $V = nR$ generates a subalgebra $\mathrm{Cliff}_C(\bar{V}, \bar{g})$, where $\bar{g}$ is the restriction of g to $\bar{V} \times \bar{V}$. A particularly interesting case is realized when $\bar{V}$ is

$$\bar{V} = \{\sum_{m=1}^{n} \alpha^m \gamma_m : \sum_{m=1}^{n} \alpha^m = 0\} \quad (61)$$

of codimension one, with the corresponding subalgebra denoted by $\bar{C}_{n-1}$ [51]. If we consider a special basis $\{t'_m\}_{m=1}^{n-1} \subset \bar{V}$ (which is not orthonormal) defined by

$$t'_{m;m+1} = \frac{1}{\sqrt{2}}(\gamma_m + \gamma_{m+1}) \quad m = 1,...n-1, \quad (62)$$

then the group generated by this basis is isomorphic to $\bar{S}_n$. This can be seen by mapping t_i to t'_i and z to -1, and by noticing that

1) For $m = 1,...,n-1$:

$$t'^2_{m;m+1} = -1; \quad (63)$$

2) For $n-2 \geq j$:

$$(t'_{j;j+1} t'_{j+1;j+2})^3 = -1; \quad (64)$$

3) For $n-1 \geq q > m+1$:

$$t'_{m;m+1} t'_{q;q+1} = -t'_{q;q+1} t'_{m;m+1}, \quad (65)$$

as can be checked by direct calculation. One choice for the matrices is provided by the following construction (Brauer and Weyl, 1935 [63]):

$$\begin{cases} \gamma_{2m-1} = \sigma_3 \otimes ... \sigma_3 \otimes (\sigma_1) \otimes 1 ... \otimes 1, \\ \gamma_{2m} = \sigma_3 \otimes ... \sigma_3 \otimes (\sigma_2) \otimes 1 ... \otimes 1, \\ m = 1, 2, 3, ... k, \end{cases} \quad (66)$$

for $n = 2k$. Here σ_1, σ_2 occur in the m-th position, the product involves M factors, and $\sigma_1, \sigma_2, \sigma_3$ are the Pauli matrices. If $n = 2k+1$, we first add one more matrix,

$$\gamma_{2k+1} = \sigma_3 \otimes \ldots \otimes \sigma_3 \quad (k \text{ factors}), \tag{67}$$

and then define:

$$\begin{cases} \Gamma_{2m-1} = \gamma_{2m-1} \oplus \gamma_{2m-1} \\ \Gamma_{2m} = \gamma_{2m} \oplus \gamma_{2m} \\ \Gamma_{2k+1} = \gamma_{2k+1} \oplus (-\gamma_{2k+1}). \end{cases} \tag{68}$$

The representation $\hat{\pi}(S_n)$ so constructed is reducible. An irreducible module of $\bar{C}_{n-1}$ restricts that representation to the irreducible representation of $\tilde{S}_n$, since $\{t'_{i;i+1}\}_{i=1}^{n-1}$ generates $\bar{C}_{n-1}$ as an algebra [51]. The simplest (irreducible) non-trivial projective representations of S_n are already surprisingly intricate and have dimensions which grow exponentially with n. They are intimately related to spinor representations of $SO(n)$ [59].

Now we try to obtain the tomography via wavelets transform in Banach space based on projective representation of permutation group (spinor representation of permutation group). In order to do so, we need to choose the tomographic set $\{\hat{\pi}_{i_1,\ldots,i_m} = \gamma_1^{i_1}\gamma_2^{i_2}, \gamma_3^{i_3}, \ldots, \gamma_m^{i_m} \; ; \; , \; i_1, i_2, \ldots, i_m = \{0,1\}\}$ as a projective representation of the wavelet transformation and the identity operator as a vacuum vector. Therefore, the corresponding wavelet transformation takes the following form:

$$\mathcal{W} : \mathcal{B} \mapsto L_2(G) : \hat{\rho} \mapsto \hat{\rho}(i_1, \ldots, i_m) =$$

$$< \hat{\rho}, l_{(i_1,\ldots,i_m)} > = < \hat{\rho}, \hat{\pi}_{i_1,\ldots,i_m} l_0 > = < \hat{\rho}\hat{\pi}^{\dagger}_{i_1,\ldots,i_m}, l_0 > = Tr(\hat{\rho}\hat{\pi}^{\dagger}_{i_1,\ldots,i_m}). \tag{69}$$

With this condition, inverse wavelet transform in $\mathcal{M}$ is a left inverse operator on B for the wavelet transform $\mathcal{W}$:

$$\mathcal{M} : L_2(G) \mapsto \mathcal{B} : \hat{\rho}(i_1, \ldots, i_m) \mapsto \mathcal{M}[\hat{\rho}] = \mathcal{M}(\hat{\rho})$$

$$\mathcal{M}\mathcal{W}(\hat{\rho}) = \frac{1}{p} \sum_{i_1,\ldots,i_m} < \hat{\rho}, l_{(i_1,\ldots,i_m)} > b_{(i_1,\ldots,i_m)} \tag{70}$$

According to relation (29), $C(b_0, b_0') = 2^{2k}$ and $P = \frac{c(b_0,b_0')}{<b_0,l_0'>} = 2^k$ and the reconstruction procedure of wavelet transform (operation of the combination of wavelet transform and its inverse one, $\mathcal{M}\mathcal{W}$ on the density operator $\hat{\rho}$) leads to the tomography relation of the following form

$$\rho = \frac{1}{2^k} \sum_{i_1,\ldots,i_m} Tr[\hat{\rho}\hat{\pi}_{i_1,\ldots,i_m}]\hat{\pi}_{i_1,\ldots,i_m}, \tag{71}$$

that is, similar to k-qubit tomography relation [56]. By the same choice as above for vacuum vectors and test functions, we can get the atomic decomposition and Banach frame for this example with finite bound $A = B = 2^{\frac{k}{2}}$.

6. Quantum Tomography with Wavelet Transform Based on Projective Representation of Group(Singular Case)

6.1. Homodyne Tomography

The problem of measuring the density matrix $\hat{\rho}$ of radiation has been extensively considered both experimentally and theortically [38]. Homodyne tomography is presently the only method that can be used to achieve such measurement. This method is based on the idea that the density matrix can be evaluated in optical Homodyne experiments from the collection of quadrature probability distribution for the radiation state. As shown in [49], the matrix can be obtained after calculating the Wigner function as the inverse Radon transform of such quadrature distributions [50]. Quantum homodyne tomography is used in quantum optic in the measurement of the quantum state of light. In this case, we get [16,35]:

$$\hat{\rho} = \int_C \frac{d^2\alpha}{\hat{\pi}} Tr[\hat{\rho}\hat{U}^\dagger(\alpha)]\hat{U}(\alpha), \tag{1}$$

where $\hat{U}(\alpha) = exp(\alpha a^\dagger - \alpha^* a)$ is a displacement operator. By Changing polar variable $\alpha = \frac{i}{2}ke^{i\phi}$ this formula becomes

$$\hat{\rho} = \int_0^{2\pi} \frac{d\phi}{\pi} \int_{-\infty}^{\infty} \frac{dk\mid k\mid}{4} Tr[\hat{\rho}e^{ikX_\phi}]e^{ikX_\phi}, \tag{2}$$

where $X_\phi = \frac{(a^\dagger e^{i\phi} + ae^{-i\phi})}{2}$ is field-quadrature operators that is measured by balance Homodyne [19].

Now we try to obtain the tomography equation (1) via wavelets transform in Banach space based on projective representation. Obviously the group is Heisenberg. Since the representation of H^R fails to be square-integrable, according to Stone-Von Neumann [39], we can factor out the center H^R and consider only the factor space, that is, projective representation of Heisenberg group.

For the operators $\hat{\pi}(\alpha)$ the following properties are crucial. For $\alpha_1, \alpha_2 \in C$

$$\hat{\pi}(\alpha_1)\hat{\pi}(\alpha_2) = \mathrm{e}^{\alpha_1\bar{\alpha_2} - \bar{\alpha_1}\alpha_1}\ \hat{\pi}(\alpha_2)\hat{\pi}(\alpha_1), \tag{3}$$

$$\hat{\pi}(\alpha_1 + \alpha_2) = \mathrm{e}^{-\frac{1}{2}(\alpha_1\bar{\alpha_2} - \bar{\alpha_2}\alpha_1)}\ \hat{\pi}(\alpha_1)\hat{\pi}\alpha_2). \tag{4}$$

Comparison with projective representation (18), obviously the factor set $\chi(\alpha_1, \alpha_2)$ is defined by

$$\chi(\alpha_1, \alpha_2) = \mathrm{e}^{-\frac{1}{2}(\alpha_1\bar{\alpha_2} - \bar{\alpha_2}\alpha_1)}\ . \tag{5}$$

For the vacuum vector and test functional, we need to choose the identity operator and $l_0(O) = Tr[O]$ for any operator O, respectively. Then the wavelet formula is given by:

$$\mathcal{W} : \mathcal{B} \mapsto L_2(\alpha) : \hat{O} \mapsto \hat{\rho}(\alpha) = <\hat{\rho}, l_\alpha> = <\hat{\rho}, \hat{\pi}(\alpha)l_0> = <\hat{\rho}\hat{\pi}(\alpha)^\dagger, l_0> = Tr(\hat{\rho}\hat{\pi}(\alpha)^\dagger), \tag{6}$$

But above reference state is not admissible. Thus according the singular cases, we must select a probe vector $p_0 \in \mathcal{B}$ in which equation (32) is non-zero and finite. In this case, the probe vector is selected by:

$$p_0 = \int \mid \alpha >< \alpha \mid e^{(\frac{-|\alpha|^2}{\Delta})} \frac{d^2\alpha}{\pi}, \tag{7}$$

where Δ is non-zero and finite and $b_0 \in \hat{B}$ is identity. Since the representation is irreducible Since the representation is irreducible and

$$C(b_0, p_0) = < \int_\alpha < U(\alpha)^\dagger \int |\alpha' >< \alpha' | e^{(\frac{-|\alpha'|^2}{\Delta})} \frac{d^2\alpha'}{\pi}, l_0 > U(\alpha) \frac{d^2\alpha}{\pi}, l_0 >$$

$$= \int_\alpha \int_{\alpha'} tr(U^\dagger(\alpha) |\alpha' >< \alpha' | e^{(\frac{-|\alpha'|^2}{\Delta})} \frac{d^2\alpha'}{\pi}) tr(U(\alpha)) \frac{d^2(\alpha)}{\pi} = \Delta,$$

where $tr(U(\alpha)) = \pi\delta^2(\alpha)$ then the inverse wavelet transform in $\mathcal{M}$ is a left inverse operator on $\mathcal{B}$ for the wavelet transform $\mathcal{W}$:

$$\mathcal{M}\mathcal{W} = I \Rightarrow \mathcal{M} : L_2(\alpha) \mapsto B : \hat{\rho}(\alpha) \mapsto \mathcal{M}[\hat{\rho}] = \mathcal{M}\mathcal{W}(\hat{\rho}), \tag{8}$$

then;

$$\hat{\rho} = \int d\mu(\alpha) < \hat{\rho}, l_\alpha > b_\alpha = \int d\mu(\alpha) Tr(\hat{\rho}\hat{\pi}^\dagger(\alpha))\hat{\pi}(\alpha) b_0, \tag{9}$$

where $d\mu(\alpha) = \frac{d^2\alpha}{\pi}$ is the invariant measure of the group of translation and group is unimodular. For $b_0 = I$, the reconstruction procedure of wavelet transform (9) leads to the tomography relation (1).

In this relation $Tr(\hat{\rho}\hat{\pi}^\dagger(\alpha))$ is Wigner characteristic function. We also can obtain another quasidistribution characteristic functions by choosing different representations. For example, we need to choose the representations, $\hat{\pi}_{an}(\alpha) = e^{\alpha\hat{a}^\dagger} e^{-\alpha^*\hat{a}}$, $\hat{\pi}_n(\alpha) = e^{-\alpha^* a} e^{\alpha a^\dagger}$, $\hat{\pi}_h(\nu) = e^{-\nu^* b} e^{\nu b^\dagger}$ ($\hat{b} = \mu\hat{a} + \nu\hat{a}^\dagger$ and $\mu^2 - \nu^2 = 1$), $\hat{\pi}_s(\xi.\eta) = e^{i\xi\hat{q}} e^{i\eta\hat{p}}$ and $\hat{\pi}_{as}(\xi.\eta) = e^{i\eta\hat{p}} e^{i\xi\hat{q}}$, for P-function characteristic function [43], Q-function characteristic function [43], Husimi characteristic function [44], Standard-ordered characteristic function [45] and Antistandard-ordered characteristic function [46], respectively.

For the complex Fourier transform of the displacement operator $\hat{\pi}$ [34]

$$\hat{\pi}(\alpha) = \int \frac{d^2\xi}{\pi} \hat{\pi}(\xi) exp(\alpha\xi^* - \alpha^*\xi), \tag{10}$$

the expansion of the operator in terms of the operator $\hat{\pi}(\alpha)$ is given by

$$\hat{\rho} = \int \frac{d^2\alpha}{\pi} W(\alpha)\hat{\pi}(\alpha), \tag{11}$$

where $W(\alpha)$ is Wigner function. Also by defining complex Fourier transform for each of above representation, we can get its tomography relation for each quasidisrbution. Now we will try to obtain the atomic decomposition and Banach frame for this example. Let M_2 be a Banach space and let L_2 be an associated Banach space of scalar-valued sequences and let $\{\{\hat{\pi}(\alpha) l_0\} \subset \mathcal{B}^*\}$. Finally by the same choice of vacuum vector, test functional and index sequence, we can show that required conditions (a), (b) and (c) are satisfied by atomic bounds A=B=1. Since we have

$$||\{Tr(\hat{\pi}^\dagger(\alpha)\hat{O})\}|| = [\int Tr(\hat{\pi}^\dagger(\alpha)\hat{O})\overline{Tr(\hat{\pi}^\dagger(\alpha)\hat{O})} d\mu(\alpha)]^{\frac{1}{2}}$$

$$= |\int tr(\hat{\pi}^\dagger(\alpha)\hat{O}) tr(\hat{\pi}(\alpha)\hat{O}^\dagger) d\mu(\alpha)|^{1/2} = |tr(\int tr(\hat{\pi}^\dagger(\alpha)\hat{O})\hat{\pi}(\alpha) d\mu(\alpha)\hat{O}^\dagger)|^{1/2},$$

if we use the tomography relation (1) we obtain

$$||\{Tr(\hat{\pi}^{\dagger}(\alpha)\hat{O})\}|| = |tr(\hat{O}\hat{O}^{\dagger})|^{1/2} = ||\hat{O}||.$$

Therefore, $\{\hat{\pi}(\alpha)b_0, \hat{\pi}(\alpha)l_0\}$ is a linear atomic decomposition of M_2 with respect to L_2. Similarly, by using the relation (1) and definition T, $\{\hat{\pi}(\alpha)b_0, T\}$ is Banach frame for M_2 with respect to L_2 with frame bounds A=B=1. We can generalize single mode Homodyne tomography to multimode state, too. In the wavelet notation, the irreducible representation is $\hat{\pi} = \hat{\pi}_0 \bigotimes \hat{\pi}_1 \bigotimes \ldots \bigotimes \hat{\pi}_m$, which $\hat{\pi}_j = exp(z_j \hat{a}_j^{\dagger} - z_j^* \hat{a}_j)$, and reduced wavelets formula with the choice of $b_0 = \hat{I} \bigotimes \hat{I} \bigotimes \ldots \bigotimes \hat{I}$ is given by:

$$\mathcal{W} : B \mapsto L_2(z_0, z_1, \ldots, z_m) : \rho \mapsto \hat{\rho}(z_0, z_1, \ldots, z_m)$$

$$= < \hat{\rho}, l_{z_0, z_1, \ldots, z_m} > = < \hat{\rho}, \hat{\pi}(z_1, z_2, \ldots, z_m) l_0 > = < \hat{\rho}\hat{\pi}^{\dagger}(z_0, z_1, \ldots, z_m), l_0 >$$
$$= tr(\hat{\rho}\hat{\pi}^{\dagger}(z_0, z_1, \ldots, z_m)). \tag{12}$$

But this reference state is not admissible. Thus according the singular cases, we must select a probe vector $p_0 \in \mathcal{B}$ in which that equation (32) is non-zero and finite. In this case, the probe vector is selected by:

$$p_0 = \int | z_0, z_1, \ldots, z_m >< z_0, z_1, \ldots, z_m | e^{(\frac{-\sum_{j=0}^{m} |z_j|^2}{\Delta})} d\mu(z_0, z_1, \ldots, z_m), \tag{13}$$

where

$$| z_0, z_1, \ldots, z_m > = | z_0 > \otimes | z_1 > \otimes, \ldots, \otimes | z_m >, \tag{14}$$

and

$$d\mu(z_0, z_1, \ldots, z_m) = \frac{d^2 z_0}{\pi} \frac{d^2 z_1}{\pi} \cdots \frac{d^2 z_m}{\pi}, \tag{15}$$

where Δ is non-zero and finite, and $b_0 \in \hat{B}$ is identity. Since the representation is irreducible and $c(b_0, p_0) = \Delta^{m+1}$, the inverse wavelet transform in $\mathcal{M}$ is a left inverse operator on $\mathcal{B}$ for the wavelet transform $\mathcal{W}$:

$$\mathcal{M}\mathcal{W} = I \Rightarrow \mathcal{M} : L_2(z_0, z_1, \ldots, z_m) \mapsto B : \hat{\rho}(z_0, z_1, \ldots, z_m) \mapsto \mathcal{M}[\hat{\rho}]$$

$$= \mathcal{M}\mathcal{W}(\hat{\rho}) = \hat{\rho} = \int d\mu(z_0, z_1, \ldots, z_m) < \hat{\rho}, l_{z_0, z_1, \ldots, z_m} > b_{z_0, z_1, \ldots, z_m}, \tag{16}$$

Then;

$$\hat{\rho} = \int_C \frac{d^2 z_0}{\hat{\pi}} \int_C \frac{d^2 z_1}{\hat{\pi}} \cdots \int_C \frac{d^2 z_m}{\hat{\pi}} Tr[\hat{\rho}\hat{\pi}^{\dagger}(z_0, z_1, \ldots, z_m)]\hat{\pi}(z_0, z_1, \ldots, z_m). \tag{17}$$

The atomic decomposition and Banach frame is similar to one mode Homodyne, and A , B are equal to identity.

6.2. $SU(1,1)$ Tomography

The Lie algebra $su(1,1)$ of the $SU(1,1)$ group is spanned by the operators $\hat{K}_+, \hat{K}_-, \hat{K}_z$. The Casimir invariant operator that labels all the unitary irreducible representations of the group is given by $(\hat{K}_z)^2 - 1/2(\hat{K}_+\hat{K}_- + \hat{K}_-\hat{K}_+) = k(k+1)\hat{I}$, where the eigenvalue k is also called the Bargeman index.
In the single mode setup for the tomographic reconstruction of $SU(1,1)$ D'Ariano et. al. [37] employs the following $SU(1,1)$ realization

$$K_+ = \frac{1}{2}(a^\dagger)^2,\ K_- = K_+^\dagger,\ K_z = \frac{1}{2}(a^\dagger a + \frac{1}{2}) \tag{18}$$

where **a** is the annihilation operator of the mode, and $k = \frac{1}{4}, \frac{3}{4}$. Notice that this realization is actually a projective representation of $SU(1,1)$. It allows to reconstruct only the operators that act on the subspace of the radiation space constituted by the states with fixed (even or odd) parity in the number of photons. In this case, the observable K_z is related to the number of photons in the mode, while the operator $u(\theta,\phi)$ describes a degenerate parametric amplifier (or phase sensitive amplifier) [37], i.e.

$$u(\theta,\phi) = exp(\frac{1}{4}\theta(a^{\dagger 2} e^{-i\phi} - a^2 e^{i\phi})). \tag{19}$$

In the following section, we will try to obtain the tomographic equation via wavelet transform in Banach space based on projective representation. Obviously group is $SU(1,1)$ and the representation $\hat{\pi}^*(x) = u^\dagger(\theta,\phi)\hat{K}_z u(\theta,\phi)$ of $SU(1,1)$ is the dual set of

$$\hat{\pi}(x) = \{(-1)^{\hat{K}_z} e^{\theta(e^{i\phi K_+} - e^{i\phi K_-})}, K_z\}_+,$$

where $\{*.*\}_+$ denotes the anti-commutator. The wavelet transform is given by

$$\mathcal{W}: \mathcal{B} \to L_2(x) : \hat{\rho} \to \hat{\rho}(x) = [\mathcal{W}\hat{\rho}](x) = <\hat{\pi}(x^{-1})\hat{\rho}, l_0> = <\hat{\rho}, \hat{\pi}^*(x) l_0> = Tr[\hat{\pi}^\dagger(x)\rho], \tag{20}$$

and inverse wavelet transform is given by

$$\mathcal{M}: L_2(x) \to \mathcal{B} : \hat{\rho}(x) \to \mathcal{M}[\hat{\rho}(x)] = \int_x \hat{\rho}(x) b_x d\mu(x) = \int_X \hat{\rho}(x)\hat{\pi}(x) b_0 d\mu(x), \tag{21}$$

The reference state is $b_0 = I$ but this reference state is not admissible. Thus according the singular cases, we must select a probe vector $p_0 \in \mathcal{B}$ in which equation (32) is non-zero and finite. In this case, the probe vector is selected by

$$p_0 = \sum_r b^r \mid r >< r \mid, \tag{22}$$

where this probe vector is similar to thermal states described by the density operator ρ_T

$$\rho_T = \frac{1}{1+\widetilde{N}} \sum_r (\frac{\widetilde{N}}{1+\widetilde{N}})^r \mid r >< r \mid, \tag{23}$$

where $\widetilde{N} \equiv <\rho_T N> = \frac{1}{exp(\hbar\omega/KT)-1}$, and $N = a^\dagger a$. In the high temperature this thermal state is proportional to identity. Since the representation is irreducible and $C(b_0, p_0) = \frac{1}{1-b}$, the inverse wavelet transform in $\mathcal{M}$ is a left inverse operator on $\mathcal{B}$. Then the tomography formula for $SU(1,1)$ group is given by

$$\hat{\rho} = \frac{1}{\pi}\int_0^{2\pi} d\phi \int_0^{\pi} d\theta \tanh(\theta) Tr[\{(-\mathbf{1})^{\hat{\mathbf{K}}_z}\mathbf{e}^{\theta(\mathbf{e}^{-i\phi}\hat{\mathbf{K}}_- - \mathbf{e}^{i\phi}\hat{\mathbf{K}}_+)}, \hat{\mathbf{K}}_z\}_+ \hat{\rho}] \times \quad (24)$$

$$e^{i\theta/2(e^{-i\phi}\hat{K}_+ + e^{i\phi}\hat{K}_-)}\hat{K}_z e^{-i\theta/2(e^{-i\phi}\hat{K}_+ + e^{i\phi}\hat{K}_-)}.$$

Now we will obtain atomic decomposition and Banach frame for this example. Let M_2 be a coorbit space and let L_2 be an associated Banach space of scalar-valued sequences. Let $\{\hat{\pi}(x)l_0\} \subset \mathcal{B}^*$, then we can show that:

a) $\{< \hat{\rho}, \hat{\pi}(x)l_0 >\} = \{Tr(\hat{\rho}\hat{\pi}^\dagger(x))\} \in L_2$ for each $f \in M_2$,

b) The norms $||\hat{\rho}||_{M_2}$ and $||\{Tr(\hat{\rho}\hat{B}^\dagger(x))\}||$ are equivalent in the sense that they satisfy the inequality (34) with the atomic bounds A=B=1, provided that we use the Hilbert-Schmidt norm for the operator $\hat{\rho}$

$$||Tr(\hat{\rho}\hat{\pi}^\dagger(x))||^2 = \int d\mu(x) Tr(\hat{\rho}\hat{\pi}^\dagger(x))\overline{Tr(\hat{\rho}\hat{\pi}(x))}, \quad (25)$$

Since the dual couple $\hat{\pi}^*(x)$ and $\hat{\pi}(x)$ satisfy the orthogonality relation [37]:

$$\delta_{mk}\delta_{nl} = \int d\mu(x) < m|B^\dagger(x)|n >< l|C(x)| >,$$

then;

$$||Tr(\hat{\rho}\hat{\pi}^\dagger(x))||^2 = \int d\mu(x)\rho_{mn}\hat{\pi}^*_{mn}(x)\rho^*_{kl}\hat{\pi}_{kl}(x) = ||\hat{\rho}||^2,$$

and if we use the relation (24), we have:

c) $\hat{\rho} = \int d\mu(x) Tr(\hat{\rho}\hat{\pi}^\dagger(x))\hat{\pi}(x)$,

Therefore, $\{\hat{\pi}(x)b_0, \hat{\pi}(x)l_0\}$ is an atomic decomposition of M_2 with respect to L_2 with atomic bounds A=B=1. Similar to atomic decomposition, $\{\hat{\pi}(x)l_0, S\}$ is a Banach frame for coorbit space of operators with respect to L_2 with frame bounds A,B are equal to identity.

7. Conclusions

The Banach space wavelets transformation nature of quantum tomography of mixed quantum states has been revealed. Also by considering various well known examples of quantum tomography , it is shown that, the quantum tomography of mixed quantum states, are almost the same as the Banach space wavelets reconstruction formalism associated with some unitary representation of finite or infinite groups. Using this fact, the frame and atomic decomposition nature of quantum tomography of mixed states is also explained.

References

[1] S. G. Mallat. A Wavelet Tour of Signal Processing, Academic Press, New York, (1998).

[2] Randy K. Young. Wavelet Theory and its Applications, Kluwer Academic Publishers (1993).

[3] Y. Meyer. Wavelets: Algorithms and Applications (SIAM), Philadelphia (1993).

[4] P. Goupillaud, A. Grossmann, and J. Morlet, Geoxploration **23**, 85, (1984).

[5] A. Grossmann, J. Morlet and T. Paul, J. Math. Phys **26**, 2473, (1985).

[6] A. Grossmann, J. Morlet and T. Paul, Ann. Inst. H. Poincare **45**, 293, (1986).

[7] I. Daubechies. Ten Lectures on Wavelets, SIAM, Philadelphia, (1992).

[8] S. T. Ali, J-P Antoine, J-P.Gazeau. Coherent States, Wavelets and their Generalizations Springer (2000).

[9] O. Christensen, C. Heil, Math. Nachr. **185**, 33, (1997).

[10] H.G. Feichtinger and K.H. Grochenig, J. Functional Anal, **86**, 307, (1989).

[11] M. A. Manko, V. I. Manko, R. V. Mendes, J. of Physics A: Math. and Gen. **34**, 8321, (2001).

[12] V. V. Kisil, Acta Appl. Math. **59**, 1, 79, (1999).

[13] D'Ariano, Phys Lett A. **268**, 151, (2000).

[14] D'Ariano, Acta. Physca Slovaca,**49**, 513, (1999).

[15] G. S. Agarwal, Phys.Rev.A **24**, 2889, (1981).

[16] G. M. D'Ariano, L. Maccone and M. G. A. Paris, J.Phys.A: Math, Gen.**34**, 1, 93, (2001).

[17] V. V. Dodonov and V.I. Manko, Phys.Lett.A **229**, 335, (1997).

[18] V. I. Manko and O.V. Manko, GEFT, **112**, 796, 33, (1997).

[19] G. M. D'Ariano,L. Maccone, M. Paini, J.Opt.B: Quantum Semicalss.Opt. **5**, 77, (2003).

[20] R. T. Thew, K. Nemoto, A. G. White and W. J. Munro, Phys.Rev.A, **66**, 012303, (2002).

[21] G. M. D'Ariano, L. Maccone and M. G. A. Paris, Physics Letters A, **276**, Issues 1-4, 25, (2000).

[22] E. B. Davis. Quantum Theory of Open System, Academic Press (1976).

[23] W. Miller. Topics in Harmonic Analysis With Applycations To Radar and Sonar Lecture note (2002) www.ima.umn.edu/ miller/radarla.pdf.

[24] P. G. Gasazza, Advances in computational Mathematics, special issue on frames (2002).

[25] M. Paini, quant-ph/0002078.

[26] J. L. Park, W. Band, Found. Phys. **1**, 211, 339, (1971).

[27] H. F. Jones, Groups, Representation and physics, institute of physics publishing Bristol and Philadelphia (1998).

[28] L. C. Biedenharn and J. D. Louck, Angular Momentum in Quantum Physics: Theory and Applications (Reading, MA: Addison-Wesley) (1981).

[29] S. S. Safonov, quant-ph/9807031.

[30] M. Mirzaee, M. Rezaei, M. A. Jafarizadeh, Eur. Phys. J. B **60**, 193, (2007).

[31] R. Schack and C. Caves, Optics Communications **179**, 439, (2000).

[32] S. Heiss, S. Weigert, Phys. Rev. A **63**, 012105, (2001).

[33] C. Miquel, J, P. Paz, M. Saraceno, Phy. Rev A **65**, 062309,(2002).

[34] G. M. D'Ariano, S. Mancini, V. I. Manko, P. Tombesi, J. Opt. B: Quantum and Semiclassical opt. **8**, 1017, (1996).

[35] G. M. D'Ariano, M. G. A. Paris, and M. F. Sacchi, **128**, 205, (2003).

[36] S. Mancini, O. V. Manko, V. I. Manko, P. Tombesi, J. Phys. A: Math. Gen. **34**, 16 , 3461, (2001).

[37] G. M. D'Ariano, E. De Vito and L. Maccone, Phys. Rev A **64**, 033805, (2001).

[38] U. Leonhardt. Measuring the Quantum State of Light (Cambridge University Press, Cambridge, England (1997)).

[39] A. A. Kirillov. Elements of the Theory of Representations. (Springer-Verlag, Berlin, (1976)).

[40] S. Dahlke, G. Steidl and G. Teschke. Coorbit Spaces and Banach Frames on Homogeneous Spaces with Application to Analizing Function on Sphers , Advances in Computational Mathematics **21**(1-2) (2004) 147. S. Dahlke, G. Steidl and G. Teschke. Weighted Coorbit spaces and Banach Frames on Homogeneous Spaces. to appear in Journal of Fourier Analysis and Applications.

[41] G. W. Wei, Y. B. Zhao and Y. Xiang, Int. J. Numer. Math. Eng. **55**, 913, (2002).

[42] J. Várilly and J. M. Gracia-Bondía, Ann. Phys. **190**, 107, (1989).

[43] R. J. Glauber, Phys. Rev. **130**, 2529, (1963).

[44] K. Husimi, Proc. Phys. Mat. Soc. Jpn, **22**, 264, (1940).

[45] C. L. Mehta, J. Math. Phys. **5**, 69, (1940).

[46] J. G. Kirkwood, Phys. Rev, **44**, 31, (1933).

[47] H. G. Feichtinger and G. zimmermann , An exotic minimal Banach space of functions.

[48] G. James and M. Liebeck. Representations and Characters of Groups, Cambridge Mathematical Textbooks Cambridge, (1995).

[49] K. Vogel and Risken, Phys. Rev. A **40**, 2847, (1989).

[50] Giacomo mauro D'Ariano and Nicoletta Sterpi, J. modern Opt. **44**, 2227, (1997).

[51] P. N. Hoffman, J. F. Humphereys, Projective Representation of the Symmetric Groups, Clarendon press. Oxford, (1992).

[52] G. Karpilovesky, Projective Representation of Finite Groups, Marcel Dekker, Ink. New York and Basel, (1985).

[53] L. Dornhoff, Group Representation Theory Volume A), Marcel Dekker, New York, (1971). MR 50:458a.

[54] A. Klappenecker, M. Roetteler . Beyond Stabilizer Codes I: Nice Error Bases, Beyond Stabilizer Codes II: Clifford Codes, IEEE Trans.

[55] A. O. MORRIS, H. I. JONES, PROJECTIVE REPRESENTATIONS OF GENERALIZED SYMMETRIC. GROUPS, St'eminaire Lotharingien de Combinatoire 50 (2003).

[56] M. Mirzaee, M. Rezaei,M. A. Jafarizadeh, Int J Theor Phys, **46**, 10, (2007).

[57] F. Mezzadri , Nonlinearity ,**15** , 905, (2002).

[58] J. H. Hannay and M. V. Berry, Physica D **1**, 267,(1980).

[59] F. Wilczek, hep-th/9806228.

[60] M. E. Degli, Quantization of the orientation preserving automorphisms of the torus Ann. Inst. H. Poincare, **58**, 323, (1993).

[61] S. Knabe, J. Phys. A: Math. Gen. **23**, 2013, (1990).

[62] I. Schur, Journal flur die reine und angewandte Mathematik **139**, 155, (1911).

[63] R. Brauer, and H. Weyl, Amer. J. Math. LVII(2), 425, (1935).

In: Wavelets: Classification, Theory and Applications
Editors: Manel del Valle et al.
ISBN 978-1-62100-252-9

Chapter 9

AUTOMATIC MODELING OF DISCRETE WAVELET TRANSFORMS USING PARTICLE SWARM OPTIMIZATION

Marcus V.M. Figueredo*, A. Beckert Neto, S.R. Rogal Jr and J.C. Nievola
Programa de Pós-Graduação em Informática - PPGIa
Pontifícia Universidade Católica do Paraná - PUCPR
Curitiba, Paraná, Brazil

Abstract

Discrete wavelet transforms have been successfully utilized in digital signals processing, such as data compression or classification. One of the key points in these applications is the choice of the wavelet function that best fits the domain. In many situations, one of the various available wavelet families can be applied with satisfactory results. However, there are several contexts where the modeling of new kinds of wavelet functions may be useful. Traditionally, the creation of a new function is a manual process that involves the detailed analysis of the studied domain. A better solution would be to automatically model the coefficients of the best wavelet for each situation. This chapter presents an automatic process that models the wavelet coefficients using Particle Swarm Optimization (PSO). The algorithm starts from a random population of coefficients that are indexed by two variables: velocity and position. These parameters are updated iteratively based on best global and best local results, causing their evolution. At the end, the algorithm finds the wavelet coefficients that most suit the specific application. In tasks involving classification, the method tries to optimize the accuracy of the classifier. On the other hand, the algorithm minimizes the residual error in lossy data compression applications. So, the method can be applied on both compression and classification (combined with machine learning techniques) tasks. To verify that, two situations involving the electrocardiogram (ECG) are presented. In the first application, a wavelet function is optimized to improve the efficiency of ECG data compression. In the last one, another wavelet is created to process the waveforms that are analyzed by an arrhythmia detection algorithm. Once the presented method is generic, these two examples can be replicated in other domains.

*E-mail address: marcus@ppgia.pucpr.br

Introduction

In the last years, several applications involving digital signals processing have utilized discrete wavelet transforms (DWT) with considerable success. In this kind of application, the choice of the most adequate wavelet function is crucial to the success. Usually, one of the normal wavelet families (Haar, Symlet, Daubechies, Coiflet and others) can be applied with satisfactory results. However, the utilization of specialized wavelets may achieve a better performance.

This reason stimulated many authors to create specific wavelet functions for different domains. Normally, it is a mathematical process that involves the detailed analysis of the domain. Another approach is the use of automatic optimization techniques to adjust the wavelet coefficients.

Particularly, this chapter presents an automatic process that models the wavelet coefficients using Particle Swarm Optimization (PSO). The PSO algorithm can be described as a simplified social model of flocks or fish shoals looking for food. Initially, the birds do not know where the food is. So, each bird starts flying without any pattern. Eventually, they start to follow that birds that are near to food, which creates several small flocks. Finally, one bird finds the food and attracts the others, creating an unique large flock.

In the same way, the proposed method randomly initializes a population of coefficients arrays. Each array is associated to two variables: velocity and position. Iteratively, the algorithm updates the coefficients based on these individual and global variables. During the process, it tries to find the wavelet coefficients that are most suited to the application. For example, in a classification task, the algorithm tries to find the wavelet function that maximizes the accuracy of the classifier. In other situation, the technique maximizes the performance of a data compressor.

To validate the efficiency of this method, it is utilized in two applications with the electrocardiogram (ECG) waveform. The first experiment involves the optimization of wavelet coefficients that are utilized in the ECG data compression. The other intends to generate wavelet coefficients for arrhythmias classification.

This chapter is organized as follows. The next section reviews the foundations of wavelets, which are essential to understand which coefficients are optimized. So, the following section presents the PSO algorithm, that is utilized to optimize the coefficients. Then, the proposed methodology is described, including the ECG domain and the both experiments. The results are listed and discussed after that. Finally, the last section presents the conclusions about the method.

Discrete Wavelet Transform

Because the continuous wavelet transform (CWT) calculation must be discretely performed in computers, it generates a high redundancy of information, creating a high demand of computational resources like time and memory. This reason stimulated the use of discrete wavelets transforms (DWT) instead [1]. Unlike the CWT, the DWT provides enough information to analyze and reconstruct the analyzed signal in a reduced computational time.

The DWT is implemented by a sequence of digital filtering on the original signal according to an algorithm known as sub-band coding, where a pair of digital filters (defined as quadrature filters) is represented by the functions $h(n)$ and $g(n)$, which are, respectively, the impulse response functions of a low-pass and a high-pass filter, each with half the bandwidth of the original signal [2]. In figure 1, one can observe the DWT application in a one-dimensional signal composed of a 20Hz sinusoid and a high frequency noise.

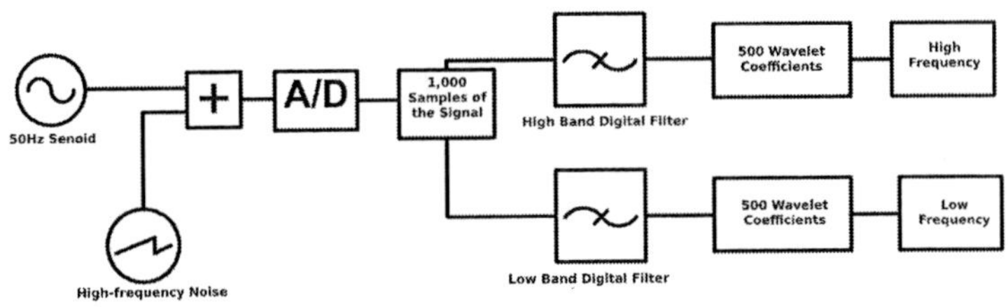

Figure 1. The DWT application in a one-dimensional signal composed of a 20Hz sinusoid and a high frequency noise.

The filtering can be performed recursively, as shown in figure 2. An original signal $x[n]$ with 1,000 samples and a frequency of π rad/s [1] is processed by a high-pass filter and a low-pass filter. The first filter generates a signal that has frequencies between $\pi/2$ rad/s and π rad/s and is represented by 500 samples [2] . These samples will represent the first detail coefficients of DWT. The low-pass filter, in turn, generates a signal with also 500 samples, containing frequencies between 0 rad/s and π rad/s. These are the approximation coefficients of DWT. It should be noted that, in total, 1,000 samples are kept of the original signal. Approximation coefficients can be filtered again, generating new approximation and detail coefficients. This operation can be repeated several times [3] , creating multiple layers of analysis, with different levels of resolution.

Coefficients Optimization

Once a DWT is a digital filter, it is described as an array of coefficients. The coefficients of usual wavelets like Daubechies or Coiflets are already known. However, in certain domains, it could be necessary to utilize particular wavelets with unknown coefficients. Some optimization technique can be utilized to solve this kind of problem. This chapter investigates the particle swarm optimization (PSO) as a feasible solution.

James Kennedy and Russel Eberhart had proposed the PSO in 1995 [3]. The technique was created as a simplified social model of flocks or fish shoals looking for food [4]. Although the method was first intended for simulating social behaviour, it was discovered that it could perform optimization.

By having a population of candidate solutions (or particles) that are moving around the search-space, the PSO can optimize a problem. The particle movement is based on simple

[1]Although most of this work the unit Hz is used to express the frequencies, it is very important to note that the appropriate unit of frequency for discrete signals is rad/s.

[2]The Nyquist criterion says that an analog signal can be reproduced without distortions (errors) from one set of samples, provided that the sampling frequency exceeds twice the highest frequency existing in the original continuous signal. As the frequency range of the signal was reduced by half, it becomes possible to halve the number of samples.

[3]In a sampled signal of size $2n$, the filtering operations can be repeated n times.

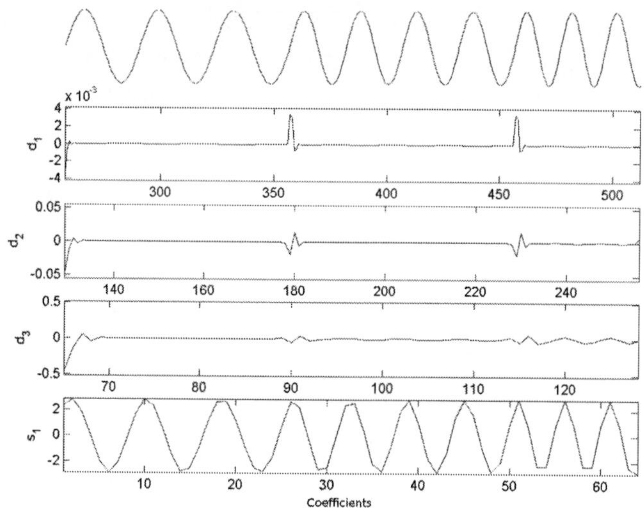

Figure 2. Application of DWT and generation of detail and approximation coefficients.

mathematical formulae which guide them using the best local and global positions.

Differently from classical optimization algorithms, PSO can be applied to any kind of problem. It happens because it does not use gradients during the process, so the optimized problem does not need to be differentiable. This way, PSO can be defined as an efficient, robust and simple non-deterministic optimization algorithm [5].

As commented before, PSO can be conceptually explained using a "flock metaphor":

1. Let be a flock looking for food.
2. Initially, the birds do not know where the food is.
3. So, each bird starts flying without any pattern.
4. Eventually, they start to follow that birds that are near to food, which creates several small flocks.
5. Finally, one bird finds the food and attracts the others, creating an unique large flock.

This "flock metaphor" can be translated to an algorithm like this:

1. Let be a fitness function $f(x)$ that must be minimized and x_i a candidate solution in the form of a vector of real numbers. So, $f(x_i)$ will output a real number that is the fitness of x_i. The best global particle (the one with the lowest fitness) is represented as g.
2. Produce a population of n particles (candidate solutions) with positions x_i and velocities v_i, where $0 <= i < n$.
3. Initialize the position of each particle as a uniformly distributed random vector.

4. Initialize the velocity of each particle as a uniformly distributed random vector.
5. Repeat until a termination criterion is met[4]:
   (a) Update the velocity of each particle.
   (b) Use each velocity to update its respective particle's position.
   (c) If $f(x_i) < f(g)$, then $g \rightarrow x_i$.
6. The best solution is g.

As commented before, a DWT can be described as an array of coefficients. This way, the PSO algorithm can be applied to optimize them. To do this, the algorithm generates an initial population of candidates. Each candidate is an random array of coefficients and is associated to two variables (position and velocity). Incrementally, the coefficients are evolved during the early described process. Finally, the best candidate is selected.

To determine what is the best set of coefficients, one should determine a fitness criterion. This criterion must be objective enough so it can be described as an algorithm, i.e. a fitness function. The fitness function will prescribe the optimality of each candidate, allowing the algorithm to find the best solution.

The key point in the proposed method is to find an objective way to determine how well a wavelet fits the studied domain. It is a hard task, because the criterion may drastically change according to the context. Particularly, this chapter studies two usual applications of DWTs: data compression and patterns classification. Thus, the fitness functions may vary, once the objectives for each case are quite different.

Electrocardiogram

Despite the differences between data compression and patterns classification, the two studied cases rely on the same domain: the electrocardiogram. Because of that, this section describes the foundations of this vital signal.

According to Décio Faraco Azevedo [6], about 60% of the human body consists of water, which is distributed by blood vessels (intravascular compartment), the cells (intracellular compartment) and between the cells (extracellular compartment). Dissolved in this water, there are many chemicals, including ions such as sodium (Na +), the chlorine (Cl-), potassium (K +) and calcium (Ca +). When those ions move through the cardiac cell membranes, electrical current is produced in nature electrochemistry. According to Gerson P. Goldwassser [7], sarcolema is the membrane surrounding the heart cell, being formed by two layers of phospholipids and a layer of protein. This membrane has the property of the selective permeability, allowing the passage of ions through its layers according to some factors. The component of sarcolema that handles this passage is called ionic channel and there are channels suitable for each type of ion.

This channel serves as a gate, opening up or closing them. When at rest, the heart muscle cells have their positive and negative loads separated by the sarcolema. The positive charges are extended externally while the negative charges concentrate internally. The cell

[4]Normally, a maximum number of iterations or a fitness threshold.

is in the state of electrical diastole or polarized. In this state, its internal negative tension equals -90 mV while the external positive tension is equivalent to +10 mV. The muscle cells remain so indefinitely, unless they receive some kind of stimulus. When this occurs, the electrical balance is amended and there is a reversal of electric charges, producing an electric current. The nature of this current is electrochemical and the cell is in the state of electrical systole or depolarized. Once depolarized, the cells start automatically to re-polarize themselves, at the expense of electrochemical gradients and reorganizing the electric charges over their membranes. There are two types of heart cells: the PTR stables and the PTR unstable. The first ones depolarize just if stimulated while the last ones depolarize automatically, controlling the cardiac rhythm.

Azevedo [6] defined the electrocardiography as the study of the electrical activity of the heart, the record of this activity is the electrocardiogram (ECG) and the equipment that records it is the electrocardiograph. The ECG is one of the most important resources in the cardiology because it is cheap, very informative and non-invasive. Indeed, the Nobel Prize winner of medicine Bernardo A. Houssay affirmed in 1946: a physician does not consider an exam complete without analyzing the electrical activity of the heart.

Figure 3 shows the schematic representation of a normal electrocardiogram highlighting its main components and patterns. It is relevant to consider that ECG is a non-stationary signal, where the high frequencies take place locally and the low frequencies happen during the entire signal. The first wave of the cardiac cycle is known as P wave and expresses the depolarization of the halls. The re-polarization is expressed in the halls of the T wave, which is generally not viewed in the ECG because it is generally absorbed by the P wave. The PR segment begins at the end of the P wave and ends at the beginning of the QRS complex, corresponding to the passage of the stimulus by the atrio-ventricular junction, where it undergoes a delay to allow atrial contraction and ventricular filling. The PR interval is the distance between the beginning of the P wave and the beginning of the QRS complex, measuring the time of atrio-ventricular conduction. The QRS complex corresponds to the depolarization time of the ventricles, from the mid-third of the septum to portions of the basal ventricles and of the septum. Inside the QRS complex, the first negative wave that appears is called the Q wave, the first positive wave is called the R wave and the second negative wave is called S wave. The J section is located in the end of the QRS complex and the beginning of the ST segment, that starts in end of the QRS complex and before the T wave. The QT interval is the electrical systole of the ventricles, starting at the beginning of the QRS complex and ending at the end of the T wave. The RR interval is the distance between two successive R waves. The interval PP, in turn, is the distance between two successive P waves. Finally, the W wave does not always appear because its beginning confuses with the T wave.

ECG Data Compression

The telecardiology involves the real-time broadcast of the ECG waveform or the transmission of single exams. It is one of the most mature telemedicine applications, being highly instigated by the development of real-time applications [8].

An ECG exam can generate a large amount of data, what inspired many authors to develop efficient ECG data compression techniques. Many of these applications made use

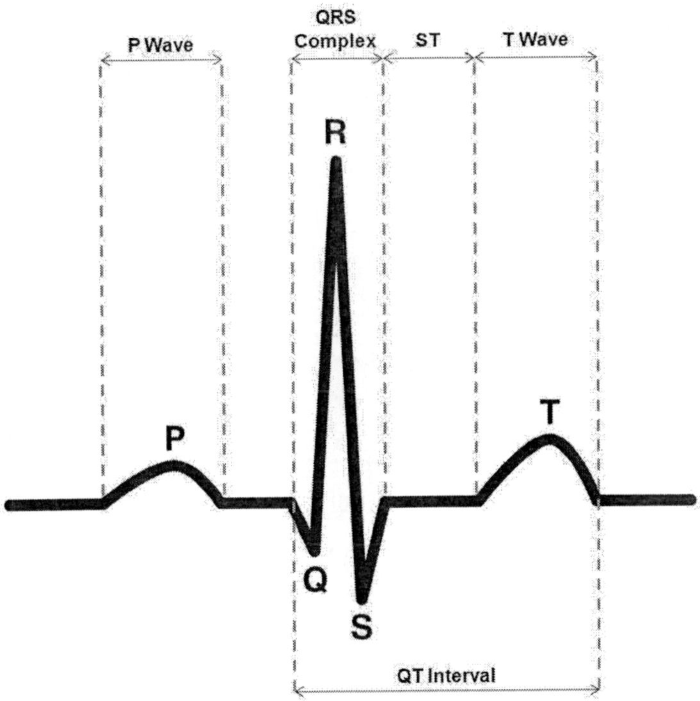

Figure 3. A schematic representation of a regular electrocardiogram signal. The main components of the signal are highlighted: P wave, QRS complex, ST segment and T wave.

of wavelets.

For example, Alesanco presented an ECG compression method based in wavelet transforms [9]. The signal is segmented, separating each QRS complex. Each segment is subtracted from a template signal selected from a pattern dictionary. The residual signal and the template are coded with a Coiflet wavelet expansion. The compression is obtained with the thresholding of the wavelet coefficients.

Normally, the ECG compression techniques utilize thresholding to remove the coefficients near to zero. Then, the significant coefficients are compressed using Huffman coding, for example. The employment of Huffman coding, however, needs the previous knowledge of the symbols that will be compressed. This characteristic is negative in situations where the ECG waveform is much different from the normal.

To avoid the problems of Huffman coding, Chen, Zhang and Shi proposed a compression schema with a uniform scalar dead zone quantiser for encoding the significant coefficients [10]. The method utilizes the Golomb coding for the zero sequences and the Golomb-Rice coding for the non-zero sequences. As these algorithms do not need a previous knowledge of the symbols, the method becomes more robust. Additionally, the great computational efficiency of Golomb coding algorithm turns it into a good approach to real-time applications.

A ECG compression method without thresholding was proposed by Ku and others [11]. They utilized a reversible round-off no recursive one-dimensional (1-D) discrete periodized wavelet transform, executing a staged decomposition that is more resistant to errors. Normally, the increase of the compression rate implicates in the increase of the error. To avoid this, the authors tried to stay in the linear operation zone, reaching the desired compression rate with many stages.

In real-time applications, the transmission delay is a very relevant characteristic to be analyzed. Generally, high compression rates present high delays, because they utilize large portions of the waveform. The use of large time windows, however, makes this kind of technique inviable to applications where time is crucial; such is the case of telecardiology in ambulances.

Kim, Yoo and Lee proposed a method that utilizes Coiflet transforms and presents a low delay [12]. The algorithm reduces the frame size as much as possible to achieve a low delay, while maintaining reconstructed signal quality. To attain both low delay and high quality, it employs waveform partitioning, adaptive frame size adjustment, wavelet compression, flexible bit allocation, and header compression.

The present authors also investigated the ECG data compression before [13]. In that work, neural networks were utilized with wavelet transforms. An algorithm sliced the ECG in segments, which start 250 ms before the QRS complex. Each segment was sliced in the middle, generating two blocks with the same size that were classified as "complex" or "simple" according to their standard deviances. A discrete wavelet transform was utilized to process each block with ECG samples. Two different neural networks were trained to compress and expand the ECG signals. The first network was trained only with "complex" blocks, while the last one worked just with plain blocks. The adopted methodology obtained good results, which are similar to the state of art. The "plain" blocks were compressed in 40 times while the "complex" blocks were compressed in 5 times. The use of both compressors guaranteed a compression rate of 28:1, approximately. The process obtained good grades in the quality aspect: percent root mean squared difference (1.846%), maximum error (0.1789) and standard derivation of errors (0.1044).

In a general way, all the presented method follow the same principle. First, they convert the ECG to time-frequency domain by using wavelet transforms. Then, they reduce the quantity of time-frequency coefficients according to a chosen approach. The key point is to select the correct coefficients, so the reconverted waveform may be very similar (with a small error) to the original version.

Therefore, it is reasonable to assume that, in the ECG data compression context, a good fitness criterion could be related to the distribution of the time-frequency coefficients. In other words, the PSO algorithm could optimize wavelets coefficients looking for a candidate that generates more "condensed" time-frequency coefficients, which means that few coefficients concentrate much information.

In fact, this principle was previously utilized in many ECG data compression applications that make use of wavelets. For example, the present authors proposed a compression method that utilizes the 40 most important coefficients, where the importance of a coefficient is measured by its absolute value [13].

This way, a good fitness function is presented as the algorithm .1.

Algorithm .1 Fitness Criterion of the ECG Data Compression Problem

Require: A candidate wavelet, composed by an array coefficients $W = w_i, i = 1, ..., N$.
Require: An example ECG waveform $E = e_i, i = 1, ..., M$.
Require: A number of coefficients to be utilized ξ.
Ensure: The fitness value of the candidate wavelet.

```
for i = 1 to N do
   if W_i = 0 then
      -1
   end if
end for
timeFrequency ← dwt(E, W)
for i = 1 to M do
   if timeFrequency_i < 0 then
      timeFrequency_i ← timeFrequency_i * (−1)
   end if
end for
timeFrequency ← sort(timeFrequency)
fitness ← 0
for i = M to M − ξ do
   fitness ← fitness + timeFrequency_i
end for
fitness
```

Automatic Detection of Cardiac Arrhythmias

The early discovery of cardiac arrhythmias can prevent the sudden death of many patients. Once the number of patients tends to grow, the automatic methods become a very interesting tool to physicians. In a general way, these algorithms analyze the ECG waveform in a classical application of patterns detection.

Because the ECG waveforms are highly sensible to noise, several methods utilize transforms to overcome the problem. For example, Jung and Tompkins presented an algorithm based on wavelets that can detect and classify four types of ventricular arrhythmias: ventricular fibrillation, ventricular tachycardia, supraventricular tachycardia and ventricular flutter [14]. After testing four different types of wavelets, they found out that the Daubechies-4 presented the best results. The wavelets reduced the utilized data in ten times and the method achieved an average sensitivity of 92.3%.

Fourier transforms were utilized by Al-Fahoum and Khadra for classifying ECG waveforms as normal, atrial fibrillation, ventricular tachycardia and ventricular fibrillation [15]. The proposal has achieved a hit rate of around 93.6%.

Inan, Giovangrandi and Kovacs applied wavelet transforms together with temporal information [16]. Thus, the accuracy of method was over 95% with more than 93,000 analyzed heartbeats.

Applications involving the detection of cardiac arrhythmias demand wavelets that can highlight differences among the different classes. This principle could be implemented by a geometric operation called Euclidean distance. This operation measures the absolute distance among two vectors. This way, the fitness function could calculate the potential of a candidate wavelet by transforming the different classes of arrhythmia in time-frequency coefficients and measuring the total distance among them.

The algorithm .2 presents this implementation.

Algorithm .2 Fitness Criterion of the Arrhythmias Classification Problem

Require: A candidate wavelet, composed by an array coefficients $W = w_i, i = 1, ..., N$.
Require: Examples of ECG waveforms $E = e_i, i = 1, ..., M$, where each waveform belongs to an arrhythmia class: $E_j, j = 1, ..., N$.
Ensure: The fitness value of the candidate wavelet.

```
for i = 1 to N do
  if W_i = 0 then
    -1
  end if
end for
tf ← E
for i = 1 to N do
  tf_i ← dwt(E_i, W)
end for
fitness ← 0
for i = 1 to N do
  for j = 1 to N do
    if i <> j then
      fitness ← fitness + calculateDistance(tf_i, tfj)
    end if
  end for
end for
fitness
```

Results

The Java programming language and the JSwarm-PSO framework [17] were utilized to implement both cases. The obtained results are shown and discussed in this section.

The JSwarm-PSO framework implements a complete engine of particle swarm optimization. To utilize it, one should implement the fitness criterion, which is particular for each domain. Additionally, the developer should define some PSO parameters, such as number of particles, number of iterations and optimization objective (maximization or minimization). The fitness criterion is implemented as a Java class that extends an abstract class called FitnessFunction. The other parameters can be defined in the main function of the program.

The optimization process produced a six-sized array of wavelet coefficients, which is graphically represented in the figure 4. The figure 5 presents the effects of applying

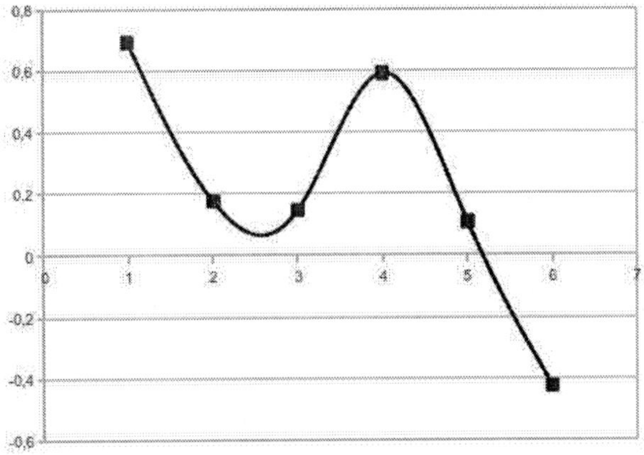

Figure 4. The optimized wavelet in the ECG data compression application.

the optimized wavelet in ECG waveform, comparing to traditional wavelets. The optimized mother-wavelet obtained a better compression performance when compared to other wavelets, as follows:

- Optimized: 17.07;
- Daubechies-8: 14.60;
- Daubechies-6: 14.17;
- Daubechies-4: 13.76;
- Coiflet-6: 13.62;
- Coiflet-8: 13.62;
- Symlet-6: 13.59;
- Daubechies-2: 13.24.

These results represents the sum of the 40 most significant coefficients in each application. Largest values are better, once they mean that the compression process is losing less data.

The analysis of the results shows that the optimization process achieved a performance 16.92% better than the Daubechies-8 (worst case) and 28.93% better when compared to the Daubechies-2 (best case).

Obviously, these results are not fully transfered to the compression performance, since the wavelets are just one step of the process. Its final impact cannot be measured, because it depends of the compression method that was chosen. However, it is possible to affirm that the optimized wavelet should enhance the results.

On the other hand. four different kinds of ECG signal were utilized in the classification experiment: normal (no arrhythmia), atrial fibrillation, ventricular tachycardia and ventricular flutter.

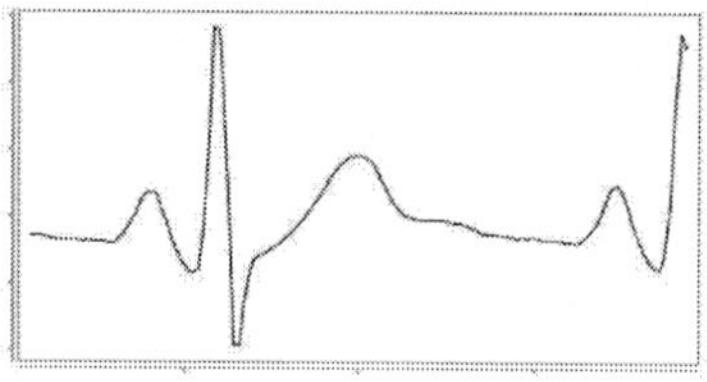

(a) Original waveform.

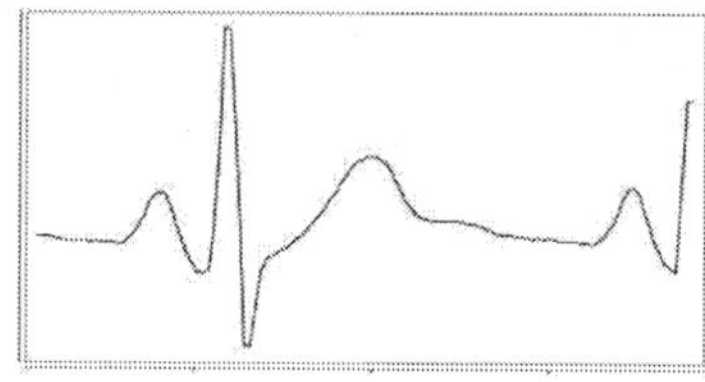

(b) Waveform processed with the optimized wavelet.

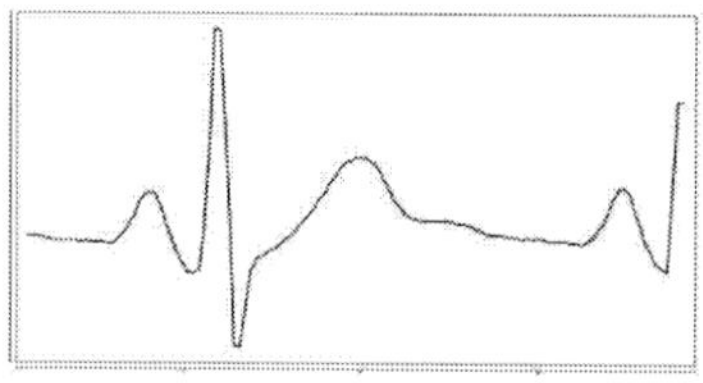

(c) Waveform processed with a Daubechies-2 wavelet.

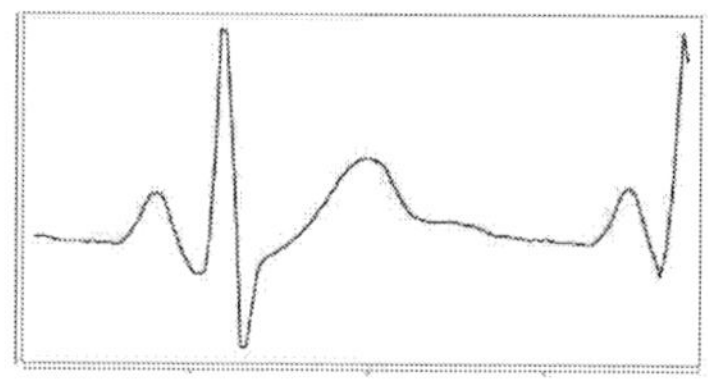

(d) Waveform processed with a Daubechies-4 wavelet.

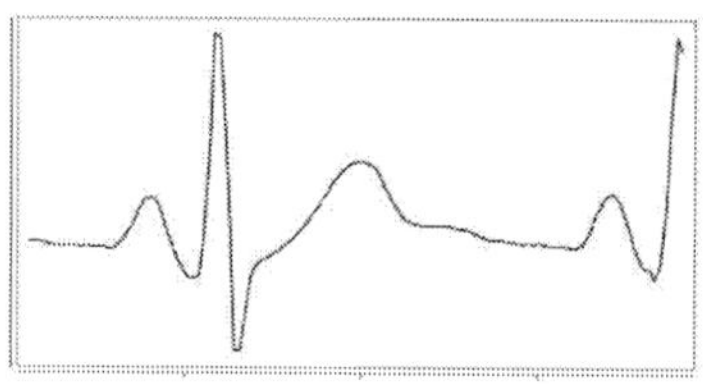

(e) Waveform processed with a Daubechies-8 wavelet.

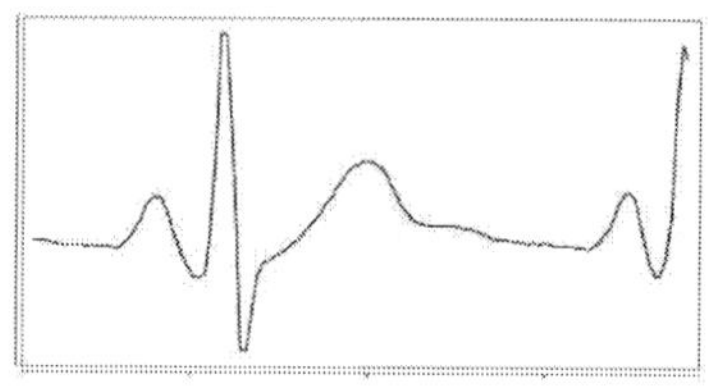

(f) Waveform processed with a Coiflet-6 wavelet.

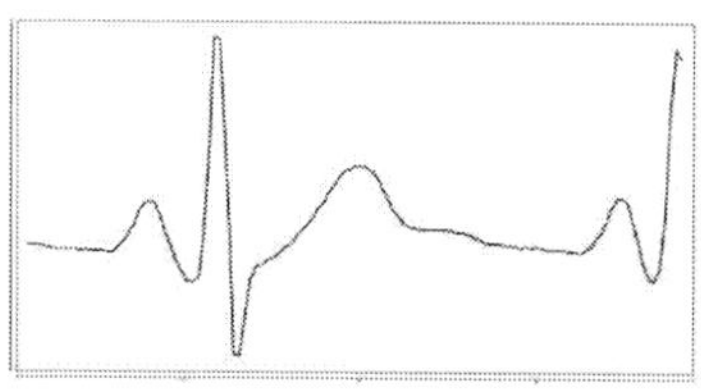

(g) Waveform processed with a Coiflet-8 wavelet.

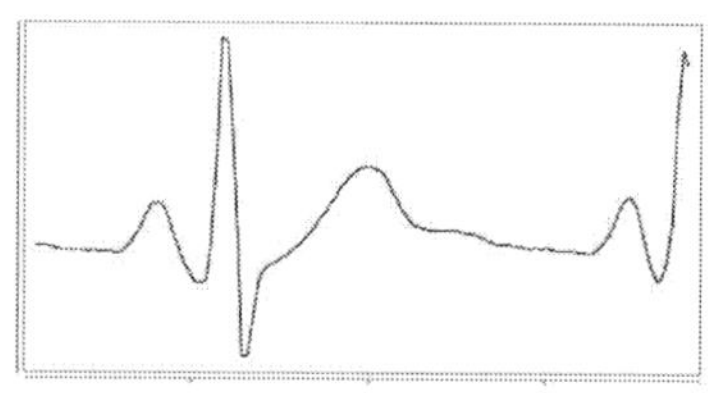

(h) Waveform processed with a Symlet-6 wavelet.

Figure 5. Each figure presents the same waveform processed by a different wavelet.

This way, the PSO execution produced a eight-sized array of wavelet coefficients, which is graphically represented in the figure 6. This optimized mother-wavelet presented better results when compared to other wavelets, as follows:

- Optimized: 44.01;
- Daubechies-4: 22.81;
- Daubechies-8: 22.03;
- Coiflet-8: 19.01;
- Symlet-6: 18.82;
- Coiflet-6: 17.89;
- Daubechies-6: 14.71;
- Daubechies-2: 12.41.

These results represents the sum of the Euclidean distances among the vectors of transformed coefficients in each application. Largest values are better, once they mean that the differences between the ECG types are more evident, enhancing the classifier performance.

The analysis of the results shows that the optimization process achieved a performance 92.94% better than the Daubechies-4 (worst case) and 255.63% better when compared to the Daubechies-2 (best case).

It is relevant to observe that, as in the previous case, the classification performance is not affected in the same magnitude by these results. It happens because the wavelets are a preprocessing step of the process and the final results depend of the classification method. Anyway, it is evident that the optimized wavelets may have a positive effect in the classifier performance, once they highlight the differences among the signal types.

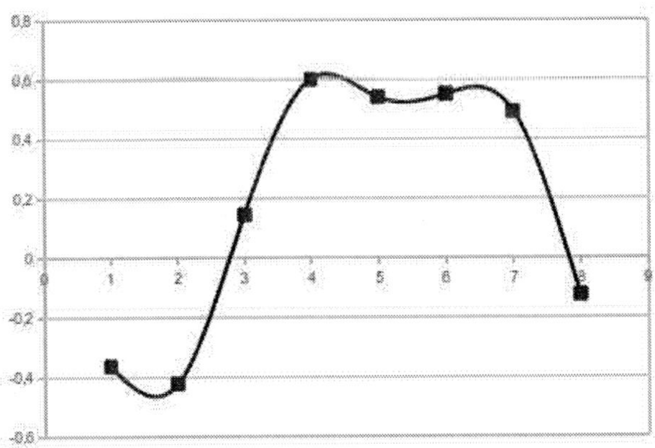

Figure 6. The optimized wavelet in the arrhythmias detection application.

Source-Code

The ECG data compression problem was implemented in two Java classes: CompressionOptimization and CompressionFitnessFunction. The Java codes are presented in the listings 1 and 2.

The arrhythmias detection problem was also implemented in two Java classes: ClassificationOptimization and ClassificationFitnessFunction. The Java codes are presented in the listings 3 and 4.

```java
package br.pucpr.ppgia.figueredo.wavelets;

import net.sourceforge.jswarm_pso.Swarm;

public class CompressionOptimization {

    static public void main(String[] args) {
        // starts the PSO engine with a default number of particles
        // the engine tries to optimize an array of wavelet
        // coefficients that produces more zeros

        // (actually, values < 0.1)
        Swarm swarm = new Swarm(Swarm.DEFAULT_NUMBER_OF_PARTICLES,
                new CoefficientsParticle(),
                new CompressionFitnessFunction(40));
        swarm.setMaxPosition(1);
        swarm.setMinPosition(0);


        // the engine is run for 500 epochs
        for (int i = 0; i < 500; i++) {
            swarm.evolve();
        }
        // the results are presented
        System.out.println(swarm.toStringStats());
    }
}
```

Listing 1: Main function of ECG data compression case.

```java
package br.pucpr.ppgia.figueredo.wavelets;

import net.sourceforge.jswarm_pso.FitnessFunction;

public class CompressionFitnessFunction extends FitnessFunction {

    // an example of ECG
    final static public double[] EXAMPLE = {
            -0.060, -0.065, -0.060, -0.075, -0.065,
            -0.070, -0.070, -0.090, -0.080, -0.095,
            -0.080, -0.095, -0.080, -0.095, -0.085,
            -0.090, -0.090, -0.100, -0.085, -0.105,
            -0.090, -0.045, 0.005, 0.015, 0.045,
            0.155, 0.140, 0.045, 0.005, -0.040, -0.085,
            -0.200, -0.195, -0.200, -0.200, -0.240,
            -0.130, 0.340, 1.155, 1.470, -0.155,
            -0.825, -0.590, -0.350, -0.155, -0.170,

            -0.140, -0.155, -0.115, -0.125, -0.090,
            -0.095, -0.065, -0.055, -0.015, -0.005,
            0.035, 0.045, 0.090, 0.110, 0.150,
```

```
        0.180, 0.205, 0.225, 0.230, 0.220,
        0.235, 0.230, 0.200, 0.170, 0.120, 0.075,
        0.040, 0.020, 0.005, -0.005, -0.005,
        -0.010, -0.015, -0.010, 0.000, -0.005,

        -0.020, -0.020, -0.015, -0.025, -0.035,
        -0.045, -0.045, -0.060, -0.065, -0.070,
        -0.055, -0.060, -0.080, -0.075, -0.090,
        -0.065, -0.075, -0.080, -0.080, -0.085,
        -0.090, -0.075, -0.085, -0.095, -0.105,
        -0.090, -0.095, -0.070, -0.045, 0.005,
        0.030, 0.070, 0.195, 0.145, 0.050,
        -0.015, -0.025, -0.115, -0.180, -0.190,
        -0.185, -0.185, -0.215, 0.010, 0.575,
        1.185, 0.960, -0.160
    };
    // the example in the framework model
    private Wavelet example;
    // quantity of selected coefficients

    private int limit;


    public CompressionFitnessFunction(int limit) {
        super();
        example = new Wavelet(EXAMPLE);
        this.limit = limit;
        // it is a maximization process
        setMaximize(true);
    }

    @Override
    public double evaluate(double[] position) {
        // avoids invalid coefficients
        for (int i = 0; i < position.length; i++) {
            if (position[i] == 0) {
                return -1;
            }
        }

        // transforms the example using the candidate (position)
        Wavelet wavelet = new Wavelet(position);
        Wavelet transformed = Wavelet.dwt(example, wavelet, 2);
        double[] coeffs = new double[transformed.length()];
        for (int i = 0; i < coeffs.length; i++) {
            coeffs[i] = Math.abs(transformed.valueAt(i));
        }
        Arrays.sort(coeffs);


        // calculates the fitness
        double fitness = 0;
        for(int i=coeffs.length-1, j=0; j<limit; i--, j++) {

            fitness += coeffs[i];
```

```
        }

        // just checks the error to avoid invalid wavelets
        double error = checkError(transformed, wavelet);
        if(error>5) {
            fitness -= error*3;
        }


        // returns the fitness
        return (double) fitness;
    }

    public double checkError(Wavelet transformed, Wavelet wavelet) {
        // tests the validity of the candidate
        Wavelet inversed = Wavelet.idwt(transformed, wavelet, 2);
        double error = 0;
        for (int i = 0;
                i < inversed.length() && i < EXAMPLE.length;
                i++) {
            error += Math.abs(
                Math.abs(inversed.valueAt(i)) - Math.abs(EXAMPLE[i]));
        }

        return error;

    }

}
```

Listing 2: Fitness function of ECG data compression case.

```

package br.pucpr.ppgia.figueredo.wavelets;

import net.sourceforge.jswarm_pso.Swarm;



public class ClassificationOptimization {



        static public void main(String[] args) {

        // starts the PSO engine with a default number of particles

        // the engine tries to optimize an array of wavelet coefficients

        // that produces more zeros (actually, values < 0.1)

        Swarm swarm = new Swarm(Swarm.DEFAULT_NUMBER_OF_PARTICLES,

                new CoefficientsParticle(8),

                new ClassificationFitnessFunction());
```

```
        swarm.setMaxPosition(1);

        swarm.setMinPosition(-1);



        // the engine is run for 500 epochs

        for (int i = 0; i < 5000; i++) {
            swarm.evolve();
        }
        // the results are presented
        System.out.println(swarm.toStringStats());
    }

}
```

Listing 3: Main function of arrhythmias detection case.

```

package br.pucpr.ppgia.figueredo.wavelets;

import net.sourceforge.jswarm_pso.FitnessFunction;

public class ClassificationFitnessFunction extends FitnessFunction {

    // the samples were removed to facilitate the understanding
    final static public double[][] EXAMPLES = {
        new double[] {/*AFIB*/},
        new double[] {/*VT*/},

        new double[] {/*NORMAL*/},
        new double[] {/*FLUTTER*/} };

    public ClassificationFitnessFunction() {
        super();
        setMaximize(true);

    }

    @Override
    public double evaluate(double[] position) {

        for(int i=0; i<position.length; i++) {

            if(position[i] == 0) {

                return -1;

            }

        }



```

```
        double t[][] = new double[EXAMPLES.length][0];

        for(int i=0; i<EXAMPLES.length; i++) {

            Wavelet transformed = Wavelet.dwt(

                new Wavelet(EXAMPLES[i]), new Wavelet(position), 2);

            t[i] = new double[transformed.length()];

            for(int j=0; j<t[i].length; j++) {
                t[i][j] = transformed.valueAt(i);
            }
        }

        double fitness = 0;
        for(int i=0; i<t.length-1; i++) {
            for(int j=i+1; j<t.length; j++) {
                fitness += calculateDistance(t[i], t[j]);
            }
        }

        double error = calculateError(position);
        if(error>4) {
            fitness -= error*3;

        }

        return fitness;

    }

    public double calculateDistance(double[] v1, double[] v2) {
        if(v1.length != v2.length) {
            return -1;
        }

        double distance = 0;
        for(int i=0; i<v1.length; i++) {
            distance += Math.pow(v1[i]-v2[i], 2);
        }
        distance = Math.sqrt(distance);

        return distance;
    }

    public double calculateError(double[] candidate) {
        Wavelet wavelet = new Wavelet(candidate);
```

```
        Wavelet transformed = Wavelet.dwt(
                new Wavelet(EXAMPLES[0]), wavelet, 2);
        Wavelet inversed = Wavelet.idwt(
                transformed, wavelet, 2);

        double error = 0;
        for(int i=0;
                i<EXAMPLES[0].length && i<inversed.length();
                i++) {

            error += Math.abs(
                EXAMPLES[0][i] - inversed.valueAt(i));

        }

        return error;
    }
}
```

Listing 4: Fitness function of arrhythmias detection case.

Conclusion

This chapter proposed an innovative way to optimize wavelets for several domains. Instead of dealing with complicated mathematical formulae, the method utilizes the PSO algorithm to evolve the wavelet coefficients.

Two different applications of the present method were described. Although the both belong to the same domain (ECG), they give a good vision of two usual wavelets applications: data compression and patterns recognition.

It is important to notice that the method is generic, so it can be applied in several domains. However, the developer must implement a specific fitness function for each application, which will guide the algorithm during the optimization process.

The obtained results are quite optimistic and the optimized wavelets overpassed the traditional ones. Nevertheless, it is necessary to remind that the final performance of an application (compression or classification) depends mostly of the chosen method. This way, the impact of optimized wavelets may vary for each circumstance.

Anyway, the presented method reveals itself as an useful tool for the researchers involved with wavelets. It represents a computationally inexpensive way of finding more adequate wavelets for each domain. Thus, it can enhance the final performance of several applications, without many requirements.

References

[1] J.S. Walker. *A Primer on Wavelets and Their Scientific Applications*. Chapman & Hall, Boca Raton, FL, 1999.

[2] C.R. Castelano. *Estudo Comparativo da Transforma Wavelet no Reconhecimento de Padrões da Íris Humana, Dissertação (Mestrado em Engenharia Elétrica)*. Depar-

tamento de Engenharia Elétrica, Universidade de São Paulo, São Carlos, São Paulo, 2006.

[3] J. Kennedy and R.C. Eberhart. Particle swarm optimization. In *Proceedings of the IEEE International Conference on Neural Networks*, pages 1942–1948, 1995.

[4] J. Vesterstroem and J. Riget. Particle swarms: Extensions for improved local, multimodal, dynamic search in numerical optimization, 2002.

[5] E.C. Biscaia-Jr., M. Schwaab, and J.C. Pinto. Um novo enfoque do método do enxame de partículas. In *Anais do NanoBio2004 - I Workshop em Nanotecnologia e Computação Inspirada na Biologia*, pages 1–7, Rio de Janeiro, RJ, Brasil, 2004.

[6] D.F. Azevedo. *Iniciação à Eletrocardiografia.* Artmed, Porto Alegre, 1999.

[7] G.P. Goldwassser. *O Eletrocardiograma Orientado para o Clínico Geral.* Revinter, Rio de Janeiro, 1997.

[8] M.L. Sedgewick, K. Dalziel, J. Watson, D.J. Carrington, and S.M. Cobbe. Performance of an established system of first responder out-of- hospital defibrillation: the results of the second year of the heartstart scotland project in the utslein style. *Resuscitation*, **6**:75–78, 1993.

[9] A. Alesanco, S. Olmos, R.S.H. Istepanian, and J. García. Enhanced real-time ecg coder for packetized telecardiology applications. *IEEE Trans. Inf. Technol. B.*, **10**(2):229–236, 2006.

[10] J. Chen, Y. Zhang, and X. Shi. Ecg compression based on wavelet transform and golomb coding. *El. Lett.*, **42**(6):322–324, 2006.

[11] C.-T. KU, H.-S. Wang, K.-C. Hung, and Y.-S. Hung. A novel ecg data compression method based on nonrecursive discrete periodized wavelet transform. *IEEE Trans. Bio-Med. Eng.*, **53**(12):2577–2583, 2006.

[12] B. Kim, S. K. Yoo, and H. Moon. Wavelet-based low-delay ecg compression algorithm for continuous ecg transmission. *IEEE Trans. Inf. Technol. B.*, **10**(1):77–83, 2006.

[13] M. Figueredo, J. Nievola, S. Rogal, and A. Neto. Compression of electrocardiogram using neural networks and wavelets. In R. Lee and H.K. Kim, editors, *Computer and Information Science*, volume 131 of *Studies in Computational Intelligence*, pages 27–40. Springer Berlin - Heidelberg, 2008.

[14] Y. Jung and W.J. Tompkins. Detecting and classifying life-threatening ecg ventricular arrhythmias using wavelet decomposition. In *25th Annual International Conference of the IEEE*, pages 2390–2393, Cancun, Mexico, 2003. IEEE, IEEE Press.

[15] A. Al-Fahoum and L. Khadra. Combined bispectral and bicoherency approach for catastrophic arrhythmia classification. In *27th Annual Conference of IEEE Engineering in Medicine and Biology*, pages 332–336, Shanghai, China, 2005. IEEE, IEEE Press.

[16] O. T. Inan, L. Giovangrandi, and G.T.A. Kovacs. Robust neural-network-based classification of premature ventricular contractions using wavelet transform and timing interval features. *IEEE Trans. Bio-Med. Eng.*, **53**(12):2507–2515, 2006.

[17] P. Cingolani. Jswarm-pso - particle swarm optimization package written in java, 2011.

In: Wavelets: Classification, Theory and Applications
Editors: Manel del Valle et al.
ISBN: 978-1-62100-252-9

Chapter 10

WAVELET NEURAL NETWORKS ¬ INNOVATIVE PROCESSING STRATEGY FOR CLASSIFICATION AND QUANTIFICATION MODELS

Juan Manuel Gutiérrez[1,*]***, Roberto Muñoz***[1] ***and Manel del Valle***[2]
[1]Department of Electrical Engineering, Bioelectronics Section, CINVESTAV, MEXICO
[2]Sensors and Biosensors Group, Department of Chemistry,
Universitat Autònoma de Barcelona, Spain

ABSTRACT

Biologically inspired technologies for artificial sensing have been investigated in the last years using different kinds of sensors and applying novel computational paradigms, in order to emulate the behavior of their biological counterparts. In this sense, the use of advanced processing techniques is essential in interpreting the obtained information. Among the available processing tools, the combination of Wavelet Transform (WT) and Artificial Neural Networks (ANNs) has been adopted as innovate strategy for processing complex signals, especially when they are composed of a large number of measures and include non-linear content. Wavelet Neural Networks (WNNs) are a new class of networks that combine wavelets and neural networks in order to improve tasks such as classification, prediction and modeling, taking benefit of both paradigms. This chapter describes the fundamentals of WNNs and their advantages in the treatment of signals, especially when feature extraction stage and/or signal preprocessing are mandatory to build an appropriate calibration or classification model. This chapter will highlight the successful applications of these networks in areas such as electronic tongues and myoelectric analysis. To demonstrate the computing power of the WNNs, two case studies from our laboratories are presented. The first relates to the quantification of analytes using a voltammetric electronic tongue and the second involves the classification of motor unit spikes at different levels of muscle contraction from myoelectric records obtained with a needle electrode.

* E-mail address: mgutierrez@cinvestav.mx; Tel: +52 5557473800; Fax: +52 5557477002 (Corresponding Author)

INTRODUCTION

Recent advances in the development of bio-inspired technologies have attempted the improvement of the instrumentation parameters such as response time, stability, detection limit, interpretation skills, size, power consumption and portability. However, the most notable progress in this issue is related to the development of novel computational paradigms focused in the proper interpretation of data, in order to make the right decisions in different application fields. Thus, many applications of signal processing entail detecting, extracting and classifying specific elements from high-dimensional data. Classification and quantification models are widely used in various fields of science, whose goal is either to classify a large set of samples into predefined classes or estimating some parameters from a sample.

On the one hand, classification algorithms have extensively been reported, some of them are specialized tools in the analysis of particular data and domains. In this way, it is possible to find pure statistical models, genetic algorithms, adaptative spline methods, graphical methods, non-linear algorithms based on ANNs, Bayesian classifiers, Support Vector Machines (SVM), among some of the most commonly used [1]. On the other hand, quantification models are mainly related with regression analysis techniques. Usually, these models are used to describe the effect of certain variables upon anothers, in order to estimate the quantitative effect between them. The regression can be observed as an optimization problem where it is attempted to find a solution where the error is minimized. Algorithms such as Principal Component Regression (PCR), Partial Least Squares (PLS) and Multiple Linear Regression (MLR), are typically used to solve this optimization problem [2].

Nowadays, the use of modern instruments produces more accurate and higher dimensional measurements than in the past, thus allowing acquisition of more abundant information on the samples analyzed [3]. Although increasing computer performance allows the processing of raw data, increased information databases requires extracting relevant data features in order to achieve better interpretation models and reduce computational cost [4]. In this way, different strategies have been proposed to discard irrelevant content and compress original data for partially overcoming the curse of dimensionality associated with high-dimensional data modeling problems [5].

In the last decade, Wavelet Transform (WT) has gained wide application in signal processing methodologies where data compression is an important step in the data mining process. WT has been proved to be fast and reliable in computations related with non-stationary signals coming from real applications, opposite to existing popular methods, especially to Short Time Fourier transform (STFT). WT yields a series of coefficients as a result of relating an original signal with a family of functions that are scaled and translated versions of a base function known as mother wavelet. Its most attractive feature is its ability to optimally describe temporal information from the spectrum in a better way that does STFT [6]. These coefficients can be deconvolved using computer approaches to obtain information about individual samples or classes.

The development of advanced computational algorithms to face proper handling of the information, have encouraged the establishment of bio-inspired computational models with predictive capabilities and improved performance. Despite having been created more than 60 years ago, it has been recently that ANNs have demonstrated their power and versatility in data processing. These mathematical models are computational systems made up with a large number of simple highly interconnected processing elements (called neurons) that mimic the structure and operation of the brain [7, 8]. Their ability to find and interpret patterns have allowed their application to a broad spectrum of data-intensive applications.

Traditionally, ANNs and WT have been two separate disciplines, which have been employed in data processing. In the recent years a new class of networks called Wavelet Neural Network (WNN) have emerged and grown vigorously both in research and applications. Nowadays, the potential of these networks in engineering, chemistry, economics, and social science applications is rich and still growing. A WNN model generally consists of a feedforward neural network with one hidden layer, whose activation functions are drawn from an orthonormal wavelet family [9]. Most applications of WNN are related with function learning [10], compression and pattern recognition [11], speech segmentation [12], image processing for face recognition [13], characterization of real environments for use in virtual reality [14], prediction of chaotic signals [15] and classical robotics control problems [16].

Nowadays WNN are used, as processing tools in research areas such as chemistry and biomedical engineering where processing non-stationary signals is a difficult task. On the one hand, the first contributions of WNN in chemistry are focused in prediction tasks or quantitative structure-properties relationship (QSPR) where complexation constants of α-cyclodextrins and β-cyclodextrins were studied [17]; similar prediction applications have been described, like the programmed-temperature separation of naphtas in a petrochemical industry [18] and in the prediction of gas chromatography retention indexes of methyl-substitud alkanes produced by insects [19]. Other interesting works developed by research groups in Iran used WNN models to predict the solubility of polycyclic aromatic hydrocarbons [20], anthraquinone dyes [21] and azo dyes [22] in supercritical CO_2. WNN models have also been used as multivariate tool for the determination of analytes. Some of the first contributions were related with the multivariate spectrophotometric determination of Cu^{2+}, Fe^{2+} and Al^{2+} using pyrocathecol violet as the chromogenic reagent [23], or that of Fe^{2+}, Ni^{2+} and Co^{2+} [24] using pyridylazo naphtol in micellar media;both works employed UV-VIS spectra preprocessed with PCA as departure information for the WNN modeling. A similar work was reported by Ensafi et al. describing the simultaneous kinetic determination of two toxic compounds (thiocyanate and sulfide) from their spectrophotometric transient record previously treated with PCA [25]. WNNs have been also applied in electroanalysis studies, the first contribution was regarded with the oscillographic chronopotentiometric determination of mixtures of Pb^{2+}, In^{3+} and Zn^{2+} where a discrete version of wavelet network was used [26]. The resolution of highly overlapped traces commonly obtained in voltammetric signals was firstly treated by Khayamian et al. using a WNN model able to resolve a three-species heavy metals mixture, extending the calibration range of Cu^{2+} from 1-50 ng mL^{-1} to 1-5000 ng mL^{-1} [27]. Nevertheless, there are only some applications which dealt with multianalyte quantification in overlapped-signals of voltammetric applications, the simultaneous quantification of oxidizable compounds [7] the resolution of mixtures of three phenolic compounds [4] and the determination of three heavy metals (Pb^{2+}, Cd^{2+} and Cu^{2+}) in presence of Tl^{2+} and In^{2+} as interfering species [28]. Latest report of WNN is related with a QSPR work for the prediction of the half-wave potential of aldehydes and ketones [29].

On the other hand, bioengineering applications of WNNs are frequently associated with the classification and identification of biosignals, e.g. electromyographic (EMG), electrocardiography (ECG) and electroencephalography (EEG). A biosignal is an electrical signal acquired from any organ that represents a physical variable of interest. This signal is normally a function of time and can be described in terms of its amplitude, frequency and phase. Its origin is related to the electrochemical activity of a certain class of cells, known as excitable cells, which are components of nervous, muscular or glandular tissue [30]. Biosignals are in general complex signals due to the non-stationary nature related with the physiological properties [31]. Many research and efforts have been made in the area, developing better processing algorithms, upgrading existing methodologies, improving detection techniques to reduce noise, in order to acquire accurate and representative signals.

Recent advances using signal processing technologies and mathematical models have driven the development of better detection methodologies and analysis. One area that has received extensive attraction is the Artificial intelligence (AI). Its use has allowed to take advantage of the pre-processing techniques such as WT, STFT, Wigner-Ville Distribution (WVD), statistical measures and higher-order statistics [32]. AI approaches towards signal recognition and processing may include ANNs, Dynamic Recurrent Neural Networks (DRNN), fuzzy logic systems and Genetic Algorithm (GAs) [33].

The time-frequency approach offered by WT is well suited to non-stationary biosignals. WT analysis may allow the development of a real-time instrument that can be used for suppressing noise, compact information and extract significant features of data. However, although it is possible to find many reports using wavelets in the treatment of biosignals, there are only a couple of works where the use of WNN plays a major role in the processing of information.

Both reports are by Sudasi et al.; the first recorded work is related to the classification of EEG signals in order to identify epileptic or nonepileptic seizures [34]. The referred work compares three different approaches for designing classifiers: logistic regression, fedforward ANN trained with backpropagation error and WNN.

Departure information was obtained from a pre-processing stage using Fast Fourier Transform (FFT) and Autoregressive (AR) model considering maximum likelihood estimation of EEG. The second report is oriented to classify EMG signal patterns of subjects suffering neurogenic diseases (e.g. myopathy and neurogenic disorder) [35]. Similarly to the first work, ANN and WNN models were developed and fed with estimated AR parameters coming from raw information. Final results show that WNN-based classifier outperformed the ANN counterpart.

This chapter is focused in describing not only nature of WNNs to build prediction models able to interpret the complexity of electrochemical signals, but also in establishing robust classification models where the use of data pretreatment may be circumvented. Our research groups have worked closely on the development of multidisciplinary applications for these purposes. We have been pioneers in the use of WNN either as a chemometric tool for processing voltammetric signals and also as processing strategy for biosignals. Gained knowledge will be summarized along the different sections to give the reader a comprehensive and specialized information on the use of this novel networks. Two different case studies will be presented, in order to show advantages of WNN in classification and prediction tasks.

FOUNDATIONS OF WNN

Theory

The idea behind the WNN concept is related with the merging of wavelets with neural networks. This successful synthesis of theories has been generated a new class of networks called WNNs or simply wavelet networks [9]. The innovation of these networks is the use of wavelet functions as hidden neuron activation functions in the ANN architecture. Applying theoretical features of WT, different network construction methods can be proposed. The foundation of a basic WNN model is evident by observing the similarity between Strömberg's definition and the hidden layer of a Multi-layer Perceptron (MLP) [36, 37]. Thus, wavelet decomposition can be seen as a neuronal model assuming that the scale and translation (s,t) of wavelets can be replaced by a new index $i=1,...,k$, as is defined in the equation:

$$y=\sum_{i=1}^{k} w_i \psi_i(x) \qquad (1)$$

where w_i represent the wavelet coefficients of the decomposition of y , and ψ_i the daughter wavelets.

This approximation represents an important relation between the single output of a one-hidden layer MLP and one-dimensional signal decomposition using simple scaling and translations of a mother wavelet. A graphic interpretation of this concept can be seen on Figure 1.

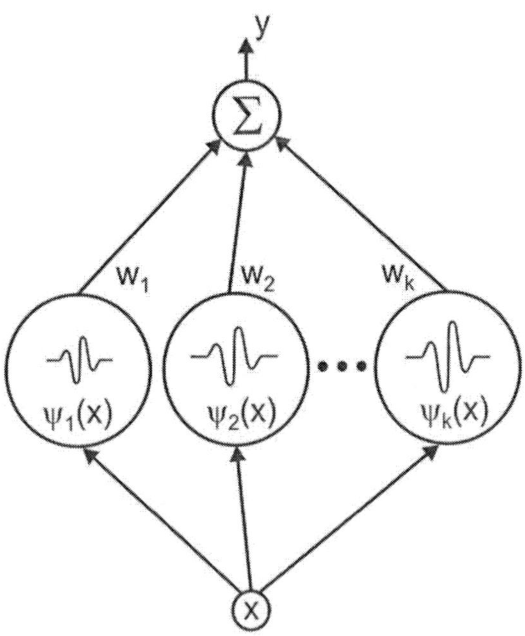

Figure 1. Wavelet expansion shown as neural network.

For developing a WNN, the use of $1+1/2$ network's architectures (one input layer, one hidden layer and a single output) is enough to approximate any arbitrary and continuous function, using an appropriate family of functions in the hidden layer [38-40]. The final accuracy in the approach depends on the used function family characteristics as well as the network's training error. The essence of WNN is to find an appropriate family of wavelets in a representative space in order that the complex relationships contained in the original signals might be properly expressed [41]. In this sense, there are two possibilities to construct a complete and useful family of wavelets for its inclusion in a neuronal model.

The first, involves an orthogonal decomposition: the obtained daughter wavelets need to be linearly independent and orthogonal to each other so that there are not any redundant functions [42]. The second is the use of wavelet frames, which can be constructed by simple operations of translation and dilation without fulfilling stringent orthogonal conditions [7, 37, 39]. Although many wavelet applications work well using orthogonal wavelet basis, many others work better with redundant wavelet families. The redundant representation offered by wavelet frames has demonstrated to be good both in signal denoising and in compaction [41, 43].

In this manner, a signal $f(x)$ can be approximated by generalizing a linear combination of daughter wavelets $\psi_{s,t}(x)$ derived from its mother wavelet $\psi(x)$, this family of functions is defined as:

$$M_c = \left\{ \frac{1}{\sqrt{|s_i|}} \psi\left(\frac{x - t_i}{s_i}\right), \quad t_i, s_i \in \Re, \quad s_i > 0 \right\} \tag{2}$$

where the translation t_i and the scaling s_i are real numbers in $\Re$.

The family of functions M_c is known as a continuous wavelet frame of $L^2(\Re)$ if there exist two constants A and B that fulfills with [44]:

$$A\|f(x)\|^2 \leq \sum_{s,t} |\langle f(x), \psi_i(x) \rangle|^2 \leq B\|f(x)\|^2 \text{ with } A > 0, B < +\infty \tag{3}$$

Nevertheless for multi-variable model's applications it is necessary to use multidimensional wavelets. Families of multidimensional wavelets can be obtained from the product of P monodimensional wavelets, $\psi(a_{ij})$, of the form:

$$\Psi_i(x^n) = \prod_{j=1}^{P} \psi(a_{ij}) \text{ where } a_{ij} = \frac{x_j^n - t_{ij}}{s_{ij}} \tag{4}$$

where now t_{ij} and s_{ij} are the translation and scaling vectors respectively.

WNN Algorithm

The most widespread WNN architecture is shown in Figure 2. This model corresponds to a one-hidden layer feedforward MLP with single output. The output y^n (where *n* is an index, not a power) depends on the connection weights c_i between the output of each neuron and the output of the network, the connection weights w_j between the input data and output, an offset value b_0 useful when adjusting functions that has a mean value other than zero, the *n*-th input vector x^n and the wavelet function Ψ_i of each neuron. The approximated signal of the model y^n can be represented by the next equation:

$$y^n = \sum_{i=1}^{K} c_i \Psi_i(x^n) + b_0 + \sum_{j=1}^{P} w_j x_j^n \text{ with } \{i, j, K, P\} \in Z \tag{5}$$

where subindexes i and j stand for the *i*-th neuron in the hidden layer and the *j*-th element in the input vector, x^n, respectively, K is the number of wavelet neurons and P is the length of input vector, x^n. With this model, a P-dimensional space can be mapped to a monodimensional space $(\Re^P \rightarrow \Re)$, allowing the network to predict a value for the output y^n when the n-th signal x^n is input to the trained network.

The basic neuron will be a multidimensional wavelet $\Psi_i(x^n)$, which is built using the definition (4), where scaling s_{ij} and translation t_{ij} coefficients are the adjustable parameters of the *i*-th wavelet neuron. With this mathematical model for the wavelet neuron, the network's output becomes a linear combination of several multidimensional wavelets [7].

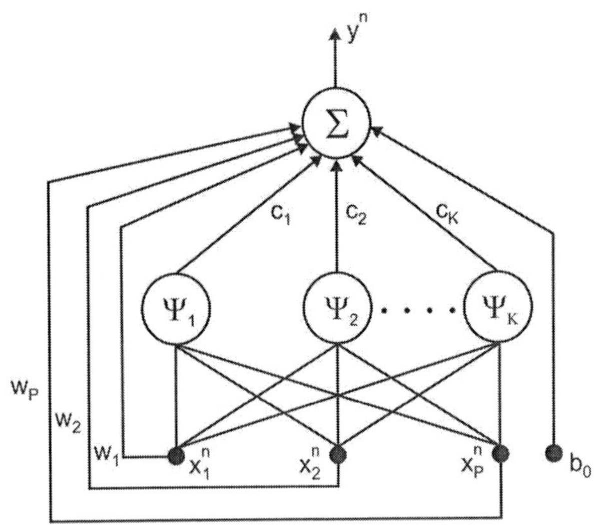

Figure 2. WNN topology architecture.

Training Stage

Different learning methods have been implemented in WNN training: some authors use stochastic gradient algorithm [9], conjugate gradient method [11], backpropagation algorithm [45] and most recently, the use of sampling theory [46]. There are also cases where backpropagation is combined with Orthogonal Least Square-Backward Elimination [9, 47], mainly used in the selection of network structures. Other algorithms besides backpropagation have been also proposed to optimize the configuration of the wavelet network, e.g., Kalman filter [48], genetic algorithms [49] and immune algorithms [50].

Nevertheless, backpropagation remains the most popular algorithm in the training process. This algorithm proposed by Rumelhart [51] consists on the derivation of the delta rule, which allows updating the weights of the network whenever a training vector is presented to the network. A variation of this rule is called least minimum square rule; in this variation the weights of the network are updated when all the training vectors have been input to the network (this incremental is named an epoch). As the aim of training is diminishing the difference between the outputs of the network and the expected values, this difference is evaluated according with the Mean Squared Error (MSE) defined by equation (6):

$$J(\Omega) = \tfrac{1}{2}\sum_{n=1}^{N}\left(y_{\exp}^{n} - y^{n}\right)^{2} = \tfrac{1}{2}\sum_{n=1}^{N}\left(e^{n}\right)^{2} \tag{6}$$

where y^n is the output of the network and $y_{\exp}^n$ is the expected output value related to the input vector x^n.

Since the proposed model is multivariate, we define Ω as the parameters set which will be adjusted during training stage so that y^n maps a value from $\Re^P \rightarrow \Re^1$:

$$\Omega = \{b_0, w_j, c_i, t_{ij}, s_{ij}\} \qquad i = 1,..,K\,;\, j = 1,...,P \tag{7}$$

These parameters are changed in the direction determined by the negative of the output error function's gradient.

$$\frac{\partial J}{\partial \Omega} = -\sum_{n=1}^{N} e^{n}\frac{\partial y^{n}}{\partial \Omega} \text{ where } \frac{\partial y^{n}}{\partial \Omega} = \left.\frac{\partial y}{\partial \Omega}\right|_{x=x^{n}} \tag{8}$$

In addition, we propose to average these changes with the number *N*, of input vectors, in order to obtain a weighted error

$$\frac{\partial J}{\partial \Omega} = -\frac{1}{N}\sum_{n=1}^{N} e^{n}\frac{\partial y^{n}}{\partial \Omega} \tag{9}$$

The partial derivaties of y^n for the set of parameters in Ω are indicated in equations (10-14)

$$\frac{\partial y^n}{\partial b_0} = 1 \tag{10}$$

$$\frac{\partial y^n}{\partial w_j} = x_j^n \tag{11}$$

$$\frac{\partial y^n}{\partial c_i} = \Psi_i(x^n) \tag{12}$$

$$\frac{\partial y^n}{\partial t_{ij}} = -\frac{c_i}{s_{ij}} \frac{\partial \Psi_i}{\partial a_{ij}}\bigg|_{x=x^n} \tag{13}$$

$$\frac{\partial y^n}{\partial s_{ij}} = -\frac{c_i}{s_{ij}} a_{ij} \frac{\partial \Psi_i}{\partial a_{ij}}\bigg|_{x=x^n} \qquad i = 1,\ldots,K \quad j = 1,\ldots,P \tag{14}$$

in the last two equations

$$\frac{\partial \Psi_i}{\partial a_{ij}}\bigg|_{x=x^n} = \psi(a_{i1}^n)\psi(a_{i2}^n)\ldots\psi'(a_{ij}^n)\ldots\psi(a_{iP}^n) \tag{15}$$

where $\psi'(a_{ij}^n)$ is the value taken by the derivative of the mother wavelet at the point a_{ij}^n.

The changes in network parameters are calculated in each iteration according to:

$$\Delta\Omega = -\mu \frac{\partial J}{\partial \Omega} \tag{16}$$

where μ is a positive real value known as the learning rate.

Whit these changes the variables contained in Ω are updated using:

$$\Omega_{new} = \Omega_{old} + \Delta\Omega \tag{17}$$

where Ω_{old} represents the current values, $\Delta\Omega$ represents the changes and Ω_{new} corresponds to the new value after each iteration.

An important point in training process is related with the proper initialization of network parameters. In this sense, many algorithms are suggested to initialize the weights [52], adjust the structures of wavelet networks [53] and accelerate the convergence when the error backpropagation is applied to the training algorithms [7]. Appropriate initialization values allow convergence of training error. Considering a range in input vectors defined by the domain $\lfloor x_{j,\min}, x_{j,\max} \rfloor$, where $x_{j,\min}$ y $x_{j,\max}$ represent the minimum and maximum values in training vectors after normalization stage to an interval of $[-1,1]$, in order to obtain fast training. The initial values of the i-th neuron for translation and scaling parameters are set to $t_{ij} = 0.5(x_{j,\min} + x_{j,\max})$ and $s_{ij} = 0.2(x_{j,\max} - x_{j,\min})$, respectively, to guarantee that wavelets are not focusing in localities of the input universe. The weights connections between layers w_j and c_i are proposed as random starting values since its initialization is less critical than the translation and scaling variables.

An unsolved problem in the design of WNN is related with the curse of dimensionality, which brings some difficulties in applying this kind of networks to high-dimensional problems. In this sense, the determination of the number of hidden neurons and the correct initialization is an important issue. As in conventional ANN models, an optimized architecture checked by systematic evaluation of possible configurations, is still important to promote generalization and avoid overfitting.

WNN Prediction Models in Chemical Sensing

During the last years, the electroanalytical methods have taken advantage for the development of sensors or biosensors usable in chemical analysis. For their proper utilization data processing methods are needed to improve tasks such as the interpretation, correlation and modeling of measured chemical responses based on properties of a set of samples, e.g., concentration, intensity, potential, mass, spectra, etc. [54, 55]. Electronic tongues are analytical chemical systems, which employ electrochemical sensors in a novel way, having as main foundation the sense of taste in animals. A widely accepted definition of electronic tongue entails an analytical instrument comprising an array of non-specific, poorly selective, chemical sensors with cross-sensitivity to different compounds in a solution, and an appropriate chemometric tool for data processing [56]. The low selectivity of sensors is what precisely produces complex cross-response signals containing information about different compounds plus other features. Due to the lack of selectivity, an important part in a multisensor approach is the signal processing stage.

In the recent years, there have been published contributions on the use of electronic tongues for environmental monitoring purposes [57-62]. Our research group proposed recently the use of voltammetric electronic tongues and WNN data processing as potential monitor of pollution episodes by heavy metals [28].

The study presented here, corresponds to the direct multivariate determination of three oxidizable compounds namely ascorbic acid, 4-aminophenol and paracetamol which present overlapping signals coming from a linear sweep voltammetric technique obtained with a laboratory made sensor electrode (consisting of a composite of epoxy-graphite).

Chemical Procedure

The three oxidizable compounds Ascorbic Acid (AA), 4-Aminophenol (4-Aph) and Parcetamol (Pct) are analyzed by the proposed voltammetric e-tongue. Despite the sensitivity of the method, the obtained voltammetric signals have a high degree of overlapped response, making difficult the determination of each component. A total of 60 samples with different concentrations were prepared in the intervals (mM) [0.012,0.373] for AA, [0.017,0529] for 4-Aph and [0.010,0.424] for Pct, in order to train the system and generate its response model. The different samples were prepared using automated analytical system based on the Sequential Injection Analysis (SIA) developed and validated in our laboratories. This system employs a bi-directional microburette (Precision Syringe drive/2 Module, Hamilton, Switzerland) equipped with a syringe of 2.5 ml (Hamilton) plus a motorized MVP selection valve (Hamilton) with a 6 way valve head, HVXM 6-5 (Hamilton). PTFE tubing (Bioblock, France) with 1 mm i.d. was used to connect the different SIA components and as holding coil. Connections were completed with low-pressure chromatography fittings. A specially designed mixing cell, made in Perspex, was used in the sample preparation. Automatic stirring was used to assure homogeneity and absence of diffusion effects in the sample matrix. The whole SIA system was controlled with inhouse software programmed in BASIC (Quick-Basic, Microsoft, USA). Electrochemical measurements were developed using an Autolab/PGSTAT20 system (Ecochemie, Netherlands), with a specially designed voltammetric flow cell. Reference electrode consisted in an Ag/AgCl electrode (Model 52-40 Crison Instruments, Spain). Stainless steel used as a constitutional part of the measuring cell was used as counter electrode. The working electrode consisted in a home-made epoxy–graphite transducer, of general use in our laboratories. Voltrammograms were measured applying a scan potential from 0 to 1.0V at 0.1V/s in steps of 10mV. Figure 3 summarizes the nature of the different signals generated, each one corresponding to a trinary mixture of the oxidizable compoinds is considered.

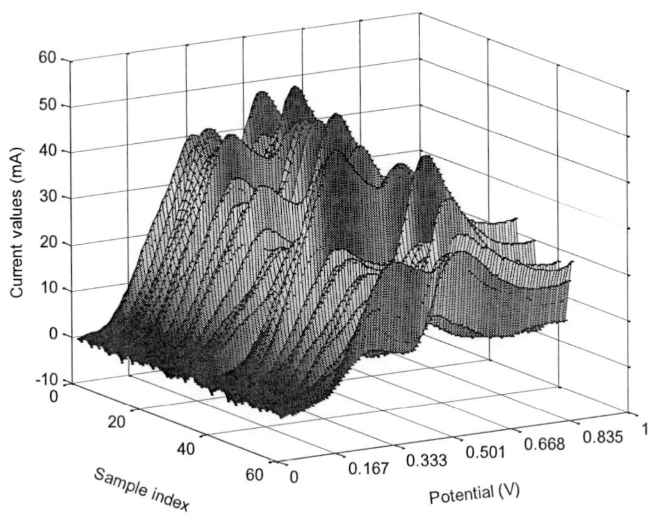

Figure 3. Measured signals obtained with the generated standards.

Data Processing

With the 60 samples determined, the input data for the WNN models is a voltammetric matrix of dimension [101x60], and the target is a [3x60] matrix (AA, 4-Aph and Pct concentrations). Prior to any calculation, the processed information was normalized to an interval of [-1,1], and randomly separated in two sets; 70% of the data were used for training and the remaining 30% for testing. Each voltammogram is represented by x^n which is mapped to a point of the five-dimensional space of concentrations identified by y^n . In this way, three WNN architectures with dimension of 101x3x1 were programmed and trained to obtain the concentration of each oxidizable compound (see Figure 4). Structures with more neurons in the hidden layer were not tested because these first architectures allowed to obtain satisfactory results, as reported in previous works [4, 28]. The expected output error was programmed to reach a value of 0.001 mM and the learning rate was set to 0.001. Prediction capability of the model was evaluated using k-fold cross validation method, choosing different training and testing sets, in order to validate the reliability of the model in an unconditioned way.

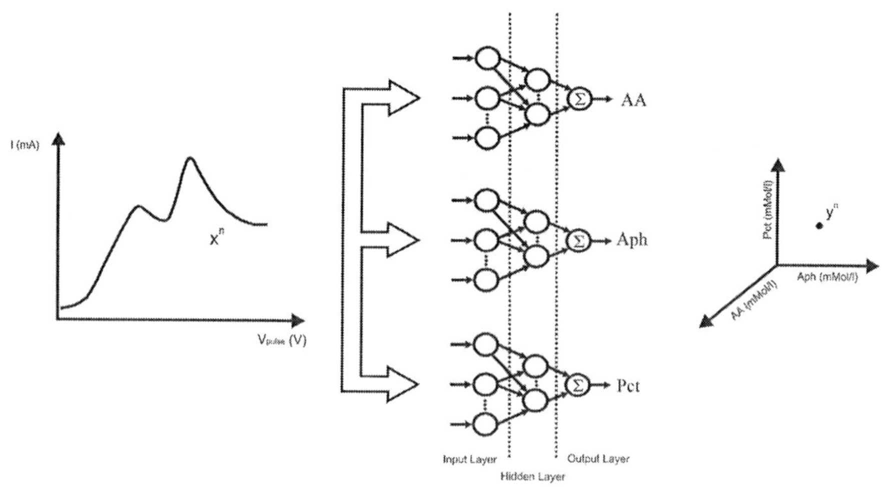

Figure 4. The voltammogram is entered in parallel in three independent WNN netwoks in order to obtain the corresponding concentration values of each compound present in the sample.

Obtained Results

To determine the WNN performance in quantification of oxidizable compounds, different comparison graphs between obtained and expected concentrations were built, both for training and for testing subsets. The linear regression of the obtained vs. expected concentration comparison graph was a measure of the goodness of the model for each compound. This fitted line, in ideal conditions, should yield the diagonal identity line (slope 1 and intercept 0) and a correlation coefficient close to 1. Figure 4 shows the comparative graphs between real concentrations for AA, 4-Aph and Pct and those predicted with the WNN models for training and testing subsets. The slope (m) and intercept (b) which defines the comparison line y=mx+b that best fits the data (along with the uncertainly intervals) are shown for one of the study cases in Table 1, where the ideal situation is fulfilled in all cases (at the 95% confidence level). Simultaneously, the high level of linearity accomplished, where all correlation coefficients were larger than 0.947, permits to express the low dispersion of obtained data.

Table 1. Linear regression parameters for tested oxidizable compounds with the training and testing data sets using WNN architectures whith 3 neurons in the hidden layer

Compound	Training			Testing		
	m	*b*	*R*	*m*	*b*	*R*
AA	0.951±0.056	0.010±0.013	0.983	1.076±0.200	0.008±0.0.042	0.947
4-Aph	0.956±0.053	0.012±0.016	0.985	1.002±0.170	0.011±0.047	0.965
Pct	0.964±0.053	0.008±0.013	0.985	0.906±0.105	0.019±0.024	0.979

Two extra training procedures, using the same WNN architectures, were programmed with random initialization of weights to verify the final model's consistency. Employing a 17-fold cross validation method, a testing set was selected each time randomly from the total set of the available data. To compare the accuracy of the predicted information, an average Recovery Percentage (RP) was calculated for each WNN model. RP is defined by equation (18):

$$RP = \frac{\sum_{i=1}^{N} 100 \cdot \left(1 + \frac{y_i - y_{\exp_i}}{y_{\exp_i}} \right)}{N} \quad (18)$$

where y_i is the *i*-th obtained concentration value, $y_{\exp_i}$ is the *i*-th expected concentration value and N is the size of the external test set.

Table 2. Recovery percentage for the three trainings of the WNN architectures whit three neurons in the hidden layer

Compound	Training					Testing				
	Case 1	Case 2	Case 3	Mean	RSD(%)	Case 1	Case 2	Case 3	Mean	RSD(%)
AA	101.94	101.94	102.72	102.20	0.44	103.82	103.82	103.56	103.73	0.14
4-Aph	103.67	103.43	104.03	103.71	0.29	106.33	104.74	105.74	105.67	0.78
Pct	96.82	97.32	97.70	97.28	0.45	105.83	106.59	102.88	104.10	1.48

RP value is a parameter which indicates the ability of the model to determine quantitatively a chemical analyte present in a sample; the ideal value is 100% [7]. The obtained results for this case study are summarized in Table 2, including the mean and Relative Standard Deviation (RSD) for each compound. It is possible to observe the fluctuation of the RP for each analyte in the different WNN models. RP's value is a consequence of the random selection of the data, which in the worst case never exceeded the 5% of the ideal RP. At the same time, the accuracy of the predicted information is always below a 5% RSD uncertainly, and the ideal 100% situation is always attained. This jacknife approach ensures the models' validity regardless of the data selection used or its construction.

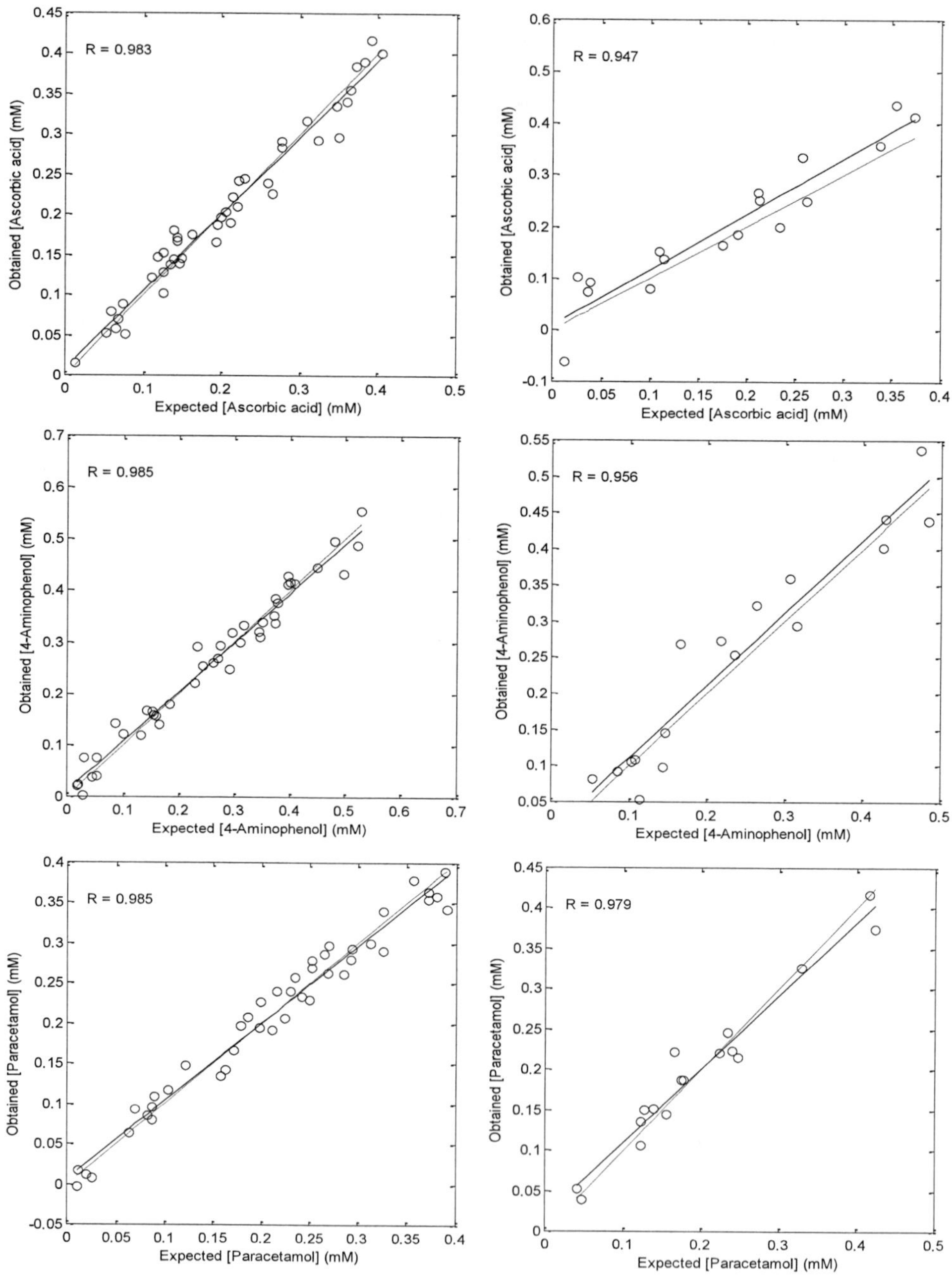

Figure 5. Comparison graphs between expected results and those obtained with the WNN models. The graphs correspond to the three species under study. The dashed line corresponds to ideality ($y=x$) and the solid line is the regression of the comparison data. Plots on the left correspond to training and plots on the right to testing.

WNN Models for EMG Classification

Electromyography is the detection of the electrical activity associated with muscle contraction. An electromyographic (EMG) signal is obtained by measurement of the electrical activity of a muscle during contraction. In this sense, EMG reflects the electrical depolarization of excitable muscle fiber membranes which create electrical signals called muscle fiber potentials (MFPs) [63].

The main problem in electromyography analysis is related with the composition of the electrical signal detected during a muscle contraction. In normal mammalian skeletal muscle, the fiber never contract as individuals. Instead, small groups of them contract in concert. All the fibers of each group of muscle fibers are controlled by the terminal branches of one nerve fiber or axon whose cell body is in the anterior horn of the spinal cord grey matter. These cells are known as motoneurons and directly innervate skeletal muscle fibers.

Motoneurons consisting of the nerve cell body, the long axon running down the motor nerve with its terminal branches, and all the muscle fibers controlled by these branches constitute a motor unit (MU). The summation of all of a MU's spatially and temporally dispersed results in a signal called a motor unit potential (MUP). In order to sustain a muscle contraction, motor units must be repeatedly activated and each motor unit generates multiple MUPs. The collection of MUPs generated by one motor unit, positioned at their times of occurrence or separated by their inter-discharge intervals (IDIs) is called a motor unit potential train (MUPT). The superposition of the MUPTs of all recruited motor units and background noise comprises an EMG signal [63].

Clinically, EMG signal analysis, in the form of EMG signal decomposition and MUP classification into groups of similar shapes, is used to assist in the diagnosis of neuromuscular disorders, to analyze the neuromuscular system, and in biofeedback training.

The characteristics of an EMG signal are largely affected by anatomical and physiological properties of the muscle. For example, when the force in muscular contractions increases, the number of active MUs and the rate at which they are active increases [64]. The EMG signal therefore becomes more complex with increasing force of contraction. Furthermore, the fundamental structure of a muscle such as the size, distribution and number of motor units and how they are controlled can also be reflected in the characteristics of an EMG signal. For these reasons, many researchers are interested in devising techniques for the quantitative analysis of EMG signals.

The study presented here, is related with the use of WNN classifier responsible for the identification of three different force levels in an EMG signal during isometric muscular contraction. Considering the ability of WNN models to extract significant features of signals, this classification strategy will employ raw EMG records segmented in analysis windows that assure the presence of at least two MU spikes.

Acquisition EMG Procedure

The measurements were done in our laboratory following an internal clinical management. A total of 3 healthy subjects (without apparent neuromuscular problems) were selected with ages ranging from 25 to 28 years. EMG was recorded using a needle electrode inserted in the First Dorsal Interosseous (FDI) muscle and a reference Ag/AgCl disc electrode

(Nicolet, 9mm diameter) on the wrist. The obtained signal is acquired employing an isolated pre-amplifier (Axon Bioamp 100) and finally amplified whit a gain of x100 (Axon, CyberAmp 380 , USA). The obtained signal was filtered using a fourth order band pass filter (2kHz and 10kHz cutoff frequency). Digitalization stage was performed using a 16 bits analog to digital board (Digidata 1322A , USA). Once the needle electrode is inserted in the FDI, MU must be found to initiate voluntary isometric muscle contraction. The localization of a MU was done using an acoustic monitor as feedback. Demand for FDI muscle strength is done bending and ascending the index finger gradually (Figure 5). Three different force levels were established by engaging the subject's finger to a dynamometer (Ohaus 8263M, USA). The defined levels were associated with 250mg, 750mg and 1250mg force levels. Each FDI contraction at specific force level had a duration of 10 s.

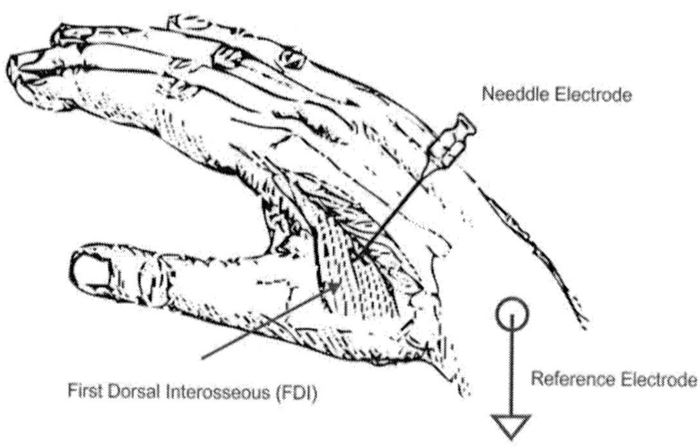

Figure 6. Needle electrode and its point of insertion in the FDI muscle.

WNN Modeling Procedure

The whole acquired information contained 255000 samples per subject. For building WNN models each sample represents an input of the structure. In order to manage a reduced number of samples a down-sampling procedure was carried out to decrease the recorded dimension. Each final force record dimension contained 10625 samples (related whit a sample rate of 1250 Hz). A WNN classifier was devised to identify between three different force levels according with the MU firing frequency. Before training a WNN model, the training data set must be defined. The objective of this step is considering the presence of at least two continuous MU in each training record. The information was segmented considering a MU firing frequency ranging from 6Hz to 20Hz [63]. In this way, 20 analysis windows with 512 samples each were formed by each force level. Final training data matrix had a dimension of [512,60] following the sequence ($Level1_1$, $Level2_1$, $Level3_1$... $Level1_{20}$, $Level2_{20}$, $Level3_{20}$). The objective vector contained the force level (i.e. 0,1,2…0,1,2…etc) until completing a dimension of [1,60].

The classification model was built using a single output WNN. The output is segmented according to the following classification ranges: [-0.25,0.25] for Level 1, [0.75,1.25] for

Level 2 and [1.75,2.25] for Level 3. Intermediate ranges were considered as uncertain, thus any value inside these ranges was identified as unclassified although might belong to any of the defined levels. Figure 6, shows the scheme of the proposed classification strategy.

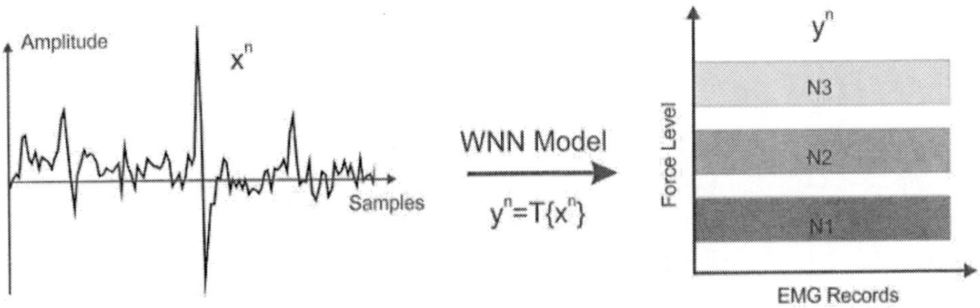

Figure 7. EMG mapping record to a force level band.

Test sets were obtained from original recordings plus white noise. Added noise is done using a percentage of 25%, 30% and 50% equivalent to the absolute signal power defined by P_{dBW}. This value is evaluated from the original record (before segmentation) using the equations (19) and (20):

$$P_{dBW} = 10\log_{10} P \qquad (19)$$

where P is

$$P = \frac{1}{N}\sum_{n=1}^{N} |x(n)|^2 \qquad (20)$$

where P is the recording average power, $x(n)$ the n-th element of the record and N the number of elements.

In this way, the signal to noise ratio is constant and added noise contamination of the recorded elements is proportional to its energy content. All percentages of noise were established with the purpose to identify appreciable changes in the original recorded composition. Either the composition of the test set or the dimension were developed as described for the training set.

Obtained Results

Preliminary tests allowed to establish the proper number of hidden neurons in WNN architecture. Some parameters, as learning rate, epochs and training error were modified in a trial and error process, in order to find the best configuration to ensure the good performance of the classification model. Final configuration of the application model had six neurons in hidden layer with a learning rate set to a value of 0.0005, a training error of 0.01 and 10000 epochs.

After the WNN training phase for each subject, designated as alpha, beta and gamma, linear regression analysis was performed in order to evaluate correlation coefficients values obtained either in training or test phases. These results are shown in Figure 7.

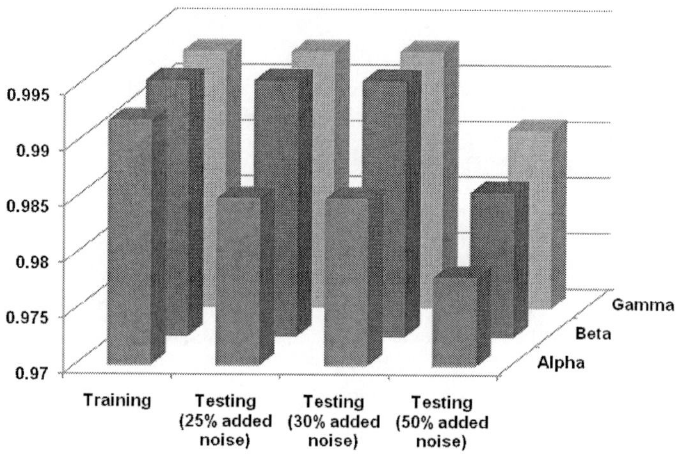

Figure 8. Correlation coefficients obtained for training and testing phases. The study presents three subjects and the classification of three different force levels.

Obtained correlation coefficients above 0.9 indicate the good performance of the WNN in the classification of the different force levels. The fact that the correlation coefficient did not show large variations with the increasing noise percentage added to original records, is a good indicator demonstrating the network's immunity to random signals and confirming the architecture robustness. Figure 8 shows the WNN behavior with the training and test vectors obtained for the subject alpha. As can be observed, records are properly classified within the three defined ranges for each force level; the vectors that will inevitably fall outside the defined ranges are the cause of the correlation coefficient deterioration.

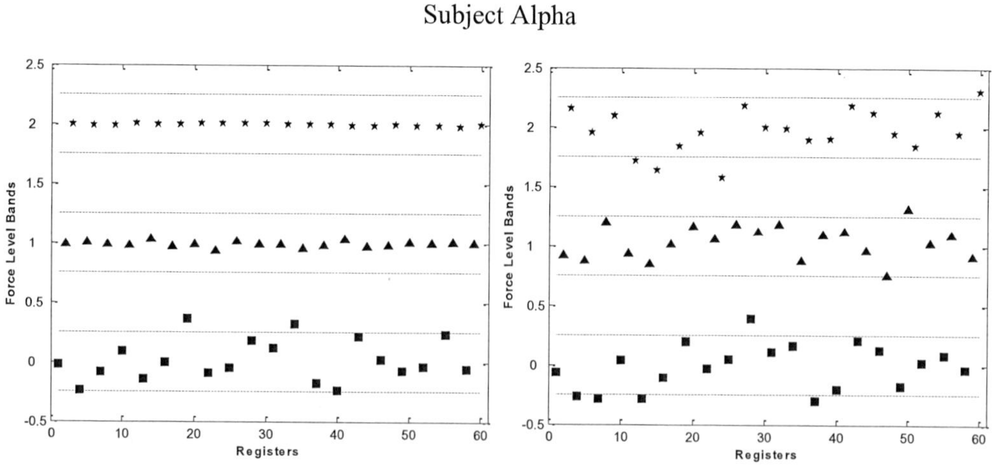

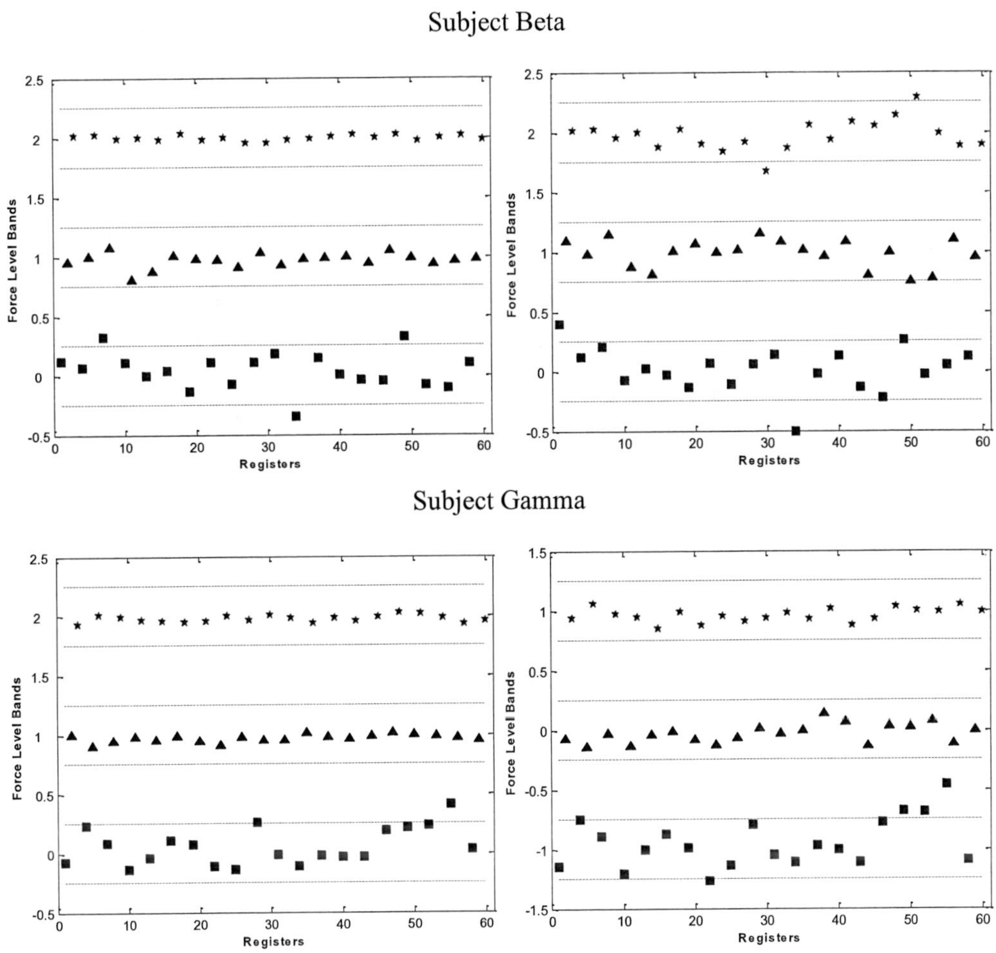

Figure 9. WNN results in the classification of EMG signal in force levels. The dotted lines correspond to the ranges chosen for each class. Plot at the left corresponds to training and plot at the right to testing. These results were obtained using original records corrupted with 50% of noise. N1, N2 and N3 correspond to eah force level, respectively.

Another way to corroborate the model's performance is to evaluate the percentage classification. Confusion matrices were built using the information of the testing subsets (25%, 30% and 50% of added noise) obtained for each subject, in order to characterize the accuracy of the identification model. Table 3 shows the results of these classifications, expressed as the corresponding confusion matrix; its content displays expected vs. found response of each force level. The cells in the table without highlighting are related to force levels which are confused with an incorrect level. Last column in the table represents the percentage of vectors identified as unclassified.

Table 3. Confusion matrix obtained using a WNN classifier with 6 neurons in the hidden layer for the three subjects. The model was able to discriminate correctly between three force levels in the test. The classification rate decreased slightly with the increment of noise

	Subject Alpha				Subject Beta				Subject Gamma			
Expected Force Level	*Testing (25% added noise)*				*Testing (25% added noise)*				*Testing (25% added noise)*			
	N1	N2	N3	Unclassified	N1	N2	N3	Unclassified	N1	N2	N3	Unclassified
N1	75	0	0	25	85	0	0	15	90	0	0	10
N2	0	95	0	5	0	100	0	0	0	100	0	0
N3	0	0	100	0	0	0	100	0	0	0	100	0
Percentage classification	90%				95%				96.97%			
Expected Force Level	*Testing (30% added noise)*				*Testing (30% added noise)*				*Testing (30% added noise)*			
	N1	N2	N3	Unclassified	N1	N2	N3	Unclassified	N1	N2	N3	Unclassified
N1	70	0	0	30	85	0	0	15	90	0	0	10
N2	0	90	0	10	0	100	0	0	0	100	0	0
N3	0	0	100	0	0	0	100	0	0	0	100	0
Percentage classification	86.67%				95%				96.97%			
Expected Force Level	*Testing (50% added noise)*				*Testing (50% added noise)*				*Testing (50% added noise)*			
	N1	N2	N3	Unclassified	N1	N2	N3	Unclassified	N1	N2	N3	Unclassified
N1	75	0	0	25	85	0	0	90	85	0	0	10
N2	0	95	0	5	0	100	0	0	0	100	0	0
N3	0	0	80	20	0	0	90	10	0	0	100	0
Percentage classification	88.33%				91.67%				96.97%			

Besides knowing percentage classification obtained by the WNN models, it is also important to evaluate the classification error; this value is calculated by the following equation:

$$e = \frac{\sum_{i=1}^{N}(t_i - \Gamma(o_i))^2}{N} \cdot 100 \tag{21}$$

with

$$\Gamma(o_i) = \begin{cases} 0 & \forall \quad -0.25 < t_i < 0.25 \\ 1 & \forall \quad 0.75 < t_i < 1.25 \\ 2 & \forall \quad 1.75 < t_i < 2.25 \end{cases} \tag{22}$$

where *N* is the number samples per record, o_i is the obtained output at the point i, t_i is the expected classification output and $\Gamma(o_i)$ is the modified output according with the classified ranges.

Thus, the classification error for the test sessions and for each force level are evaluated and presented in Table 4.

Table 4. Classification error of the three force levels

	Force Level	*Classification Error*		
		Alpha	*Beta*	*Gamma*
Training	N1	1.18	1.65	1.17
	N2	0.00	0.00	0.00
	N3	0.00	0.00	0.00
Testing (25% added noise)	N1	2.61	1.62	1.20
	N2	0.39	0.00	0.00
	N3	0.00	0.00	0.00
Testing (30% added noise)	N1	3.15	1.54	1.20
	N1	0.78	0.00	0.00
	N3	0.00	0.00	0.00
Testing (50% added noise)	N1	2.38	2.41	3.12
	N2	0.49	0.00	0.00
	N3	2.43	0.00	0.00

CONCLUSION

Wavelet Neural Networks (WNNs) are powerful processing tools thanks to the rigorous mathematical foundations of their origin. We have described two different advanced signal processing applications. The first was aimed to quantitatively determine the concentration of chemical species based on information obtained from voltammetric sensors and the second was oriented to the classification of force levels coming from EMG records. In this sense, WNN models allow the processing of nonstationarities present in the signals and capture the essential characteristics in the time-frequency domain. The use of continuous wavelet frames and backpropagation algorithm were proved to be enough to build proper quantification and classification models for aplicactions related with analytical chemistry and biosignal analysis. In the first case, the adjusted parameters of WNN and the wavelet family functions implicit in the model were able to find relationships in strongly overlapped signals and amounts of oxidizable compounds, permitting the resolution of the trinary mixture in a application typical of sophisticated analytical instrumentation such as HPLC. In the second case, even departing from recordings highly contaminated by noise, a firing rate pattern of MUs spikes could be identified clearly in the EMG signal and related with different levels of force for three different subjects.

ACKNOWLEDGMENTS

Financial support for this work was provided by Mexican National Council of Science and Technology CONACYT, the Spanish Ministry of Science and Innovation through project TEC2007-68012-C03-02 and the program ICREA Academia from Generalitat de Catalunya.

REFERENCES

[1] N. Mastrogiannis, B. Boutsinas, I. Giannikos, *Compu. Oper. Res.* 36, 2829 (2009).
[2] P.K. Hopke, *Anal. Chim. Acta*, 500, 365 (2003).
[3] X. -G. Shao, A.K,-M. Leung, F.-T. Chau, *Acc. Chem. Res.* 36, 276 (2003).
[4] J.M. Gutiérrez, A. Gutés, F. Céspedes, M. del Valle, R. Muñoz, *Talanta* 76, 373 (2008).
[5] I.A. Gheyas, L.S. Smith, *Pattern Recogn.* 43, 5 (2010).
[6] L. Moreno-Barón, R. Cartas, A. Merkoçi, S. Alegret, J.M. Gutiérrez, L. Leija, P.R. Hernandez, R. Muñoz, M. del Valle, *Anal. Lett.* 38, 2189 (2005).
[7] A. Gutés, F. Céspedes, R. Cartas, S. Alegret, M. del Valle, J.M. Gutiérrez, R. Muñoz, *Chemom. Intell. Lab. Syst.* 83, 169 (2006).
[8] J.A. Freeman and D.M: Skapura, *Neura Networks: Algorithms, Applications and Programming Techniques*, C. Koch (Ed.) Addison-Wesley Publishing Company, Reading, MA, USA,1991.
[9] Q. Zhang, A. Benveniste, *IEEE Trans. Neural Netw.* 3, 889 (1992).
[10] J. Zhang, G.G. Walter, Y. Miao, W.N.W. Lee, *IEEE Trans. Signal Proces.* 43, 1485 (1995).
[11] H. Szu, B. Telfer, J. Garcia, *Neural Netw.* 9, 695 (1996).
[12] H. Szu, C. Hsu, D.-H., Xie, *Proceedings of the SPIE,* Vol. 3391, Orlando, FL, USA,1998, pp 393-408.
[13] V. Kruger, A. Happe, G. Sommer, *Proceedings of 15th International Conference on Pattern Recognition*, Barcelona, Spain, 2000, pp 127-130.
[14] B.E. Miller, J. E. Colgate, *Proceedings of the ASME Dynamic Systems and Control Division,* Anaheim, CA, USA, 1998, 64, pp 257-264.
[15] C. Liangyue, H. Yiguang, F. Haiping, H. Guowei, *Phys. D.* 85, 225 (1995).
[16] D. Katic, M. Vukobratovic, *Proceedings of the 1997 IEEE International Symposium on Intelligent Control*, Istanbul, Turkey, 1994, pp 245-250.
[17] Q.X. Guo, L. Liu, W.S. Cai, Y. Jiang, Y.C. Liu, *Chem. Phys. Lett.* 290, 514 (1998).

[18] X. Zhang, J. Qi, R. Zhang, M. Liu, Z. Hu, H. Xue, B. Fan, *Comput. Chem.* 25, 125 (2001).
[19] M. Atabati, K. Zarei, *J. Chinese Chem. Soc.* 55, 732 (2008).
[20] T. Khayamian, M. Esteki, *J. Supercrit. Fluids*, 32, 73 (2004).
[21] R. Tabaraki, T. Khayamian, A.A. Ensafi, *J. Mol. Graph. Modell.* 25, 46 (2006).
[22] R. Tabaraki, T. Khayamian, A.A. Ensafi, *Dyes Pigm.* 73, 230 (2007).
[23] T. Khayamian, A.A. Ensafi, R. Tabaraki, M. Esteki, *Anal. Lett.* 38, 1477 (2005).
[24] K. Zarei, M. Atabati, *Anal. Lett.* 39, 2085 (2006).
[25] A.A. Ensafi, T. Khayamian, R. Tabaraki, *Talanta*, 71, 2021 (2007).
[26] H. Zhong, J. Zhang, M. Gao, J. Zheng, G. Li, L. Chen, *Chemom. Intell. Lab. Sys.* 59, 67 (2001).
[27] T. Khayamian, A.A. Ensafi, A. Benvidi, *Talanta*, 69 1176 (2006).
[28] J.M. Gutiérrez, L. Moreno-Barón, F. Céspedes, R. Muñoz, M del Valle, *Electroanal.* 21, 445 (2009).
[29] Z. Garkani-Nejad, H. Rashidi-Nodeh, *Electrochim. Acta*, 55, 2597 (2010).
[30] J.G. Webster, *Medical Instrumentation: Application and Design* 3rd Edition, J.G. Webster (Ed.), John Wiley and Sons, NY, USA, 1997.
[31] J.R. Cameron, J. G. Skofronick, R.M. Grant, *Physics of the Body* 2nd Edition. Medical Physics Publishing Corporation, 1999.
[32] O. Rioul, M. Vetterli, *IEEE Signal Proc. Mag.*, 8,14 (1991).
[33] A. Nait-Ali, *Advanced Biosignal Processing*, A. Nait-Ali (Ed.), Springer-Verlag Berlin Heidelberg, Germany, 2009.
[34] A. Subasi, A. Alkan, E. Koklukaya, M.K. Kiymik, *Neural Netw.* 18, 985 (2005).
[35] A. Subasi, M. Yilmaz, H.R. Ozcalik, *J. Neurosci. Meth.* 156, 360 (2006).
[36] Y. Meyer, Wavelets, *Algorithms and Applications. Society for Industrial and Applied Mathematics* 2nd Edition, SIAM, Philadelphia, PA, USA, 1994.
[37] M. Akay, Time frequency and wavelets; In M. Akay (Ed.) *Biomedical Signal Processing. IEEE Press Series in Biomedical Engineering,* Wiley-IEEE Press: Piscataway, NJ, 1992.
[38] F. Scarselli, A. Chung Tsoi, *Neural Netw.* 11, 15 (1998).
[39] C. Heil, D.F. Walnut, *SIAM Review* 31, 628 (1989).
[40] K. Hornik, M. Stinchcombe, H. White, *Neural Netw.* 2, 359 (1989).
[41] I. Daubechies, *Ten Lectures of wavelets. CBMS-NSF Regional Conference Series In Applied Mathematics,* Vol. 61. Society for Industrial an Applied Mathematics, Philadelphia, PA, USA,1992.
[42] G. Kaiser, A Friendy Guide to Wavelets. Ed Birkhäuser, Boston, MA, USA, 1994.
[43] I. Daubechies, A. Grossmann, Y. Meyer, *J. Math. Phys.* 27, 1271 (1986).
[44] T. Kugarajah, Q. Zhang, *IEEE Trans. Neural Netw.* 6, 1552 (1995).
[45] Y. Oussar, I. Rivals, L. Personnaz, G. Dreyfus, *Neurocomputing*, 20, 173 (1998).
[46] Z. Zhang, *Neurocomputing* 71, 244 (2007).
[47] Q. Zhang, *IEEE Trans. Neural Netw.* 8, 227 (1997).
[48] S. Qingmei, G. Yongchao, Proceedings of the International Conference on Wavelet Analysis and Its Applications (WAA), Chongqing, China, 2003, pp 633–638.
[49] L. Chang Gyoon, K. Kangchui, K.; Eungkon, *Proceedings of the Fifteenth IASTED International Conference on Modeling and Simulation*, CA, USA, 2004, pp 55-59.

[50] M. Huang, B. Cui, A Novel learning algorithm for wavelet neural networks, In L. Wang, K. Chen, Y.S. Ong (Eds.), *Advances in natural computation ICNC* 2005, Springer-Verlang, Berlin, Germany, 2005, 1-7.

[51] D.E. Rumelhart, G. E. Hilton, R. J. Williams, *Learning internal representations by error propagation.* In D.E. Rumelhart, J.L McClelland (Eds.), Parallel Distributed Processing: Exploration in the microstructure of cognition, MIT Press, vol. 1, Cambridge, MA, USA, 1986.

[52] Y. Oussar, G. Dreyfus, *Neurocomputing* 34, 131 (2000).

[53] B. Yevgeniy, L. Nataliya, P. Iryna, V. Olena, *Expert Syst.* 22, 235 (2005).

[54] S. Wold, *Chemom. Intell. Lab. Syst.*, 30, 109 (1995).

[55] Y. Ni, S. Kokot, *Anal. Chim. Acta*, 626, 130 (2008).

[56] M. Holmberg, M. Eriksson, C. Krantz-Rülcker, T. Artursson, F. Winquist, A. Lloyd-Spetz, I. Lundström, *Sens. Actuators B.*, 101, 213 (2004).

[57] A. Mimendia, J.M. Gutiérrez,L. Leija, P.R. Hernández, L. Favari, R. Muñoz, M. del Valle, *Environ. Modell. Softw.* 25, 1023 (2010).

[58] C. Di Natale, A. Macagnano, F. Davide, A. D'Amico, A.; Legin, Y. Vlasov, A. Rudnitskaya, B. Selezenev, *Sens. Actuators B.* 44, 423 (1997).

[59] A.V. Legin, Y.G. Vlasov, A.M Rudnitskaya, E.A. Bychkov, *Sens. Actuators B.* 34, 456 (1996).

[60] C. Krantz-Rülcker, M. Stenberg, F. Winquist, I. Lundström, *Anal. Chim. Acta* 426, 217 (2001).

[61] F. Winquist, R. Bjorklund, C. Krantz-Rülcker, I. Lundström, K. Östergren, T. Skoglund, *Sens. Actuators B.*, 111-112, 299 (2005).

[62] M. Gutiérrez, J. M. Gutiérrez, L. Leija, P.R. Hernández, L. Favari, R. Muñoz, M. del Valle, *Inter. J. Environ. Anal. Chem.* 88, 103 (2007).

[63] J. Basmajian, C.J. De Luca, *Muscles Alive. Their Functions Revealed by Electromyography* 5th Edition, J. Butler (Ed.), Williams and Wilkinns Baltimore, MD, USA,1985.

[64] M. Aminoff, *Electromyography in Clinical Practice* 3rd Edition; M. Aminoff (Ed.) Churchill-Livingstone NY, 1998.

In: Wavelets: Classification, Theory and Applications
Editors: Manel del Valle et al.
ISBN 978-1-62100-252-9

Chapter 11

WAVELETS AND FRAMES IN WALSH ANALYSIS

Yu. A. Farkov*
Department of Mathematics at
Russian State Geological Prospecting University, Moscow, Russia

Abstract

It is well known that Walsh functions can be identified with characters of the Cantor dyadic group. We give a review of recent results on wavelets and frames for this group. Among the main subjects to be discussed are the following:

- algorithms to construct orthogonal and biorthogonal wavelets associated with the Walsh polynomials;
- estimates of the smoothness of dyadic orthogonal wavelets of Daubechies type;
- an algorithm for constructing Parseval dyadic frames.

Orthogonal and biorthogonal wavelets for Vilenkin groups are also discussed.

Introduction

The Walsh functions can be considered as characters of the Cantor dyadic group. This fact was first recognized by Gelfand in 1940s, who offered to Vilenkin study series with respect to characters of a large class of abelian groups which includes the Cantor dyadic group as a special case (see [1, 2, 3]). The basic properties of the Cantor dyadic group are given in PontryaginŠs book [4]. Concerning its applications to the theory of trigonometric Fourier series see, e.g., [5]. At present the Walsh analysis is an actively developing domain of the harmonic analysis. In the books [6, 7] and [8] among the main subjects are the following:

- convergence and summability of Walsh-Fourier series;
- approximation by Walsh and Haar polynomials;
- Walsh-Fourier analysis of Boolean functions;

*E-mail address: farkov@list.ru

- derivatives and integrals in Walsh analysis;
- dyadic martingales and Hardy spaces;
- dyadic Hardy and Hardy-Littlewood operators.

At the Workshop on Discrete Analysis and Applications held on September 27-29, 2008 in Thessaloniki within the scope the main following topics were discussed: Walsh analysis, harmonic analysis on Vilenkin groups and particular non-commutative groups, the Gibbs derivatives, non-linear coding methods, and related subjects; see the special issue of the journal *Facta Universitatis* (Vol. 21, No. 3, 2008). Applied aspects of Walsh analysis are reflected in the books [8, 9, 10, 11] and the references therein.

Let G be the Cantor dyadic group. A continuous wavelet transform on the space $L^2(G)$ was studied by Lukashenko [12] and by Schipp and Wade [13] (see also the recent paper [14]). The Haar multiresolution analysis for Vilenkin's group was defined by Farkov and Perlov (see [15] and [16]). The idea is to construct analogous multidimensional generalized Haar bases that would have, as scaling functions, characteristic functions of appropriate sets in G with dilation replaced by an automorphism $A \in \operatorname{Aut} G$. W.C. Lang constructed in [17] the family of orthonormal compactly supported wavelets in $L^2(G)$, which is similar to the Daubechies wavelets. Then, in [18] and [19], he studied a multiresolution analysis in $L^2(G)$ and a multifractal structure of dyadic wavelets on G, as well as an orthonormal wavelet which generate an unconditional basis in $L^q(G)$ for all $1 < q < \infty$. These investigations have been continued in [20] where the sharp estimate of regularity of the scaling function constructed in [17] was obtained and some new examples of orthonormal wavelets for the p-adic Vilenkin group G_p, including the case $p = 2$ for the Cantor dyadic group, were constructed. Also, in the paper [20], for any integers $p, n \geq 2$ a compactly supported function $\varphi \in L^2(G_p)$ which satisfies a refinement equation with p^n coefficients and generates a multiresolution analysis in $L^2(G_p)$ is defined. Further results for dyadic wavelets on G are concerned with the following topics:

- methods for the efficient evaluation of orthogonal and biorthogonal wavelets associated with the Walsh functions (Farkov [21, 22, 23], Farkov and Protasov [24], Farkov and Rodionov [52]);
- estimates of the smoothness of dyadic orthogonal wavelets of Daubechies type (Farkov and Rodionov [25]);
- unconditional convergence of a dyadic wavelet expansion in the dyadic Hardy space (Farkov, Goginava and Kopaliani [26]);
- approximation properties of dyadic wavelet expansions (Protasov [27], Sendov [50,51]);
- an algorithm for constructing Parseval dyadic frames (Farkov [28]);
- periodic wavelets in the dyadic harmonic analysis (Farkov [28,53]);
- discrete wavelets related to the Vilenkin-Chrestenson transform (Farkov [29]);

- wavelet packets on the p-adic field (Shah [30], Shah and Wahid [31]).

In [32] and [33], some applications of the theory of dyadic wavelets for image processing tasks are discussed. Wavelet bases on the group G as well as on the additive group $\mathbb{Q}_p$ of the p-adic number field by means of an iterative method giving rise to so-called wavelet sets in the dual group G^* were derived by J. J. Benedetto and R. L. Benedetto [34]; for some other wavelet constructions on the group $\mathbb{Q}_p$ see [35, 36, 37, 38, 39]. Although the Haar and Shannon wavelets are naturally antipodal in the Euclidean setting, it is observed in [34] that their analogues for G and $\mathbb{Q}_p$ are equivalent.

Let us recall that the *Cantor dyadic group* G can be defined as a weak direct product of a countable set of the cyclic groups of order 2. Therefore the group G consists of all sequences of the form

$$x = (x_j) = (\ldots, x_{-2}, x_{-1}, x_0, x_1, x_2, \ldots), \quad x_j \in \{0,1\},$$

for which $x_j \to 0$ as $j \to -\infty$. We see that for each non-zero $x \in G$ there exists $s = s(x) \in \mathbb{Z}$ such that $x_s = 1$ and $x_j = 0$ for all $j < s$. The group operation on G is denoted by $\oplus$ and is defined as coordinate-wise addition modulo 2:

$$(z_j) = (x_j) \oplus (y_j) \iff z_j = x_j + y_j \,(\mathrm{mod}\, 2) \quad \text{for} \quad j \in \mathbb{Z}.$$

The topology in G can be introduced via the complete system of neighbourhoods of the zero sequence:

$$U_l = \{(x_j) \in G \,|\, x_j = 0 \ \text{for}\ j < l\}, \quad l \in \mathbb{Z}.$$

The sets U_l satisfy the following properties:

- $U_{l+1} \subset U_l$, $\bigcap U_l = \{\theta\}$, $\bigcup U_l = G$;
- each U_l is a compact open subgroup of G;
- each U_l is homeomorphic to the Cantor ternary set.

Observe that $x \oplus x = \theta$ for all $x \in G$, where θ is the zero sequence. This means that $\ominus$, the inverse operation of $\oplus$, coincides with $\oplus$.

The group G is a locally compact abelian group. Denote by μ the Haar measure on G normalized so that $\mu(U) = 1$ where $U := U_0$. As usual, $L^q(G)$, $1 \le q < \infty$, are the Lebesgue spaces on G defined by μ. Denote by $\langle \cdot, \cdot \rangle$ and $||\cdot||$ the inner product and the norm in $L^2(G)$, respectively.

The group dual to G is denoted to G^*. Each non-zero element $\omega \in G^*$ can be written as a sequence of the form

$$\omega = (\omega_j) = (\ldots, \omega_{-2}, \omega_{-2}, \omega_0, \omega_1, \omega_2, \ldots), \quad \omega_j \in \{0,1\},$$

where $\omega_s = 1$ and $\omega_j = 0$ for $j < s = s(\omega)$. The operations of addition and subtraction, the neighbourhoods $\{U_l^*\}$ and the Haar measure μ^* for G^* are introduced as above for G. Each character on G can be defined by the formula

$$\chi(x, \omega) = \exp\left(\pi i \sum_{j \in \mathbb{Z}} x_j \omega_{-j-1}\right), \qquad x \in G,$$

for some $\omega \in G^*$ (see, e.g., [6]).

Take in G a discrete subgroup $H = \{(x_j) \in G \,|\, x_j = 0 \text{ for } j > 0\}$ and define an automorphism $A \in \operatorname{Aut} G$ by the formula $(Ax)_j = x_{j+1}$. It is easy to see that the quotient group $H/A(H)$ contains two elements and the annihilator $H^\perp$ of the subgroup H consists of all sequences $(\omega_j) \in G^*$ which satisfy $\omega_j = 0$ for $j \geq 0$.

Let $\mathbb{R}_+ = [0, +\infty)$. We define a map $\lambda : G \to \mathbb{R}_+$ by

$$\lambda(x) = \sum_{j \in \mathbb{Z}} x_j 2^{-j-1}, \qquad x = (x_j) \in G.$$

The image of H under λ is the set of non-negative integers: $\lambda(H) = \mathbb{Z}_+$. For every $\alpha \in \mathbb{Z}_+$, let $h_{[\alpha]}$ denote the element of H such that $\lambda(h_{[\alpha]}) = \alpha$. Also, we define the map $\lambda^* : G^* \to \mathbb{R}_+$, the automorphism $B \in \operatorname{Aut} G^*$, the subgroup U^* and the elements $\omega_{[\alpha]}$ of $H^\perp$ similarly λ, A, U and $h_{[\alpha]}$, respectively. We note that $\chi(Ax, \omega) = \chi(x, B\omega)$ for all $x \in G$, $\omega \in G^*$.

The *Walsh functions* for G can be defined by

$$W_\alpha(x) = \chi(x, \omega_{[\alpha]}), \qquad \alpha \in \mathbb{Z}_+, x \in G.$$

These functions are continuous on G and satisfy the orthogonality relations

$$\int_U W_\alpha(x) W_\beta(x)\, d\mu(x) = \delta_{\alpha,\beta}, \qquad \alpha, \beta \in \mathbb{Z}_+,$$

where $\delta_{\alpha,\beta}$ is the Kronecker delta. It is well-known that $\{W_\alpha\}$ is a complete system in $L^2(U)$ (e.g., [6]) The corresponding system for G^* is defined by

$$W^*_\alpha(\omega) = \chi(h_{[\alpha]}, \omega), \qquad \alpha \in \mathbb{Z}_+, \omega \in G^*.$$

The system $\{W^*_\alpha\}$ is an orthonormal basis of $L^2(U^*)$.

For any function $f \in L^1(G) \cap L^2(G)$ the Fourier transform $\widehat{f}$, defined by

$$\widehat{f}(\omega) = \int_G f(x) \overline{\chi(x, \omega)}\, d\mu(x), \quad \omega \in G,$$

belongs to $L^2(G)$. The Fourier operator

$$\mathcal{F} : L^1(G) \cap L^2(G) \to L^2(G), \quad \mathcal{F} f = \widehat{f},$$

extends uniquely to the whole space $L^2(G)$. The properties of the Fourier transform on $L^2(G)$ are quite similar to those of the classical Fourier transform. In particular, if $f, g \in L^2(G)$ then $\langle f, g \rangle = \langle \widehat{f}, \widehat{g} \rangle$ (Parseval's equality).

For any positive integer n let $\mathcal{E}_n(G)$ denotes the collection of all functions on G which are constant on

$$U_{n,\alpha} := A^{-n}(h_{[\alpha]}) \oplus A^{-n}(U), \quad \alpha \in \mathbb{Z}_+.$$

The class $\mathcal{E}_n(G^*)$ is defined in a similar way. According to Proposition 2 in [21] (see also [8]), the following properties hold:

(a) if $f \in L^1(G) \cap \mathcal{E}_n(G)$, then $\operatorname{supp} \widehat{f} \subset U^*_{-n}$;

(b) if $f \in L^1(G)$ and $\operatorname{supp} f \subset U_{-n}$, then $\widehat{f} \in \mathcal{E}_n(G^*)$.

The elements of $\mathcal{E}_n(G)$ will be called *dyadic entire functions of order n*. In the sequel, $\mathbf{1}_E$ stands for the characteristic function of a subset E of G.

1. Construction of Orthonormal Compactly Supported Wavelets

In this section, we will consider orthonormal wavelets with compact supports on G which can be represented by expansions in Walsh series. Let us recall that an *orthonormal wavelet* on G is a function $\psi \in L^2(G)$ such that $\{\psi_{j,\alpha} : j \in \mathbb{Z}, \alpha \in \mathbb{Z}_+\}$ is an orthonormal basis in $L^2(G)$, where

$$\psi_{j,\alpha}(x) := 2^{j/2}\psi(A^j x \oplus h_{[\alpha]}) \quad \text{for all} \quad j \in \mathbb{Z},\ \alpha \in \mathbb{Z}_+.$$

For example, the Haar wavelet on G can be defined by

$$\psi_H(x) = \begin{cases} 1, & x \in U_1, \\ -1, & x \in U \setminus U_1, \\ 0, & \text{otherwise.} \end{cases}$$

Indeed, if $\psi = \psi_H$ then $\{\psi_{j,\alpha} : j \in \mathbb{Z},\ \alpha \in \mathbb{Z}_+\}$ is the Haar basis for $L^2(G)$ (see [15, 16] and [17]). Further, we have

$$\psi_H(x) = \varphi_H(Ax) - \varphi_H(Ax \oplus h_{[1]}),$$

where $\varphi_H := \mathbf{1}_U$ is the Haar function for G. Note also, that $\widehat{\varphi}_H = \varphi_H$. The last equality, in fact, shows [34] that the Haar and Shannon wavelets coincide on G.

We denote the support of a function $f \in L^2(G)$ by supp$f$; it is defined as the minimal (with respect to inclusion) closed set such that on its complement f vanishes almost everywhere. The set of functions in $L^2(G)$ with compact support is denoted by $L^2_c(G)$. We say that a function $\varphi \in L^2_c(G)$ is a *refinable function*, if it satisfies an equation of the type

$$\varphi(x) = \sum_{\alpha=0}^{2^n-1} a_\alpha \varphi(Ax \oplus h_{[\alpha]}). \tag{1}$$

The functional equation (1) is called the *refinement equation*. The Walsh polynomial

$$m(\omega) = \frac{1}{2}\sum_{\alpha=0}^{2^n-1} a_\alpha W^*_\alpha(\omega) \tag{2}$$

is called the *mask* of equation (1) (or its solution φ). We note that the Haar function φ_H is a solution of equation (1) when $a_0 = a_1 = 1$ and $a_\alpha = 0$ otherwise.

The sets

$$U^*_{n,s} := B^{-n}(\omega_{[s]}) \oplus B^{-n}(U^*), \quad 0 \le s \le 2^n - 1,$$

are the cosets of the subgroup $B^{-n}(U^*)$ in U^*. From the definition of W^*_α it follows that the mask m defined by (2) is constant on each $U^*_{n,s}$ and hence $m \in \mathfrak{E}_n(G^*)$. Moreover, the coefficients of equation (1) are related to the values of m on cosets $U^*_{n,s}$ by means of the Walsh discrete transforms:

$$a_\alpha = \frac{1}{2^{n-1}}\sum_{s=0}^{2^n-1} b_s W^*_\alpha(B^{-n}\omega_{[s]}), \quad 0 \le \alpha \le 2^n - 1, \tag{3}$$

$$b_s = \frac{1}{2}\sum_{\alpha=0}^{2^n-1} a_\alpha W_\alpha^*(B^{-n}\omega_{[s]}), \qquad 0 \le s \le 2^n - 1. \tag{4}$$

They can be realized by the fast algorithms (see, for instance, [6, 8]). Thus, any choice of the values b_s of m on $U_{n,s}^*$ defines also the coefficients a_α of equation (1). In [21], it was proved the following

THEOREM 1.1. *Let $\varphi \in L_c^2(G)$ satisfies the refinement equation* (1). *If* $\widehat{\varphi}(\theta) = 1$, *then*

$$\sum_{\alpha=0}^{2^n-1} a_\alpha = 1 \quad \textit{and} \quad \operatorname{supp}\varphi \subset U_{1-n}.$$

This solution is unique, is given by the formula

$$\widehat{\varphi}(\omega) = \prod_{j=1}^{\infty} m(B^{-j}\omega),$$

and possesses the following properties:

1) $\widehat{\varphi}(h^*) = 0$ *for all* $h^* \in H^\perp \setminus \{\theta\}$ (*the modified Strang-Fix condition*);

2) $\sum_{h\in H} \varphi(x \oplus h) = 1$ *for almost every* $x \in G$ (*the partition of unity property*).

Let us discuss necessary and sufficient conditions for a refinable function to generate a multiresolution analysis in $L^2(G)$. As usual, a collection of closed subspaces $V_j \subset L^2(G)$, $j \in \mathbb{Z}$, is called a *multiresolution analysis* (an *MRA*) in $L^2(G)$ if the following hold:

(i) $V_j \subset V_{j+1}$ for all $j \in \mathbb{Z}$;

(ii) $\overline{\bigcup V_j} = L^2(G)$ and $\bigcap V_j = \{0\}$;

(iii) $f(\cdot) \in V_j \Longleftrightarrow f(A\cdot) \in V_{j+1}$ for all $j \in \mathbb{Z}$;

(iv) $f(\cdot) \in V_0 \Longrightarrow f(\cdot \oplus h) \in V_0$ for all $h \in H$;

(v) there is a function $\varphi \in L^2(G)$ such that $\{\varphi(\cdot \oplus h) \mid h \in H\}$ is an orthonormal basis of V_0.

Sometimes condition (v) is relaxed by assuming that $\{\varphi(\cdot \oplus h) \mid h \in H\}$ is a Riesz basis for V_0. That is, for every $f \in V_0$ there exists a unique sequence $\{a_\alpha\} \in \ell^2$ such that

$$f(x) = \sum_{\alpha=0}^{\infty} a_\alpha f(x \oplus h_{[\alpha]})$$

with convergence in $L^2(G)$ and

$$A_0\Big(\sum_{\alpha=0}^{\infty} |a_\alpha|^2\Big)^{1/2} \le \Big\|\sum_{\alpha=0}^{\infty} a_\alpha f(\cdot \oplus h_{[\alpha]})\Big\| \le B_0\Big(\sum_{\alpha=0}^{\infty} |a_\alpha|^2\Big)^{1/2}$$

with $0 < A_0 \le B_0 < \infty$ constants independent of $f \in V_0$. If this is the case we say that we have an MRA with a Riesz basis. The fact that this weaker version of (v) is equivalent to the one announced above is proved as in [40] for the space $L^2(\mathbb{R})$.

We say that *a function φ generates an MRA in $L^2(G)$* if $\{\varphi(\cdot \oplus h) \mid h \in H\}$ is an orthonormal system in $L^2(G)$ and, in addition, the family of subspaces

$$V_j = \operatorname{clos}_{L^2(G)} \operatorname{span}\{\varphi_{j,\alpha} \mid \alpha \in \mathbb{Z}_+\}, \quad j \in \mathbb{Z},$$

is an MRA in $L^2(G)$. If a function φ generates an MRA in $L^2(G)$, then $\{\varphi_{j,\alpha} \mid \alpha \in \mathbb{Z}_+\}$ is an orthonormal basis of V_j for every $j \in \mathbb{Z}$.

Suppose that $\varphi \in L_c^2(G)$ is a solution of equation (1) such that $\widehat{\varphi}(\theta) = 1$. It was shown in [20], that if $\{\varphi(\cdot \oplus h) \mid h \in H\}$ is an orthonormal system in $L^2(G)$ then

$$m(\theta) = 1 \quad \text{and} \quad |m(\omega)|^2 + |m(\omega \oplus \delta_0)|^2 = 1, \quad \omega \in G^*, \tag{5}$$

where δ_0 denotes the sequence $\omega = (\omega_j)$ such that $\omega_0 = 1$ and $\omega_j = 0$ for $j \neq 1$. Hence the equalities

$$b_0 = 1, \quad |b_j|^2 + |b_{j+2^{n-1}}|^2 = 1, \quad 0 \leq j \leq 2^{n-1} - 1, \tag{6}$$

are necessary for the system $\{\varphi(\cdot \oplus h) | h \in H\}$ to be orthonormal in $L^2(G)$. Under which additional conditions φ generates an MRA in $L^2(G)$? Theorem 1.2 below contains the answer to this question.

A compact subset E of G^* is said to be *congruent to* U^* *modulo* $H^\perp$ if $\mu^*(E) = 1$ and, for any $\omega \in E$, there exists an $h^* \in H^\perp$ such that $\omega \oplus h^* \in U^*$. We say that the mask m satisfies the *modified Cohen condition*, if there exists a compact subset E of G^* containing a neighbourhood of θ such that:

1) E congruent to U^* modulo $H^\perp$;

2) the inequality

$$\inf_{j \in \mathbb{N}} \inf_{\omega \in E} |m(B^{-j}\omega)| > 0 \tag{7}$$

is true.

Since E is compact, we note that if $m(\theta) = 1$ then there exists a number j_0 such that $m(B^{-j}\omega) = 1$ for all $j > j_0$, $\omega \in E$. Therefore, (7) holds if m does not vanish on the sets $B^{-1}(E), \ldots, B^{-j_0}(E)$. Moreover, we can choose $j_0 \leq 2^n$, because m is completely defined by the values b_j (and m is an $H^\perp$-periodic function).

Assume, that M is a union of certain sets $U^*_{n-1,s}$, $1 \leq s \leq 2^{n-1} - 1$, or coincides with one of them. The set M is said to be *blocking* for a mask m, if it satisfies the following relation

$$\{B^{-1}(\omega) \mid \omega \in M\} \cup \{\delta_0 + B^{-1}(\omega) \mid \omega \in M\} \subset \{\omega \in U^* \mid m(\omega) = 0\}.$$

According to this definition, if M is a blocking set for m, then $M \cap U^*_{n-1,0} = \emptyset$. It is clear that each mask may have only finitely many of blocking sets.

THEOREM 1.2. *Let* $\varphi \in L_c^2(G)$ *be a solution of equation* (1) *such that* $\widehat{\varphi}(\theta) = 1$. *Suppose that the mask m satisfies* (5). *Then the following are equivalent*:

(a) φ *generates an MRA in* $L^2(G)$;

(b) *m satisfies the modified Cohen's condition*;

(c) *m has no blocking sets.*

The equivalence $(a) \Leftrightarrow (b)$ of Theorem 1.2 was noted by Lang in [18]. Generalizations of Theorems 1.1 and 1.2 for Vilenkin groups are given in [21]; some similar results for dyadic wavelets on $\mathbb{R}_+$ have been obtained in [22] and [24]. Also, as in [40] it can be verified that any compactly supported orthonormal wavelet on G must be an MRA wavelet.

Using Theorem 1.2 we obtain the following *algorithm to construct an orthonormal wavelet* ψ *in* $L^2(G)$:

1. Choose numbers b_s, $0 \le s \le 2^n - 1$, so that (6) is true.
2. Compute a_α, $0 \le \alpha \le 2^n - 1$, by (3) and verify that the mask

$$m_0(\omega) = \frac{1}{2} \sum_{\alpha=0}^{2^n-1} a_\alpha W_\alpha^*(\omega).$$

has no blocking sets.

3. Find

$$m_1(\omega) = \frac{1}{2} \sum_{\alpha \in \mathbb{Z}_+} a_\alpha^{(1)} W_\alpha^*(\omega),$$

such that $(m_l(\omega \oplus B^{-1}\omega_{[k]}))_{l,k=0}^1$ is an unitary matrix. In particular, one can choose $a_\alpha^{(1)} = (-1)^\alpha a_{\alpha\oplus 1}$ or $a_\alpha^{(1)} = (-1)^\alpha a_{2^n-1-\alpha}$ for $0 \le \alpha \le 2^n - 1$ (and $a_\alpha^{(1)} = 0$ for the rest α).

4. Define ψ by the formula

$$\psi(x) = \sum_{\alpha \in \mathbb{Z}_+} a_\alpha^{(1)} \varphi(Ax \oplus h_{[\alpha]}).$$

EXAMPLE 1.1. Let $n = 2$ and

$$b_0 = 1,\ b_1 = a,\ b_2 = 0,\ b_3 = b,$$

where $|a|^2 + |b|^2 = 1$. Put

$$a_0 = (1+a+b)/2, \quad a_1 = (1+a-b)/2, \quad a_2 = (1-a-b)/2, \quad a_3 = (1-a+b)/2$$

and denote by φ the corresponding solution of (1). For $a \ne 0$ the modified Cohen condition is fulfilled on the set $E = U^*$ and, by Theorem 2.2, φ generates an MRA in $L^2(G)$. In particular, for $a = 1$ and $a = -1$ the Haar function: $\varphi(x) = \mathbf{1}_U(x)$ and the displaced Haar function: $\varphi(x) = \mathbf{1}_U(x \oplus h_{[1]})$ are obtained. If $0 < |a| < 1$, then φ is defined by the expansion

$$\varphi(x) = (1/2)\mathbf{1}_U(A^{-1}x)(1 + a \sum_{j=0}^{\infty} b^j W_{2^{j+1}-1}(A^{-1}x)), \qquad x \in G. \tag{8}$$

As an illustration of this expansion, a few representative pictures are given in [17] (cf. [18], [23]). The corresponding wavelet ψ can be written as

$$\psi(x) = \sum_{\alpha=0}^{3} (-1)^\alpha a_{\alpha\oplus 1} \varphi(Ax \oplus h_{[\alpha]}).$$

When $a = 0$ the family $\{\varphi(\cdot \oplus h) \mid h \in H\}$ is a linear dependence system (see Example 5.4 below).

Now, let us take a compactly supported L^2-solution φ of equation (1) such that $\widehat{\varphi}(\theta) = 1$. Suppose that the values b_s of the corresponding mask m satisfy (6). Then we let $\gamma(i_1, i_2, \ldots, i_n) = b_s$, if

$$s = i_1 2^0 + i_1 2^1 + \ldots + i_n 2^n, \quad i_j \in \{0, 1\}.$$

Further, for an integer l with the binary expansion

$$l = \sum_{j=0}^{k} \mu_j 2^j, \quad \mu_j \in \{0,1\}, \quad \mu_k = 1, \quad k = k(l), \tag{9}$$

we define $d_l[m]$ as follows

$$d_l[m] = \gamma(\mu_0, 0, 0, \ldots, 0, 0) \quad \text{if} \quad k(l) = 0;$$

$$d_l[m] = \gamma(\mu_1, 0, 0, \ldots, 0, 0)\gamma(\mu_0, \mu_1, 0, \ldots, 0, 0) \quad \text{if} \quad k(l) = 1;$$

$$\ldots\ldots\ldots\ldots\ldots\ldots\ldots\ldots ; \ldots\ldots\ldots\ldots\ldots\ldots\ldots\ldots\ldots\ldots$$

$$d_l[m] = \gamma(\mu_k, 0, 0, \ldots, 0, 0)\gamma(\mu_{k-1}, \mu_k, 0, \ldots, 0, 0)\ldots\gamma(\mu_0, \mu_1, \mu_2, \ldots, \mu_{n-2}, \mu_{n-1})$$

if $k = k(l) \geq n-1$. The indices of each factor in the last product, starting with the second factor, are equal to the indices of the preceding factor shifted one position rightwards; at the free first position one puts the corresponding digit of the binary expansion (9). Let $\mathbb{N}_0(n)$ be the set of all positive integers $l \geq 2^{n-1}$ whose binary expansion (9) contains no n-tuple $(\mu_j, \mu_{j+1}, \ldots, \mu_{j+n-1})$ coinciding with the n-tuple $(0, 0, \ldots, 0, 1)$. It is proved in [20] that φ can be written as the lacunary Walsh series:

$$\varphi(x) = (1/2^{n-1})\mathbf{1}_U(A^{1-n}x)(1 + \sum_{l \in \mathbb{N}(n)} d_l[m] W_l(A^{1-n}x)), \quad x \in G, \tag{10}$$

where $\mathbb{N}(n) = \{1, 2, \ldots, 2^{n-1} - 1\} \cup \mathbb{N}_0(n)$. We see that in the case $n = 2$ decompositions (8) and (10) coincide. Below we give a generalization of (10) to the p-adic Vilenkin group (see (41)).

In the next example we shall indicate a scaling function in $L^2(G)$, which is a linear combination of finite number of the Walsh functions. In this case, the inequality (7) is true for some E different from U^*.

EXAMPLE 1.2. Let $n = 3$ and

$$b_0 = 1,\ b_1 = a,\ b_2 = b,\ b_3 = c,\ b_4 = 0,\ b_5 = \alpha,\ b_6 = \beta,\ b_7 = \gamma,$$

where

$$|a|^2 + |\alpha|^2 = |b|^2 + |\beta|^2 = |c|^2 + |\gamma|^2 = 1.$$

Then by (3) we obtain the following expressions:

$$a_0 = \frac{1}{4}(1 + a + b + c + \alpha + \beta + \gamma),$$

$$a_1 = \frac{1}{4}(1 + a + b + c - \alpha - \beta - \gamma),$$

$$a_2 = \frac{1}{4}(1 + a - b - c + \alpha - \beta - \gamma),$$

$$a_3 = \frac{1}{4}(1 + a - b - c - \alpha + \beta + \gamma),$$

$$a_4 = \frac{1}{4}(1-a+b-c-\alpha+\beta-\gamma),$$

$$a_5 = \frac{1}{4}(1-a+b-c+\alpha-\beta+\gamma),$$

$$a_6 = \frac{1}{4}(1-a-b+c-\alpha-\beta+\gamma),$$

$$a_7 = \frac{1}{4}(1-a-b+c+\alpha+\beta-\gamma).$$

Further, using (10) we have

$$\varphi(x) = (1/4)\mathbf{1}_U(y)(1+a\,W_1(y)+ab\,W_2(y)+ac\,W_3(y)+ab\alpha\,W_5(y)$$
$$+ac\beta\,W_6(y)+ac\gamma W_7(y)+ab^2\alpha\,W_{10}(y)+abc\alpha\,W_{11}(y)+ac\alpha\beta\,W_{13}(y)$$
$$+ac\beta\gamma W_{14}(y)+ac\gamma^2\,W_{15}(y)+ab^2\alpha^2\,W_{21}(y)+abc\alpha\beta\,W_{22}(y)+ab\alpha\beta\gamma W_{23}(y)$$
$$+abc\alpha\beta\,W_{26}(y)+ac^2\alpha\beta\,W_{27}(y)+ac\alpha\beta\gamma W_{29}(y)+ac\beta\gamma^2\,W_{30}(y)+\ldots), \qquad (11)$$

where $y = A^{-2}x$. The blocking sets for the mask

$$m(\omega) = \frac{1}{2}\sum_{\alpha=0}^{7} a_\alpha W_\alpha^*(\omega)$$

are: 1) $U_{1,1}^*\cup U_{2,1}^*$ for $a=0$, 2) $U_{2,1}^*\cup U_{2,2}^*$ for $a=\beta=0$, 3) $U_{2,3}^*$ for $c=0$, 4) $U_{1,1}^*$ for $b=c=0$. If $a=0$ or $c=0$, then $\{\varphi(\cdot\oplus h)\,|\,h\in H\}$ is a linear dependence system. If a and c distinct from zero, then φ generate an MRA in $L^2(G)$. In the case where $abc\neq 0$ inequality (7) is fulfilled for $E=U^*$, and when $a\neq 0, b=0, c\neq 0$ it holds for $E=B(U_{3,0}^*\cup U_{3,1}^*\cup U_{3,3}^*\cup U_{3,6}^*)$. In particular, for $a=c=1, 0\leq |b|<1$, from (11) we obtain

$$\varphi(x) = (1/4)\mathbf{1}_U(y)(1+W_1(y)+bW_2(y)+W_3(y)+\beta W_6(y)), \quad y=A^{-2}x, \qquad (12)$$

(cf. Example 4 in [42], Examples 2 and 4 in [24]).

Note that the expansion (11) was found by Lang [18] only in the following two cases: 1) $a=1, b=0, |c|<1$ ($\alpha=0, \beta=1, |\gamma|<1$) and 2) $|a|<1, b=1, c=0$ ($|\alpha|<1, \beta=0, \gamma=1$).

EXAMPLE 1.3. Suppose that for some numbers b_s, $0\leq s\leq 2^n-1$, the equalities (6) are true. Applying (3), we find the mask

$$m(\omega) = \frac{1}{2}\sum_{\alpha=0}^{2^n-1} a_\alpha W_\alpha^*(\omega),$$

which takes the values b_s on $U_{n,s}^*$, $0\leq s\leq 2^n-1$. If in addition $b_j\neq 0$ for all $1\leq j\leq 2^{n-1}-1$, then equation (1) have a solution φ, which generates an MRA in $L^2(G)$ (the modified Cohen condition is fulfilled for $E=U^*$). The expansion of this solution in a lacunary Walsh series can be obtained by (10).

Let us recall that the Walsh system $\{w_n\,|\,n\in\mathbb{Z}_+\}$ ∎ on $\mathbb{R}_+$ is defined by

$$w_0(x)\equiv 1, \qquad w_n(x)=\prod_{j=0}^{k}(w_1(2^jx))^{\nu_j}, \quad n\in\mathbb{N}, \quad x\in\mathbb{R}_+,$$

where the ν_j are the digits of the binary expansion

$$n = \sum_{j=0}^{k} \nu_j 2^j, \quad \nu_j \in \{0,1\}, \quad \nu_k = 1, \quad k = k(n),$$

and $w_1(x)$, for $x \in [0,1)$, is given by

$$w_1(x) = \begin{cases} 1, & \text{for } x \in [0,1/2), \\ -1, & \text{for } x \in [1/2,1) \end{cases}$$

and w_1 is extended to $\mathbb{R}_+$ by periodicity: $w_1(x+1) = w_1(x)$ for all $x \in \mathbb{R}_+$.

For each $x \in \mathbb{R}_+$ and any $j \in \mathbb{N}$, we define $x_j, x_{-j} \in \{0,1\}$ from the binary expansion

$$x = \sum_{j<0} x_j 2^{-j-1} + \sum_{j>0} x_j 2^{-j}$$

(in the case of a binary rational number x, we choose an expansion with a finite number non-zero terms). If we denote by $[x]$ the integer part of a number x, then

$$x_j = [2^j x] (\bmod 2), \quad x_{-j} = [2^{1-j} x] (\bmod 2), \quad x \in \mathbb{R}_+, \ j \in \mathbb{N}. \tag{13}$$

The binary addition on $\mathbb{R}_+$ is defined by the formula

$$x \dot{+} y = \sum_{j<0} |x_j - y_j| 2^{-j-1} + \sum_{j>0} |x_j - y_j| 2^{-j}, \quad x,y \in \mathbb{R}_+,$$

where x_j and y_j can be calculated by (13).

Remark 1.2. Investigation of Walsh functions may proceed [6] in two directions:

a) use the system $\{w_n\}$ and Lebesgue integration to study functions defined on $(\mathbb{R}_+, \dot{+})$,

or

b) use the system $\{W_\alpha\}$ and Haar integration to study functions defined on $(G, \oplus)$.

Thus, the above results can be reformulate to dyadic wavelets on $\mathbb{R}_+$.

The next result characterizes all dyadic entire refinable compactly supported functions in $L^2(\mathbb{R}_+)$.

THEOREM 1.3 (cf. [24]). *Let φ be a compactly supported L^2-solution of the refinement equation*

$$\varphi(x) = \sum_{k=0}^{2^n-1} a_k \varphi(2x \dot{+} k), \quad x \in \mathbb{R}_+,$$

and let m be its mask. Assume that $\widehat{\varphi}(0) = 1$. Then φ is not a dyadic entire function if and only if there exists a finite sequence $\{d_k\}_{k=1}^{N}$, $N \leq 2^{n-1} + n - 1$, $d_k \in \{0,1\}$, such that:

(a) $d_1 = \ldots = d_{n-1} = 0$, $d_n = 1$;

(b) *there exists an integer j, $n-1 \leq j \leq N-1$, such that $d_{j-s} = d_{N-s}$ for $s = 0, \ldots, n-2$;*

(c) $m(0.d_{k+1} \ldots d_{k+n}) \neq 0$ *for all $k = 0, \ldots, N-n$.*

The dyadic orthonormal wavelets from Example 1.1 can be identified with multiwavelets on $\mathbb{R}_+$ consisting of piecewise fractal functions. To illustrate this fact, consider the refinable function φ defined by the expansion

$$\varphi(x) = (1/2)\mathbf{1}_{[0,1)}(x/2)(1 + a\sum_{j=0}^{\infty} b^j w_{2^{j+1}-1}(x/2)), \quad x \in \mathbb{R}_+,$$

and let

$$a_0 = \frac{1+a+b}{2}, \qquad a_1 = \frac{1+a-b}{2}, \qquad a_2 = \frac{1-a-b}{2}, \qquad a_3 = \frac{1-a+b}{2},$$

where $0 < |a| < 1$, $|a|^2 + |b|^2 = 1$. Denote by ψ the corresponding wavelet in $L^2(\mathbb{R}_+)$. Then φ and ψ satisfy the equalities

$$\varphi(x) = \sum_{k=0}^{3} a_k \varphi(2x \dotplus k), \quad \psi(x) = \sum_{k=0}^{3} d_k \varphi(2x \dotplus k), \quad x \in \mathbb{R}_+,$$

where $d_0 = -a_1$, $d_1 = a_0$, $d_2 = -a_3$, $d_3 = a_2$. Further, we write

$$\overrightarrow{\varphi} = \begin{pmatrix} \varphi_1 \\ \varphi_2 \end{pmatrix}, \quad \overrightarrow{\psi} = \begin{pmatrix} \psi_1 \\ \psi_2 \end{pmatrix},$$

where $\varphi_1(x) := \varphi(x)$, $\varphi_2(x) := \varphi(x \dotplus 1)$, $\psi_1(x) = \psi(x)$, $\psi_2(x) = \psi(x \dotplus 1)$ for all $x \in \mathbb{R}_+$.

THEOREM 1.4. *We have* $\overrightarrow{\varphi}(x/2) = P_0\overrightarrow{\varphi}(x) + P_2\overrightarrow{\varphi}(x \dotplus 2)$ *and* $\overrightarrow{\psi}(x/2) = Q_0\overrightarrow{\varphi}(x) + Q_2\overrightarrow{\varphi}(x \dotplus 2)$ *where*

$$P_0 = \tfrac{1}{\sqrt{2}}\begin{pmatrix} a_0 & a_1 \\ a_2 & a_3 \end{pmatrix}, \quad P_2 = \tfrac{1}{\sqrt{2}}\begin{pmatrix} a_2 & a_3 \\ a_0 & a_1 \end{pmatrix}, \quad Q_0 = \tfrac{1}{\sqrt{2}}\begin{pmatrix} d_0 & d_1 \\ d_2 & d_3 \end{pmatrix}, \quad Q_2 = \tfrac{1}{\sqrt{2}}\begin{pmatrix} d_2 & d_3 \\ d_0 & d_1 \end{pmatrix}.$$

This theorem means that wavelets defined in Example 1.1 can be identified with multiwavelets on $\mathbb{R}_+$ (cf. [43]). The next result shows that these wavelets can be seen as the piecewise fractal functions in the sense of Massopust [44]:

THEOREM 1.5. *If* $f(x) = \varphi(2x)$ *then*

$$f(x) = \begin{cases} a_1 + bf(2x) & \text{if } x \in [0,1/2), \\ a_3 - bf(2x-1) & \text{if } x \in [1/2,1). \end{cases}$$

Furthermore,

$$\varphi_1(x) = \begin{cases} f_1(x) & \text{if } x \in [0,1), \\ f_2(x-1) & \text{if } x \in [1,2) \end{cases} \quad \text{and} \quad \varphi_2(x) = \begin{cases} f_2(x) & \text{if } x \in [0,1), \\ f_1(x-1) & \text{if } x \in [1,2), \end{cases}$$

where

$$f_1(x) = \begin{cases} a_1 + bf_1(2x) & \text{if } x \in [0,1/2), \\ a_0 - bf_1(2x-1) & \text{if } x \in [1/2,1), \end{cases} \quad \text{and} \quad f_2(x) = \begin{cases} a_2 + bf_2(2x) & \text{if } x \in [0,1/2), \\ a_3 - bf_2(2x-1) & \text{if } x \in [1/2,1). \end{cases}$$

Theorems 1.4 and 1.5 were proved by Lang in [19].

2. Estimates of the Smoothness of Dyadic Wavelets

It is well known that the Daubechies scaling function of order N is a solution of the functional equation

$$\varphi(x) = \sqrt{2}\sum_{k=0}^{2N-1} h_k \varphi(2x-k), \qquad x \in \mathbb{R}, \tag{14}$$

and possesses the following properties: 1) $\operatorname{supp}\varphi = [0, 2N-1]$, 2) the system $\{\varphi(\cdot - k) : k \in \mathbb{Z}\}$ is orthonormal in $L^2(\mathbb{R})$; 3) φ generates a multiresolution analysis in $L^2(\mathbb{R})$. For $N = 1$, the Daubechies construction yields the Haar function: $\varphi = \chi_{[0,1)}$ (in this case, $h_0 = h_1 = 1/\sqrt{2}$). For $2 \le N \le 10$, the coefficients of equation (1) are given in Daubechies' book (see [45]). For $N = 2$, the solution of equation (14) is continuous on $\mathbb{R}$ and satisfies the Lipschitz condition

$$|\varphi(t) - \varphi(x)| \le C|t-x|^{\alpha}, \quad t, x \in \mathbb{R},$$

where $\alpha \approx 0,5500$. The exact value of exponent α (and the corresponding quantities for $N = 3$ and $N = 4$) were obtained by the matrix method (see [46] and [47]). For $N \ge 5$, the exact values of the exponents of smoothness of the Daubechies scaling functions are not known. This section will discuss estimations for moduli of smoothness of compactly supported solutions of equation (1) with $n \le 4$. We shall assume that the a_α in (1) are real, although all arguments can be extended to the complex case.

Let us recall that the modulus of smoothness of a continuous complex-valued function f on U_{1-n} is the sequence $\{\Omega_j(f)\}$, $j \ge 1-n$, defined by

$$\Omega_j(f) = \sup\{|f(x) - f(y)| \mid x, y \in U_{1-n}, x \oplus y \in U_j\}.$$

Given any non-increasing sequence with zero limit,

$$\varepsilon_{1-n} \ge \varepsilon_{2-n} \ge \varepsilon_{3-n} \ge \dots, \qquad \lim_{j\to\infty} \varepsilon_j = 0,$$

one can find a function f such that $\Omega_j(f) = \varepsilon_j$ for all $j \ge 1-n$ (see [48]).

Put $N = 2^{n-1}$. From (5) we get

$$\sum_{\alpha=0}^{2N-1} a_\alpha = 2 \quad \text{and} \quad \sum_{\alpha=0}^{N-1} a_{2\alpha} = \sum_{\alpha=0}^{N-1} a_{2\alpha+1} = 1. \tag{15}$$

By definition, the equality $k = i \oplus_2 j$ for $i, j, k \in \{1, 2, \dots, 2N-1\}$ means that

$$k_s = i_s + j_s \pmod 2, \qquad s \in \{0, 1, \dots, 2N-1\},$$

where i_s, j_s, k_s are digits of the binary expansions

$$i = \sum_{s=0}^{2N-1} i_s 2^s, \qquad j = \sum_{s=0}^{2N-1} j_s 2^s, \qquad k = \sum_{s=0}^{2N-1} k_s 2^s.$$

We define $(N \times N)$-matrices T_0 and T_1 by

$$(T_0)_{i,j} = a_{2(i-1)\oplus_2(j-1)}, \qquad (T_1)_{i,j} = a_{(2i-1)\oplus_2(j-1)}, \tag{16}$$

where $i, j \in \{1,2,\ldots,N\}$. Let $e_1 = (1,1,\ldots,1)$ be the N-dimensional row vector all whose components equal to 1. Hence by (15)

$$e_1 T_0 = e_1 T_1 = e_1.$$

For every two N-dimensional vectors $v = (v_1,\ldots,v_N)$ and $w = (w_1,\ldots,w_N)$ we set

$$v \cdot w^t := \sum_{j=1}^{N} v_j w_j, \qquad \|v\| := \sqrt{v \cdot v^t},$$

where the column vectors v^t and w^t are obtained from the row vectors v and w by transposition.

We define E_1 to be the subspace of $\mathbb{R}^N$ orthogonal to e_1:

$$E_1 := \{u = (u_1,\ldots,u_N)^t \mid e_1 \cdot u = 0\}.$$

For any real $(N \times N)$-matrix M put

$$\|M\| := \sup\{\|Mu\|/\|u\| \mid u \in \mathbb{R}^N,\, u \neq 0\},$$

$$\|M|_{E_1}\| := \sup\{\|Mu\|/\|u\| \mid u \in E_1,\, u \neq 0\}.$$

It is well known that the value $\|M\|$ coincides with the square root of the greatest eigenvalue of the matrix $M^t M$. The following analogue of Theorem 2.2 in [48] is valid:

PROPOSITION 2.1 ([20]). *Let $\varphi \in L^2_c(G)$ be a solution of equation* (1) *such that $\widehat{\varphi}(\theta) = 1$. Suppose that the matrices T_0, T_1 are given by* (16). *If for all $m \in \mathbb{N}$*

$$\max\{\|T_{d_1} T_{d_2} \ldots T_{d_m}|_{E_1}\| \mid d_j \in \{0,1\},\, 1 \le j \le m\} \le C m^p q^m,$$

where $0 < q < 1$, $p \ge 0$ and $C > 0$, then φ is continuous on U_{1-n} and, for any integer $j \ge n-1$, we have $\Omega_j(\varphi) \le C q^j$.

Let us denote by α_φ the supremum for values of $\alpha > 0$ such that the inequality

$$\Omega_j(\varphi) \le C 2^{-\alpha j}$$

holds for all $j \ge n-1$. As first noted in [20], for the lower bounds of α_φ can be used an expansion (10) of φ in the Walsh series.

Recall that the joint spectral radius of two $(d \times d)$-matrices A_0 and A_1 is defined by the formula

$$\widehat{\rho}(A_0, A_1) := \lim_{k \to \infty} \max\{\|A_{d_1} A_{d_2} \ldots A_{d_k}\|^{1/k} : d_j \in \{0,1\},\, 1 \le j \le k\},$$

where $\|\cdot\|$ is any norm in $\mathbb{C}^{d \times d}$. It is clear that if $A_0 = A_1$ then $\widehat{\rho}(A_0, A_1)$ coincides with the spectral radius $\rho(A_0)$. The joint spectral radius of finite-dimensional linear operators L_0, L_1 is defined as the joint spectral radius of their matrices in an arbitrary fixed basis of the corresponding linear space.

Remark 2.1 Under the assumptions of Proposition 2.1, the Walsh- Fourier series (10) converges uniformly on the set $A^{n-1}(U)$. Indeed, it follows that

$$\lim_{j\to\infty} j\Omega_j(\varphi) = 0,$$

and one can apply Theorem 2.3.5 in [8].

PROPOSITION 2.2. *Let a solution φ of equation* (1) *generates an MRA in $L^2(\mathbb{R}_+)$ and let $\widehat{\rho} = \widehat{\rho}(L_0, L_1)$ be the joint spectral radius of the linear operators L_0, L_1 defined on the space $\mathbb{R}^N$ by the matrices T_0, T_1 and then restricted to the subspace E_1. Then $\alpha_\varphi = -\log_2 \widehat{\rho}$.*

The proof of Proposition 2.2 is quite similar to the proofs of the corresponding statements in [47] (see also [24]). Also note that the smoothness of the dyadic orthonormal wavelet ψ corresponding to the refinable function φ coincides with α_φ.

EXAMPLE 2.1. If φ is given as in Example 1.1, then

$$\alpha_\varphi = \log_2(1/|b|). \tag{17}$$

Indeed, in this case we have

$$T_0 = \begin{pmatrix} a_0 & a_1 \\ a_2 & a_3 \end{pmatrix}, \quad T_1 = \begin{pmatrix} a_1 & a_0 \\ a_3 & a_2 \end{pmatrix},$$

and

$$E_1 = \{v \in \mathbb{R}^2 \mid v_1 + v_2 = 0\} = \{te_1^0 \mid t \in \mathbb{R}\},$$
$$T_0 e_1^0 = b e_1^0, \qquad T_1 e_1^0 = -b e_1^0,$$

where $e_1^0 = (-1,1)^t$. Therefore, from Proposition 2.1 it follows that $\Omega_j(\varphi) \le |b|^j$ for all $j \in \mathbb{N}$. This estimate is sharp, since (8) imply

$$\varphi(h_{[0]}) - \varphi(A^{-j}h_{[1]}) = \frac{ab^j}{1-b}.$$

Hence, by Proposition 2.2 we obtain (17).

For $n = 3$ we set

$$b_0 = 1,\ b_1 = a,\ b_2 = b,\ b_3 = c,\ b_4 = 0,\ b_5 = \alpha,\ b_6 = \beta,\ b_7 = \gamma,$$

where $|a|^2 + |\alpha|^2 = |b|^2 + |\beta|^2 = |c|^2 + |\gamma|^2 = 1$. Further, we define φ and a_α as in Example 1.2. In this case, the basis of E_1 is constituted by the vectors

$$e_1 = (1,-1,1,-1)^t, \quad e_2 = (1,-1,-1,1)^t, \quad e_3 = (1,1,-1,-1)^t \tag{18}$$

and T_0, T_1 can be written as

$$T_0 = \begin{pmatrix} a_0 & a_1 & a_2 & a_3 \\ a_2 & a_3 & a_0 & a_1 \\ a_4 & a_5 & a_6 & a_7 \\ a_6 & a_7 & a_4 & a_5 \end{pmatrix}, \quad T_1 = \begin{pmatrix} a_1 & a_0 & a_3 & a_2 \\ a_3 & a_2 & a_1 & a_0 \\ a_5 & a_4 & a_7 & a_6 \\ a_7 & a_6 & a_5 & a_4 \end{pmatrix}. \tag{19}$$

The matrices of linear operators L_0 and L_1 defined on $\mathbb{R}^4$ by the matrices (19) and then restricted to E_1 with basis (18) have the following form

$$A_0 = \begin{pmatrix} 0 & \beta & b \\ 0 & \gamma & c \\ \alpha & 0 & 0 \end{pmatrix}, \quad A_1 = \begin{pmatrix} 0 & -\beta & b \\ 0 & -\gamma & c \\ -\alpha & 0 & 0 \end{pmatrix}. \tag{20}$$

We note that $\widehat{\rho}(L_0,L_1) = \widehat{\rho}(A_0,A_1)$. In [25], using Matlab 7.0 the following results are obtained:

A. The equality $\widehat{\rho}(A_0,A_1) = \rho(A_0)$ holds in the cases

1) $0 \le b \le 1,\ 0 < c \le 1,\ 0 \le \alpha < 1,\ 0 \le \beta\gamma < 1$;
2) $-1 \le b \le 0,\ -1 \le c < 0,\ -1 < \alpha \le 0,\ 0 \le \beta\gamma < 1$;
3) $0 \le b \le 1,\ -1 \le c < 0,\ 0 \le \alpha < 1,\ -1 < \beta\gamma \le 0$;
4) $-1 \le b \le 0,\ 0 < c \le 1,\ -1 < \alpha \le 0,\ -1 < \beta\gamma \le 0$.

B. The equality $\widehat{\rho}(A_0,A_1) = \rho(A_1)$ holds in the cases

1) $0 \le b \le 1,\ 0 < c \le 1,\ -1 < \alpha \le 0,\ 0 \le \beta\gamma < 1$;
2) $-1 \le b \le 0,\ -1 \le c < 0,\ 0 \le \alpha < 1,\ 0 \le \beta\gamma < 1$;
3) $0 \le b \le 1,\ -1 \le c < 0,\ -1 < \alpha \le 0,\ -1 < \beta\gamma \le 0$;
4) $-1 \le b \le 0,\ 0 < c \le 1,\ 0 \le \alpha < 1,\ -1 < \beta\gamma \le 0$.

Under a special choice of the parameters, the spectral radii of the matrices (20) can be calculated in quite a simple way. Thus, for example, if $\beta c = \gamma b$ then

$$\rho(A_0) = \max\{|\lambda| : \lambda^2 - \gamma\lambda - \alpha b = 0\}, \quad \rho(A_1) = \max\{|\lambda| : \lambda^2 + \gamma\lambda + \alpha b = 0\}.$$

THEOREM 2.1 ([25]). *Suppose that φ is as in Example 1.2 and $\widehat{\rho}$ is the joint spectral radius of the matrices* (20). *Then*

$$\widehat{\rho} = \begin{cases} \sqrt[3]{|\alpha|}, & \text{if } \ b = 0,\ |c| = 1,\ 0 \le |\alpha| < 1, \\ \max\{\sqrt{|\alpha|}, |\gamma|\}, & \text{if } \ |b| = 1,\ 0 \le |\alpha| < 1,\ 0 \le |\gamma| < 1, \\ |\gamma|, & \text{if } \ |a| = 1,\ 0 \le |\gamma| < 1 \end{cases}$$

and $\alpha_\varphi = -\log_2 \widehat{\rho}$. *In particular, if $\gamma = 0$, then φ is an dyadic entire function.*

Now, let $n = 4$. Suppose that the a_α in (1) are defined by (3) for

$$b_0 = 1,\ b_l = d_l,\ b_8 = 0,\ b_{l+8} = \gamma_l, \quad 1 \le l \le 7, \tag{21}$$

where $|d_l|^2 + |\gamma_l|^2 = 1$. Then the compactly supported L^2-solution φ of equation (1) can be expanded in the lacunar Walsh series

$$\varphi(x) = \frac{1}{8}\chi_U(A^{-3}x)\left(1 + \sum_{l \in \mathbb{N}(4)} c_l w_l(A^{-3}x)\right), \quad x \in G. \tag{22}$$

In this expansion, $\mathbb{N}(4) = \{1,2,\ldots,7\} \cup \mathbb{N}_0(4)$, where $\mathbb{N}_0(4)$ is the set of all natural numbers $l > 7$ whose binary expansion

$$l = \sum_{j=0}^{k} \mu_j 2^j, \quad \mu_j \in \{0,1\}, \quad \mu_k \ne 0 \quad k = k(l) \in \mathbb{Z}_+, \tag{23}$$

has no collection $(0,0,0,1)$, among the ordered collections $(\mu_j, \mu_{j+1}, \mu_{j+2}, \mu_{j+3})$, while the c_l are defined as in (10). We see that the mask

$$m(\omega) = \frac{1}{2}\sum_{\alpha=0}^{15} a_\alpha w_\alpha(\omega) \tag{24}$$

satisfies the conditions (5). Hence, by Theorem 1.2, the solution φ generates an MRA in $L^2(G)$ if and only if the mask m does not possess blocking sets. Let us present a complete list of collections of the parameters for which the mask m has blocking sets, and, for each of these collections, let us indicate one blocking set:

1) $d_1 = 0$, $M = U^*_{3,1} \cup U^*_{2,1} \cup U^*_{1,1}$,
2) $d_7 = 0$, $M = U^*_{3,7}$,
3) $d_2 = d_3 = 0$, $M = U^*_{2,1} \cup U^*_{3,5} \cup U^*_{2,3}$,
4) $d_2 = \gamma_5 = 0$, $M = U^*_{3,2} \cup U^*_{3,5}$,
5) $d_3 = d_5 = 0$, $M = U^*_{3,3} \cup U^*_{3,5} \cup U^*_{2,3}$.

In addition note that, for $n = 4$, the vectors

$$\begin{aligned} &e_1 = (1,-1,1,-1,1,-1,1,-1)^t, \quad e_2 = (1,1,-1,-1,-1,-1,1,1)^t,\\ &e_3 = (1,1,1,1,-1,-1,-1,-1)^t, \quad e_4 = (1,-1,1,-1,-1,1,-1,1)^t,\\ &e_5 = (1,1,-1,-1,1,1,-1,-1)^t, \quad e_6 = (1,-1,-1,1,1,-1,-1,1)^t,\\ &e_7 = (1,-1,-1,1,1,-1,1,-1)^t. \end{aligned} \tag{25}$$

constitute a basis in the space E_1.

In what follows, T_0 and T_1 stand for the matrices defined by the formulas (4) and (16) under condition (21). Also, we set

$$B_0 = \begin{pmatrix} 0 & -d_7 & 0 & 0 & d_4 & \gamma_4 & 0\\ 0 & 0 & d_3 & \gamma_3 & 0 & 0 & 0\\ \gamma_1 & 0 & 0 & 0 & 0 & 0 & 0\\ 0 & d_7 & 0 & 0 & d_5 & \gamma_5 & 0\\ 0 & 0 & d_2 & \gamma_2 & 0 & 0 & 0\\ 0 & d_6-d_7 & 0 & 0 & 0 & 0 & 0\\ 0 & 2d_7 & 0 & 0 & 0 & 0 & 0 \end{pmatrix}, \quad B_1 = \begin{pmatrix} 0 & -d_7 & 0 & 0 & d_4 & -\gamma_4 & 0\\ 0 & 0 & d_3 & -\gamma_3 & 0 & 0 & 0\\ \gamma_1 & 0 & 0 & 0 & 0 & 0 & 0\\ 0 & d_7 & 0 & 0 & d_5 & -\gamma_5 & 0\\ 0 & 0 & d_2 & -\gamma_2 & 0 & 0 & 0\\ 0 & d_6-d_7 & 0 & 0 & 0 & 0 & 0\\ 0 & 2d_7 & 0 & 0 & 0 & 0 & 0 \end{pmatrix}$$

and denote by Γ the matrix of dimension 7×7 whose first six columns consist of zeros and the last column coincides with the vector $(\gamma_4 - \gamma_7, -\gamma_3, \gamma_1, \gamma_5 + \gamma_7, -\gamma_2, \gamma_6 - \gamma_7, \gamma_7)^t$. Therefore, the linear transforms

$$y = T_0 x, \quad y = T_1 x, \quad x \in E_1,$$

have the matrices

$$A_0 = B_0 + \frac{1}{2}\Gamma, \quad A_1 = B_1 - \frac{1}{2}\Gamma,$$

respectively, in the basis (25). Under the assumptions of Proposition 2.2, the formula $\widehat{\rho}(A_0, A_1) = 2^{-\alpha_\varphi}$ is valid. Let us point out three cases in which the quantity α_φ can be calculated exactly.

THEOREM 2.2. *Suppose that the function* φ *is given by the expansion* (22). *Then*

1) $\alpha_\varphi = (1/2)\log_2(1/|\gamma_2|)$, *if* $\gamma_2 \neq 0$ *and* $\gamma_1 = \gamma_5 = \gamma_7 = 0$;

2) $\alpha_\varphi = (1/2)\log_2(1/|d_5\gamma_2|)$, *if* $d_5\gamma_2 \neq 0$ *and* $\gamma_1 = \gamma_3 = \gamma_7 = 0$;

3) $\alpha_\varphi = \log_2(1/|\gamma_7|)$, *if* $\gamma_7 \neq 0$ *and one of the following conditions holds*: $\gamma_1 = \gamma_2 = \gamma_3 = 0$, $\gamma_1 = \gamma_2 = \gamma_5 = 0$, $d_5 = \gamma_1 = \gamma_3 = 0$.

A similar result for refinable functions in $L^2(\mathbb{R}_+)$ was proved by Farkov and Rodionov in [25].

As consequence of Theorem 2.2, let us point out three cases in which the solution φ is a binary entire function. Suppose that $x \in U_{-3}$ and put $y = A^{-3}(x)$. Then, setting,

$$W_7(y) := w_1(y) + d_2w_2(y) + d_3w_3(y) + d_2d_4w_4(y) + d_3d_6w_6(y) + d_3d_7w_7(y).$$

we obtain the following three expansions:

1. If $\gamma_1 = \gamma_2 = \gamma_3 = \gamma_7 = 0$, then

$$8\varphi(x) = 1 + d_1\{W_7(y) + d_2d_5w_5(y) + d_3d_6\gamma_4w_{12}(y) + d_3d_6\gamma_5w_{13}(y)$$
$$+ d_3d_7\gamma_6w_{14}(y) + d_3d_7\gamma_4\gamma_6w_{28}(y) + d_3d_7\gamma_5\gamma_6w_{29}(y)\}.$$

2. If $\gamma_1 = \gamma_2 = \gamma_5 = \gamma_7 = 0$, then

$$8\varphi(x) = 1 + d_1\{W_7(y) + d_2d_5w_5(y) + d_2d_5\gamma_3w_{11}(y) + d_3d_6\gamma_4w_{12}(y)$$
$$+ d_3d_7\gamma_6w_{14}(y) + d_2d_5d_6\gamma_3w_{22}(y) + d_2d_5d_7\gamma_3w_{23}(y) + d_3d_7\gamma_4\gamma_6w_{28}(y)$$
$$+ d_2d_5d_6\gamma_3\gamma_4w_{44}(y) + d_2d_5d_7\gamma_3\gamma_6w_{46}(y) + d_2d_5d_7\gamma_3\gamma_4\gamma_6w_{92}(y)\}.$$

3. If $d_5 = \gamma_1 = \gamma_3 = \gamma_7 = 0$, then

$$8\varphi(x) = 1 + d_1\{W_7(y) + d_3d_6\gamma_4w_{12}(y) + d_3d_7\gamma_6w_{14}(y) + d_3d_6\gamma_2\gamma_5w_{26}(y)$$
$$+ d_3d_7\gamma_4\gamma_6w_{28}(y) + d_3d_7\gamma_5\gamma_6w_{29}(y) + d_3d_4d_6\gamma_2\gamma_5w_{52}(y)$$
$$+ d_3d_7\gamma_2\gamma_5\gamma_6w_{58}(y) + d_3d_4d_7\gamma_2\gamma_5\gamma_6w_{116}(y)\}.$$

It turn out that if an L^2-solution φ of equation (1) is not dyadic entire, then it has a finite smoothness. Moreover, the α_φ can be estimated by the smallest non-zero value of the modulus of the mask. Indeed, let

$$b(m) := \min\{\,|m(\omega)|\,|\,\omega \in G,\ m(\omega) \neq 0\};$$

since the mask m is an H-periodic function and belongs to $\mathcal{E}_n(G^*)$, the value of $b(m)$ coincides with the smallest non-zero number in the set $\{\,|b_s|\,|\,s = 0, 1, \ldots, 2^n - 1\}$, where the b_s are defined by (4).

THEOREM 2.3 ([24]). *Suppose that* φ *is a compactly supported* L^2*- solution of the equation* (1) *such that* $\widehat{\varphi}(\theta) = 1$. *Then the following alternative holds: either* φ *is a dyadic entire function, or* $b(m) < 1$ *and* $\alpha_\varphi \leq -\log_2 b(m)$.

There exist infinitely smooth functions in G that are not dyadic entire. For refinable functions, however, by Theorem 2.3 this situation is impossible. The estimate $\alpha_\varphi \le -\log_2 b(m)$ is sharp for refinable functions (it is attained, for example, for φ defined by the expansion (8)). By contrast with the situation on the real line $\mathbb{R}$, where for each n there exists a unique refinable function of maximum smoothness (the cardinal B-spline of order $n-1$), the smoothness of solutions of equation (1) is unbounded for fixed n. However, it is bounded for functions that are not dyadic entire for fixed value of $b(m)$. This estimate is always attained for fixed n and $b(m)$ on compactly supported dyadic refinable functions. Another useful consequence of Theorem 2.3 concerns the case where the modulus of the mask m takes only two values: 0 and 1; in this case, the solution φ is dyadic entire.

3. Construction of Biorthogonal Wavelets

As we have seen in Section 1, for each refinable function φ generating an MRA in $L^2(G)$, we can construct an orthogonal wavelet ψ such that $\{\psi_{j,h}\}$ is an orthonormal basis in $L^2(G)$. Moreover, for any positive integer n there exists a refinement equation of the form (1) such that its solution φ generates an MRA in $L^2(G)$.

In this section, we consider the following problem: given two masks

$$m(\omega) = \frac{1}{2}\sum_{\alpha=0}^{2^n\ 1} a_\alpha W_\alpha^*(\omega), \qquad \widetilde{m}(\omega) = \frac{1}{2}\sum_{\alpha=0}^{2^{\tilde{n}}-1} \widetilde{a}_\alpha W_\alpha^*(\omega) \tag{26}$$

how can we find biorthogonal wavelet bases in $L^2(G)$?

By analogy with [41], we have the following

PROPOSITION 3.1. *Let $\varphi, \widetilde{\varphi} \in L^2(G)$. The systems $\{\varphi(\cdot \oplus h) \,|\, h \in H\}$ and $\{\widetilde{\varphi}(\cdot \oplus h) \,|\, h \in H\}$ are biorthonormal in $L^2(G)$ if and only if*

$$\sum_{h\in H} \widehat{\varphi}(\omega \oplus h)\overline{\widehat{\widetilde{\varphi}}(\omega \oplus h)} = 1 \quad \textit{for a.e.} \quad \omega \in G^*.$$

Given a refinement equation (1), let us define two linear operators L_0, L_1 as in Proposition 2.2. By analogy with similar results in [20] (see also [41] and [47]) can be proved the following

LEMMA 3.1. *Let the mask m of the refinement equation* (1) *satisfies $m(\theta) = 1$, $m(\delta_0) = 0$ and $\widehat{\rho}\,[m] := \widehat{\rho}(L_0, L_1) < 1$. Then the function φ defined by* (10) *satisfies the equation* (1) *and belongs to $L^2(G)$. Moreover, we have*

$$\sum_{\alpha=0}^{2^{n-1}-1} a_{2\alpha} = \sum_{\alpha=0}^{2^{n-1}-1} a_{2\alpha+1} = 1.$$

Now, let us define two scaling functions $\varphi, \widetilde{\varphi}$ in terms of the Fourier transform:

$$\widehat{\varphi}(\omega) = \prod_{j=1}^{\infty} m(B^{-j}\omega), \quad \widehat{\widetilde{\varphi}}(\omega) = \prod_{j=1}^{\infty} \widetilde{m}(B^{-j}\omega), \tag{27}$$

where m, $\widetilde{m}$ are given by (26). Suppose that

$$m(\theta)=\widetilde{m}(\theta)=1, \quad m(\delta_0)=\widetilde{m}(\delta_0)=0. \tag{28}$$

We set $m^*(\omega)=m(\omega)\overline{\widetilde{m}(\omega)}$, $N=\max\{n,\widetilde{n}\}$ and $a_\alpha=\widetilde{a}_\beta=0$ for $\alpha\geq 2^n, \beta\geq 2^{\widetilde{n}}$. Using the equality $W_\alpha(\omega)W_\beta(\omega)=W_{\alpha\oplus\beta}(\omega)$, we obtain

$$m^*(\omega)=\frac{1}{2}\sum_{\alpha=0}^{2^N-1}a^*_\alpha W_\alpha(\omega), \quad a^*_\alpha=\frac{1}{2}\sum_{\beta=0}^{2^N-1}a_\beta\overline{\widetilde{a}}_{\alpha\oplus\beta}. \tag{29}$$

PROPOSITION 3.2. *If the systems* $\{\varphi(\cdot\oplus h)\,|\,h\in H\}$, $\{\widetilde{\varphi}(\cdot\oplus h)\,|\,h\in H\}$ *are biorthonormal in* $L^2(G)$*, then*

$$m^*(\omega)+m^*(\omega\oplus\delta_0)=1 \quad \textit{for all} \quad \omega\in B^{-1}(U). \tag{30}$$

PROOF. As noted in Introduction, if $f\in L^1(G)$ and $\operatorname{supp} f\subset U_{-n}$, then $\widehat{f}\in\mathcal{E}_n(G^*)$. Therefore, the Fourier transforms of φ and $\widetilde{\varphi}$ belongs to $\mathcal{E}_{N-1}(G^*)$ (see Theorem 2.1). From this, by Lemma 3.1, it follows that for any $\omega\in U$

$$1=\sum_{h\in H}\widehat{\varphi}(\omega\oplus h)\overline{\widehat{\widetilde{\varphi}}(\omega\oplus h)}=\sum_{h\in H}m^*(B^{-1}\omega\oplus B^{-1}h)\,\widehat{\varphi}(B^{-1}\omega\oplus B^{-1}h)\overline{\widehat{\widetilde{\varphi}}(B^{-1}\omega\oplus B^{-1}h)}.$$

Further, using periodicity of m^*, we obtain

$$1=m^*(B^{-1}\omega)\sum_{h\in H}\widehat{\varphi}(B^{-1}\omega\oplus h)\overline{\widehat{\widetilde{\varphi}}(B^{-1}\omega\oplus h)}+m^*(B^{-1}\omega\oplus\delta_0)\sum_{h\in H}\widehat{\varphi}((\omega\oplus(\delta_0\oplus h)))\overline{\widehat{\widetilde{\varphi}}(\omega\oplus(\delta_0\oplus h))},$$

where both sums equal to 1. This completes the proof of Proposition 3.2.

Now, let us define φ^* by the formula

$$\varphi^*(x):=\int_G\varphi(t\oplus x)\overline{\widetilde{\varphi}(t)}\,d\mu(t).$$

The Fourier transform of φ^* can be written as $\widehat{\varphi}^*(\omega)=\widehat{\varphi}(\omega)\overline{\widehat{\widetilde{\varphi}}(\omega)}$ (cf. [41]). Hence, $\widehat{\varphi}^*(\omega)=m^*(B^{-1}\omega)\widehat{\varphi}^*(B^{-1}\omega)$. Using the inverse Fourier transform, we can deduce from (29) that φ^* satisfies

$$\varphi^*(x)=\sum_{\alpha=0}^{2^N-1}a^*_\alpha\varphi^*(Ax\oplus h_\alpha), \quad x\in G.$$

Thus, the Walsh polynomial m^* is the mask of the refinable function φ^*.

PROPOSITION 3.3. *If one of the masks* m, $\widetilde{m}$, m^* *has blocking set, then* $\{\varphi(\cdot\oplus h)\,|\,h\in H\}$, $\{\widetilde{\varphi}(\cdot\oplus h)\,|\,h\in H\}$ *are not biorthonormal systems in* $L^2(G)$.

The proof of a similar result for the space $L^2(\mathbb{R}_+)$ is given in [22].

From identity

$$m^*(\omega)+m^*(\omega\oplus\delta_0)=\frac{1}{2}\sum_\alpha\sum_l a_\alpha\overline{\widetilde{a}}_{\alpha\oplus 2\beta}W_{\alpha\oplus 2\beta}(\omega)$$

we see that (30) is equivalent to the following

$$\sum_{\alpha} a_\alpha \overline{\widetilde{a}}_{\alpha\oplus 2\beta} = 2\delta_{0,\beta}, \quad \beta \in \mathbb{Z}_+.$$

Besides, it is possible write (30) in the form

$$b_\alpha \overline{\widetilde{b}}_\alpha + b_{\alpha+2^{N-1}} \overline{\widetilde{b}}_{\alpha+2^{N-1}} = 1, \quad 0 \le \alpha \le 2^{N-1}-1, \tag{31}$$

where $b_\alpha = m(B^{-N}h^*_{[\alpha]})$, $\widetilde{b}_\alpha = \widetilde{m}(B^{-N}h^*_{[\alpha]})$.

THEOREM 3.1. *Let* m^* *satisfies* (30). *Then systems* $\{\varphi(\cdot \oplus h) | h \in H\}$, $\{\widetilde{\varphi}(\cdot \oplus h) | h \in H\}$ *are biorthonormal in* $L^2(G)$ *if and only if there exists a compact set E congruent to U modulo H, and containing a neighbourhood of the zero-element* θ, *such that*

$$\inf_{j\in\mathbb{N}} \inf_{\omega\in E} |m(B^{-j}\omega)| > 0, \qquad \inf_{j\in\mathbb{N}} \inf_{\omega\in E} |\widetilde{m}(B^{-j}\omega)| > 0. \tag{32}$$

Notice that if $m^*(\omega) \neq 0$ for all $\omega \in B^{-1}(U)$, then (32) is valid for $E = U$.

Suppose that $\{V_j\}$, $\{\widetilde{V}_j\}$ are two MRA's in $L^2(G)$. We say that two functions $\psi \in V_1$ and $\widetilde{\psi} \in \widetilde{V}_1$ form a *biorthogonal wavelet pair* if $\psi \perp \widetilde{V}_0$, $\widetilde{\psi} \perp V_0$ and $(\psi(\cdot \oplus k), \widetilde{\psi}(\cdot \oplus l)) = \delta_{k,l}$, $k, l \in H$. As usual, let I be the identity matrix and let $\mathcal{M}^*$ denote the conjugate matrix of a matrix $\mathcal{M}$.

THEOREM 3.2. *Let* $\{V_j\}$, $\{\widetilde{V}_j\}$ *be two MRA's generated by refinable functions* φ, $\widetilde{\varphi}$, *respectively, and let* $\{\varphi(\cdot \oplus h) | h \in H\}$, $\{\widetilde{\varphi}(\cdot \oplus h) | h \in H\}$ *are biorthonormal systems. Suppose that the matrices*

$$\mathcal{M} = \begin{pmatrix} m(\omega) & m(\omega \oplus \delta_0) \\ m_1(\omega) & m_1(\omega \oplus \delta_0) \end{pmatrix}, \qquad \widetilde{\mathcal{M}} = \begin{pmatrix} \widetilde{m}(\omega) & \widetilde{m}(\omega \oplus \delta_0) \\ \widetilde{m}_1(\omega) & \widetilde{m}_1(\omega \oplus \delta_0) \end{pmatrix},$$

satisfy the condition $\mathcal{M}\widetilde{\mathcal{M}}^* = I$ *for a.e.* $\omega \in U^*_{1,0}$. *Then* ψ *and* $\widetilde{\psi}$ *given by the equalities*

$$\widehat{\psi}(\omega) = m_1(B^{-1}\omega)\widehat{\varphi}(B^{-1}\omega), \qquad \widehat{\widetilde{\psi}}(\omega) = \widetilde{m}_1(B^{-1}\omega)\widehat{\widetilde{\varphi}}(B^{-1}\omega),$$

form a biorthogonal wavelet pair. In particular, we can choose

$$m_1(\omega) = -w_1(\omega)\overline{m(\omega \oplus \delta_0)}, \qquad \widetilde{m}_1(\omega) = -w_1(\omega)\overline{\widetilde{m}(\omega \oplus \delta_0)}.$$

The analogues of Theorems 3.1 and 3.2 for the space $L^2(\mathbb{R})$ are well known, their proofs for the case $\varphi = \widetilde{\varphi}$ are given in [24] and can be easily modified for the biorthogonal case.

The above formulated results justify the following *algorithm of biorthogonal wavelet pair construction*.

1. Choose parameters b_l, $\widetilde{b}_s$ with $0 \le l \le 2^{n-1}-1$, $0 \le s \le 2^{\widetilde{n}-1}-1$, so that equalities (31) are true.

2. Calculate a_α, $\widetilde{a}_\alpha$ by (3) and check that the corresponding masks m, $\widetilde{m}$ have no blocking sets.

3. Check that these masks m, $\widetilde{m}$ satisfy the condition (32) (search E as a finite union of sets $U^*_{n,\alpha}$) and by means of Lemma 3.1 check that the corresponding refinable functions φ, $\widetilde{\varphi}$ belong to $L^2(G)$.

4. Define ψ, $\widetilde{\psi}$ by the formulas

$$\psi(x) = \sum_{\alpha=0}^{2^n-1} (-1)^\alpha \overline{\widetilde{a}}_{\alpha\oplus 1}\varphi(Ax\oplus h_{[\alpha]}), \qquad \widetilde{\psi}(x) = \sum_{\alpha=0}^{2^{\widetilde{n}}-1} (-1)^\alpha \overline{a}_{\alpha\oplus 1}\widetilde{\varphi}(Ax\oplus h_{[\alpha]}).$$

EXAMPLE 3.1. Let $n = \widetilde{n} = 2$. Suppose that the masks of refinable functions φ, $\widetilde{\varphi}$ are given by

$$m(\omega) = \begin{cases} 1, & \omega \in U^*_{2,0}, \\ a, & \omega \in U^*_{2,1}, \\ 0, & \omega \in U^*_{2,2}, \\ b, & \omega \in U^*_{2,3}, \end{cases} \qquad \widetilde{m}(\omega) = \begin{cases} 1, & \omega \in U^*_{2,0}, \\ \widetilde{a}, & \omega \in U^*_{2,1}, \\ 0, & \omega \in U^*_{2,2}, \\ \widetilde{b}, & \omega \in U^*_{2,3}, \end{cases}$$

where $a\overline{\widetilde{a}} + b\overline{\widetilde{b}} = 1$. If $a = 0$ or $\widetilde{a} = 0$, then $U^*_{1,1}$ is blocking set for m or $\widetilde{m}$, otherwise blocking sets for m, $\widetilde{m}$, m^* do not exist. Now, let $|b| < 1$ and $|\widetilde{b}| < 1$. Then $a\overline{\widetilde{a}} \neq 0$ and the blocking sets are absent. Moreover, the condition (32) is satisfied for $E = U$. Besides, both functions φ, $\widetilde{\varphi}$ belong to $L^2(G)$. Really, as in Example 2.1 we see that $\widehat{\rho}\,[m] = |b|$, $\widehat{\rho}\,[\widetilde{m}] = |\widetilde{b}|$; hence, Lemma 1.3 is applicable. The following decomposition takes place

$$\varphi(x) = (1/2)\mathbf{1}_U(A^{-1}x)(1 + a\sum_{j=0}^{\infty} b^j W_{2^{j+1}-1}(A^{-1}x))$$

and a similar formula is true for $\widetilde{\varphi}$; cf. (8). Thus, under conditions $a\overline{\widetilde{a}} + b\overline{\widetilde{b}} = 1$, $|b| < 1$, $|\widetilde{b}| < 1$, the H-shifts of functions φ, $\widetilde{\varphi}$ forms an biorthonormal system in $L^2(G)$.

EXAMPLE 3.2. Let $n = 3$, $\widetilde{n} = 2$. Suppose $\widetilde{m}$ is the same as in Example 3.1 and the mask m is defined by

$$\begin{array}{ll} m(\omega) = 1 \ \text{for} \ \omega \in U^*_{3,0}, & m(\omega) = a \ \text{for} \ \omega \in U^*_{3,1}, \\ m(\omega) = b \ \text{for} \ \omega \in U^*_{3,2}, & m(\omega) = c \ \text{for} \ \omega \in U^*_{3,3}, \\ m(\omega) = 0 \ \text{for} \ \omega \in U^*_{3,4}, & m(\omega) = \alpha \ \text{for} \ \omega \in U^*_{3,5}, \\ m(\omega) = \beta \ \text{for} \ \omega \in U^*_{3,6}, & m(\omega) = \gamma \ \text{for} \ \omega \in U^*_{3,7}. \end{array}$$

The set $U^*_{1,1}$ is a blocking set for m, $\widetilde{m}$, m^* in the following cases 1) $b = c = 0$, 2) $\widetilde{a} = 0$, 3) $b\overline{\widetilde{a}} = c\overline{\widetilde{a}} = 0$ respectively. Besides, $U^*_{2,1}$ is a blocking set for m and m^* in the cases $c = 0$ and $c\overline{\widetilde{a}} = 0$, respectively. For the other values of parameters blocking sets for $m, \widetilde{m}, m^*$ do not exist. Note that, in the case $\widetilde{a} = 0$, $bc \neq 0$, the set $U^*_{2,3}$ is a blocking set for m^* and is not a blocking set for $m, \widetilde{m}$. Now assume that

$$a = 1, \quad b\overline{\widetilde{a}} + \beta\overline{\widetilde{b}} = c\overline{\widetilde{a}} + \gamma\overline{\widetilde{b}} = 1$$

(in particular, for $b = 0$, $c = 1$ we have $\beta = 1/\overline{\widetilde{b}}$, $\gamma = (1 - \overline{\widetilde{a}})/\overline{\widetilde{b}}$). The blocking sets for $m, \widetilde{m}, m^*$ exist only if $\widetilde{a} = 0$ or $c = 0$. In the case $\widetilde{a} \neq 0$ the condition (32) is valid for $E = U^*_{1,0} \cup U^*_{2,3} \cup U^*_{2,6}$. According to Examples 2.1 and 2.2 we have $\widehat{\rho}\,[m] = |\gamma|$ and $\widehat{\rho}\,[\widetilde{m}] = |\widetilde{b}|$. Thus, if $\widetilde{a} \neq 0$, $|\gamma| < 1$ and $|\widetilde{b}| < 1$, the families of H-shifts of φ, $\widetilde{\varphi}$ are biorthonormal systems in $L^2(G)$.

4. Generalisation to the p-adic Vilenkin Group

Let us recall the definition of the p-adic Vilenkin group G_p and some related concepts. Given a prime number p, we denote by $\mathbb{F}_p$ the field of order p, with elements $\{0,1,\dots,p-1\}$. Then the *p-adic field* $\mathbb{F}$ is the subset of $\prod_{j\in\mathbb{Z}}\mathbb{F}_p$ consisting of sequences

$$x=(x_j)=(\dots,x_{-2},x_{-1},x_0,x_1,x_2,\dots),$$

for which $x_j\to 0$ as $j\to-\infty$. Addition on $\mathbb{F}$ is the coordinate- wise addition modulo p:

$$(z_j)=(x_j)\oplus(y_j)\iff z_j=x_j+y_j\,(\operatorname{mod} p),$$

while multiplication on $\mathbb{F}$ follows the rule

$$(z_j)=(x_j)\cdot(y_j)\iff z_j=\sum_{l+k=j}x_l y_k\,(\operatorname{mod} p).$$

Define projections $\pi_j:\ \mathbb{F}\to\{0,1,\dots,p-1\}$ by $\pi_j(x)=x_j$ where $x=(x_j)$ and, for each $n\in\mathbb{Z}$, choose an unit element $e_n\in\mathbb{F}$ such that

$$\pi_j(e_n)=\begin{cases}1, & j=n,\\ 0, & \text{otherwise.}\end{cases}$$

Observe that $e_n\cdot x=(x_{j-n})$ for all $x=(x_j)\in\mathbb{F}$.

Denote by θ the zero sequence in $\mathbb{F}$. We see that for each $x\in\mathbb{F}$ with $x\neq\theta$ there exists $s(x)\in\mathbb{Z}$ such that

$$x_{s(x)}\neq 0 \quad\text{and}\quad x_j=0 \quad\text{for}\quad j<s(x).$$

There is a non-Archimedean norm on $\mathbb{F}$. Indeed, set $\|\theta\|=0$ and for each $x\in\mathbb{F}$ with $x\neq\theta$ set $\|x\|=2^{-s(x)}$. Then

$$\|x+y\|\le\max\{\|x\|,\|y\|\}\quad\text{and}\quad\|x\cdot y\|=\|x\|\,\|y\|$$

for $x,y\in\mathbb{F}$; cf. [6, p.414].

The *p-adic Vilenkin group* G_p can be defined as the additive group of the p-adic field $\mathbb{F}$, with the topology induced by $\|\cdot\|$. In the case $p=2$, we obtain the Cantor dyadic group: $G_2=G$, where G was defined in Introduction. The above given notations A, B, G^*, λ, μ, U, H, U_l, $h_{[\alpha]}$ and so on can be extended into more general case $p\ge 2$. So, the automorphism A coincides with the multiplication-by-e_{-1}, $H=\{(x_j)\in G_p\,|\,x_j=0 \text{ for } j\ge 0\}$ is the discrete subgroup of G_p, and the sets

$$U_l=\{(x_j)\in G_p\,|\,x_j=0\text{ for } j<l\},\quad l\in\mathbb{Z},$$

form a complete system neighbourhoods of θ. Further, the duality pairing takes $x=(x_j)\in G_p$, $\omega=(\omega_j)\in G_p^*$ to

$$\chi(x,\omega)=\exp\left(\frac{2\pi i}{p}\sum_{j\in\mathbb{Z}}x_j\,\omega_{-j-1}\right),$$

a map $\lambda : G_p \to \mathbb{R}_+$ can be defined by

$$\lambda(x) = \sum_{j\in\mathbb{Z}} x_j p^{-j-1}, \qquad x = (x_j) \in G_p,$$

and the *generalized Walsh functions* on G_p are given by

$$W_\alpha(x) = \chi(x, h_{[\alpha]}), \quad x \in G_p, \quad \alpha \in \mathbb{Z}_+.$$

As in the case $p = 2$, we denote by $L^2_c(G_p)$ the set of all functions in $L^2(G_p)$ with compact support. A function $\varphi \in L^2(G_p)$ is called a *refinable function* if it satisfies a refinement equation

$$\varphi(x) = p \sum_{\alpha=0}^{p^n-1} c_\alpha (Ax \ominus h_{[\alpha]}), \quad x \in G_p, \tag{33}$$

where c_α are complex coefficients. Applying the Fourier transform, we can write

$$\widehat{\varphi}(\omega) = m(B^{-1}\omega)\widehat{\varphi}(B^{-1}\omega), \tag{34}$$

where

$$m(\omega) = \sum_{\alpha=0}^{p^n-1} c_\alpha \overline{W^*_\alpha(\omega)} \tag{35}$$

is a generalized Walsh polynomial, which is called the *mask* of the refinable function φ.

Each function $W^*_\alpha(\omega)$ with $0 \le \alpha \le p^n - 1$ is constant on the sets

$$U^*_{n,s} := B^{-n}(\omega_{[s]}) \oplus B^{-n}(U^*), \quad 0 \le s \le p^n - 1. \tag{36}$$

The coefficients c_α are related to the values b_s of the mask m on $U^*_{n,s}$ by the direct and inverse VilenkinŰChrestenson discrete transforms:

$$c_\alpha = \frac{1}{p^n} \sum_{s=0}^{p^n-1} b_s W^*_\alpha(B^{-n}\omega_{[s]}), \quad 0 \le \alpha \le p^n - 1, \tag{37}$$

$$b_s = \sum_{\alpha=0}^{p^n-1} c_\alpha \overline{W^*_\alpha(B^{-n}\omega_{[s]})}, \qquad 0 \le s \le p^n - 1. \tag{38}$$

ParsevalŠs formula for these transforms takes the form

$$\sum_{\alpha=0}^{p^n-1} c_\alpha \overline{\widetilde{c}_\alpha} = \frac{1}{p^n} \sum_{s=0}^{p^n-1} b_s \overline{\widetilde{b}_s}, \tag{39}$$

where c_α, b_s and $\widetilde{c}_\alpha, \widetilde{b}_s$ are arbitrary sets of numbers satisfying equalities (37) and (38).

A function $f \in L^2(G)$ is said to be *stable* if there exist positive constants A_0 and B_0 such that

$$A_0 \Big(\sum_{\alpha=0}^{\infty} |a_\alpha|^2 \Big)^{1/2} \le \Big\| \sum_{\alpha=0}^{\infty} a_\alpha f(\cdot \oplus h_{[\alpha]}) \Big\| \le B_0 \Big(\sum_{\alpha=0}^{\infty} |a_\alpha|^2 \Big)^{1/2}$$

for each sequence $\{a_\alpha\} \in \ell^2$. In other words, a function f is stable in $L^2(G)$ if its shifts $f(\cdot \oplus h)$ with $h \in H$ form a Riesz system in $L^2(G)$. It is shown in [24] (see also [22]) that if $\varphi \in L_c^2(G)$ is a refinable function in $L^2(G)$ and $\widehat{\varphi}(\theta) = 1$, then φ is not stable if and only if its mask m possesses a blocking set.

We say that a function $g : G^* \to \mathbb{C}$ has a *periodic zero* at a point $\omega \in G^*$ if $g(\omega \oplus h^*) = 0$ for all $h^* \in H^\perp$.

The following theorem [49] characterizes all compactly supported stable functions in $L^2(G)$:

THEOREM 4.1 *For any function $f \in L_c^2(G)$ the following conditions are equivalent*:

(a) *f is stable in $L^2(G)$*;

(b) $\{f(\cdot \ominus h) | h \in H\}$ *is linearly independent*;

(c) *the Fourier transform of f has no periodic zeros.*

For arbitrary $p, n \in \mathbb{N}$, $p \geq 2$ there exist c_α, such that equation (33) has a solution $\varphi \in L_c^2(G_p)$ which generates an MRA for $L^2(G_p)$ (see [21],[49]). Moreover, for each refinable function φ, generating an MRA in $L^2(G_p)$, it is possible to construct wavelets $\psi^{(1)}, \ldots, \psi^{(p-1)}$ such that

$$\psi_{j,h}^{(\nu)}(x) = p^{j/2} \psi^{(\nu)}(A^j x \ominus h), \quad 1 \leq \nu \leq p-1, \, j \in \mathbb{Z}, \, h \in H,$$

form an orthonormal basis in $L^2(G_p)$. More precisely, we have the following algorithm:

1. Choose numbers b_s, $0 \leq s \leq p^n - 1$, so that

$$b_0 = 1, \quad |b_j|^2 + |b_{j+p^{n-1}}|^2 + \ldots + |b_{j+(p-1)p^{n-1}}|^2 = 1, \quad 0 \leq j \leq p^{n-1} - 1.$$

2. Compute c_α by (37) and verify that the mask

$$m_0(\omega) = \sum_{\alpha=0}^{p^n-1} c_\alpha \overline{W_\alpha^*(\omega)}.$$

has no blocking sets.

3. Find

$$m_\nu(\omega) = \sum_{\alpha \in \mathbb{Z}_+} c_\alpha^{(\nu)} \overline{W_\alpha^*(\omega)}, \quad 1 \leq \nu \leq p-1,$$

such that $M(\omega) := (m_\nu(\omega \oplus B^{-1}\omega_{[k]}))_{\nu,k=0}^{p-1}$ is an unitary matrix.

4. Define $\psi^{(1)}, \ldots, \psi^{(p-1)}$ by the formula

$$\psi^{(\nu)}(x) = p \sum_{\alpha \in \mathbb{Z}_+} c_\alpha^{(\nu)} \varphi(Ax \ominus h_{[\alpha]}), \quad 1 \leq \nu \leq p-1.$$

Examples of realization of this algorithm for the case $p = 3$ are given in [22], [49] and [52] (see also Example 4.2 below).

We shall now show how to deduce an analogue of expansion (10) for the case $p > 2$. Let $l \in \{0, 1, \ldots, p-1\}$ and let δ_l denote the sequence $\omega = (\omega_j)$ in which $\omega_1 = l$ and $\omega_j = 0$ for

$j \neq 1$ (in particular, $\delta_0 = \theta$). Note that $W_\alpha^*(\delta_l) = \varepsilon_p^{\alpha l}$ and $\delta_l = B^{-1}\omega_{[l]}$ for $0 \le l \le p-1$. Suppose that $\varphi \in L_c^2(G_p)$ is a solution of equation (33) and its mask satisfies the conditions

$$m(\theta) = 1 \quad \text{and} \quad \sum_{l=0}^{p-1} |m(\omega \oplus \delta_l)|^2 = 1, \quad \omega \in G_p^*. \tag{40}$$

We shall prove that

$$\varphi(x) = (1/p^{n-1})\mathbf{1}_U(A^{1-n}x)(1 + \sum_{l \in \mathbb{N}(p,n)} d_l[m] W_l(A^{1-n}x)), \quad x \in G_p, \tag{41}$$

where $\mathbb{N}(p,n)$ and $c_l[m]$ are defined as follows. Let us represent each $l \in \mathbb{Z}_+$ in the form of a p-ary expansion

$$l = \sum_{j=0}^{k} \mu_j p^j, \quad \mu_j \in \{0,1,\dots,p-1\}, \quad \mu_k \neq 0, \quad k = k(l) \in \mathbb{Z}_+, \tag{42}$$

and denote the set of all positive integers $l \ge p^{n-1}$ for which the ordered sets $(\mu_j, \mu_{j+1}, \dots, \mu_{j+n-1})$ of the coefficients of (42) do not contain

$$(0,0,\dots,0,1), (0,0,\dots,0,2), \dots, (0,0,\dots,0,p-1)$$

by $\mathbb{N}_0(p,n)$. Then $\mathbb{N}(p,n) = \{1,2,\dots,p^{n-1}-1\} \cup \mathbb{N}_0(p,n)$. Let

$$\gamma(i_1,i_2,\dots,i_n) = b_s, \qquad s = i_1 p^0 + i_2 p^1 + \dots + i_n p^{n-1}, \quad i_j \in \{0,1,\dots,p-1\},$$

where b_s are defined by (38). Then

$$d_l[m] = \gamma(\mu_0,0,0,\dots,0,0) \quad \text{if} \quad k(l) = 0;$$

$$d_l[m] = \gamma(\mu_1,0,0,\dots,0,0)\gamma(\mu_0,\mu_1,0,\dots,0,0) \quad \text{if} \quad k(l) = 1;$$

$$\dots\dots\dots\dots\dots\dots\dots\dots\dots\dots,\dots\dots\dots\dots\dots\dots\dots\dots\dots\dots\dots$$

$$d_l[m] = \gamma(\mu_k,0,0,\dots,0,0)\gamma(\mu_{k-1},\mu_k,0,\dots,0,0)\dots\gamma(\mu_0,\mu_1,\mu_2,\dots,\mu_{n-2},\mu_{n-1})$$

if $k = k(l) \ge n-1$. Now, if we set $X_{n-1} = \mathbf{1}_{U_{n-1}^*}$ and apply the formula

$$\int_G X_{n-1}(\omega \ominus B^{1-n}\omega_{[l]})\chi(x,\omega)\,d\mu^*(\omega) = p^{1-n}\mathbf{1}_U(A^{1-n}x)W_l(A^{1-n}x),$$

then (41), in Fourier domain, become

$$\widehat{\varphi}(\omega) = X_{n-1}(\omega) + \sum_{l \in \mathbb{N}(p,n)} d_l[m] X_{n-1}(\omega \ominus B^{1-n}\omega_{[l]}).$$

The last equality follows from (40) since, for any $l \in \mathbb{N}(p,n)$,

$$\widehat{\varphi}(\omega) = \prod_{j=1}^{\infty} m(B^{-j}\omega) = \begin{cases} 1, & \omega \in U_{n-1,0}^* \\ d_l[m], & \omega \in U_{n-1,l}^* \\ 0 & \text{otherwise.} \end{cases}$$

Therefore, (41) is proved.

Returning to the biorthogonal case, let us take two refinable functions φ and $\widetilde{\varphi}$ with masks

$$m(\omega)=\sum_{\alpha=0}^{p^n-1} c_\alpha \overline{W_\alpha^*(\omega)}, \qquad \widetilde{m}(\omega)=\sum_{\alpha=0}^{p^{\widetilde{n}}-1} \widetilde{c}_\alpha \overline{W_\alpha^*(\omega)}, \tag{43}$$

respectively.

It is easy to generalize Propositions 3.1 and 3.2 to the case $p>2$:

PROPOSITION 4.1. *Let $\varphi, \widetilde{\varphi}\in L^2(G)$. The systems $\{\varphi(\cdot\ominus h)\,|\,h\in H\}$ and $\{\widetilde{\varphi}(\cdot\ominus h)\,|\,h\in H\}$ are biorthonormal in $L^2(G)$ if and only if*

$$\sum_{h^*\in H^\perp} \widehat{\varphi}(\omega\oplus h^*)\,\overline{\widehat{\widetilde{\varphi}}(\omega\oplus h^*)}=1 \quad \textit{for a.e.} \quad \omega\in G^*.$$

PROPOSITION 4.2. *Let φ and $\widetilde{\varphi}$ be refinable functions with masks m and $\widetilde{m}$, respectively. If the systems $\{\varphi(\cdot\ominus h)\,|\,h\in H\}$ and $\{\widetilde{\varphi}(\cdot\ominus h)\,|\,h\in H\}$ are biorthonormal in $L^2(G)$, then*

$$\sum_{l=0}^{p-1} m(\omega\oplus\delta_l)\,\overline{\widetilde{m}(\omega\oplus\delta_l)}=1 \quad \textit{for all} \quad \omega\in G^*. \tag{44}$$

Given masks (43), we set $m^*(\omega)=m(\omega)\overline{\widetilde{m}(\omega)}$ and $N=\max\{n,\widetilde{n}\}$. Condition (44) is then written in the form

$$\sum_{l=0}^{p-1} m^*(\omega\oplus\delta_l)=1, \qquad \omega\in G^*,$$

and is equivalent to the equalities

$$\sum_{\nu=0}^{p-1} b^{(N)}_{l+\nu p^{N-1}}\,\overline{\widetilde{b}}^{(N)}_{l+\nu p^{N-1}}=1, \quad 0\le l\le p^{N-1}-1, \tag{45}$$

where $b_l^{(N)}=m(B^{-N}\omega_{[l]})$ and $\widetilde{b}_l^{(N)}=\widetilde{m}(B^{-N}\omega_{[l]})$. Let $\oplus_p$ and $\ominus_p$ denote, respectively, addition and subtraction of integers modulo p. The equality $W_\alpha^*(\omega)W_\beta^*(\omega)=W^*_{\alpha\oplus_p\beta}(\omega)$, implies

$$m^*(\omega)=\sum_{\alpha=0}^{p^n-1}\sum_{\beta=0}^{p^{\widetilde{n}}-1} c_\alpha\overline{\widetilde{c}}_\beta\overline{W^*_{\alpha\oplus_p\beta}(\omega)}.$$

Setting $c_\alpha=\widetilde{c}_\beta=0$ for $\alpha\ge p^n$ and $\beta\ge p^{\widetilde{n}}$, we obtain

$$m^*(\omega)=\sum_{\alpha=0}^{p^N-1} c^*_\alpha\overline{W_\alpha^*(\omega)}, \qquad c^*_\alpha=\sum_{\gamma=0}^{p^N-1} c_\gamma\overline{\widetilde{c}}_{\gamma\ominus_p\alpha}.$$

Consider the function φ^* defined by

$$\varphi^*(x):=\int_G \varphi(t\oplus x)\overline{\widetilde{\varphi}(t)}\,d\mu(t), \quad x\in G.$$

The Fourier transform of this function is related to those of φ and $\widetilde{\varphi}$ by $\widehat{\varphi}^*(\omega)=\widehat{\varphi}(\omega)\overline{\widehat{\widetilde{\varphi}}(\omega)}$. Moreover, φ^* is a refinable function satisfying the equation

$$\varphi^*(x)=p\sum_{\alpha=0}^{p^N-1}c^*_\alpha\varphi^*(Ax\ominus\alpha),\quad x\in G.$$

Thus, the polynomial m^* is the mask of φ^*.

For $M\subset U^*$ let

$$S_pM:=\bigcup_{l=0}^{p-1}\Big\{B^{-1}\omega_{[l]}+B^{-1}(\omega)\,|\,\omega\in M\Big\}.$$

Suppose that M either coincides with one of the sets $U^*_{n-1,s}$, $1\le s\le p^{n-1}-1$, or is the union of some of these sets. Such a set M is said to be *blocking* for a mask m if $m(\omega)=0$ for all $\omega\in S_pM\setminus M$.

PROPOSITION 4.3. *If one of the masks m, $\widetilde{m}$, m^* has a blocking set, then the systems $\{\varphi(\cdot\ominus h)\,|\,h\in H\}$ and $\{\widetilde{\varphi}(\cdot\ominus h)\,|\,h\in H\}$ are not biorthonormal in $L^2(G)$.*

This proposition is given in [23], as well as the following Cohen's type criterion:

THEOREM 4.2. *If φ and $\widetilde{\varphi}$ are refinable functions whose masks $m,\widetilde{m}$ satisfy condition* (44) *and $\widehat{\varphi}(\theta)=\widehat{\widetilde{\varphi}}(\theta)=1$, then the following conditions are equivalent*:

(a) *$\{\varphi(\cdot\ominus h)\,|\,h\in H\}$ and $\{\widetilde{\varphi}(\cdot\ominus h)\,|\,h\in H\}$ are biorthonormal systems in $L^2(G)$*;

(b) *there exists a set E that is congruent to U^* modulo $H^\perp$, contains a neighborhood of zero in G^*, and is such that*

$$\inf_{j\in\mathbb{N}}\inf_{\omega\in E}|m(B^{-j}\omega)|>0,\qquad \inf_{j\in\mathbb{N}}\inf_{\omega\in E}|\widetilde{m}(B^{-j}\omega)|>0. \tag{46}$$

One of the main questions in constructing a biorthogonal wavelet system is: whether φ and $\widetilde{\varphi}$ belong to L^2? In some cases, this question can be answered by using the joint spectral radius of special linear operators defined from the refinement equations for φ and $\widetilde{\varphi}$ (cf. [41]).

Let $r=p^{n-1}$. The joint spectral radius of $r\times r$ complex matrices $A_0, A_1,\dots,A_{p-1}$ is defined as

$$\widehat{\rho}(A_0,A_1,\dots,A_{p-1}):=\lim_{k\to\infty}\max\{\|A_{d_1}A_{d_2}\dots A_{d_k}\|^{1/k}\,:\,d_j\in\{0,1,\dots,p-1\},\ 1\le j\le k\},$$

where $\|\cdot\|$ is an arbitrary norm on $\mathbb{C}^{r\times r}$. The joint spectral radius of finite-dimensional linear operators $L_0,L_1,\dots,L_{p-1}$ is defined as the joint spectral radius of their matrices in an arbitrary fixed basis of the corresponding linear space.

Given a refinement equation of the form (33), we set $a_\alpha=pc_\alpha$ and define $(r\times r)$-matrices $T_0,T_1,\dots,T_{p-1}$ by

$$(T_0)_{i,j}=a_{(pi-p)\ominus_p(j-1)},\ (T_1)_{i,j}=a_{(pi-p+1)\ominus_p(j-1)},\dots,(T_{p-1})_{i,j}=a_{(pi-1)\ominus_p(j-1)}$$

for $i,j\in\{1,2,\dots,r\}$. Consider the subspace

$$V:=\{u=(u_1,\dots,u_r)^t\,|\,u_1+\dots+u_r=0\}$$

and denote by $L_0, L_1, \ldots, L_{p-1}$ the restrictions to V of the linear operators defined on the whole space $\mathbb{C}^r$ by the matrices $T_0, T_1, \ldots, T_{p-1}$, respectively.

PROPOSITION 4.4 *If the mask m of the refinement equation* (33) *satisfies the conditions*

$$m(\theta) = 1, \quad m(\delta_1) = m(\delta_2) = \ldots = m(\delta_{p-1}) = 0,$$

and $\widehat{\rho}\,[m] := \widehat{\rho}(L_0, L_1, \ldots, L_{p-1}) < 1$, *then the function* φ, *defined by* (41), *satisfies equation* (33) *and is continuous on* G.

This proposition can be proved by analogy with similar results in [20, Sections 2 and 4]; it turns out that the series in (41) converges uniformly on G (see also Sect. 5 in [24]). It is easy to see that under the conditions of Proposition 4.4, φ has compact support and belongs to the space $L^2(G)$. Moreover, in this case

$$\sum_{\alpha=0}^{p^{n-1}-1} a_{p\alpha} = \sum_{\alpha=0}^{p^{n-1}-1} a_{p\alpha+1} = \ldots = \sum_{\alpha=0}^{p^{n-1}-1} a_{p\alpha+p-1} = 1.$$

As above, expansions for φ and $\widetilde{\varphi}$ of the form (41) are derived from the formulas

$$\widehat{\varphi}(\omega) = \prod_{j=1}^{\infty} m(B^{-j}\omega), \qquad \widehat{\widetilde{\varphi}}(\omega) = \prod_{j=1}^{\infty} \widetilde{m}(B^{-j}\omega)$$

by means of the Fourier transform.

EXAMPLE 4.1 Let $p = 3$, $n = \widetilde{n} = 2$, and the masks m and $\widetilde{m}$ take the value 1 on $U^*_{2,0}$, vanish on ■$U^*_{2,3} \cup U^*_{2,6}$, and be defined on the remaining part of U^* by the equalities

$m(\omega) = a$ and $\widetilde{m}(\omega) = \widetilde{a}$ for $\omega \in U^*_{2,1}$, $m(\omega) = \alpha$ and $\widetilde{m}(\omega) = \widetilde{\alpha}$ for $\omega \in U^*_{2,2}$,

$m(\omega) = b$ and $\widetilde{m}(\omega) = \widetilde{b}$ for $\omega \in U^*_{2,4}$, $m(\omega) = \beta$ and $\widetilde{m}(\omega) = \widetilde{\beta}$ for $\omega \in U^*_{2,5}$,

$m(\omega) = c$ and $\widetilde{m}(\omega) = \widetilde{c}$ for $\omega \in U^*_{2,7}$, $m(\omega) = \gamma$ and $\widetilde{m}(\omega) = \widetilde{\gamma}$ for $\omega \in U^*_{2,8}$,

where the parameters satisfy the condition

$$a\overline{\widetilde{a}} + b\overline{\widetilde{b}} + c\overline{\widetilde{c}} = \alpha\overline{\widetilde{\alpha}} + \beta\overline{\widetilde{\beta}} + \gamma\overline{\widetilde{\gamma}} = 1. \tag{47}$$

Then, in the cases $a\overline{\widetilde{a}} = \alpha\overline{\widetilde{\alpha}} = 0$, $a\overline{\widetilde{a}} = c\overline{\widetilde{c}} = 0$, and $\alpha\overline{\widetilde{\alpha}} = \beta\overline{\widetilde{\beta}} = 0$ blocking sets for m^* are $U^*_{1,1} \cup U^*_{1,2}$, $U^*_{1,1}$, $U^*_{1,2}$, respectively. Condition (46) holds in the following three cases:

1) $a\overline{\widetilde{a}} \neq 0$, $\alpha\overline{\widetilde{\alpha}} \neq 0$, and $E = U^*$;
2) $a\overline{\widetilde{a}} \neq 0$, $\beta\overline{\widetilde{\beta}} \neq 0$, and $E = U^*_{1,0} \cup U^*_{1,1} \cup U^*_{1,5}$;
3) $c\overline{\widetilde{c}} \neq 0$, $\alpha\overline{\widetilde{\alpha}} \neq 0$, and $E = U^*_{1,0} \cup U^*_{1,2} \cup U^*_{1,7}$.

Thus, if $\widehat{\rho}\,[m] < 1$ and $\widehat{\rho}\,[\widetilde{m}] < 1$ (this condition can be verified numerically for the specific parameter values), then, in the above three cases, the H-shifts of the refinable functions φ and $\widetilde{\varphi}$ form a biorthonormal system in $L^2(G)$.

Let $\{V_j\}$ and $\{\widetilde{V}_j\}$ be two MRAs in $L^2(G)$. We say that functions $\psi^{(\nu)} \in V_1$, $\widetilde{\psi}^{(\nu)} \in \widetilde{V}_1$, $\nu = 1,\dots,p-1$, form a *biorthogonal wavelet set* with respect to the pair $\{V_j\}$, $\{\widetilde{V}_j\}$, if $\psi^{(\nu)} \perp \widetilde{V}_0$ and $\widetilde{\psi}^{(\nu)} \perp V_0$ for all $\nu = 1,\dots,p-1$ and

$$(\psi^{(\nu)}(\cdot \oplus h_{[\alpha]}), \widetilde{\psi}^{(\varkappa)}(\cdot \oplus h_{[\alpha]})) = \delta_{\nu,\varkappa}\,\delta_{\alpha,\beta}, \qquad \nu,\varkappa \in \{1,\dots,p-1\}, \quad \alpha,\beta \in \mathbb{Z}_+.$$

We have the following generalization of Theorem 3.2:

THEOREM 4.3 ([23]). *Let $\{V_j\}$ and $\{\widetilde{V}_j\}$ be two MRAs generated by refinable functions φ and $\widetilde{\varphi}$ with masks $m = m_0$ and $\widetilde{m} = \widetilde{m}_0$, respectively. Suppose that $\{\varphi(\cdot \ominus h) \mid h \in H\}$ and $\{\widetilde{\varphi}(\cdot \ominus h) \mid h \in H\}$ are the biorthonormal systems. If the matrices*

$$\mathcal{M} = \{m_\nu(\omega + \delta_k)\}_{\nu,k=0}^{p-1}, \quad \widetilde{\mathcal{M}} = \{\widetilde{m}_\nu(\omega + \delta_k)\}_{\nu,k=0}^{p-1}, \tag{48}$$

where $m_\nu, \widetilde{m}_\nu \in L^2(U^*)$, satisfy the condition

$$\mathcal{M}\widetilde{\mathcal{M}}^* = I, \tag{49}$$

for almost every $\omega \in U^*$, then $\psi^{(\nu)}$, $\widetilde{\psi}^{(\nu)}$, $\nu = 1,\dots,p-1$, defined by the equalities

$$\widehat{\psi}^{(\nu)}(\omega) = m_\nu(B^{-1}\omega)\widehat{\varphi}(B^{-1}\omega), \qquad \widehat{\widetilde{\psi}^{(\nu)}}(\omega) = \widetilde{m}_\nu(B^{-1}\omega)\widehat{\widetilde{\varphi}}(B^{-1}\omega), \tag{50}$$

form a biorthogonal wavelet set with respect to the pair $\{V_j\}$, $\{\widetilde{V}_j\}$.

EXAMPLE 4.2. Let m and $\widetilde{m}$ are defined for $p = 3$ as in the preceding example under the condition (47). Since $b_0 = \widetilde{b}_0 = 1$, it follows from (39) and (47) that

$$\sum_{\alpha=0}^{8} c_\alpha \overline{\widetilde{c}_\alpha} = \frac{1}{3}.$$

Further, we have

$$\mathcal{M}\widetilde{\mathcal{M}}^* = I \iff \sum_{l=0}^{2} m_\nu(\omega \oplus \delta_l)\overline{\widetilde{m}_\mu(\omega \oplus \delta_l)} = \delta_{\nu\mu}, \quad 0 \le \nu,\mu \le 2, \tag{51}$$

where $m_0 = m$, $\widetilde{m}_0 = \widetilde{m}$, and

$$\mathcal{M} = \begin{pmatrix} m_0(\omega) & m_0(\omega\oplus\delta_1) & m_0(\omega\oplus\delta_2) \\ m_1(\omega) & m_1(\omega\oplus\delta_1) & m_1(\omega\oplus\delta_2) \\ m_2(\omega) & m_2(\omega\oplus\delta_1) & m_2(\omega\oplus\delta_2) \end{pmatrix}, \qquad \widetilde{\mathcal{M}} = \begin{pmatrix} \widetilde{m}_0(\omega) & \widetilde{m}_0(\omega\oplus\delta_1) & \widetilde{m}_0(\omega\oplus\delta_2) \\ \widetilde{m}_1(\omega) & \widetilde{m}_1(\omega\oplus\delta_1) & \widetilde{m}_1(\omega\oplus\delta_2) \\ \widetilde{m}_2(\omega) & \widetilde{m}_2(\omega\oplus\delta_1) & \widetilde{m}_2(\omega\oplus\delta_2) \end{pmatrix}$$

From (51) it follows that m_ν and $\widetilde{m}_\nu$ must be satisfy

$$\begin{cases} \sum_{\nu=0}^{2} m_\nu(\omega)\overline{\widetilde{m}_\nu(\omega)} = 1, \\ \sum_{\nu=0}^{2} m_\nu(\omega)\overline{\widetilde{m}_\nu(\omega\oplus\delta_1)} = 0, \\ \sum_{\nu=0}^{2} m_\nu(\omega)\overline{\widetilde{m}_\nu(\omega\oplus\delta_2)} = 0. \end{cases} \tag{52}$$

Let us recall that m_0 and $\widetilde{m}_0$ satisfy the conditions

$$m_0(\theta) = \overline{\widetilde{m}_0(\theta)} = 1, \quad m_0(\delta_1) = \overline{\widetilde{m}_0(\delta_1)} = 0, \quad m_0(\delta_2) = \overline{\widetilde{m}_0(\delta_2)} = 0. \tag{53}$$

In addition, we assume that

$$m_1(\theta) = \overline{\widetilde{m}_1(\theta)} = 0, \quad m_1(\delta_1) = \overline{\widetilde{m}_1(\delta_1)} = 1, \quad m_1(\delta_2) = \overline{\widetilde{m}_1(\delta_2)} = 0 \tag{54}$$

and

$$m_2(\theta) = \overline{\widetilde{m}_2(\theta)} = 0, \quad m_2(\delta_1) = \overline{\widetilde{m}_2(\delta_1)} = 0, \quad m_2(\delta_2) = \overline{\widetilde{m}_2(\delta_2)} = 1. \tag{55}$$

By substitution $B^{-2}\omega_{[j]}$ for ω in (52), we obtain

$$\begin{cases} \sum\limits_{\nu=0}^{2} m_\nu(B^{-2}\omega_{[j]})\overline{\widetilde{m}_\nu(B^{-2}\omega_{[j]})} = 1, \\ \sum\limits_{\nu=0}^{2} m_\nu(B^{-2}\omega_{[j]})\overline{\widetilde{m}_\nu(B^{-2}\omega_{[j]} \oplus \delta_1)} = 0, \\ \sum\limits_{\nu=0}^{2} m_\nu(B^{-2}\omega_{[j]})\overline{\widetilde{m}_\nu(B^{-2}\omega_{[j]} \oplus \delta_2)} = 0, \end{cases} \tag{56}$$

where $j = 0, 1, \ldots, 8$. Using Mathlab and (53) - (56) we can find $b_{\nu,j} = m_\nu(B^{-2}\omega_{[j]})$ and $\widetilde{b}_{\nu,j} = \widetilde{m}_\nu(B^{-2}\omega_{[j]})$ for all $\nu = 0, 1, 2$; $j = 0, 1, \ldots, 8$. Finally, we can use (37) and (50) to obtain $\psi^{(1)}$, $\widetilde{\psi}^{(1)}$, $\psi^{(2)}$, and $\widetilde{\psi}^{(2)}$ (see also Example 2 in [52]).

To conclude this section, we remark that, for the functions $\psi^{(\nu)}$ and $\widetilde{\psi}^{(\nu)}$, $\nu = 1, \ldots, p-1$, defined by (50), conditions to generate frames or Riesz bases in $L^2(G)$ can be stated by analogy with Theorem 2.7.5 in [41]. For example, it suffices to assume that $\widehat{\rho}\,[m] = \widehat{\rho}\,[\widetilde{m}] = 0$; then, for the refinable functions φ and $\widetilde{\varphi}$, expansions of the form (41) contain only finitely many non-zero coefficients, and their Fourier transforms $\widehat{\varphi}$ and $\widehat{\widetilde{\varphi}}$ have compact support. In this case, under conditions (44), (49), and

$$m_\nu(\theta) = \widetilde{m}_\nu(\theta) = 0, \quad \nu = 1, \ldots, p-1,$$

each of the systems $\{\psi_{j,h}^{(\nu)}\}$ and $\{\widetilde{\psi}_{j,h}^{(\nu)}\}$ is a frame in $L^2(G)$. If the systems $\{\varphi(\cdot \ominus h) \mid h \in H\}$ and $\{\widetilde{\varphi}(\cdot \ominus h) \mid h \in H\}$ are in addition biorthonormal, then the functions $\psi^{(\nu)}$ and $\widetilde{\psi}^{(\nu)}$, $\nu = 1, \ldots, p-1$, form a biorthogonal wavelet set with respect to the pair $\{V_j\}$, $\{\widetilde{V}_j\}$, and each of the systems $\{\psi_{j,h}^{(\nu)}\}$ and $\{\widetilde{\psi}_{j,h}^{(\nu)}\}$ is a Riesz basis in $L^2(G)$.

5. Examples of Parseval Dyadic Frames

In this section, we present some new examples of frames for $L^2(G)$, based on the results similar to those published in [28]. A Parseval frame for $L^2(G)$ related to the generalized Walsh- Dirichlet kernel is also constructed. As above, for any $\psi \in L^2(G)$ we write

$$\psi_{j,\alpha}(x) := 2^{j/2}\psi(A^j x \oplus h_{[\alpha]}), \quad j \in \mathbb{Z},\ \alpha \in \mathbb{Z}_+.$$

By the Plancherel theorem,

$$\langle f, \psi_{j,\alpha}\rangle = \langle \widehat{f}, \widehat{\psi}_{j,\alpha}\rangle = 2^{-j/2}\int_G \widehat{f}(\omega)\overline{\widehat{\psi}(A^{-j}\omega)}\, w_\alpha(A^{-j}\omega)\, d\omega, \quad f \in L^2(G).$$

Let us recall that $\{\psi_{j,\alpha}\}$ is a *frame* for $L^2(G)$ if there exist positive constants c_0, c_1 such that, for every $f \in L^2(G)$,

$$c_0\|f\|^2 \leq \sum_{j,\alpha} |\langle f, \psi_{j,\alpha}\rangle|^2 \leq c_1\|f\|^2.$$

The constants c_0 and c_1 are known respectively as lower and upper frame bounds. When $c_0 = c_1$ we have a tight frame; in this case,

$$f = {c_0}^{-1}\sum_{j,\alpha}\langle f, \psi_{j,\alpha}\rangle\psi_{j,\alpha}, \qquad f \in L^2(G).$$

A frame $\{\psi_{j,\alpha}\}$ will be called a *Parseval frame* if $c_0 = c_1 = 1$.

Now, we set

$$D_\psi(\omega) := \sum_{j\in\mathbb{Z}} |\widehat{\psi}(A^{-j}\omega)|^2, \quad M_{s,\psi} := \sup_{\omega\in G}\sum_{j\in\mathbb{Z}} |\widehat{\psi}(A^{-j}\omega)||\widehat{\psi}(A^{-j}\omega\oplus h_{[s]})|, \quad s \in \mathbb{N}.$$

Observe that $D_\psi(\omega) = D_\psi(A\omega)$ for all $\omega \in G$, and that suprema in the definition of $M_{s,\psi}$ is invariant under the transform $\omega \to A\omega$, so that this suprema can be taken over $\omega \in U_{1,0}$.

THEOREM 5.1. *Let $\psi \in L^2(G)$ be such that*

$$c_0^{(1)} := \operatorname{ess\,inf}_{\omega\in U_{1,0}} D_\psi(\omega) - \sum_{s\in\mathbb{N}} M_{s,\psi} > 0,$$

and

$$c_1^{(1)} := \operatorname{ess\,sup}_{\omega\in U_{1,0}} D_\psi(\omega) + \sum_{s\in\mathbb{N}} M_{s,\psi} < \infty.$$

Then $\{\psi_{j,\alpha}\}$ is a frame with frame bounds $c_0^{(1)}$ and $c_1^{(1)}$.

Notice that Theorem 5.1 includes the Daubechies type "admissible condition" (see, e.g., [40, Sect. 9.3], [45, Sect. 3.3]). We apply this theorem to obtain several examples of Parseval frames for $L^2(G)$.

Let

$$g_{l,s}(x) := 2^{-s}\mathbf{1}_U(A^{-s}x)W_l(A^{-s}x), \quad l, s \in \mathbb{Z}_+.$$

It is easy to check that $\widehat{g}_{l,s} = \mathbf{1}_{U_{l,s}}$, where $U_{l,s} = A^{-l}((h_{[s]} \oplus U)$. In particular, $g_{1,0}$ coincides with the Haar wavelet on G. Moreover, as noted in Section 1 any compactly supported orthogonal wavelet in $L^2(G)$ can be expanded in the gap series with respect to $\{g_{l,s}\}$ (see (10)).

EXAMPLE 5.1. Suppose that $\psi = g_{l,s}$ with $l \in \mathbb{N}$, $s \in \mathbb{Z}_+$. Then for any $\alpha \in \mathbb{N}$ the supports of $\widehat{\psi}(A^{-j}\omega)$ and $\widehat{\psi}(A^{-j}\omega \oplus h_{[\alpha]})$ are disjoint. Since

$$\operatorname{ess\,inf}_{\omega\in U_{1,0}} D_\psi(\omega) = \operatorname{ess\,sup}_{\omega\in U_{1,0}} D_\psi(\omega) = 1,$$

we see that $c_1^{(0)} = c_1^{(1)} = 1$. Therefore, by Theorem 5.1, $\{\psi_{j,\alpha}\}$ is a Parseval frame for $L^2(G)$.

EXAMPLE 5.2. Let us assume that

$$\psi(x) = g_{1,4}(x) + \nu g_{1,1}(x),$$

where ν is a positive parameter. Then, for every $\omega \in G$,

$$\widehat{\psi}(\omega) = \mathbf{1}_{U_{1,4}}(\omega) + \nu \mathbf{1}_{U_{1,1}}(\omega), \quad |\widehat{\psi}(\omega)|^2 = \mathbf{1}_{U_{1,4}}(\omega) + \nu^2 \mathbf{1}_{U_{1,1}}(\omega),$$

and

$$\widehat{\psi}(A^{-j}\omega)\widehat{\psi}(A^{-j}\omega \oplus h_{[\alpha]}) = 0 \quad \text{for} \quad j \in \mathbb{Z},\ \alpha \in \mathbb{N}.$$

From these equalities we deduce that

$$c_1^{(0)} = \nu^2, \quad c_1^{(1)} = 1 + \nu^2, \quad c_1^{(1)}/c_1^{(0)} = 1 + \frac{1}{\nu^2}.$$

Thus, $\{\psi_{j,\alpha}\}$ tends to the tight frame when $\nu \to \infty$.

It is known that the Cantor dyadic group G can be considered as the additive group of the field $\mathbb{F}_2((t))$ of formal Laurent series with coefficients 0 or 1; see, e.g. [34]. Thus, all our above formulated results can be rewritten for the the additive group of $\mathbb{F}_2((t))$. Let us recall the definition of the field $\mathbb{F}_2((t))$. Denote by $\mathbb{K}$ the field of order 2, with elements $\{0,1\}$. The field $\mathbb{F}_2((t))$ consists of formal Laurent series in the formal variable t, that is,

$$\mathbb{F}_2((t)) = \left\{ \sum_{j \geq j_0} a_j t^j \ :\ j_0 \in \mathbb{Z} \quad \text{and} \quad a_j \in \mathbb{K} \right\},$$

with the usual addition and multiplication of Laurent series. (Equivalently, $\mathbb{F}_2((t))$ is the completion of the field $\mathbb{F}_2(t)$ of rational functions with coefficients in $\mathbb{K}$ with respect to the absolute value $|f|_0 = 2^{-\operatorname{ord}_0(f)}$, where $\operatorname{ord}_0(f)$ is the order of the zero (or negative the order of the pole) of f at 0). Then $\mathbb{F}_2((t))$ is a LCAG under addition, with the topology induced by $|\cdot|_0$, and with compact open subgroup $U = \mathbb{F}_2[[t]]$ consisting of Taylor series in t; equivalently, U is the closure of the ring of polynomials $\mathbb{F}_2[t]$ in $\mathbb{F}_2((t))$.

For any nonzero a, the multiplication-by-a map is an automorphism of $G = \mathbb{F}_2((t))$, with adjoint also multiplication-by-a. We see that the automorphism A, defined earlier, coincides with the multiplication-by-$1/t$ and it takes U to the larger subgroup $t^{-1}U$.

EXAMPLE 5.3. Given nonzero $b \in G$ we set $U_{l,s}^{(b)} := b^{-s}(h_{[l]} + U)$. Take $g_{l,s}^{(\beta)} \in L^2(G)$ such that $\widehat{g}_{l,s}^{(\beta)} = \mathbf{1}_{U_{l,s}^{(b)}}$, where $\beta = \lambda(b)$, $l \in \mathbb{N}$, $s \in \mathbb{Z}_+$. In particular, if $b = 1/t$ then $U_{l,s}^{(b)} = U_{l,s}$, $g_{l,s}^{(\beta)} = g_{l,s)}^{(1)} = g_{l,s}$. As in Example 5.1, we obtain that if $\psi = g_{l,s}^{(\beta)}$, $\beta > 1/2$, then $\{\psi_{j,\alpha}\}$ is a Parseval frame for $L^2(G)$. Moreover, using Theorem 5.1, we derive the following frames for $L^2(G)$ with frame bounds $c_1^{(0)}$ and $c_1^{(1)}$:

1) $\psi(x) = g_{1,4}^{(3/2)}(x) + 2g_{2,1}^{(3/2)}(x), \quad c_1^{(0)} = 4, \quad c_1^{(1)} = 5, \quad c_1^{(1)}/c_1^{(0)} = 1{,}25,$

2) $\psi(x) = g_{3,1}^{(1)}(x) + 4g_{1,1}^{(0,85)}(x), \quad c_1^{(0)} = 9, \quad c_1^{(1)} = 25, \quad c_1^{(1)}/c_1^{(0)} = 2{,}777\ldots,$

3) $\psi(x) = g_{2,1}^{(0,5)}(x) + 3g_{1,1}^{(0,8)}(x), \quad c_1^{(0)} = 4, \quad c_1^{(1)} = 22, \quad c_1^{(1)}/c_1^{(0)} = 5,5.$

Given functions $\psi^1, \dots, \psi^N \in L^2(G)$ and a number $s \in \{0,1,2\}$ we let

$$\psi_{j,\alpha}^{\nu,s}(x) := 2^{j/2}\psi^\nu(A^j x \oplus A^{-s}h_{[\alpha]}), \quad \nu \in \{1,\dots,N\},\ j \in \mathbb{Z},\ \alpha \in \mathbb{Z}_+.$$

The set $\{\psi_{j,\alpha}^{\nu,s}\}$ is called *framelet* for $L^2(G)$ if there exist positive constants c_0, c_1 such that, for every $f \in L^2(G)$,

$$c_0\|f\|^2 \le \sum_{\nu=1}^{N}\sum_{j,\alpha} |\langle f, \psi_{j,\alpha}^{\nu,s}\rangle|^2 \le c_1\|f\|^2.$$

Let us denote

$$M^\nu(h,s) := \sup_{\omega\in G}\sum_j |\widehat{\psi}^\nu(A^{-j}\omega)||\widehat{\psi}^{\nu,s}(A^{-j}\omega \oplus A^s h)|, \quad R^{(N)}(s) := \sum_{\nu=1}^{N}\sum_{h\in H} M^\nu(h,s),$$

$$c_0^{(N)}(s) := \operatorname*{ess\,inf}_{\omega\in G}\sum_{\nu=1}^{N}\sum_j |\widehat{\psi}^{\nu,s}(A^{-j}\omega)|^2 - R^{(N)}(s),$$

$$c_1^{(N)}(s) := \operatorname*{ess\,sup}_{\omega\in G}\sum_{\nu=1}^{N}\sum_j |\widehat{\psi}^\nu(A^{-j}\omega)|^2 + R^{(N)}(s).$$

THEOREM 5.2. *Let $\psi^1, \dots, \psi^N \in L^2(G)$ be such that $c_0^{(N)}(s) > 0$ and $c_1^{(N)}(s) < \infty$. Then $\{\psi_{j,\alpha}^{\nu,s}\}$ is framelet for $L^2(G)$ and*

$$2^s c_0^{(N)}(s)\|f\|^2 \le \sum_{\nu=1}^{N}\sum_{j,\alpha} |\langle f, \psi_{j,\alpha}^{\nu,s}\rangle|^2 \le 2^s c_1^{(N)}(s)\|f\|^2$$

for all $f \in L^2(G)$.

As above, we denote by

$$m_0(\omega) = \frac{1}{2}\sum_{\alpha=0}^{2^n-1} a_\alpha W_\alpha^*(\omega)$$

the mask of equation (1)

THEOREM 5.3. *Let $\varphi \in L_c^2(G)$ be a solution of refinement equation* (1) *and let m_0 satisfies the condition*

$$|m_0(\omega)|^2 + |m_0(\omega \oplus \delta_1)|^2 = 1 \quad \textit{for all} \quad \omega \in G. \tag{57}$$

Suppose that φ is continuous at θ, $\widehat{\varphi}(\theta) \ne 0$, and that ψ is defined by $\widehat{\psi}(\omega) = m_1(A^{-1}\omega)\widehat{\varphi}(A^{-1}\omega)$, where m_1 is a Walsh polynomial such that the matrix

$$\begin{pmatrix} m_0(\omega) & m_1(\omega) \\ m_0(\omega\oplus\delta_1) & m_1(\omega\oplus\delta_1) \end{pmatrix}$$

is unitary for all $\omega \in G$. *Then* $\{\psi_{j,\alpha}\}$ *is a tight frame with frame bound* $|\widehat{\varphi}(\theta)|^2$.

In particular, we can choose

$$\psi(x) = \sum_{\alpha=0}^{2^n-1} (-1)^{\alpha} a_{2^n-1-\alpha} \varphi(Ax \oplus h_{[\alpha]}), \quad x \in G. \tag{58}$$

Observe that when (57) holds and m_0 has no blocking sets, Theorem 1.2 implies that ψ is an orthonormal wavelet for $L^2(G)$. Thus, if m_0 has a blocking set, than $\{\psi_{j,\alpha}\}$ will be a tight frame which is not basis in $L^2(G)$.

EXAMPLE 5.4. The function

$$\varphi(x) = 2^{-1}\mathbf{1}_U(A^{-1}x), \quad x \in G,$$

satisfies two equations

$$\varphi(x) = \varphi(Ax) + \varphi(Ax \oplus h_{[3]}) \quad \text{and} \quad \varphi(x) = \varphi(Ax \oplus h_{[1]}) + \varphi(Ax \oplus h_{[2]}).$$

The set $U_{1,1} = A^{-1}(U \oplus h_{[1]})$ is a blocking set for the corresponding masks. Since $\widehat{\varphi} = \mathbf{1}_{A^{-1}U}$, we have $\widehat{\varphi}(\theta) = 1$. Therefore, each function

$$\psi(x) = -\varphi(Ax \oplus h_{[1]}) + \varphi(Ax \oplus h_{[2]}) \quad \text{and} \quad \psi(x) = \varphi(Ax) - \varphi(Ax \oplus h_{[3]}),$$

gives by (58) a Parseval frame for $L^2(G)$ (compare these functions with ψ in Example 1.1 when $a = 0$).

EXAMPLE 5.5. Let us choose numbers a, b c and α, β, γ such that

$$|a|^2 + |\alpha|^2 = |b|^2 + |\beta|^2 = |c|^2 + |\gamma|^2 = 1.$$

Suppose that $a_0, a_1, \ldots, a_7$ are defined as in Example 1.2. Then for ψ, given by (58) with $n = 3$, we have: if $ac \neq 0$ then $\{\psi_{j,\alpha}\}$ is an orthonormal basis in $L^2(G)$, and if $ac = 0$ then $\{\psi_{j,\alpha}\}$ is a Parseval frame for $L^2(G)$.

Let γ be an nonzero element in G. The function $\mathcal{D}_\gamma : G \to \mathbb{R}$ defined by

$$\mathcal{D}_\gamma(x) := \int_{\gamma U} (x, \omega)\, d\omega, \quad x \in G,$$

is called the *generalized Walsh-Dirichlet kernel*. For $\varphi = \mathcal{D}_\gamma$ we have $\widehat{\varphi} = \mathbf{1}_{\gamma U}$; compare this function with the scaling function for the Shennon wavelet on the line $\mathbb{R}$ (see, e.g., [40]). We note that if $\gamma = h_{[1]}$ then φ is the Haar function $\mathbf{1}_U$, and if $\gamma == A^{-1}h_{[1]}$ then we can write φ as in Example 5.4: $\varphi(x) = 2^{-1}\mathbf{1}_U(A^{-1}x), \quad x \in G$.

Suppose now, that $\gamma \in U_1 \setminus U_2$, and let

$$V_j(\gamma) := \{f \in L^2(G) : \widehat{f}(\omega) = 0 \quad \text{for} \quad \omega \in G \setminus A^j(\gamma U)\}, \quad j \in \mathbb{Z}. \tag{59}$$

It is clear that the family of subspaces $\{V_j(\gamma)\}$ for each γ satisfies the following conditions:

$$V_j(\gamma) \subset V_{j+1}(\gamma), \quad \overline{\bigcup V_j(\gamma)} = L^2(G), \quad \bigcap V_j(\gamma) = \{\theta\}$$

(compare with the definition of an MRA in Section 1).

THEOREM 5.4. *Let* $\varphi = \mathcal{D}_\gamma$ *and* $\psi = \mathcal{D}_{A\gamma} - \mathcal{D}_\gamma$, *where* $\gamma \in U_1 \setminus U_2$. *Suppose that the subspaces* $V_j(\gamma)$ *are defined as in* (59). *Then the following are true*:

(a) $\{\varphi(\cdot \oplus h) : h \in H\}$ *is a Parseval frame for* $V_0(\gamma)$;

(b) $\{\varphi_{j,\alpha} : \alpha \in \mathbb{Z}_+\}$ *is a Parseval frame for* $V_j(\gamma)$;

(c) *if* $f \in V_j(\gamma)$ *then*

$$f(x) = \sum_{h \in H} f(A^{-j}h)\,\varphi(A^j x \oplus h), \quad x \in G; \tag{60}$$

(d) $\{\psi_{j,\alpha} : j \in \mathbb{Z}, \alpha \in \mathbb{Z}_+\}$ *is a Parseval frame for* $L^2(G)$.

As a consequence, we obtain the following equality for $\mathcal{D}_\gamma$:

$$\mathcal{D}_\gamma(x) = \sum_{h \in H} \mathcal{D}_\gamma(A^{-j}h)\,\mathcal{D}_\gamma(A^j x \oplus h), \quad x \in G.$$

Finally, we note that equality (60) is similar to the classical Shannon-Whittacker- Kotel'nikov formula (cf. [6, 16] and [28]).

References

[1] N. Ya. Vilenkin, *Izv. Akad. Nauk SSSR, Ser. Mat.* **11,** 363 (1947). Translation by Boas: *Amer. Math. Soc. Transl. Ser.* 2. **28,** 1 (1963).

[2] G. H. Agaev, N. Ya. Vilenkin,G. M. Dzhafarli, and A.I. Rubinstein, *Multiplicative Systems of Functions and Harmonic Analysis on 0- Dimensional Groups*, ELM, Baku (1981), in Russian.

[3] N. J. Fine, *Trans. Amer. Math. Soc.* **65,** 372 (1949).

[4] L. S. Pontryagin, *Topological Groups*, Nauka, Moscow (1984). English transl. of 1st ed., Oxford Univ. Press, London (1939).

[5] R. E. Edwards, *Fourier Series. A Modern Introduction.* Vol. 2. Springer Verlag, Berlin (1982).

[6] F. Schipp, W. R. Wade, and P. Simon, *Walsh Series: An Introduction to Dyadic Harmonic Analysis*, Adam Hilger, Bristol and New York (1990).

[7] B. I. Golubov, *Elements of Dyadic Analysis*, URSS, Moscow (2007), in Russian..

[8] B. I. Golubov, A. V. Efimov, and V. A. Skvortsov, *Walsh Series and Transforms*, URSS, Moscow (2008). English transl. of 1st ed., Kluwer, Dordrecht (1991).

[9] L. A. Zalmanzon, *Fourier, Walsh and Haar Transforms and Their Application in Control, Communication and Other Domains*, Nauka, Moscow (1989), in Russian.

[10] J. Astola and L. Yaroslavsky (eds.), *Advances in Signal Transforms*, Hindawi Publishing Corporation, New York (2007).

[11] R. S. Stankovic and J. Astola, *From Boolean Logic to Switching Circuits and Automata*, Springer, Berlin (2011).

[12] T. P. Lukashenko, *Izv. Akad. Nauk SSSR, Ser. Mat.* **58**, 88 (1994). English transl., *Russian Academy of Sciences. Izvestiya Mathematics* **44**, 515 (1995).

[13] F. Schipp and W. R. Wade, *Transforms on Normed Fields*, Eotvos Lorand University, Budapest (1995).

[14] Yu. A. Farkov, "Walsh functions and continuous wavelet transform", The 5th International symposium "Fourier Series and their Applications", May 2008 (Rostov-on-Don, Russia) pp 27-32, in Russian.

[15] Yu. A. Farkov and D. Yu. Perlov, "On wavelets on locally compact groups", Int. Conf. "New Achievements in Earth Sciences", Apr 1996 (Moscow, Russia), in Russian.

[16] Yu. A. Farkov, *Funktsional. Amal. i Prilozhen* **31,** 86 (1997). English transl., *Functional Anal. Appl.* **31**, 294 (1997).

[17] W. C. Lang, *SIAM J. Math. Anal.* **27,** 305 (1996).

[18] W. C. Lang, *Houston J. Math.* **24,** 533 (1998).

[19] W. C. Lang, *Int. J. Math. and Math. Sci.* **21,** 307 (1998).

[20] Yu. A. Farkov, *Izv. Ross. Akad. Nauk Ser. Mat.* **69,** 193 (2005). English transl., *Izv. Math.* **69**, 623 (2005).

[21] Yu. A. Farkov, *Mat. Zametki* **82,** 934 (2007). English transl., *Math. Notes* **82,** 843 (2007).

[22] Yu. A. Farkov, *J. Approx. Theory* **161,** 259 (2009).

[23] Yu. A. Farkov, *Tr. Mat. Inst. Steklova* **265,** 110 (2009). English transl., *Proc. Steklov Inst. Math.* **265,** 101 (2009).

[24] Yu. A. Farkov and V. Yu. Protasov, *Mat. Sbornik* **197,** 129 (2006). English transl., *Sbornik: Mathematics* **197,** 1529 (2006).

[25] Yu. A. Farkov and E. A. Rodionov, *Mat. Zametki* **86,** 429 (2009). English transl., *Math. Notes* **86,** 407 (2009).

[26] Yu. A. Farkov, U. Goginava and T. Kopaliani, *Jaen J. Approx.*, **3,** 117 (2011).

[27] V. Yu. Protasov, *Mat. Sbornik* **198,** 135 (2007). English transl., *Sbornik: Mathematics* **198,** 1665 (2007).

[28] Yu. A. Farkov, *Commun. Math. Appl.* **1,** 27 (2010).

[29] Yu. A. Farkov, *Mat. Zametki* **89,** 914 (2011), English transl., *Math. Notes* **89** (2011).

[30] F. A. Shah, *Int. J. Wavelets Multiresolut. Inf. Process.* **7,** 553 (2009).

[31] F. A. Shah and A. Wahid, *An. St. Univ. Ovidius Constanta Ser. Mat.* **18,** 223 (2010).

[32] Yu. A. Farkov, A. Yu. Maksimov and S. A. Stroganov, *Int. J. Wavelets Multiresolut. Inf. Process.* **9**, 44 (2011).

[33] Yu. A. Farkov and S. A. Stroganov, *Izv. VUZ. Matematika* (7), 57 (2011), English transl., *Russian Mathematics (Iz. VUZ)* (2011).

[34] J. J. Benedetto and R. L. Benedetto, *J. Geometric Analysis* **14,** 423 (2004).

[35] S. Albeverio and S. V. Kozyrev, *p-Adic Numbers Ultrametric Anal. Appl.* **1,** 181 (2009).

[36] S. Albeverio, S. Evdokimov and M. Skopina, *J. Fourier Anal. Appl.* **16,** 693 (2010).

[37] A. Yu. Khrennikov, V. M. Shelkovich and M. Skopina, Preprint, Nov 2007 at http://arxiv.org/abs/0711.2820

[38] S. F. Lukomskii, *Mat. Sbornik* **201,** 41 (2010). English transl., *Sb. Math.* **201,** 669 (2010).

[39] P. Manchanda, A. Meenakshi and A. H. Siddiqi, *Aligarh Bull. Math.* **27,** 59 (2008).

[40] E. Hernandez and G. Weiss, *A First Course on Wavelets*, FL: CRC Press, Boca Raton (1996).

[41] I. Ya. Novikov, V. Yu. Protasov and M. A. Skopina, *Wavelet Theory*, Translations of Mathematical Monographs 239. Providence, RI: American Mathematical Society (2011).

[42] Yu. A. Farkov, "Orthogonal p-wavelets on $\mathbb{R}_+$", Int. Conf. "Wavelets and Splines" July 3-8, 2003 (St. Petersburg, Russia), pp. 4-26.

[43] F. Keinert, *Wavelets and Multiwavelets*, Chapman & Hall/CRC, Boca Raton (2004).

[44] P. R. Massopust, *Fractal Functions, Fractal Surface, and Wavelets*, Academic Press, San Diego (1994).

[45] I. Daubechies, *Ten Lectures on Wavelets*, SIAM, Philadelphia (1992).

[46] I. Daubechies and J. C. Lagarias, *SIAM J. Math. Anal.* **123,** 1031 (1992).

[47] V. Yu. Protasov, *Izv. Ross. Akad. Nauk Ser. Mat.* **70,** 123 (2006). English transl., *Izv. Math.* **70**, 975 (2006).

[48] A. I. Rubinshtein, *Mat. Zametki* **23,** 379 (1978). English transl., *Math. Notes* **23,** 205 (1978).

[49] Yu. A. Farkov, *Facta Univ., Ser. Electron. Energ.* **21,** 309 (2008).

[50] Bl. Sendov, *East J. Approx.* **3,** 225 (1997).

[51] Bl. Sendov, *J. Approximation Theory* **96,** 258 (1999).

[52] Yu. A. Farkov and E. A. Rodionov, *p -Adic Numbers, Ultrametric Analysis and Applications* **3,** 181 (2011).

[53] Yu. A. Farkov, *p -Adic Numbers, Ultrametric Analysis and Applications*, in press.

In: Wavelets: Classification, Theory and Applications
Editors: Manel del Valle et al.
ISBN: 978-1-62100-252-9

Chapter 12

STABLE ALGORITHM FOR COMPUTATION OF HANKEL TRANSFORM USING CHEBYSHEV WAVELETS

Rajesh K. Pandey*, Vineet K. Singh and Om P. Singh
Indian Institute of Information Technology Design and Manufacturing Jabalpur India
Department of Mathematics,
Birla Institute of Technology and Science Pilani-Goa campus, Goa, India
Department of Applied Mathematics, Institute of Technology,
Banaras Hindu University, Varanasi, India

ABSTRACT

A new stable numerical method, based on Chebyshev wavelets for numerical evaluation of Hankel transform is proposed in this paper. Chebyshev wavelets are used as a basis to expand a part of the integrand, $rf(r)$, appearing in the Hankel transform integral. This transforms the Hankel transform integral into a Fourier-Bessel series. Truncating the series, an efficient, stable algorithm is obtained for the numerical evaluations of the Hankel transforms of order $\nu > -1$. The method is quite accurate and stable, as illustrated by given numerical examples with varying degree of random noise terms $\varepsilon\theta_i$ added to the data function $f(r)$, where θ_i is a uniform random variable with values in [-1,1]. Finally, an application of the proposed method is given in solving the heat equation in an infinite cylinder with a radiation condition.

1. INTRODUCTION

There are several integral transforms which are frequently used as a tool for solving numerous scientific problems. It is well known that the Fourier transform (FT) is used to obtain spatial spectrum of optical light [1]. Fourier optics is widely used in optical instrument

* E-mail addresses: rkpandey@iiitdmj.ac.in, wavelet

design, optical propagation through lenses and in quadratics graded index mediums. Most classical optical systems like mirrors or lenses are axially symmetrical devices. In many practical problems, data are often acquired in such a form that is desirable to perform a two-dimensional polar Fourier transform that is a Hankel transform (HT) rather than the Cartesian forms. So, we transform the Cartesian coordinates into the polar coordinates.

Let $f(x,y)$ be an input field such that it can be separated as $f(x,y)=f_1(x)f_2(y)$, where f_1 and f_2 are independent functions. Then its two-dimensional Fourier transform $\widehat{f}$ is also separable as the same symmetry property is transposed through a linear FT. Hence, $\widehat{f}(u,v)=\widehat{f}_1(u)\cdot\widehat{f}_2(v)$.

Changing to the polar coordinates and if $f(r,\theta)=f(r)$ is axially symmetrical, then in [2], it was shown that

$$\widehat{f}(k,\varphi)=\frac{1}{2}\int_0^\infty d(r^2)f(r)J_0(kr) \equiv F_0(k) \tag{1}$$

which is also axially symmetrical in the Fourier frequency domain, where F_0 is the Hankel transform of order zero. The general Hankel transform pair with the kernel being J_ν is defined as [3]

$$F_\nu(p)= \int_0^\infty rf(r)J_\nu(pr)dr, \tag{2}$$

and HT being self reciprocal, its inverse is given by

$$f(r)= \int_0^\infty pF_\nu(p)J_\nu(pr)dp, \tag{3}$$

where J_ν is the ν th-order Bessel function of first kind.

The Hankel transform arises naturally in the discussion of problems posed in cylindrical coordinates and hence, as a result of separation of variables, involving Bessel functions. The Hankel transform is frequently used as a tool for solving numerous scientific problems. It is widely used in several fields like, elasticity [4], optics [5-6], fluid mechanics [7], seismology [8], astronomy and image processing [9-16]. The Hankel transform becomes very useful in analysis of wave fields where it is used in mathematical handling of radiation, diffraction, and field projection. Recently, it has been utilized to study pseudo-differential operators. Singh and Pandey [17] used HT of order ν, $\nu\in\mathbf{R}$ to study a special class of pseudo-differential operator (PDO) $(-x^{-1}D)^\nu$, $D=\dfrac{d}{dx}$ and proved that the (PDO) is almost an inverse of HT operator h_ν in the sense that $h_\nu o(-x^{-1}D)^\nu(\varphi)=h_0(\varphi)$ over certain Freshet space F, thus representing the PDO as a Fourier-Bessel series. Further, in 1995, Singh[18], using the HT representation of the PDO, proved that $e^{-\alpha x^2}$, Re $\alpha > 0$ are the eigenfunctions and $e^{-x^2/2}$ is a fixed point of $(-x^{-1}D)^{\nu}$, $\nu \in C$.

Several papers have been written to the evaluation of the Hankel transform in general and the zeroth order in particular. Analytical evaluations of (2) and (3) are rare and their numerical computations are difficult because of the oscillatory behavior of the Bessel function and the infinite length of the interval. Since seminal work by Siegman [19] in 1977, a number of algorithms for the numerical evaluation of the Hankel transform have been published for both zero-order [5-6,9-12,20-24] and high-order [25-33] Hankel transform. Unfortunately, the efficiency of a method for computing Hankel transform is highly dependent on the function to be transformed, and thus it is difficult to choose the optimal algorithm for given function. In [21], the authors used Filon quadrature Philosophy to evaluate zero-order Hankel transform. They separated the integrand into the product of (assumed) slowly varying component and a rapidly oscillating one (in this case, former is $rf(r)$ and the later is $J_\nu(pr)$). This methods works quite well for computing $F_0(p)$, for $p \geq 1$, but the calculation of inverse Hankel transform is more difficult, as $F_0(p)$ is no longer a smooth function but a rapidly oscillating one. Moreover, the error is appreciable between $0 < p < 1$. In 1998, Yu et al. [23] gave another method to compute zero-order quasi discrete HT by approximating the input function by a Fourier-Bessel series over a finite integration interval. It led to a symmetric transformation matrix for the HT and the IHT that satisfies the discrete form of the Parseval theorem. Later in 2004, Guizar-Sicairos et al. [34] obtained a powerful scheme to calculate the HT of integer order $\nu \geq 0$ by extending the zero-order HT algorithm of Yu [23] to higher orders. Their algorithm is based on the orthogonality properties of Bessel functions. Postnikov [32], proposed, for the first time, a novel and powerful method for computing zero and first order HT by using Haar wavelets. Refining the idea of Postnikov [32], we [35-36] obtained two more algorithms for numerical evaluation of HT of order $\nu > -1$ using linear Legendre multi-wavelets and rationalized Haar wavelets which were shown to be superior to the other mentioned algorithms. The data function $f(r)$ when measured experimentally may contain some noise terms affecting the accuracy of the algorithms for computing the HT. Thus, it is desirable to have algorithms stable under small random perturbation in the data function. The purpose of this communication is to present a stable algorithm that is quite accurate and fast for numerical evaluation of the HT using Chebyshev wavelets. Test functions with known analytic HT are used with random noise term $\varepsilon\theta_i$ added to the data function $r\,f(r)$, where θ_i is a uniform random variable with values in [-1,1], to illustrate the stability and efficiency of the proposed algorithm. As an application of the theory developed, we solve the heat flow problem in an infinite cylinder with radiation condition.

2. Chebyshev Wavelets

Wavelets are a class of functions constructed from dilation and translation of a single function called the mother wavelet. When the dilation and translation parameters a and b vary continuously, the following family of continuous wavelets are obtained

$$\psi_{ab}(t) = |a|^{-1/2}\psi\left(\frac{t-b}{a}\right), \quad a, b \in \mathbf{R}, \quad a \neq 0.$$

When the parameters a and b are restricted to discrete values as $a = 2^{-k}, b = n2^{-k}$, then, we have the following family of discrete wavelets

$$\psi_{kn}(t) = 2^{k/2}\psi(2^k t - n), \ k, n \in Z,$$

where the function ψ, the mother wavelet, satisfies $\int_R \psi(t)dt = 0$.

We are interested in the case where ψ_{kn} constitutes an orthonormal basis of $L^2(\mathbf{R})$. A systematic way to do this is by means of multiresolution analysis (MRA).

In 1910, Haar [37] constructed the first orthonormal basis of compactly supported wavelets for $L^2(\mathbf{R})$. It has the form $\{2^{j/2}\psi(2^j t - k): j, k \in Z\}$ where the fundamental wavelet ψ is constructed as follows: Construct a compactly supported scaling function φ by the two-scale scaling relation $\varphi(t) = \varphi(2t) + \varphi(2t-1)$ together with the normalization constraint $\int \varphi(t)dt = 1$. A solution of this recursion that represents φ in $L^2(\mathbf{R})$ is $\chi_{[0,1)}$. Then $\psi(t) = \varphi(2t) - \varphi(2t-1)$. The Haar wavelets are piecewise continuous and have discontinuities at certain dyadic rational numbers.

In seminal papers; Daubechies [38-39], constructed the first orthonormal basis of continuous compactly supported wavelets for $L^2(\mathbf{R})$. They have led to a significant literature and development, both in theoretical and applied arenas.

Later in 1989, Mallat [40] studied the properties of multiresolution approximation and proved that it is characterized by a 2π-periodic function. From any MRA, one can derive a function $\psi(t)$ called a wavelet such that $\{2^{j/2}\psi(2^j t - k): j, k \in Z\}$ is an orthonormal basis of $L^2(\mathbf{R})$. The MRA showed the full computational power that this new basis for $L^2(\mathbf{R})$ possessed. In the same year, Mallat [41] applied MRA for analyzing the information content of the images.

Note that a system $\{\varphi_k : k \in Z\}$ is called a Riesz basis if it is obtained from an orthonormal basis by means of a bounded invertible operator [42].

Definition: The increasing sequence $\{V_k\}_{k\in Z}$ of closed subspaces of $L^2(\mathbf{R})$ with scaling function $\varphi \in V_0$ is called MRA if

(i) $\bigcup_k V_k$ is dense in $L^2(\mathbf{R})$ and $\bigcap_k V_k = \{0\}$,

(ii) $f(t) \in V_k$ iff $f(2^{-k}t) \in V_0$,

(iii) $\{\varphi(t-n)\}_{n\in Z}$ is a Riesz basis for V_0.

Note that (iii) implies that the sequence $\{2^{k/2}\varphi(2^k t-n)\}_{n\in Z}$ is an orthonormal basis for V_k.

Let $\psi(t)$ be the mother wavelet, then $\psi(t)=\sum_{n\in Z} a_n\varphi(2t-n)$ and $\{2^{k/2}\psi(2^k t-n)\}_{k,n\in Z}$ forms an orthonormal basis for L^2 (**R**) under suitable conditions [38].

For $k=1,2,3,\ldots$ and $n=1,2,3,\ldots,2^{k-1}$, the Chebyshev wavelets $\psi_{nm}(t)$ are defined as [43]

$$\psi_{nm}(t)=\begin{cases} 2^{k/2}\,\widetilde{T}_m(2^k t-2n+1), \; for \; \dfrac{n-1}{2^{k-1}}\le t<\dfrac{n}{2^{k-1}}, \\ 0, \quad otherwise, \end{cases} \tag{4}$$

where

$$\widetilde{T}_m(t)=\begin{cases} \dfrac{1}{\sqrt{\pi}}, & m=0, \\ \sqrt{\dfrac{2}{\pi}}\,T_m(t), & m>0, \end{cases} \tag{5}$$

and t is the time. The well known Chebyshev polynomials $T_m(t)$ of the first kind and of degree m are orthogonal with respect to the weight function $w(t)=1/\sqrt{1-t^2}$. They are defined on $[-1,1]$ by the following recurrence relation:

$$T_0(t)=1, \quad T_1(t)=t \quad and \quad T_{m+1}(t)=2tT_m(t)-T_{m-1}(t), \quad m=1,2,3,\ldots$$

Note that in dealing with Chebyshev wavelets the weight function $w(t)$ have to be dilated and translated as $w_n(t)=w(2^k t-2n+1)$, to get orthogonal wavelets forming an orthonormal basis for $L_w^2[0,1]$.

3. Outline of Algorithm

The function $f(r)$ representing physical fields are either zero or have an infinitely long decaying tail outside a disk of finite radius R. Hence, in most practical applications either the signal $f(r)$ has a compact support or for a given $\varepsilon > 0$ there exists a $R > 0$ such that $\left|\int_R^\infty r\, f(r) J_\nu(pr)\, dr\right| < \varepsilon$, which is the case if $f(r) = O(r^\eta)$, as $r \to \infty$, where $\eta < -\frac{3}{2}$.

Therefore, in either case,

$$\hat{F}_\nu(p) = \int_0^R r\, f(r) J_\nu(pr)\, dr = \int_0^1 r\, f(r) J_\nu(pr)\, dr \text{, (by scaling)} \tag{6}$$

known as the finite Hankel transform(FHT) is a good approximation of the HT as given by (2). The algorithm is efficient if f decays faster than $r^{-3/2}$ outside some disk of finite radius R which includes the majority of input signals of physical interest.

Writing $rf(r) = g(r)$ in equation (6), we get:

$$\hat{F}_\nu(p) = \int_0^1 g(r) J_\nu(pr)\, dr. \tag{7}$$

We may expand $g(r)$ as follows

$$g(r) = \sum_{n=1}^{\infty}\sum_{m=0}^{\infty} c_{nm}\, \psi_{nm}(r), \tag{8}$$

where $c_{nm} = \langle g(r), \psi_{nm}(r)\rangle_{w_n} = \int_0^1 w_n(r) g(r) \psi_{nm}(r)\, dr$.

By truncating the infinite series (8) at levels $n = 2^{k-1}$ and $m = M$, we obtain an approximate representation for $g(r)$ as

$$g(r) \approx \sum_{n=1}^{2^{k-1}}\sum_{m=0}^{M} c_{nm}\, \psi_{nm}(r) = C^T \Psi(r), \tag{9}$$

where the matrices C and Ψ are given by

$$C = [c_{10}, c_{11}, \ldots, c_{1M}, c_{20}, \ldots, c_{2M}, \ldots, c_{2^{k-1}0}, \ldots, c_{2^{k-1}M}]^T \tag{10}$$

and

$$\Psi(r) = [\psi_{10}(r), \psi_{11}(r), \ldots, \psi_{20}(r), \ldots, \psi_{2M}(r), \ldots, \psi_{2^{k-1}0}(r), \ldots, \psi_{2^{k-1}M}(r)]^T. \tag{11}$$

Substituting (9) in (7), we get

$$\hat{F}_\nu(p) \approx C^T \int_0^1 \Psi(r) J_\nu(pr)\, dr. \tag{12}$$

Taking $M = 2$ and $k = 2$, Eq.(12) reduces to

$$\hat{F}_\nu(p) \approx C^T \left[\int_0^1 \psi_{10}(r) J_\nu(pr) dr, \int_0^1 \psi_{11}(r) J_\nu(pr) dr, \ldots, \int_0^1 \psi_{22}(r) J_\nu(pr) dr \right]^T, \tag{13}$$

where the six basis functions are given by

$$\psi_{10}(r) = \frac{2}{\sqrt{\pi}}, \quad \psi_{11}(r) = \frac{2\sqrt{2}}{\sqrt{\pi}}(4r-1), \quad \psi_{12}(r) = \frac{2\sqrt{2}}{\sqrt{\pi}}\left[2(4r-1)^2 - 1\right], \qquad 0 \le r < \frac{1}{2},$$

and

$$\psi_{20}(r) = \frac{2}{\sqrt{\pi}}, \quad \psi_{21}(r) = \frac{2\sqrt{2}}{\sqrt{\pi}}(4r-3), \quad \psi_{22}(r) = \frac{2\sqrt{2}}{\sqrt{\pi}}\left[2(4r-3)^2 - 1\right], \qquad \frac{1}{2} \le r < 1 \tag{14}$$

We re-label and write Eq. (13) as

$$\hat{F}_\nu(p) \approx [c_0, c_1, c_2, c_3, c_4, c_5][I_\nu^0, I_\nu^1, I_\nu^2, I_\nu^3, I_\nu^4, I_\nu^5]^T, \tag{15}$$

where I_ν^k' s are $(k+1)th$ place integral in Eq.(13) and are evaluated by using the following formulae:

$$\int_0^a J_\nu(t) dt = 2 \lim_{N\to\infty} \sum_{n=0}^{N} J_{\nu+2n+1}(a), \ \operatorname{Re}\nu > -1 \ \text{[Ref 44, p. 333]}$$

$$\int_0^a t^{1-\nu} J_\nu(t) dt = \frac{1}{2^{\nu-1}\Gamma(\nu)} - a^{1-\nu} J_{\nu-1}(a), \ \text{[Ref 44, p. 333]}$$

$$\int_0^a t^\mu J_\nu(t) dt = \frac{a^\mu \Gamma(\frac{\nu+\mu+1}{2})}{\Gamma(\frac{\nu-\mu+1}{2})} \lim_{N\to\infty} \sum_{n=0}^{N} \frac{(\nu+2n+1)\Gamma(\frac{\nu-\mu+1}{2}+n)}{\Gamma(\frac{\nu+\mu+3}{2}+n)} J_{\nu+2n+1}(a), \ \operatorname{Re}(\nu+\mu+1) > 0,$$

[Ref 45, p.480]. (16)

In 2004, Piessens[46] approximated the input field $g(r)$ on [0,1] as

$$g(r) \approx r^{\alpha} \sum_{k=o}^{N} c_n T_k^{\,*}(r), \tag{17}$$

where $T_k^*(r)$, is the shifted Chebyshev polynomials of degree k and $\alpha > 0$ is a real parameter. Thus approximating the Hankel transform $\hat{F}_v(p)$ as,

$$\hat{F}_v(p) \approx \sum_{k=0}^{N} c_k \int_0^1 r^{\alpha} J_v(pr)\, T_k^{\,*}(r)\, dr$$

$$= C^T \left[\int_0^1 r^{\alpha} T_0^{\,*}(r) J_v(pr), \int_0^1 r^{\alpha} T_1^{\,*}(r) dr, \ldots, \int_0^1 r^{\alpha} T_N^{\,*} J_v(pr) dr \right]^T$$

$$= [c_0, c_1, c_2, \ldots\ldots c_k][M_0, M_1, M_2, \ldots\ldots\ldots M_N]^T \tag{18}$$

where $$M_l = \int_0^1 r^{\alpha}\, T^*{}_l(r) J_v(pr) dr,\ l = 0,1,\ldots,N. \tag{19}$$

These modified moments M_l's (similar to $I_v^{\,l}$'s in Eq. (15)) satisfy a homogenous, linear, nine-term recurrence relation. But the forward and backward recursion are asymptotically unstable. However, the instability of forward recursion is less pronounced if $l \le \frac{Rp}{2}$, but for $l > \frac{Rp}{2}$ the loss of significant figures increases and forward recursion is no longer applicable. In that case, these recurrence relations have to be solved as a boundary value problem with six initial values and two end values, making the evaluation of M_l's more difficult than that of $I_v^{\,l}$'s.

In our proposed method, the integrals $I_v^{\,l}$ $l = 0,1,2,\ldots,5$ are evaluated easily. For example,

$$I_v^{\,0} = \frac{2}{\sqrt{\pi}} \int_0^1 J_v(pr) dr \approx \frac{4}{p\sqrt{\pi}} \sum_{n=0}^{N} J_{v+2n+1}(p).$$

Thus, we see that these integrals $I_v^{\,l}$ are approximated by finite sums of different orders Bessel functions J_{μ}'s evaluated at the particular point p. Besides the simplicity of our algorithm in computation over [46], it is also stable under the noise in the input field where as the algorithm proposed in [46] is not tested for the stability.

4. NUMERICAL RESULTS

In this section we discuss, the implementation of our numerical method and investigate its accuracy and stability by applying it on numerical examples with known analytical HT. In all the examples, the exact data function is denoted by $g(r)$ and the noisy data function $g^{\varepsilon}(r)$ is obtained by adding an ε random error to $g(r)$ such that $g^{\varepsilon}(r_i) = g(r_i) + \varepsilon\theta_i$, where $r_i = ih, i = 1,2, \dots L$, $Lh = 100$ and θ_i is a uniform random variable with values in [-1,1] such that $\max_{0 \le i \le L} \left| g^{\varepsilon}(r_i) - g(r_i) \right| \le \varepsilon$.

The following examples are solved with and without random perturbations to illustrate the efficiency and stability of our method by choosing four different values of the random error ε as $\varepsilon = 0,\ 0.001, 0.002$ and 0.005 and computing the error Ej(p) = Approximate HT obtained from (15) with random error ε_j - the exact HT, $j = 0,1,2,3$. The various Ej(p)'s are shown in Figures 2-3,5-6,8-9,11-12,14-15 and 17-18. Note that the various graphs in the following examples are plotted by choosing the sample points as as $p = 0.01(0.01)P$, where $P = 20$ in Figures 1,4,7,10,13,16 and $P = 100$ in Figures 2-3,5-6,8-9,11-12,14-15, 17-18.

We also use the discrete l^2 norm and the continuous L^2 norm in $I = [0, P]$ to measure errors as well. They are defined as:

$$\|f\|_{2,I} = \left(\frac{1}{N} \sum_{i=1}^{N} |f(r_i)|^2 \right)^{1/2} \tag{20}$$

and

$$\|f\|_2 = \left(\int_0^P |f(r)|^2 dr \right)^{1/2}, \tag{21}$$

respectively.

Example 1: Sombrero function

A very important, and often used function, is the Circ function that can be defined as

$$\mathrm{Circ}(r/a) = \begin{cases} 1, & r \le a, \\ 0, & r > a. \end{cases} \tag{22}$$

The zeroth-order HT of Circ(r/a) is the Sombrero function [47], given by:

$$S_0(p) = a^2 \frac{J_1(ap)}{ap}.$$

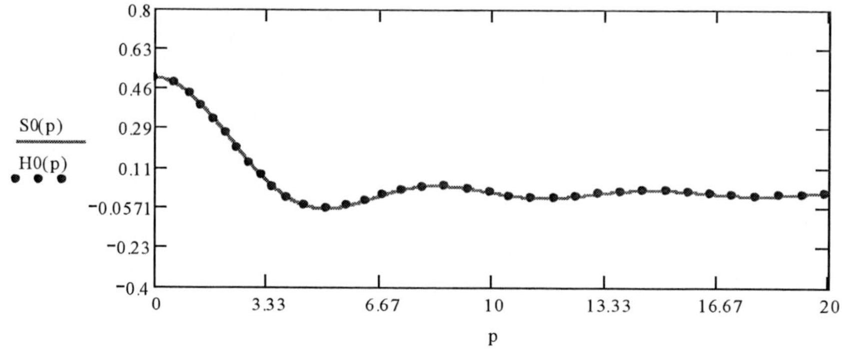

Figure 1. The exact transform, $S0(p)$ (solid line) and the approximate transform, $H0(p)$ (dotted- line) truncated at level N=20 and M=2.

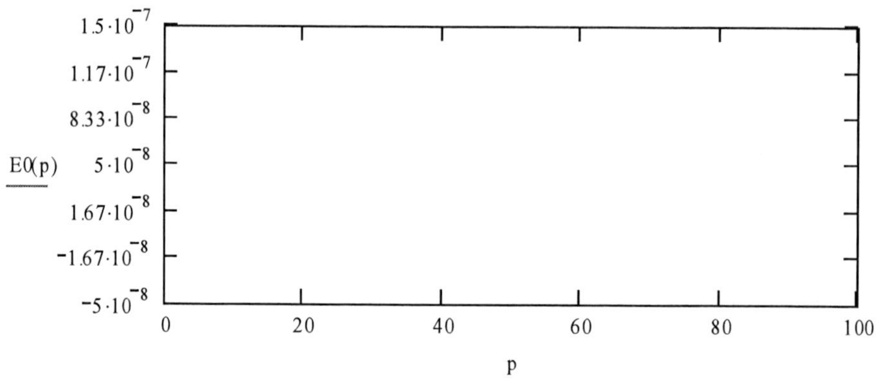

Figure 2. Error between exact transform, $S0(p)$ and approximated transform $H0(p)$ (without random noise).

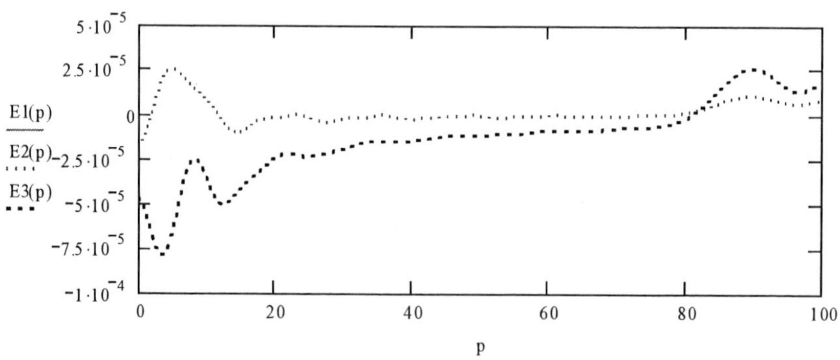

Figure 3. Errors between exact transform and approximated transform with different random perturbations.

We use Eq. (15) to obtain the approximation for the FHT $\hat{F}_0(p)$ of the Circ(r/a). This approximation is compared with the exact HT $S_0(p)$ and is shown in Figure1. Figures 2 and 3 represent the corresponding error E0(p) and Ej(p), j=1,2,3, respectively. Note that $S_0(p)$ and

$\widehat{F}_0(p)$ are indicated by $S0(p)$ (solid line) and $H0(p)$ (dotted line) in the Figures 1 and 2 respectively. Figure 3 summarizes the errors obtained for different random perturbations.

Example 2: Let $f(r) = \frac{2}{\pi}\left[\arccos(r) - r(1-r^2)^{1/2}\right]$, $0 \le r \le 1$,

$$F_0(p) = 2\frac{J_1^2(p/2)}{p^2}, \qquad 0 \le p \le \infty. \text{ [21]} \tag{23}$$

Barakat et al. [21], evaluated $F_0(p)$ numerically using Filon quadrature philosophy but the associated error is appreciable for $p < 1$; whereas our method gives almost zero error in that range. Note that $F_0(p)$ and $\widehat{F}_0(p)$ are indicated by $F0(p)$ (solid line) and $H0(p)$ (dotted line) in the Figure 4 and the corresponding error graphs $Ej(p)$ in Figures 5 and 6.

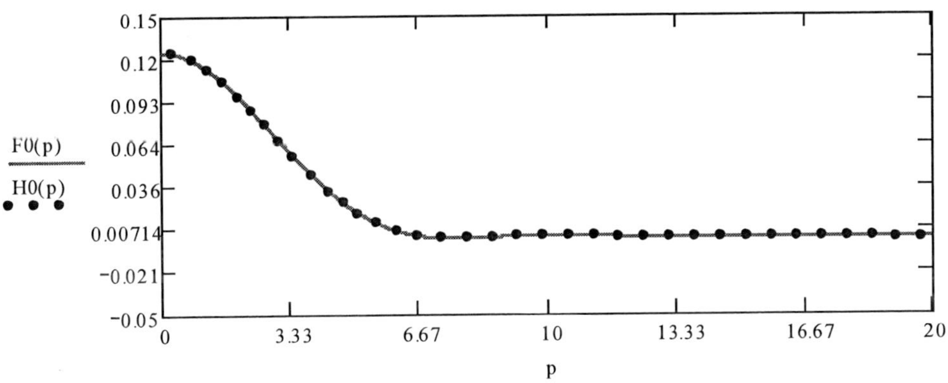

Figure 4. The exact transform, $F0(p)$ (solid line) and the approximate transform, $H0(p)$ (dotted-line) truncated at level N=50 and M=2.

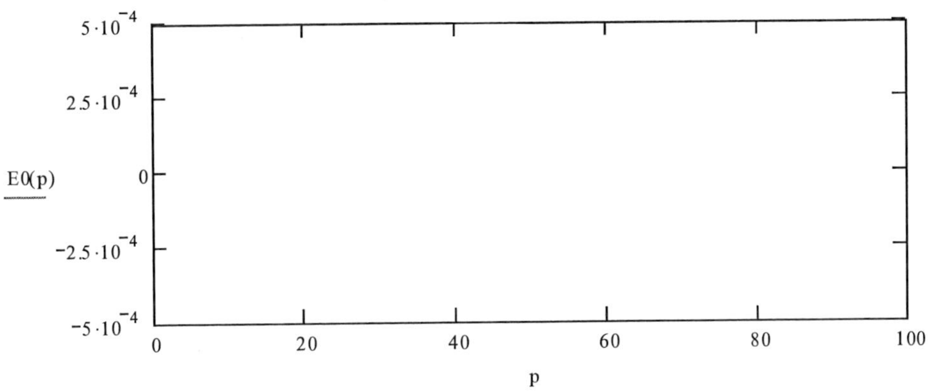

Figure 5. Error between exact transform, $F0(p)$ and approximated transform $H0(p)$ (without random noise).

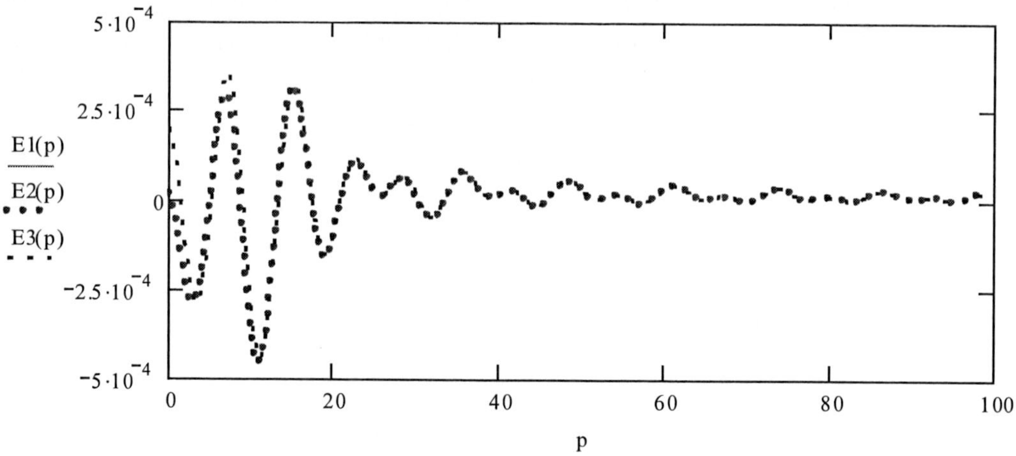

Figure 6. Errors between exact transform and approximated transform with different random perturbations.

Example 3: Let $f(r)=(1-r^2)^{1/2}$, $0\le r\le 1$, then,

$$F_1(p)=\begin{cases}\pi\dfrac{J_1^2(p/2)}{2p}, & 0<p<\infty \\ 0, & p=0\end{cases}. \tag{24}$$

Barakat et al. [22], evaluated $F_1(p)$ numerically using Filon quadrature philosophy but again the associated error is appreciable for $p<1$; whereas our method give almost zero error in that range. The comparison of the approximation $H1(p)$ (dotted line) with the exact HT $F1(p)$ (solid line) is shown in Figure 7 and the corresponding errors $Ej(p)$ in Figures 8-9.

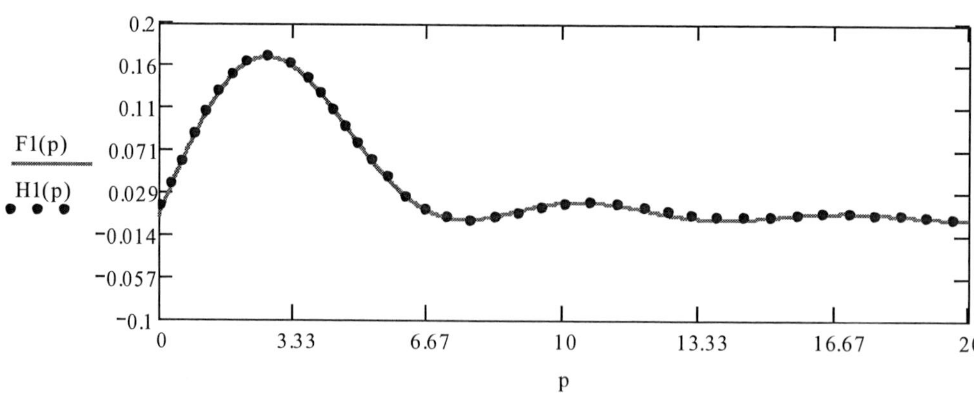

Figure 7. The exact transform, $F1(p)$ (solid line) and the approximate transform, $H1(p)$ (dotted- line) truncated at level N=50 and M=2.

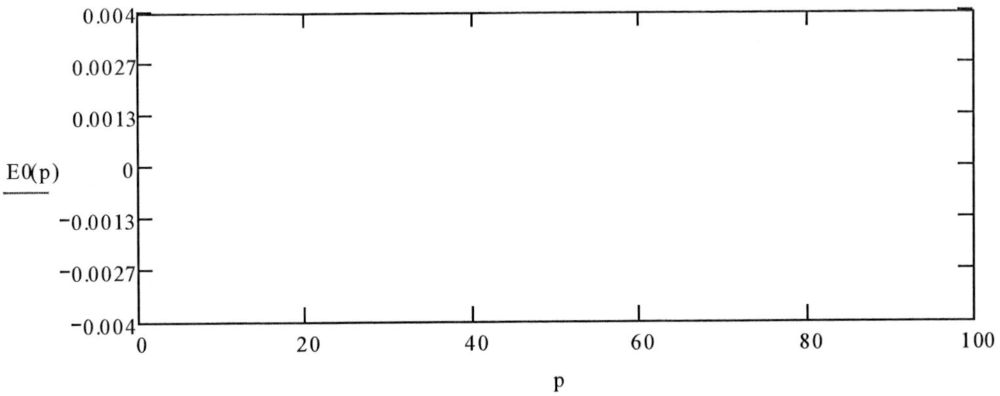

Figure 8. Error between exact transform, $F1(p)$ and approximated transform $H1(p)$ (without random noise).

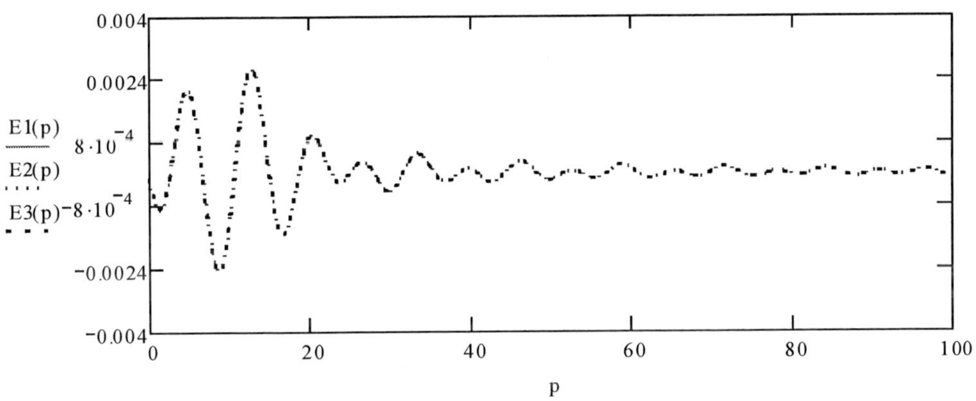

Figure 9. Errors between exact transform and approximated transform with different random perturbations.

Example 4: In this example, we choose as a test function the generalized version of the top-hat function, given as:

$f(r) = r^{\nu}[H(r) - H(r-a)], a > 0$ and $H(r)$ is the step function $H(r) = \begin{cases} 1, & r \geq 0 \\ 0, & r < 0 \end{cases}$. Then,

$$F_{\nu}(p) = \frac{J_{\nu+1}(p)}{p}.$$

In [34], authors took $a = 1$ and $\nu = 4$ for numerical calculations. We take $a = 1$, $\nu = 1/10, 5$ and observe that the errors are quite small as shown in Figures 11 and 14 respectively. The corresponding errors $Ej(p)$ with different level of random noises are shown in Figures 12 and 15 and Figures 10 and 13 show the comparison between the exact transform and the approximate transform for $\nu = 1/10, 5$ respectively.

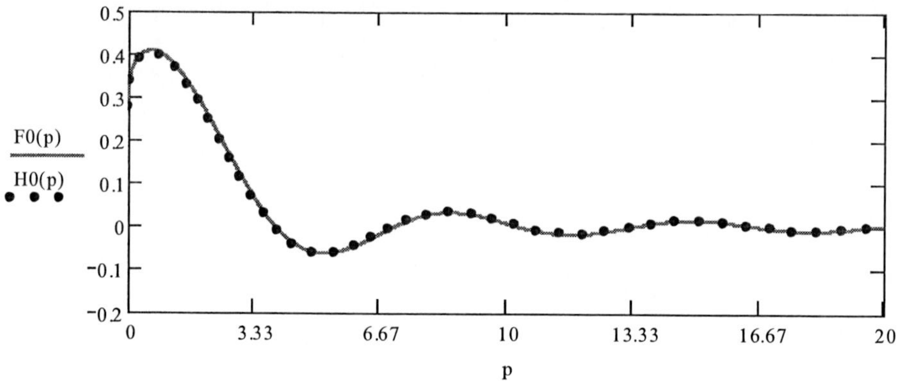

Figure 10. The exact $F0(p)$ transform, (solid line) and the approximate transform, $H0(p)$ (dotted- line) truncated at level N=50, M=2 (v=1/10).

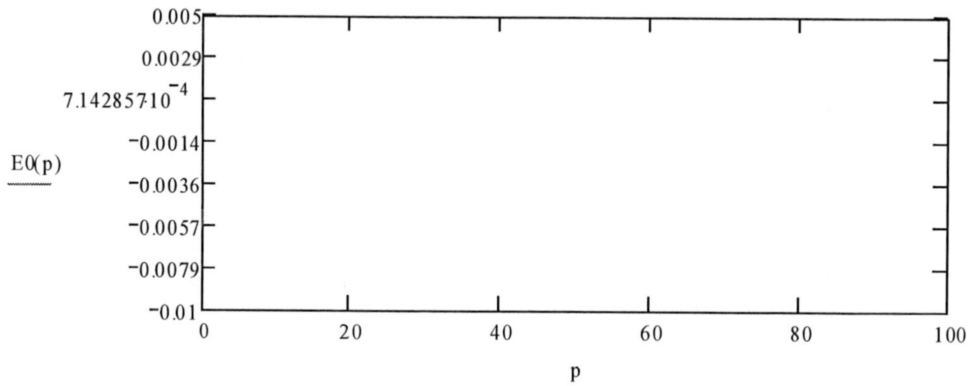

Figure 11. Error between exact transform, $F0(p)$ and approximated transform $H0(p)$, (without random noise) (v=1/10).

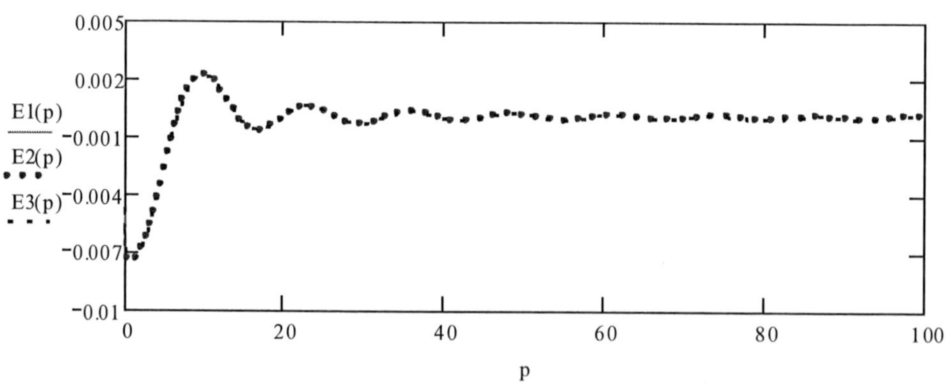

Figure 12. Errors between exact transform and approximated transform with different random perturbations (v=1/10).

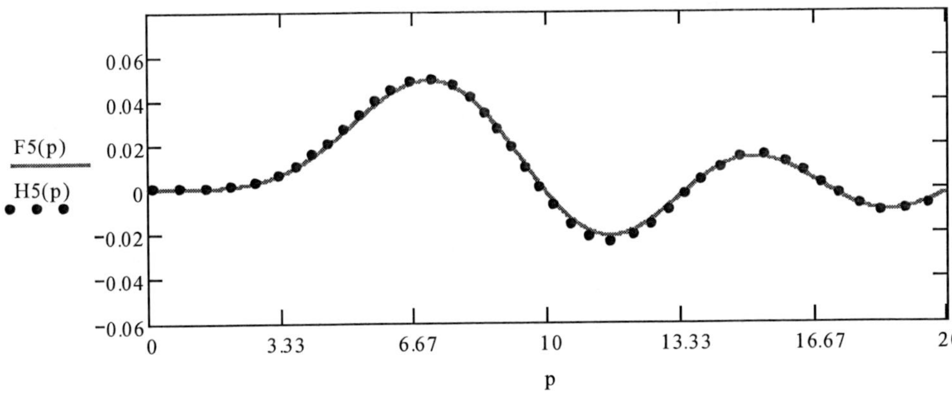

Figure 13. The exact transform, $F5(p)$ (solid line) and the approximate transform, $H5(p)$ (dotted- line) truncated at level N=50 and M=2 (v=5).

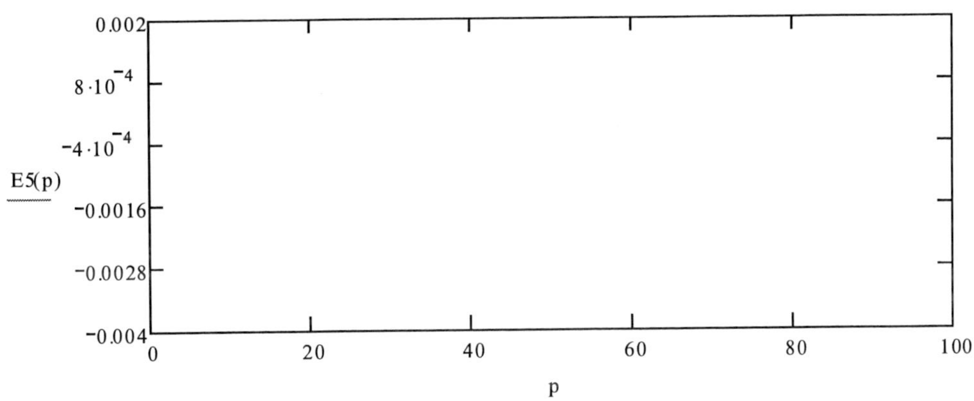

Figure 14. Error between exact transform, $F5(p)$ and approximated transform $H5(p)$, (without random noise). (v=5).

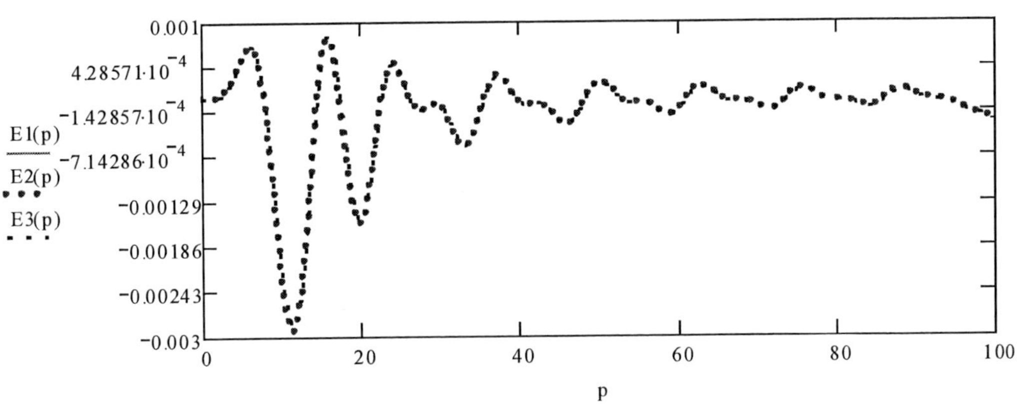

Figure 15. Errors between exact transform and approximated transform with different random perturbations. (v=5).

Example 5: The following example was solved numerically by Knockaert [15]:

For $f(r)=e^{-r}$, its HT is $F_0(p)=\dfrac{1}{(1+p^2)^{3/2}}$.

We solve the above problem by the proposed algorithm and observe that our method give result comparable to [15]. Note that $F_0(p)$ and $\widehat{F}_0(p)$ are indicated by $F0(p)$ (solid line) and $H0(p)$ (dotted line) in the Figures 16 and 17 respectively. Figure 18 depicts the associated errors $Ej(p)$ for $j=0$ and $j=1,2,3$ respectively.

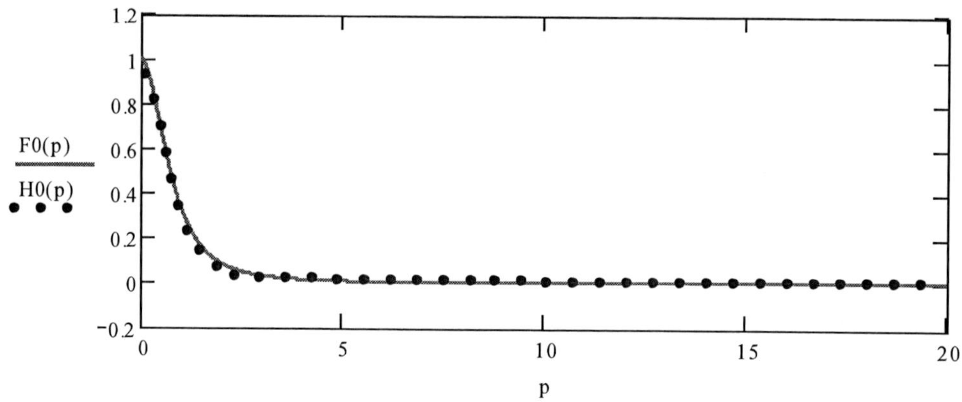

Figure 16. The exact transform, $F0(p)$ (solid line) and the approximate transform, $H0(p)$ (dotted-line) truncated at level $N=65$ and $M=2$.

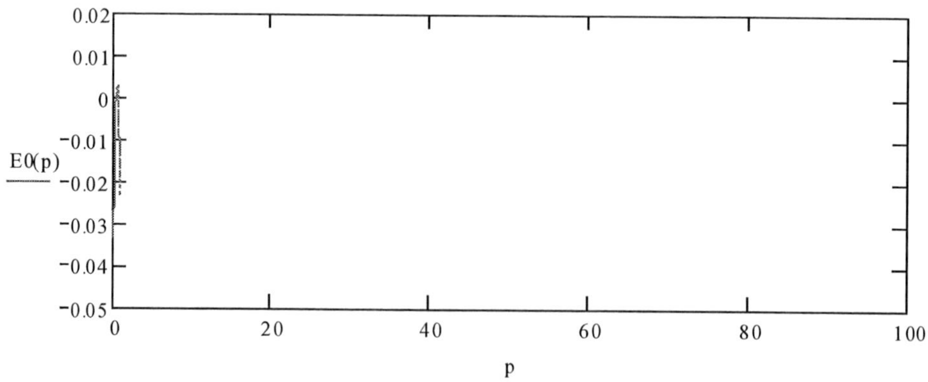

Figure 17. Error between exact transform, $S0(p)$ and approximated transform $H0(p)$ (without random noise).

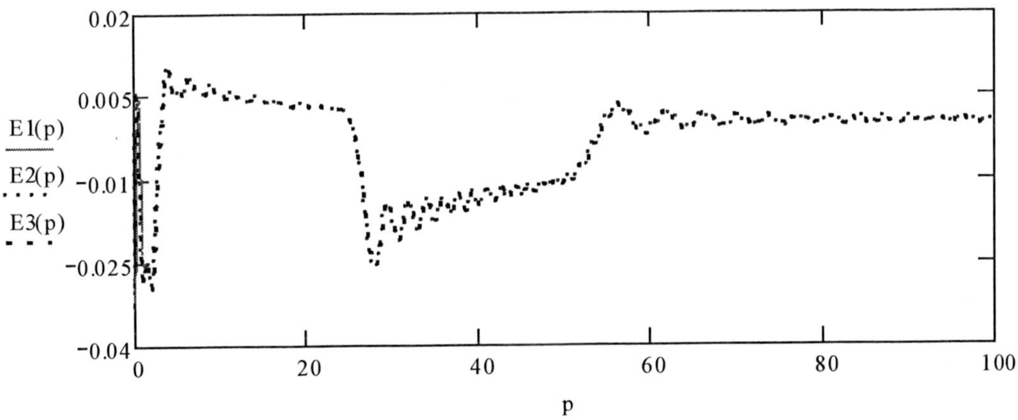

Figure 18. Errors between exact transform and approximated transform with different random perturbations.

5. APPLICATION

As an application, we solve the heat equation in cylindrical coordinates inside an infinitely long cylinder of radius unity, by using the theory of Hankel transform developed in the preceding pages. We seek a function $u(r, t)$; where r is radius and t is time, (u does not depend on θ and t) satisfying the differential equation

$$D_r^2 u + \frac{1}{r} D_r u = D_t u \quad (0 < r < 1, 0 < t < \infty) \tag{25}$$

and the following initial and boundary conditions:

As $t \to 0^+$, $u(r, t) \to f(r) = \frac{2}{\pi}\left[\arccos(r) - r(1-r^2)^{1/2}\right],\ 0 \le r \le 1$.

As $r \to 1^-$, $D_r u + H u \to 0$ for each fixed $t > 0$, where $H > 0$.

When u denotes the temperature within the cylinder, $H > 0$ means that heat is being radiated away from the surface of the cylinder.

Let $\Omega_{\nu,r}$ denotes the differential operator $D_r^2 + \frac{1}{r} D_r - \frac{\nu^2}{r^2}$, then the differential equation (25) can be written as

$$\Omega_{0,r} u = \frac{\partial u}{\partial t}. \tag{26}$$

The Dini expansion associated with $f(r)$ is

$$B_0(r)+\sum_{m=1}^{\infty} b_m J_\nu(\lambda_m r) \quad [48, \text{p.}596] \tag{27}$$

where λ_m , $m=1,2,3,\cdots$, are the positive rots (arranged in ascending order of magnitude) of the transcendental equation

$$z J_\nu'(z)+H J_\nu(z)=0 \ , \ \nu \geq -\frac{1}{2} \ , \tag{28}$$

b_m, $m=1, 2, 3,\cdots$, are given by

$$b_m = \frac{2\lambda_m^2 \int_0^1 r f(r) J_\nu(\lambda_m r)\, dr}{(\lambda_m^2 - \nu^2) J_\nu^2(\lambda_m) + \lambda_m^2 J_\nu'^2(\lambda_m)} \tag{29}$$

and $B_0(r)=0$ if $H+\nu>0$, which is the case for the present problem as $\nu=0$ and $H>0$.

The condition of validity of (27) are given in the following theorem [Ref 48, p. 601].

Theorem 1. Let $f(r)$ be a function defined over the interval $(0,1)$, and let $\int_0^1 r^{\frac{1}{2}} f(r) dr$ exist and (if it is improper integral) let it be absolutely convergent. If $f(r)$ has limited total fluctuation in (a, b) where $0 \leq a < b \leq 1$ then the series (27) converges to the sum $\frac{1}{2}[f(r+0)+f(r-0)]$ at all points r such that $a+\Delta \leq r \leq b-\Delta$ where Δ is arbitrarily small; and the convergence is uniform if $f(r)$ is continuous in (a, b).

Substituting equations (6) and (29) in (27) and using the above theorem, we obtain the following inversion theorem for the finite Hankel transform

Theorem 2 (inversion). Let $f(r)$ satisfies the condition of Theorem 1, then

$$f(r)=\lim_{N\to\infty} \sum_{m=1}^{N} \frac{2\lambda_m^2 \hat{F}_\nu(\lambda_m) J_\nu(\lambda_m r)}{(\lambda_m^2-\nu^2)\, J_\nu^2(\lambda_m)+J_\nu'^2(\lambda_m)} \tag{30}$$

From the well known formula

$$\Omega_{0,r} J_0(\lambda_m r) = -\lambda_m^2 J_0(\lambda_m r) \ , \tag{31}$$

it follows from integration by parts that

$$\int_0^1 [\Omega_{0,r} f(r)] r J_0(\lambda_m r)\,dr = \int_0^1 f(r) r\, \Omega_{0,r} J_0(\lambda_m r)\,dr,$$
$$= -\lambda_m^2 \int_0^1 f(r) r J_0(\lambda_m r)\,dr, \tag{32}$$

if we put some suitable condition on $f(r)$ such that the limit terms in integration by parts in (32) vanish.

Applying the finite Hankel transform operator to equation (26) and using (32) , we obtain

$$-\lambda_m^2 U(\lambda_m, t) = \frac{\partial U(\lambda_m, t)}{\partial t} \quad ,$$

where

$$U(\lambda_m, t) = \int_0^1 U(r,t) r\, J_0(\lambda_m r)\,dr \; ,$$

so that

$$U(\lambda_m, t) = A(\lambda_m) e^{-\lambda_m^2 t} .$$

The initial condition determines the constant A . Thus

$$A(\lambda_m) = \hat{F}_0(\lambda_m) = \int_0^1 f(r) r J_0(\lambda_m r)\,dr \; .$$

Hence

$$U(\lambda_m, t) = \hat{F}_0(\lambda_m) e^{-\lambda_m^2 t} .$$

Therefore, by inversion theorem 2, we have

$$U(r,t) = \lim_{N\to\infty} \sum_{m=1}^{N} \frac{2\hat{F}_0(\lambda_m) e^{-\lambda_m^2 t} J_0(\lambda_m r)}{J_0^2(\lambda_m) + J_1^2(\lambda_m)} \quad , \tag{33}$$

since $J_0'(r) = -J_1(r)$.

We want to prove that $u(r,t)$, given by (33) is truly a solution of (25) that satisfies the given initial and boundary conditions. To achieve this, we need the following well known estimates:

$$F_0(\lambda_m) = O(\lambda_m^{-\frac{3}{2}}) \text{ as } m \to \infty, \text{ [Ref 48, p. 595]}$$

$$\lambda_m \sim \pi\,(m + \frac{1}{4}) \text{ as } m \to \infty,$$

$$J_0^2(\lambda_m) + J_1^2(\lambda_m) \sim \frac{2}{\pi\,\lambda_m} \text{ as } m \to \infty.$$

Hence,

$$F_0(\lambda_m)\,[J_0^2(\lambda_m) + J_1^2(\lambda_m)]^{-1} = O(m^{-1/2}) \text{ as } m \to \infty.$$

Using the above estimates, we see that the series (33) and the series obtained by applying $\Omega_{0,r}$ and D_t separately under the summation sign of (33) converges uniformly on $0 < r < 1$ and $t > 0$. Hence by applying $\Omega_{0,r} - D_t$ and using the fact $\Omega_{0,r}[J_0(\lambda_m r)] = -\lambda_m^2 J_0(\lambda_m r)$, we see that (33) satisfies the differential equation (25).

Let us verify the boundary condition (ii), we have

$$\lim_{r\to 1^-}[D_r u + H u] = \lim_{r\to 1^-}\left[\sum_{m=1}^{\infty} D_r\left\{\frac{2\hat{F}_0(\lambda_m)e^{-\lambda_m^2 t}J_0(\lambda_m r)}{J_0^2(\lambda_m)+J_1^2(\lambda_m)}\right\} + H\sum_{m=1}^{\infty}\frac{2\hat{F}_0(\lambda_m)e^{-\lambda_m^2 t}J_0(\lambda_m r)}{J_0^2(\lambda_m)+J_1^2(\lambda_m)}\right]$$

and since the convergence is uniform, we can take the $\lim r \to 1^-$ inside the summation sign and arrive at the conclusion, since λ_m's are the roots of the equation $\lambda J_0'(\lambda) + H J_0(\lambda) = 0$.

The initial condition (i) is already taken care of as we evaluated the constant A by using it. Through Figures 19-23, we establish the accuracy of the propose method. All figures are drawn by truncating the series (33) at $N = 10$. The presence of $e^{-\lambda_m^2 t}$ ensures that even ten terms give satisfactory solution for $t > 0$. While evaluating the solution $u(r,t)$ from (33) , we have evaluated $\hat{F}_0(\lambda_m)$ first from its analytical expression given by (23) and denote the solution thus obtained by $u(r, t)$ in Figures 19-23 and then evaluating $\hat{F}_0(\lambda_m)$ by using the proposed algorithm for evaluation of the finite Hankel transform as given by equation (13). This solution is denoted by $u_a(r, t)$ in the above mentioned figures.

Figure 19, compares the given initial condition $f(r)$ with $u(r, t)$ as $t \to 0^+$ and Figure 20 shows the error corresponding error $E(r) = u(r, 0) - f(r)$. Figure 21 depicts the various profiles of $u_i(r, t)$ at times $t = 0, \frac{1}{100}, \frac{1}{50}$ *and* $\frac{1}{25}$, the various profiles are denoted by

u_0, u_1, u_2 *and* u_3. As the maximum possible error occurs in the neighbourhood of 0 *and* 0.001, we have restricted t in (0,1] in Figures 22 and 23 representing $u(r, t)$ and $u_a(r, t)$ respectively and note that they are in good agreement in the range.

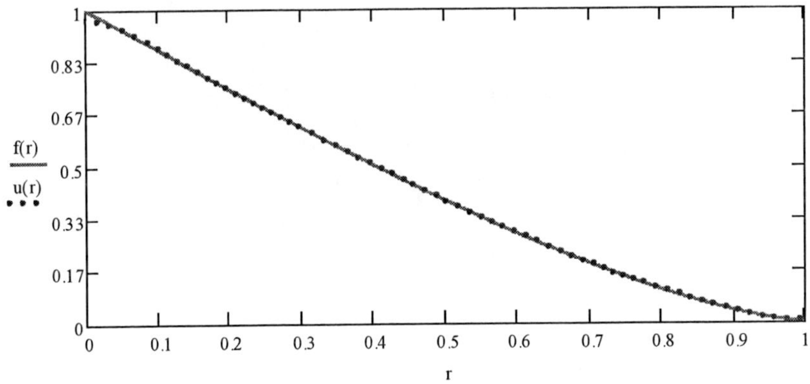

Figure 19. Initial condition function $f(r)$ (solid) and $u(r)\left(=\lim_{t\to 0^+} u(r,t)\right)$ (dotted).

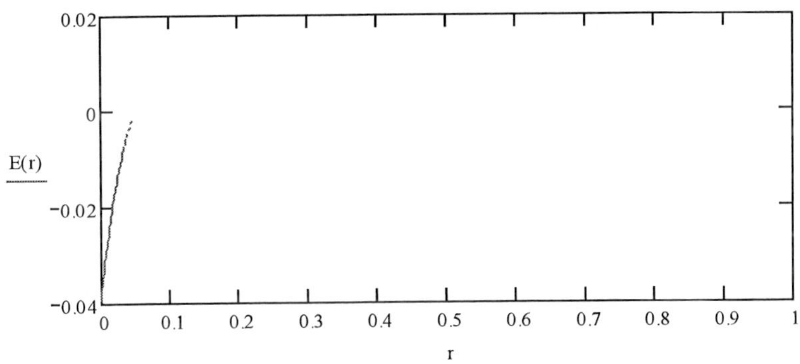

Figure 20. Error between $f(r)$ and $u(r)$.

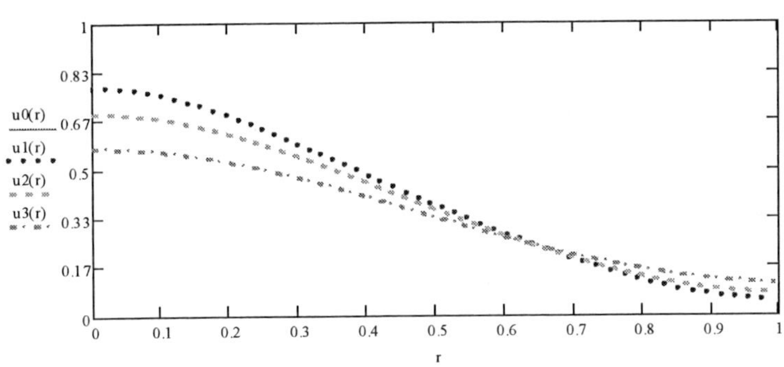

Figure 21. The various profiles of the solutions $u(r, t)$ at fixed times.

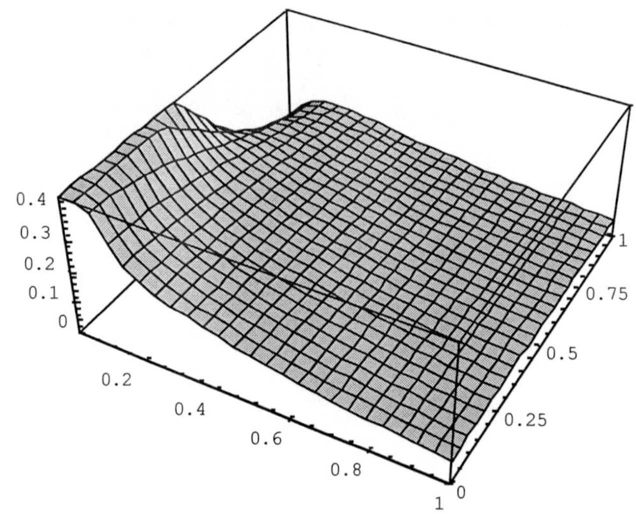

Figure 22. The solution $u(r,t)$ for $0<t<1,\ 0<r<1$.

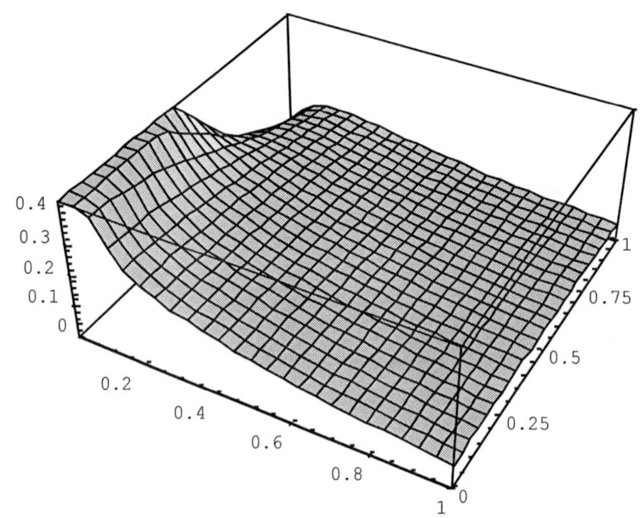

Figure 23. The solution $u_a(r,t)$ for $0<t<1,\ 0<r<1$.

6. Error Analysis

The numerical stability property of the algorithm is illustrated in Tables 1-6 where the discrete l^2 norm as well as L^2 norm of the error is shown as a function of the amount of noise ε in the data function, for Examples 1- 5 respectively (Example 4 has two tables 4 and 5 for different order of HT for the same test function). We notice that in all the cases, the numerical stability of the proposed algorithm is confirmed. Moreover, in the ε range 0.000 to 0.005 the discrete error norms are barely sensitive to changes in $h=1/N$ in examples 1, 3, 4 ($\nu=1/10$) and 5 and only slightly more sensitive to changes in h in examples 2 and 4 ($\nu=5$).

Table 1. Error norm as function of ε in Example 1

EXAMPLE 1	ERROR NORMS (DISCRETE l^2)			ERROR NORMS (L^2)
ε	N=10000	N=5000	N=1000	
0.000	$5.00\text{ x}10^{-8}$	$3.00\text{ x}10^{-8}$	$1.00\text{ x}10^{-8}$	$4.60\text{ x}10^{-7}$
0.001	$9.55\text{ x}10^{-6}$	$6.75\text{ x}10^{-6}$	$2.96\text{ x}10^{-6}$	$9.60\text{ x}10^{-5}$
0.002	$6.77\text{ x}10^{-6}$	$5.76\text{ x}10^{-6}$	$2.57\text{ x}10^{-6}$	$6.79\text{ x}10^{-5}$
0.005	$2.57\text{ x}10^{-5}$	$1.96\text{ x}10^{-5}$	$8.79\text{ x}10^{-6}$	$2.57\text{ x}10^{-4}$

Table 2. Error norm as function of ε in Example 2

EXAMPLE 2	ERROR NORMS (DISCRETE l^2)			ERROR NORMS (L^2)
ε	N=10000	N=5000	N=1000	
0.000	$1.05\text{ x}10^{-4}$	$1.36\text{ x }10^{-3}$	$6.66\text{ x}10^{-4}$	$1.05\text{ x}10^{-3}$
0.001	$1.05\text{ x}10^{-4}$	$1.35\text{ x}10^{-3}$	$6.59\text{ x}10^{-4}$	$1.05\text{ x}10^{-3}$
0.002	$1.06\text{ x}10^{-4}$	$1.37\text{ x}10^{-3}$	$6.68\text{ x}10^{-4}$	$1.07\text{ x}10^{-3}$
0.005	$1.07\text{ x}10^{-4}$	$1.34\text{ x}10^{-3}$	$6.56\text{ x}10^{-4}$	$1.09\text{ x}10^{-3}$

Table 3. Error norm as function of ε in Example 3

EXAMPLE 3	ERROR NORMS (DISCRETE l^2)			ERROR NORMS (L^2)
ε	N=10000	N=5000	N=1000	
0.000	$6.69\text{ x}10^{-4}$	$5.13\text{ x}10^{-4}$	$2.30\text{ x}10^{-4}$	$6.22\text{ x}10^{-3}$
0.001	$6.68\text{ x}10^{-4}$	$5.12\text{ x}10^{-4}$	$2.29\text{ x}10^{-4}$	$6.20\text{ x}10^{-3}$
0.002	$6.69\text{ x}10^{-4}$	$5.12\text{ x}10^{-4}$	$2.30.\text{x}10^{-4}$	$6.23\text{ x}10^{-3}$
0.005	$6.71\text{ x}10^{-4}$	$5.15\text{ x}10^{-4}$	$2.31\text{ x}10^{-4}$	$6.24\text{ x}10^{-3}$

Table 4. Error norm as function of ε in Example 4

EXAMPLE 4 $\nu = 1/10$	ERROR NORMS (DISCRETE l^2)			ERROR NORMS (L^2)
ε	N=10000	N=5000	N=1000	
0.000	$1.50\text{ x}10^{-3}$	$1.07\text{ x}10^{-3}$	$4.77\text{ x}10^{-4}$	$1.50\text{ x}10^{-2}$
0.001	$1.53\text{ x}10^{-3}$	$1.09\text{ x}10^{-3}$	$4.88\text{ x}10^{-4}$	$1.53\text{ x}10^{-2}$
0.002	$1.48\text{ x}10^{-3}$	$1.05\text{ x}10^{-3}$	$4.71\text{ x}10^{-4}$	$1.48\text{ x}10^{-2}$
0.005	$1.47\text{ x}10^{-3}$	$1.04\text{ x}10^{-3}$	$4.67\text{ x}10^{-4}$	$1.47\text{ x}10^{-2}$

Table 5. Error norm as function of ε in Example 4

EXAMPLE 4 $\nu = 5$	ERROR NORMS (DISCRETE l^2)			ERROR NORMS (L^2)
ε	N=10000	N=5000	N=1000	
0.000	$6.01 \text{x} 10^{-4}$	$7.22 \text{x} 10^{-3}$	$3.35 \text{x} 10^{-3}$	$5.73 \text{x} 10^{-3}$
0.001	$5.95 \text{x} 10^{-4}$	$7.23 \text{x} 10^{-3}$	$3.36 \text{x} 10^{-3}$	$5.67 \text{x} 10^{-3}$
0.002	$6.07 \text{x} 10^{-4}$	$7.21 \text{x} 10^{-3}$	$3.35 \text{x} 10^{-3}$	$5.80 \text{x} 10^{-3}$
0.005	$5.95 \text{x} 10^{-4}$	$7.19 \text{x} 10^{-3}$	$3.34 \text{x} 10^{-3}$	$5.73 \text{x} 10^{-3}$

Table 6. Error norm as function of ε in Example 5

EXAMPLE 5	ERROR NORMS (DISCRETE l^2)			ERROR NORMS (L^2)
ε	N=10000	N=5000	N=1000	
0.000	$8.91 \text{x} 10^{-3}$	$6.30 \text{x} 10^{-3}$	$2.82 \text{x} 10^{-3}$	$8.89 \text{x} 10^{-2}$
0.001	$8.92 \text{x} 10^{-3}$	$6.31 \text{x} 10^{-3}$	$2.82 \text{x} 10^{-3}$	$8.90 \text{x} 10^{-2}$
0.002	$8.83 \text{x} 10^{-3}$	$6.24 \text{x} 10^{-3}$	$2.79 \text{x} 10^{-3}$	$8.81 \text{x} 10^{-2}$
0.005	$8.78 \text{x} 10^{-3}$	$6.20 \text{x} 10^{-3}$	$2.78 \text{x} 10^{-3}$	$8.76 \text{x} 10^{-2}$

SUMMARY AND CONCLUSIONS

A new method based on the Chebyshev wavelets for the numerical evaluation of HT is proposed and analyzed. As the basis functions used to construct the Chebyshev wavelets are orthogonal and have compact supports, it makes them more useful and simple in actual computations compared to the algorithm based on Chebyshev polynomials [46]. Also, since the number of mother wavelet's components is restricted to one, they do not lead to the growth of complexity of calculations. Our choice of wavelets make it more attractive in their application in the applied physical problems as they eliminate the problems connected with the Gibbs phenomenon taking place in [32, 34]. The error associated with Filon quadrature philosophy [6], [21-22] is appreciable for small $p < 1$ compared to our algorithm.

The stability with respect to the data is restored and excellent accuracy is obtained even for small sample interval and high noise levels in the data. Several text cases are investigated by varying the number of sampling points N and amount of noise ε in the data. We notice that in all cases, the numerical accuracy and stability of the proposed algorithm is confirmed.

REFERENCES

[1] J.W. Goodman, *Introduction to Fourier Optics*, McGraw-Hill, New York, (1968).
[2] H.-Y. Fan, *Phys. Lett. A.*, 313, 343 (2003).

[3] I.N. Sneddon, *The use of Integral Transforms*, McGraw-Hill, New York (1972).
[4] V.S. Kulkarni and K.C. Deshmukh, *J. Frank. Inst.,* 345 (1) 29 (2008).
[5] V. Magni, G. Cerullo and De Silvestri, *J. Opt. Soc. Am. A.*, 12, 2031 (1992).
[6] R. Barakat, E. Parshall and B.H. Sandler, *J. Opt. Soc. Am. A.*, 15, 652 (1998).
[7] N.T. Eldabe, M. El-Shahed , M. Shawkey, *Appl. Math. Comput.*, 151, 713 (2004).
[8] D. Patella, *Geophysical Prospecting*, 28, 744 (1980).
[9] D.R. Mook, *IEEE Trans. Acoust. Speech Signal Process.*, ASSP-31, 979 (1983).
[10] E.C. Cavanagh and B.D. Cook, *IEEE Trans. Acoust. Speech Signal Process.*, ASSP-27, 361 (1979).
[11] E.V. Hansen, *IEEE Trans. Acoust. Speech Signal Process.*, ASSP-33, 666 (1985).
[12] E.V. Hansen, *IEEE Trans. Acoust. Speech Signal Process.*, ASSP-34, 623 (1986).
[13] S.M. Candel, *IEEE Trans. Acoust. Speech Signal Process.*, ASSP-29, 963 (1981).
[14] B.W. Suter, *IEEE Trans. Signal Process.,* 39, 532(1991).
[15] L. Knockaert, *IEEE Trans. Signal Process.*, 48, 1695 (2000) -1701.
[16] W.E. Higgins and D.C. Munson, *IEEE Trans. Med. Imag.*, 7, 59 (1988).
[17] O.P. Singh and J.N. Pandey, Proc. *Amer. Math. Soc.*, 115 (4), 969 (1992).
[18] O.P. Singh, *J. Math. Anal. Appl.*, 191, 450 (1995).
[19] A.E. Siegman, *J. Optics. Lett.*, 1, 13 (1977).
[20] J.A. Ferrari, *J. Opt. Soc. Am. A.*, 12, 1812 (1995).
[21] R. Barakat and E. Parshall, *Appl. Math. Lett.*, 9 (5), 21 (1996).
[22] R. Barakat and B.H. Sandler, *Appl. Math. Lett.*, 11(1), 127 (1998).
[23] Li Yu, M. Huang, M. Chen, W. Huang and Z. Zhu, *Opt. Lett.*, 23, 409 (1998).
[24] J. Markham and J.A. Conchello, *J. Opt. Soc. Am. A.*, 20 621 (2003).
[25] P.K. Murphy and N.C. Gallagher, *J. Opt. Soc. Am.*, 73 1130 (2003).
[26] A.V. Oppenheim, G.V. Frish and D.R. Martinez, *Proc. IEEE.*, 66 264 (1980).
[27] A.V. Oppenheim, G.V. Frish and D.R. Martinez, *J. Acoust. Soc. Am.*, 68, 523 (1980).
[28] W.E. Higgins and D.C. Munsons Jr , *IEEE Trans. Acoust. Speech Signal Process.*, ASSP-35, 86 (1987).
[29] J.A. Ferrari, D. Peciante and A. Durba, *J. Opt. Soc. Am. A.*, 16, 2581 (1999).
[30] A.Agnesi, G.C. Reali, G. Patrini and A. Tomaselli, *J. Opt. Soc. Am. A.*, 10, 1872 (1993).
[31] B.W. Suter and R.A. Hedges, *J. Opt. Soc. Am. A.*, 18, 717 (2001).
[32] E.B. Postnikov, *J. Applied Math.*, 6, 319 (2003).
[33] R. Piessens and M. Branders, *BIT Numerical Mathematics*, 23, 370 (1983).
[34] M. Guizar-Sicairos and J.C. Gutierrez-Vega, *J. Opt. Soc. Am. A.*, 21(1), 53 (2004).
[35] V.K. Singh, O.P. Singh and R.K. Pandey, *Comput. Phys. Commun*. 179, 424 (2008).
[36] R.K. Pandey, V.K. Singh and O.P. Singh, *J. Franklin Inst.*, 346, 102 (2009).
[37] A. Haar, *Math. Ann.*, 69, 331 (1910).
[38] I. Daubechies, *Commun. Pure Appl. Math.*, 41, 909 (1988).
[39] I. Daubechies, *Ten Lectures on wavelets*, CBMS-NSF, (1992).
[40] S. Mallat, *Trans. Amer. Math. Soc.*, 315, 69 (1989).
[41] S. Mallat, *IEEE Trans. Pattern Recogn. Mach. Intell.,* 11 (7), 674 (1989).
[42] R.M. Young, *An Introduction to Nonharmonic Fourier Series*, Academic Press, New York, (1980).
[43] E. Babolian and F. Fattahzadeh, *Appl. Math. Comput.*, 188, 417 (2007).
[44] A. Erdelyi (Ed.), *Tables of Integral Transforms*, McGraw-Hill, New York, (1954).

[45] M. Abramowitz and I.A. Stegun, *Handbook of Mathematical Functions, With Formulas, Graphs, and Mathematical Tables*, Dover Publications, Inc., New York, (1965).

[46] R. Piessens, *J. Comput. Appl. Math.*, 121, 113(2000).

[47] S. M. Candel, *Comput. Phys. Commun.*, 23, 343, (1981).

[48] G.N. Watson, *Theory of Bessel functions,* 2nd ed., Cambridge University Press, Cambridge (1958).

In: Wavelets: Classification, Theory and Applications
Editors: Manel del Valle et al.
ISBN: 978-1-62100-252-9

Chapter 13

APPLICATIONS OF WAVELETS IN FINANCIAL PORTFOLIO OPTIMIZATION

N. C. Suganya * ***and G. A. Vijayalakshmi Pai***
Department of Mathematics and Computer Applications,
PSG College of Technology,Coimbatore, India

ABSTRACT

Constrained portfolio optimization determines the optimal proportions of capital that the investor should invest on each asset of the portfolio to meet the bi-criterion objective of maximizing return and minimizing risk, subject to constraints such as basic, bounding, cardinality, asset class and short sales. Initially, the application of wavelets as noise filters for the financial covariance matrix, which is one of the key input to the portfolio optimization problem is discussed. The reliability of the portfolios obtained by the wavelet filtered method is measured and statistical analyses of the performance results, when compared with its counterpart's viz., Random Matrix Theory based filters and k-means cluster analysis are undertaken.

Later, the application of wavelets in finding a neural network based solution to a complex constrained portfolio optimization problem is discussed. Usually, the bi-objective problem is converted into a single objective problem by introducing risk aversion parameter. When the constraints reflecting investor preferences and/or market frictions are included, the problem turns into a complex mixed integer-quadratic programming problem, making it difficult for direct solving by analytical or numerical methods. Although there are many heuristic algorithms, Evolution based Wavelet Hopfield Neural Network (EWHNN) is proposed for its solution, because of the following advantages a) handles diversification in both large and small portfolios. b) Intrinsically reports faster convergence and yields reliable optimal portfolios when compared with other heuristic algorithms. c) Incorporates more constraints that are of practical use, in one lot. The major limitations of single objective formulation are 1) need to form minimum N different problems to find N different Pareto-optimal solutions that lie on the efficient frontier. 2) uniformly spaced set of risk aversion parameter values may not produce uniformly spaced Pareto-optimal solutions. 3) different parameter values

* E-mail address: eliotsuganya@gmail.com

may produce an identical solution. These drawbacks are rectified in a multi objective formulation.

Hence, Pareto-archived Evolutionary Wavelet Network (PEWN) is introduced for the solution of complex constrained multi objective portfolio optimization problem. The results obtained using the aforementioned two wavelet network based algorithms are evaluated using different quantitative measures. All the experimental studies are undertaken on the Bombay Stock Exchange, India (BSE200 index, period: July 2001 – July 2006) and Tokyo Stock Exchange (Nikkei225 index, period: March 2002 – March 2007) data sets.

INTRODUCTION

Investment management is an integral component of financial planning and personalized wealth management. This needs a deep economic understanding of the markets with strong mathematical models. A *financial portfolio* is an appropriate collection of assets hold by a private individual or institutions. The assets discussed in this work are conventional assets which are different from alternative assets such as real estates, hedge funds and venture capital. The problem of *portfolio optimization* is a process of determining the assets and its optimal proportion in the portfolio that meets the bi-criterion objective of *minimizing risk* and *maximizing return*. The solution to this problem is obtained by drafting the optimal risk and the corresponding return which forms the risk-return trade-off curve called *efficient frontier*, giving the maximum return for a given level of risk, or, alternatively minimum risk for an expected return.

Harry Markowitz [1] first laid a quantitative framework through mean-variance model termed as *Modern Portfolio Theory* (MPT) with the assumption that asset returns follow a multivariate normal distribution. Since, unconstrained portfolio optimization problem formulation is quadratic in nature, analytical methods can be efficiently used to find the optimal portfolios that lie on the efficient frontier. The key inputs of Markowitz model are the *expected mean* and *covariance (correlation) matrix* obtained from financial return series of assets.

Empirical covariance matrix is of great importance in asset allocation and investment theory and has gained industry-wide applications in risk management. It is necessary to have reliable estimates of covariance of asset returns in the portfolio. Since, the length of the empirical time series T is limited due to practical reasons, inevitably leading to the measurement error in the covariance matrix estimate. This estimation error induces a high amount of statistical uncertainty which causes uncertainty in portfolio estimation, dominates the probability of large losses for a certain portfolio, leads to serious instability and a high degree of degeneracy in optimal portfolios. Finally, it makes the whole theory more dangerous and questionable. 'Markowitz portfolio optimization scheme based on a purely historical determination of the correlation matrix is inadequate...' is concluded in [2]. Hence, better estimation models are required to remove noise from the covariance matrix.

The study on Random Matrix Theory (RMT) filtered covariance matrices for portfolio optimization is done by [3-5]. There are also other filtering procedures based on clustering in the econophysics literature as proposed in [6-11]. Following them, single linkage and average linkage clustering algorithms for noise filtering in covariance matrix is demonstrated in [12] and proved that clustering algorithms act as better reliable filters when compared to RMT based filtering methods in terms of the ratio between predicted and realized risk for a given expected portfolio return. Later Pai and Michel [13] applied k-means clustering for portfolio optimization and concluded that k-means cluster analysis outperforms Markowitz and RMT filtered models.

Since, wavelets are proved to be a good noise filter in a wide spectrum of disciplines, Suganya and Pai [14,15] attempted exploring the efficacy of wavelet based filters and compared its efficiency with RMT based filter and k-means clustered counterparts. Several experiments were carried out on various wavelet functions such as Haar, Daubechies, Symlet, Discrete Meyer and Coiflets for the efficient noise removal from the covariance matrix using wavelet shrinkage denoising technique. Finally 'symlet 3' wavelet function was chosen to act as mother wavelet because symlet based covariance matrix showed higher correlation with the empirical covariance matrix. The reliability of portfolios obtained using wavelet shrinkage denoising technique based on predicted and realized risk of the portfolio set was compared with that of k-means cluster analysis.

The mathematical model of the portfolio optimization problem should include more constraints to obtain more appropriate solutions close to reality. Therefore, the current constrained problem involves basic, bounding, cardinality, class and short sales constraints. In reality, portfolio optimization problem is a constrained optimization problem. When an investor decides to diversify his or her amount in only K out of the universe of N assets, the *cardinality* constraint is enforced. Several theoretical and practical approaches are introduced to handle the cardinality constraint. Pai and Michel [13] showed that k-means cluster analysis not only act as noise filter, but also eliminates cardinality constraint by forming K groups based on denoised covariance matrix and expected return and selecting randomly or preferably one asset from each group to form an *investable universe*. Hence, k-means cluster analysis is included in this chapter to eliminate the cardinality constraint.

There are heuristic algorithms such as memetic algorithm, simulated annealing, evolutionary algorithms, tabu search, Ant colony optimization and Neural network [16-20] which are sought for its solution. All these algorithms reported near optimal or mixed results. However, most of the works have primarily confined itself to tackle only bounding and cardinality constraints on diversification in small portfolios (30 or less). Later Suganya and Pai [21] made use of an Evolution based Hopfield Neural Network (EHNN) strategy, a hybrid model that adopts evolutionary operators for convergence and Hopfield Neural network as its host. This strategy was a refinement of Fernandez and Gomez's Hopfield Neural Network (HNN) strategy which handled only a restricted set of constraints viz. basic, bounding and cardinality constraints for the portfolio optimization problem. Also, the HNN strategy which resorted to pruning of network to tackle the cardinality constraint resulted in increased computational load and slow convergence for large portfolio sets. However, the EHNN strategy handled class and short sales constraints apart from other constraints described earlier and efficiently handled large portfolio sets.

Wavelet Networks are known to be one of the best heuristic solution strategies in many applications. Hence, Suganya and Pai undertook many experiments and finally embedded wavelets into the EHNN solution strategy. This Evolution based Wavelet Hopfield Neural Network (EWHNN) strategy adopts Gaussian derivative wavelet function as activation function instead of sigmoid function and with modification in EHNN, reports faster convergence and gives better optimal solutions when compared to other heuristic algorithms in literature.

All the above discussed strategies provide solution for the bi-objective problem of constrained portfolio optimization that is converted into single objective problem by weighted formulation. There are several limitations in solution strategies of weighted single objective problem. Hence, Suganya and Pai [22] proposed a Pareto-archived Evolutionary Wavelet Network (PEWN) for constrained multi objective portfolio optimization problem. The key features of PEWN solution strategy is the inclusion of wavelet network and mapping of *'concept of dominance'* to the specified constrained multi objective portfolio optimization problem.

Different performance metrics such as error ratio, set coverage metric and generational distance are employed to evaluate the closeness of the optimal set obtained using PEWN to the Pareto-optimal front. Spacing, spread and maximum spread are some of the other metrics evolved to find the diversity among non-dominated solutions. Due to the absence of heuristic strategies that handles all the constraints included in this chapter, the performance of PEWN is compared with EHNN and EWHNN, the only existing counterparts, which are proved to be better strategies when compared with other heuristic algorithms including evolutionary algorithms and neural networks [23, 24] that have been meant for the solution of portfolio optimization problem. All the algorithms are implemented, simulated and compared using MATLAB software. The experimental studies are undertaken on the Bombay Stock Exchange, India (BSE200 index, Period: July 2001-July 2006) and Tokyo Stock Exchange (Nikkei225 index, Period: March 2002 – March 2007) data sets.

The objectives of the chapter are therefore to,

(i) describe the financial constrained portfolio optimization problem which needs the use of wavelet concept to obtain more efficient portfolio sets.
(ii) give a brief account of the application of wavelet shrinkage denoising technique for the removal of noise from the financial covariance matrix, which is one of the key inputs to the portfolio optimization problem.
(iii) explain the working of Evolution based Wavelet Hopfield Neural Network (EWHNN) for constrained portfolio optimization.
(iv) discuss the Pareto-archived Evolutionary Wavelet Network (PEWN) for constrained multi objective portfolio optimization.
(v) present the conclusions of the study.

FINANCIAL CONSTRAINED PORTFOLIO OPTIMIZATION PROBLEM

The Markowitz Mean-Variance deterministic model is the simplest and basic mathematical model of portfolio optimization problem, which assumes that asset returns follow a multivariate normal distribution. Therefore, expected mean and covariance of the daily expected returns of the assets is adopted to represent the portfolio return and risk respectively. According to Modern Portfolio Theory, an investment strategy that seeks to construct an optimal portfolio should diversify the wealth among many different assets instead of selecting the dominant asset.

Markowitz's model does not incorporate aspects of real-world trading such as cardinality constraint, transaction costs, short sales, management costs etc. Hence, in the absence of these constraints, the mathematical formulation becomes a quadratic optimization problem with a positive semi definite matrix (covariance matrix), solvable in polynomial time and traditional methods like quadratic programming is directly applied to trace the efficient frontier.

Let N be the total number of available assets, μ_i be the expected return (per period) of asset i (i=1, 2…N), C_{ij} be the covariance between the assets i and j (i =1, 2…N, j =1, 2 …N), μ_p be the desired expected portfolio return, p_i be the proportion ($0 \le p_i \le 1$) invested in each asset i of the portfolio (i =1, 2…N). Therefore the standard mean-variance approach introduced by Markowitz is as follows.

$$\text{Minimize} \quad \sum_{i=1}^{N}\sum_{j=1}^{N} p_i C_{ij} p_j$$

$$\text{Subject to} \quad \sum_{i=1}^{N} p_i \mu_i = \mu_p \tag{1}$$

$$\sum_{i=1}^{N} p_i = 1$$

$$0 \le p_i \le 1$$

In this problem formulation, the objective is to minimize the total variance associated with the portfolio. The constraints ensure that the total expected return equals μ_p and the total investment is 1 with each asset proportion lying between 0 and 1. A non-decreasing curve called efficient frontier which consists of optimal portfolios is the solution of this problem.

The major disadvantages of this model is that it assumes markets to be perfect, excluding taxes and transaction costs, disallowing short sales, ignoring investor's preferences and market frictions. Prior to the application of solution strategy, finding the range of values for the expected return (μ_p) is done only by trial and error approach.

In reality, more appropriate solutions can be found only by the inclusion of constraints that depicts the original problem. Hence, better models and better solutions are needed. The portfolio optimization problem becomes more meaningful when the objective consists of two competing goals of minimizing the risk and maximizing the return of the portfolio.

Single Objective Portfolio Optimization

The single-period portfolio optimization deals with holdings in K risky assets. The portfolio is adjusted at the beginning of the time period. The single period portfolio optimization problem can be formulated as both single objective and multi objective portfolio optimization problems.

The risk aversion parameter λ $(0 \leq \lambda \leq 1)$ is introduced in single objective problem to trace the efficient frontier. This model is intuitive and easy to use. Hence the equations (1) becomes

$$\text{Minimize} \quad \lambda\left(\sum_{i-1}^{N}\sum_{j=1}^{N} p_i C_{ij} p_j\right) - (1-\lambda)\left(\sum_{i=1}^{N} p_i \mu_i\right) \tag{2}$$

$$\text{Subject to} \quad \sum_{i=1}^{N} p_i = 1$$

$$0 \leq p_i \leq 1$$

This weighted formulation transforms the bi-criterion objective function into a single criterion objective function to simplify the problem model. When $\lambda=0$, the objective function reduces to maximize (return) by completely eliminating the portfolio risk. Alternatively, when $\lambda=1$, the objective function reduces to minimize (risk). Thus, for a specific value of λ, solution of equations (2) yields the optimal weights $p_{i.}$ The number of values for λ is arbitrarily chosen to trace the shape of the curve representing the efficient frontier, though a large number of points could obviously lead to smooth curves. Therefore, the efficient frontier which is the graphical solution is traced using the risk-return couples for different values of $\lambda \in [0,1]$.

Some of the disadvantages of single objective weighted formulation are

a) In each simulation run of the solution strategies, only one optimal solution can be obtained. Therefore, in order to find N different optimal solutions minimum N different single objective optimization problems need to be formed.
b) Uniformly spaced set of risk aversion parameters will not produce uniformly spaced optimal solutions.
c) Identical Pareto-optimal solutions are obtained for different risk aversion parameter values.

The multi objective formulation of portfolio optimization problem rectifies all these drawbacks.

Multi objective Portfolio Optimization Problem

Multi objective optimization problems have objective functions which are to be minimized or maximized. The two objectives in multi objective formulation of portfolio optimization problem are as follow:

$$\text{Minimize} \sum_{i=1}^{N}\sum_{j=1}^{N} p_i C_{ij} p_j \tag{3}$$

$$\text{Maximize} \sum_{i=1}^{N} p_i \mu_i \quad (4)$$

The two conflicting goals (minimize risk and maximize return) give rise to a number of optimal solutions, known as Pareto-optimal solutions or non-inferior solutions. The concept of dominance is the key feature which helps in finding as many Pareto-optimal solutions as possible.

The merits of multi objective optimization are availability of efficient population based optimization algorithms, avoidance of multiple simulation runs and above all, the concept of dominance gives a practical meaning to handle multiple objectives irrespective of its complexity.

A non-dominated set must be collectively compared with other solutions using dominance concept for establishing whether the latter solution belongs to the non-dominated set or not. The two conditions that a non-dominated set should satisfy are a) any two solutions in the set must be non-dominated with respect to each other b) any solution not belonging to the set is dominated by at least one member of the set. Hence, in multi objective optimization problem, it is difficult to prefer one solution over the other without any further information about the problem. Therefore, it is important to find as many Pareto-optimal solutions as possible.

Complex Additional Constraints

In order to obtain more appropriate solution, there is a need to include more constraints, which in turn makes the problem complex. The list of constraints formulated and solved in this chapter that realistically model investor preferences are as follows.

Basic constraint:
The *'basic'* constraint verifies whether the total proportion invested in assets adds up to 1.

Bounding constraint:
This constraint specifies the maximal and minimal proportion of possible investments in each asset in the portfolio.

Cardinality constraint:
The maximum number of assets in a portfolio is fixed to a constant.

Class constraint:
The proportion of amount that can be invested in assets of each class is limited. Some of the typical classes might be financial services, healthcare, oil and gas, industrial goods and services, banking and telecommunication.

Linear Short sales constraint:
The technique of borrowing a security from the broker and selling it, with the understanding that it must later be bought back and returned to the broker is called short sales.

Let R^- be the minimum proportion represented by negative value, allowed for short selling each asset in the portfolio, R^+ is the maximum proportion allowed due to short selling of R^- for each asset in the portfolio $\left(R^+ = 1 + \left|R^-\right|\right)$, x_i is the constant that takes the value 1, if asset i is included in the portfolio, else it takes value 0, K is the maximum number of assets allowed in a portfolio, L_i is the minimum proportion allowed for investing in asset i, U_i is the maximum proportion allowed for investing in asset i, α_a is the minimum investment limit in the ath asset class, β_a is the maximum investment limit in the ath asset class. Therefore, the mathematical formulations of all the above mentioned constraints are as follows:

$$R^- \le p_i \le R^+ , \ \forall i \in N \text{ (Short sales constraint)} \tag{5}$$

$$\sum_{i=1}^{N} x_i \le K \text{ (Cardinality constraint)} \tag{6}$$

$$L_i x_i \le p_i \le U_i x_i \text{ (Bounding constraint)} \tag{7}$$

$$x_i \in \{0,1\} \quad i = 1,2,3 \cdot\cdot N$$

For each class of assets R_a,

$$\alpha_a \le \sum_i p_i \le \beta_a ,\ i \in R_a \text{ (Class Constraint)(8)}$$

$$0 \le \alpha_a < \beta_a \le 1$$

Efficient Frontier

The efficient frontier is conventionally plotted with the expected return and the standard deviation (risk) of the portfolio on the vertical and horizontal axis respectively. Hence, the set of portfolios that offers a minimum risk for an expected level of portfolio return traces a curve called the efficient frontier. The line created from the risk-reward graph, comprises of optimal portfolios as given in Figure 1.

The expected portfolio return increases with increase in risk. Therefore, low levels of risk (uncertainty) are associated with low potential returns, whereas high levels of risk are associated with high expected portfolio returns. According to the risk-return trade off curve (Efficient Frontier), higher profits are rendered for the invested amount, only if it is subjected to the possibility of being lost. Investors must be aware of the personal risk tolerance when choosing the portfolio.

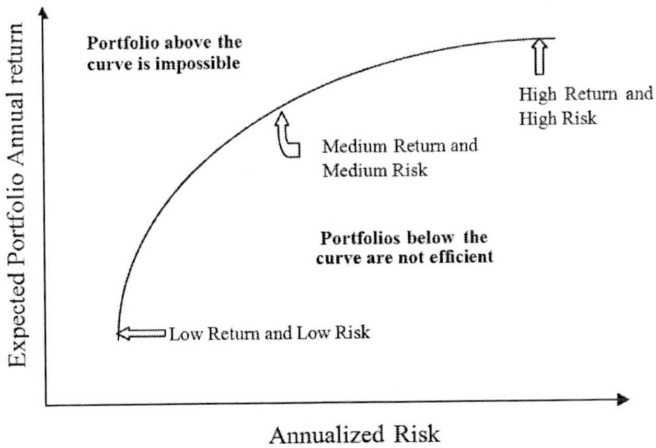

Figure 1. Efficient frontier.

HISTORICAL ESTIMATES

The key inputs of portfolio optimization problem are expected return and covariance matrix obtained from the historical return series of assets. Hence, these estimates must be reliable to obtain appropriate solution. Since, the length of the available time series is limited for practical reasons, the estimation of covariance matrix introduces high amount of noise (measurement noise). The covariance matrix has $N(N-1)/2=N^2/2$ distinct entries, where N is the number of assets. The time series of length T>>N is needed to minimize error on the covariance matrix, which is not practical. This introduces high degree of degeneracy and serious instability in optimal portfolios. Therefore, filtering methods capable of denoising the covariance matrix are needed, then less likely to be affected by statistical uncertainty.

Existing Denoising Techniques for Financial Covariance Matrix

There are several contributions in the literature that are devoted to noise filtering in the covariance matrix. Many models based on single and multi-index models are introduced on economic grounds to impose some structure on the covariance matrix to decrease the effective number of parameters to be estimated. However, pure statistical estimation methods such as Bayesian shrinkage estimation or principal component analysis are used in [25-28].

Later Random Matrix Theory (RMT) based filters and clustering algorithms are employed by few researchers to denoise the covariance matrix.

Since clustering algorithms report improved portfolio reliability in terms of predicted and realized risk for a given expected portfolio return, Pai and Michel claimed that k-means clustering based on the expected return and covariance matrix is also a better noise filter, yielding portfolios with better reliability when compared to RMT based filters and Markowitz model.

Wavelet Shrinkage Denoising Technique for Financial Covariance Matrix

In the area of applied mathematics, wavelet theory is one of the latest developments emerging as one of the fastest growing field with applications ranging from astrophysics to seismology. The presence of noise in the real time application is studied by many engineers and scientists. This corrupts the original data. Hence, this noise should be eliminated to recover the original data and to proceed with further reliable data analysis. The study on wavelet shrinkage denoising technique shows that this technique attempts to remove whatever noise is present and retains the data regardless of the frequency content of the data.

The raw market data consists of market prices are irregularly spaced in time. Due to faulty input or other factors, errors may be present in the data set. Hence filters are required to clean and present the raw price information as a regularly spaced time series for analysis. Wavelet analysis act as an important tool for calculation of variance in relation to specific time scales. Wavelet analysis also extracts information from financial market data with applications ranging from short term prediction to the testing of market models. One of the fundamental advantages of wavelet analysis is the capability to decompose time series into different components which helps in multi resolution analysis of high frequency market data.

Although wavelet theory is employed in many areas and applications, there is no literature that adopts wavelet shrinkage denoising technique to denoise financial covariance matrix. Since, wavelet based filters are proved to be best in other fields, Suganya and Pai [14, 15] tested wavelet shrinkage denoising technique with different wavelet functions to denoise financial covariance matrix. The denoised covariance matrix when used in portfolio optimization problem produces better optimal portfolio sets. The efficiency and reliability of portfolios obtained using wavelet based filtered covariance matrix and expected return is compared with its counterparts k-means cluster analysis, RMT based filters and Markowitz model in terms of predicted and realized risk for a given expected portfolio return.

Let $S_{ij}(t)$ be the noisy covariance obtained from return series of assets, $\hat{C}_{ij}(t)$ be the original data and $\varepsilon_{ij}(t)$ be the independent standard normal random variable representing the noise. The procedure involved in wavelet shrinkage denoising technique is as follows:

1. Initially, wavelet coefficient matrix ω is calculated for the covariance matrix of assets by the application of 'symlet 3' wavelet transform ψ.

$$S_{ij}(t)=\hat{C}_{ij}(t)+\varepsilon_{ij}(t), \quad i=1,2...N, j=1,2...N \tag{9}$$

$$\omega = \psi(S) = \psi(\hat{C}) + \psi(\varepsilon) \tag{10}$$

The wavelet coefficient matrix (ω) consists of both approximation and detail coefficients.

2. The detail coefficients are modified by the application of global threshold operator $\lambda = \sqrt{2\log N}$, where N is the size of the covariance matrix of assets.
3. The estimate $\hat{\omega}$ of the wavelet coefficient matrix ω is obtained.

$$\omega \rightarrow \hat{\omega} \tag{11}$$

4. Finally, inverse 'symlet 3' wavelet function is employed to estimate $\hat{\omega}$ to obtain the denoised covariance matrix (C).

$$C = \psi^{-1}(\hat{\omega}) \tag{12}$$

Several studies are performed with different wavelet functions for the efficient denoising of covariance matrix. The 'symlet 3' wavelet function is chosen to act as both forward (ψ) and inverse (ψ^{-1}) transform functions respectively, due to the high correlation between the 'symlet 3' based denoised covariance matrix and original covariance matrix. The number of resolution levels is chosen as five because no improvement in reliability of optimal portfolios is felt after this level.

Measuring the Reliability of Wavelet Filtered Portfolio Sets

Measures used in [12,13] are adopted to compute the reliability of portfolios obtained using wavelet based filter. Predicted and realized risks of the portfolio sets are necessary for reliability analysis. Since Pai and Michel already proved that k-means cluster analysis outperforms Markowitz model and RMT based filters, wavelet shrinkage denoising technique is compared with k-means cluster analysis in terms of predicted and realized risks.

Predicted and Realized Risk

Let T be the total trading days and t_0 be the expected optimum time. The observed mean returns $\mu_i^{(t_0+T)}$ and variance $\sigma_i^{(t_0+T)}$ of the (t_0+T) period and the covariance matrix of the (t_0-T) period of N assets are used to compute the optimal weights p_i^* for the (t_0-T) period. Thus, predicted risk-return curve is traced for the optimal weights p_i^*. The realized risk-return curve is obtained by computing the risk-return couple obtained using the optimal weights p_i^*, the mean returns $\mu_i^{(t_0+T)}$, volatilities $\sigma_i^{(t_0+T)}$ and the covariance matrix of the (t_0+T) period.

Since, Pai and Michel proved that k-means cluster analysis which is one of the clustering algorithm outperforms RMT based filter and Markowitz model, the current work compares wavelet based filter with k-means cluster analysis. Figures 2(a) and 2(b) gives the experimental analysis and comparison of the predicted and realized risk-return graphs obtained using wavelet shrinkage denoising technique with k-means cluster analysis using investable universe of K=20 (small portfolio set) and K=50 (large portfolio set) for BSE200 and Nikkei225 data sets respectively. The figures show less deviation between the predicted and realized risk-return curves obtained using wavelet-based filter model when compared with that of k-means cluster analysis.

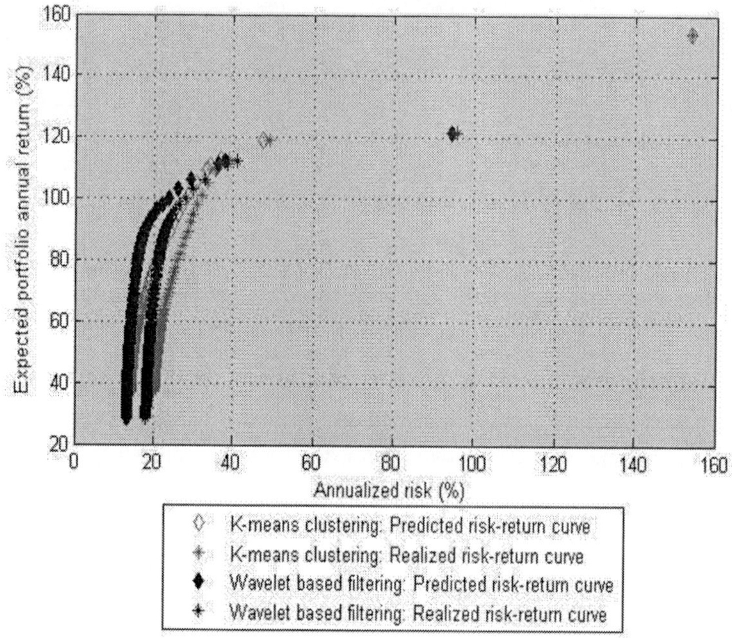

Figure 2(a). Risk-return curves (predicted and realized) for the k-means cluster based model (K=20) and wavelet based filter model for BSE200 (July 2001 – July 2006) data set.

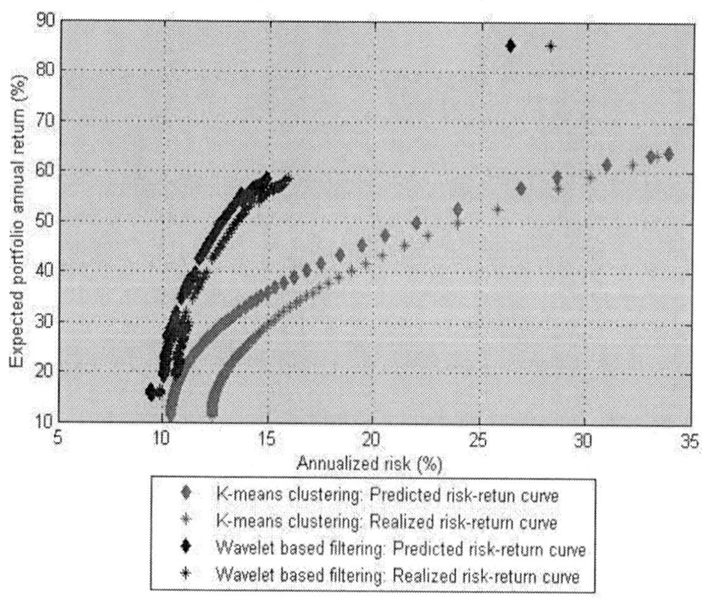

Figure 2(b). Risk-return curves (predicted and realized) for the k-means cluster based model (K=50) and wavelet based filter model for Nikkei225 (March 2002 – March 2007) data set.

The predicted σ_p and realized risks σ_R is obtained for a given expected portfolio return by the linear interpolation of the known values reflected on the risk-return curves. The reliability coefficient R is calculated using the formula:

$$R = \frac{|\sigma_p - \sigma_R|}{\sigma_p} \tag{13}$$

In Figures 3(a) and 3(b), the reliability of the portfolio sets obtained using wavelet-shrinkage denoising technique is compared with k-means cluster analysis. The small value of reliability R reported by portfolio sets obtained using wavelet shrinkage denoising technique in both the data sets indicate better reliability.

K-means Cluster Analysis for Cardinality Constrained Portfolio Optimization

Although, there are many clustering algorithms, k-means cluster analysis is employed because it eliminates cardinality constraint in portfolio optimization problem and already proved to be an inherent noise filter. The cardinality constraint is one of the complex constraints that turn the problem into a mixed integer quadratic programming problem, making difficult for direct solving by analytical or numerical methods. K-means clustering algorithm divides the universe of N assets into K clusters for a specified value of K.

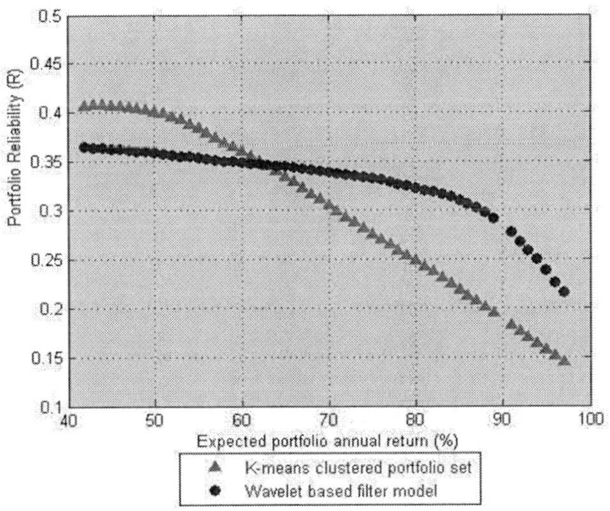

Figure 3(a). Reliability analysis of portfolios obtained using wavelet based filtering model and k-means clustered analysis (K=20) for the BSE200 data sets.

The traditional approach of selecting K assets from N assets, i.e. NC_K leads to a combinatorial explosion for a large value of K. In [29], the assets are grouped into classes based on the size, sector, geographical aspects and selects K assets from each of these classes, obtaining inferior solutions due to the ignorance of correlation between the assets. Fernandez and Gomez adopted a network pruning technique which is complex to handle the cardinality constraint. A more meaningful investable assets are obtained when N assets are partitioned based on the expected return and wavelet-based filtered covariance matrix. In portfolio optimization problem, if the investor prefers to have a maximum of K assets, then K clusters are formed. Then, an investable universe of K assets is obtained by selecting an asset of the investor's choice from each of the K clusters, which serves to eliminate the cardinality constraint. The sequential combination of wavelet shrinkage denoising technique and k-means cluster analysis reports better portfolio reliability.

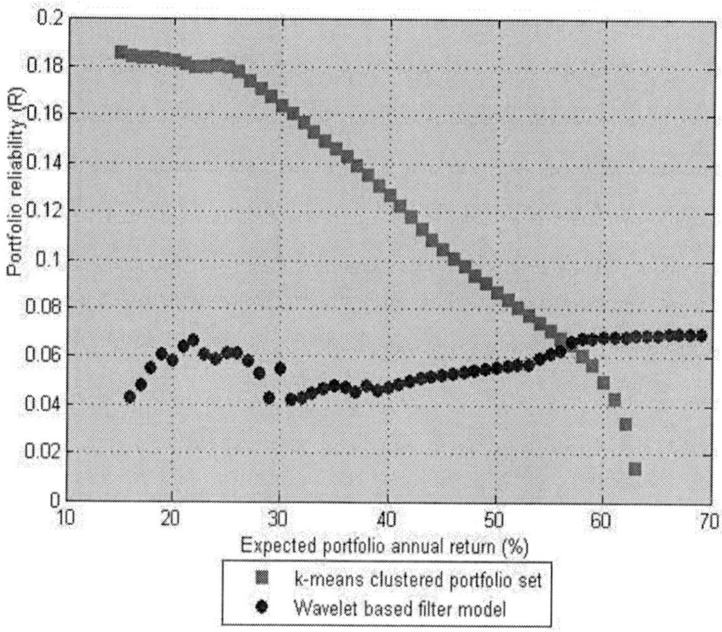

Figure 3(b). Reliability analysis of portfolios obtained using wavelet based filtering model and k-means clustered analysis (K=50) for the Nikkei225 data sets.

The merits of k-means cluster analysis in portfolio optimization problem are as follows:

a) simplifies the mathematical model by the elimination of cardinality constraint.
b) acts as an inherent noise filter.
c) improves the reliability of portfolios when wavelet-based denoised covariance matrix is adopted as one of the inputs to k-means cluster analysis.
d) reduces the number of decision variables which leads to faster convergence.

Evolution Based Wavelet Hopfield Neural Network (EWHNN)

The Evolution based Hopfield Neural Network (EHNN) algorithm which is introduced by Suganya and Pai [21] for the proposed constrained portfolio optimization is modified to obtain evolution based wavelet Hopfield neural network (EWHNN). In the previous works, the portfolio sets obtained using EHNN is compared with other heuristic algorithms such as evolutionary algorithm and Hopfield neural network. Moreover, there is no heuristic strategy in literature that solves the portfolio optimization problem with all the constraints considered in this chapter. Some of the features of EWHNN solution strategy are as follows:

a) The EWHNN based solution strategy adopts wavelet function (Gaussian derivative function), evolution based strategy, Hopfield dynamics, k-means cluster analysis and wavelet filter. Hence, it is computationally less expensive due to the inclusion of wavelet function as activation function when compared with EHNN.
b) The EWHNN provides more efficient portfolios when compared with portfolios obtained using EHNN.
c) The EWHNN handles diversification in both large and small portfolios.
d) The EWHNN intrinsically reports faster convergence when compared with other heuristic algorithms.

Architecture of Evolution Based Wavelet Hopfield Neural Network – A Hybrid Model

The EWHNN strategy is computationally very efficient only due to the adaptation of wavelet function as activation function. Furthermore, global optimal portfolio sets are obtained and the results are experimented for different runs of BSE200 and Nikkei225 data sets. The Evolution based Wavelet Hopfield Neural Network is a hybrid model which adopts concepts from neural network, wavelet network and evolutionary strategies.

Neural Networks – A Review

Human brain consists of highly interconnected network of simple processing elements called neurons. Natural neuron is composed of soma, dendrites, axon and synapses. The neuronal activities such as learning and memory are related to the number of synaptic inputs and synaptic outputs, which are believed to excite or inhibit other neurons. In fact, the neuronal activities are complex, but can be turned as reasonable approximation by a simple process of summation of the input received.

In order to understand complex systems, they can be decomposed into simple elements and solved efficiently. Networks are one such approach characterized by a set of nodes and connections between the nodes. An artificial neural network is defined as a parallel processing system with a large number of relatively simple individual processing elements (artificial neurons) in a network inspired by the structure of the human brain. Since the human brains show many desirable characteristics such as massive parallelism, learning ability, distributed representation and computation, inherent contextual information processing, generalization ability, fault tolerance and low energy consumption, the artificial neural networks are desired to possess some of these desirable characteristics.

A simple model of an artificial neuron consists of inputs, weights, activation function and an output as shown in Figure 4.

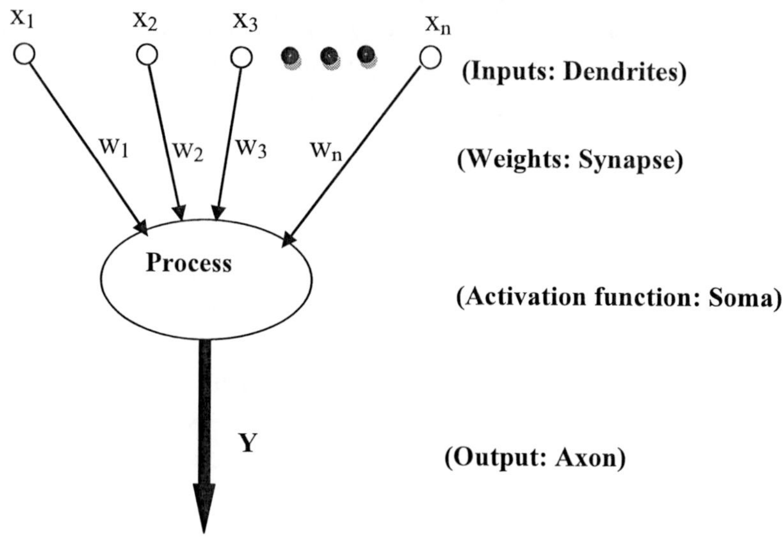

Figure 4. An artificial neuron.

Basically, inputs are multiplied by their corresponding weights and then computed by a mathematical function (activation function) which determines the activation of the neuron. Some of the activation functions are threshold function, signum function, sigmoid function and hyperbolic tangent function. After the application of activation function, the weights of an artificial neuron are adjusted by using some desired algorithms in order to obtain the desired output from the network. This process of adjusting the weights of the artificial neural network is called learning or training. According to learning mechanisms, neural networks are classified into several classes. However, there are three fundamentally different classes of networks: single layer feed forward network, multilayer feed forward network and recurrent networks. Single layer feed forward network comprises of input layer and an output layer. The input layer transmits the signals to the output layer and the computations are performed only in the output layer. The weights are connected from every input neuron to the output neuron but not vice-versa. Thus, the name single layer feed forward network. In multi layer feed forward network, apart from input and output layer there are also intermediate layers called hidden layers. Unlike single layer and multi layer feed forward network, recurrent network have the output values fed back and augment the existing output. Recurrent neural networks have been very successful in network architecture that adapts time-dependent nonlinear mappings.

Hopfield Neural Network

The Hopfield Neural Network belongs to recurrent neural network which consists of a single layer of neurons, where the output of each of the neuron is given as an input to all the other neurons. Different models can be defined as a result of the particular choice of the three elements that Hopfield Neural Network includes: activation function of the neurons, dynamical equations and energy function. Naturally, these elements are interrelated and an independent definition of each one is not possible. Figure 5 illustrates the architecture of Hopfield neural network.

This system implements a recurrent neural system in order to model the behavior of a natural associative memory. According to this model, neuron activation function and energy function are defined as,

$$f(U_i) = \frac{1}{1+e^{-\beta U_i}} \tag{14}$$

$$U_i = \sum_{j=1}^{N} w_{ji} p_i(t) + Z_i \tag{15}$$

The energy function of the Hopfield neural network given by

$$E(P) = -\frac{1}{2}\sum_{i=1}^{N}\sum_{j=1}^{N} p_i w_{ij} p_j - \sum_{i=1}^{N} Z_i p_i \tag{16}$$

where $P = (p_1, p_2, \ldots, p_N)$ defines a vector of state variables, $w_{ij} = w_{ji}$ (symmetric) is the weight of the connection between the neurons j and i, β is the gaining parameter and Z_i is the external input to the neuron i. Hopfield neural network proposed the energy function or Lyapunov function of the system, as long as the following conditions are satisfied: absence of self connections i.e. the w_{ii} are null; and connection symmetry i.e. $w_{ij}=w_{ji}$, symmetrical weight matrix of the connection from neuron i to j. f is the activation function (sigmoid function). Hence the model is described by equations (14), (15) and (16). In order to solve multi objective portfolio optimization problem by means of a Hopfield network, the network must be designed such a way that the objective function of the problem matches with the Lyapunov function.

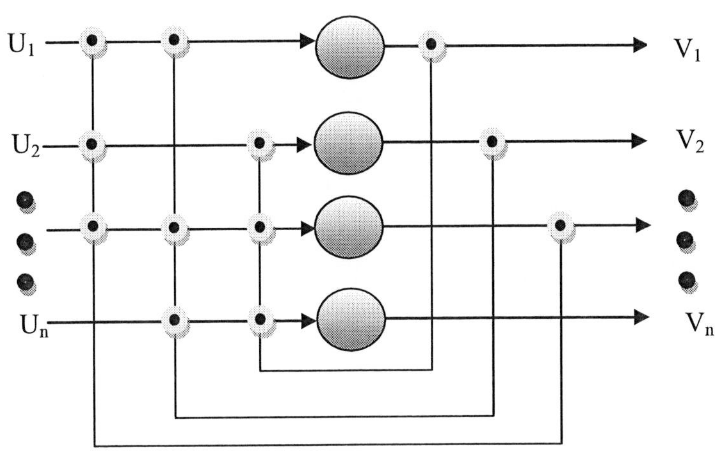

Figure 5. Hopfield Neural Network.

Wavelet Network

Wavelets are a special kind of functions which exhibits a damped oscillation for a very limited duration of time. Wavelet theory is a versatile tool with very rich mathematical content employed in many fields and applications. Some of the desirable properties of wavelets are multi resolution representation, invariance, uniqueness, stability and special localization. A wavelet is a small wave which grows and decays in a limited time frame. These are the functions called the mother wavelets denoted by $\psi(t)$ defined on the real axis and satisfying the following conditions.

$$\int_{-\infty}^{\infty} \psi(t)dt = 0 \tag{17}$$

$$\int_{-\infty}^{\infty} |\psi(t)|^2 dt = 1 \tag{18}$$

A variety of functions such as Haar, Daubechies, Symlet, Discrete Meyer, Mexicon Hat and Coiflets meets these conditions. The dynamical neural networks such as Hopfield neural network are widely used in optimization, controls and identification. A neural network structure with wavelet as the node activation function is called a *wave net* or *wavelet neural network*. Wavelet based networks have been introduced as an alternative to neural networks. Since wavelet functions are piecewise differentiable and monotone non-decreasing like the activation function in neural networks, these functions can also be used as activation functions in a Hopfield neural network. One of the commonly used wavelet function is Gaussian derivative function ($\omega(U) = -Ue^{-\frac{U^2}{2}}$) with zero crossing at $U = 0$.

The Hopfield neural network structure shown in Figure 5 is adopted for wavelet network. The only difference is that sigmoid activation function is replaced by Gaussian derivative function. Some of the major properties of wavelet networks are preserve the universal approximation property, exhibit an explicit link between the network coefficients and some appropriate transform and possibly achieve the same quality of approximation with a network of reduced size. Moreover, due to the existence of ortho normal wavelet bases, the rates of convergence for approximation is available in wavelet based networks. Therefore, EWHNN employs the same as the activation function.

Evolutionary Strategy in EWHNN

A class of stochastic search algorithms which are developed from ideas and principles of natural evolution and population based is called evolutionary algorithm. Evolutionary algorithm consists of evolutionary strategy and genetic algorithm. In reproductive phase of genetic algorithm, the parent population is completely replaced by its offspring and the same is adopted in EWHNN. Real coded representation of chromosomes used in evolutionary strategy is employed in EWHNN. The weights of the assets in the investable universe are represented by genes in the chromosomes. Each chromosome in the population comprises of

K genes, if the investor prefers to invest in a maximum of K assets. In every generation, the chromosome undergoes a set of weight standardization process (logical procedures) to handle basic, bounding, class and short sales constraints. The objective function is the fitness function used to rank chromosomes.

Wavelet Hopfield Dynamics in EWHNN

The objective function of the optimization problem should be matched with the energy function of Hopfield neural network to obtain the solution for the proposed problem. In portfolio optimization problem, the objective function given in equation (2) is compared with the energy function of the Hopfield neural network given in equation (16). This comparison yields the mapping where $w_{ij} = -2\lambda C_{ij}$ and $Z_i = (1-\lambda)\mu_i$. The wavelet Hopfield dynamics used to obtain optimal weights in EWHNN are as follows.

If p_i (t) is the state of a neuron at time t, then

$$p_i(t+1) = p_i(t) + 9 * \omega_i(U_i) \text{ [Short sales not allowed]} \tag{19}$$

$$p_i(t+1) = p_i(t) - 9 * \omega_i(U_i) \text{ [Short sales allowed]} \tag{20}$$

Where

$$\omega_i(U_i) = -U_i * e^{-\beta \frac{U_i}{2}} \tag{21}$$

```
Algorithm EWHNN ( )
   Let the population size be M and maximum portfolio size be K;
   For R iterations /* R = desired number of λ values */
     Generate parent population P with M chromosomes containing K genes;
     /* each K genes initialized in the range [R-, R+] */
     If (constrained problem)
     Include logical procedures to handle constraints;
     End if
     Choose chromosomes C_min having minimum fitness value (objective function value) f_min;
     Initialize the gain value β = |f_min| / 10000 ;
     For T repetition /* T = desired number of repetitions */
      Randomly choose a gene from each chromosome;
      Apply wavelet Hopfield dynamics to obtain optimal weights;
      Gradually increase β value; /* β = β / 0.99 */
      If (constrained problem)
      Include logical procedures to handle the constraints for the new population P';
      End if
      Obtain the best optimal chromosome C'_min with minimum fitness value f'_min for P';
      If (f'_min<f_min), then f_min = f'_min; C_min = C'_min; End if
     End T
     Update P=P' for next value of λ ;
     End R
 End EWHNN ( )
```

The number '9' is chosen because it is the maximum value after which no faster improvement in convergence is felt. β is the positive scaling parameter which is initialized to very small value (i.e. $\beta = \frac{|f_{min}|}{10000}$) and is slowly increased to obtain global optimum solution. Since cardinality constraint is already eliminated by k-means cluster analysis, the rest of the constraints are handled in the process of obtaining the global optimum solution using Algorithm EWHNN ( ) solution strategy.

ARCHITECTURE OF PARETO-ARCHIVED EVOLUTIONARY WAVELET NETWORK

```
Algorithm PEWN( )
 Initialize the population size (M) and maximum portfolio size (K);
 For R iterations
 Randomly create population P with M chromosomes containing K genes;
  /* Each K genes initialized in the range [R-, R+] */
 If (Constrained problem)
 Include logical procedures to handle constraints;
 End if
 Obtain Pareto-optimal set O using concept of dominance;
 Calculate the risk (O_risk) and return (O_return) for all chromosomes in O;
 Initialize β1 = absolute (min(O_risk)/1000); /* Initialize to very small value */
 Initialize β2 = absolute (max(O_return)/1000); /* Initialize to very small value */
 For T repetitions
 Randomly choose half of the chromosomes; (i.e. M/2)
 For a randomly chosen gene in each chromosome
 Apply wavelet Hopfield dynamics with β1 ;
 /*obtain optimal portfolios satisfying minimize risk objective function */
 End for
 For another half of chromosomes
 Randomly choose a gene in each chromosome;
 Apply wavelet Hopfield dynamics with β2 ;
 /* obtain optimal portfolios satisfying maximize return objective function */
 End for
 Gradually increase β1 = β1/0.95 and β2 = β2/0.95 ;
 If (Constrained problem)
 Include logical procedures to handle constraints;
 End if
 Update the set O using dominance concept by including P';
 End T
 End R
 Obtain the efficient frontier using the final Pareto-optimal set O;
End PEWN ( )
```

Although, the weighted single objective portfolio optimization problem is easier in implementation than multi objective portfolio optimization problem, there are several limitations in the solution strategies of single objective portfolio optimization problem. The introduction of wavelets in evolutionary strategy and the concept of dominance help to overcome these drawbacks in multi objective optimization problem. Hence, Suganya and Pai [22] came out with a new heuristic algorithm named Pareto-archived evolutionary wavelet network for the solution of constrained multi objective portfolio optimization problem.

Initially both the objective functions (minimize risk and maximize return) are compared with the energy function of Hopfield neural network given in equation (16) to obtain the value of weights (w_{ij}) and extern inputs (Z_i). In PEWN, half of the chromosomes which are randomly chosen from the population uses the minimize risk as the fitness function with $w_{ij} = -2C_{ij}$ and $Z_i=0$. Similarly, the remaining half of the chromosomes employs maximize return as the fitness function with $w_{ij}=0$ and $Z_i = -\mu_i$.

The Gaussian derivative function used in EWHNN which is an oscillating function is employed as node activation function in PEWN strategy. This solution strategy converges towards the optimal solution faster than other heuristic algorithms only because of the introduction of wavelet function as activation function. The concept of dominance is another factor in PEWN solution strategy which handles the multiple objectives.

The working of the Pareto-archived evolutionary wavelet network for the constrained multi objective portfolio optimization problem is given in the Algorithm PEWN ( )

PERFORMANCE ANALYSIS OF SOLUTION STRATEGIES

The performance of newly proposed wavelet based solution strategies EWHNN and PEWN are compared with EHNN. There are two distinct goals for the optimization problem: (i) obtain optimal solutions close to 'ideal' frontier (ii) diversify the solutions in the non-dominated front. Since, no single metric can measure the performance of the algorithm in an absolute sense; two different types of metrics [30] are employed to study the performance of algorithms.

The error ratio (ER), set coverage ratio (C) and generational distance (GD) are used to measure the first goal. The spacing (S), spread (Δ) and maximum spread (D) are employed to measure the second goal. Hence, these measures are adopted to test the performance of the three heuristic solution strategies (EHNN, EWHNN and PEWN).

These strategies are evaluated for the cardinality constrained portfolio optimization problem with K=20 and K=50 for BSE200 and Nikkei225 data sets respectively and the portfolio sets are compared with the 'ideal' frontier obtained using Markowitz traditional quadratic programming problem. The comparison of efficient frontiers obtained using these strategies with ideal frontier for BSE200 and Nikkei225 data sets are shown in Figures 6(a) and 6(b) respectively.

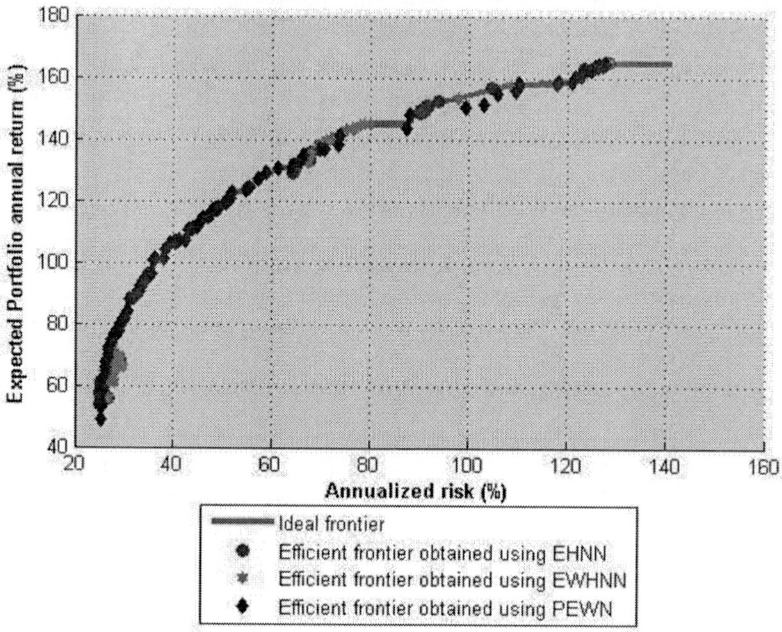

Figure 6(a). Comparison of the efficient frontiers obtained using three strategies (EHNN, EWHNN and PEWN) with 'ideal' frontier on BSE200 data set with K=20.

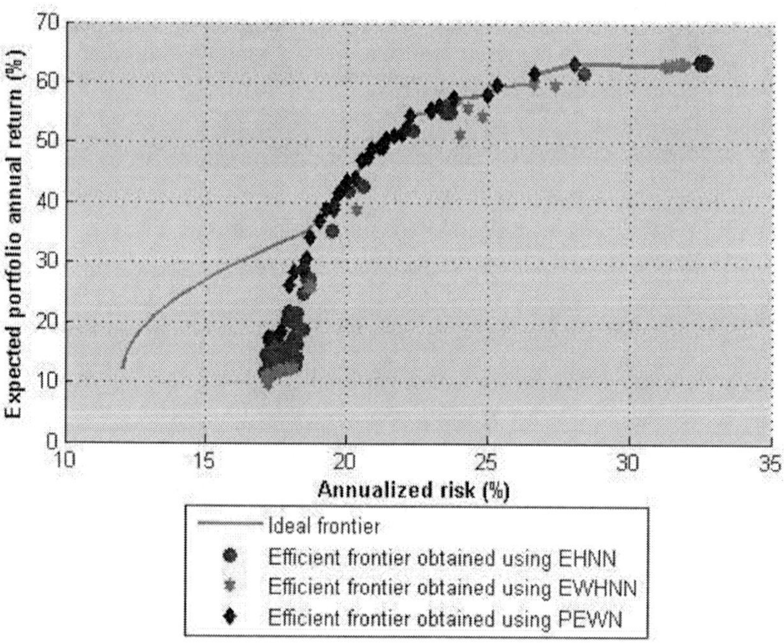

Figure 6(b). Comparison of the efficient frontiers obtained using three strategies (EHNN, EWHNN and PEWN) with 'ideal' frontier on Nikkei225 data set with K=50.

Testing the Optimality of Portfolio Sets

The three metrics employed to compute the measure of the closeness of the optimal portfolio set obtained using the test solution strategies and a known set of values of 'ideal' frontier are shown below.

Error Ratio (ER)

The number of solutions obtained using the test strategy which are not members of the 'ideal' frontier is counted using this metrics. If P* is the ideal portfolio set, Q is the solution set obtained using test strategy, e_i is the decision variable which takes the value 1, for all the solutions not belonging to P*, then

$$ER = \frac{\sum_{i=1}^{|Q|} e_i}{|Q|} \tag{22}$$

The better convergence to 'ideal' frontier is revealed by the smaller value of ER. If ER=1, then no solution in Q is a member of P*. Similarly, if ER=0, then all solutions in Q are members of P*. The smaller error ratio for PEWN indicates a better convergence to 'ideal' frontier when compared with EHNN and EWHNN strategies.

Set Coverage Metric (C)

The relative spread of solutions between two sets A and B is obtained using set coverage metric. C(A,B) represents the proportion of solutions in B that are weakly dominated by solutions in A.

$$C(A,B) = \frac{\left|\{b \in B / \exists a \in A : a \text{ dominates } b\}\right|}{|B|} \tag{23}$$

The key feature is that the cardinality of both the sets need not be same. If Q1, Q2 and Q3 are set of solutions obtained using EHNN, EWHNN and PEWN, then the smaller value obtained for C(Q1,Q3) and C(Q2,Q3) indicates the better performance of PEWN when compared with both EHNN and EWHNN in terms of the set coverage metric.

Generational Distance (GD)

The average distance of the solutions in Q from P* is measured using generational distance. If d_i is the Euclidean distance between the solution in Q and the nearest member of P* with p=2, N is the number of objective functions, $f_n{}^*(k)$ is the nth objective function value of the kth member of P* and $f_n(i)$ is the nth objective function value of the ith member in Q, then

$$GD = \frac{\left(\sum_{i=1}^{|Q|} d_i^p\right)^{1/p}}{|Q|} \tag{24}$$

$$d_i = \min_{k=1}^{|P^*|} \sqrt{\sum_{n=1}^{N} \left(f_n(i) - f_n^*(k)\right)^2} \tag{25}$$

The solutions obtained using PEWN yields better optimal solutions when compared with EHNN and EWHNN. Table 1 illustrates the comparison of error ratio, set coverage ratio and generational distance for the solutions obtained using three strategies (EHNN, EWHNN and PEWN) with respect to BSE200 and Nikkei 225 data sets.

Table 1. Comparison of error ratio, set coverage and generational distance for the solutions obtained using three algorithms (EHNN, EWHNN and PEWN) with respect to BSE200 and Nikkei225 data sets

Metrics		BSE200 index	Nikkei225 index
Error ratio (ER)	ER(Q1)	0.3947	0.3333
	ER(Q2)	0.2821	0.2800
	ER(Q3)	0.1368	0.2381
Set coverage metrics (C)	C(Q1,Q2)	0.1795	0.2200
	C(Q2,Q1)	0.2105	0.2745
	C(Q1,Q3)	0.1282	0.0476
	C(Q3,Q1)	0.3684	0.2941
	C(Q3,Q2)	0.4103	0.3200
	C(Q2,Q3)	0.0940	0.0238
Generational distance (GD)	GD(Q1)	0.6467	4.3009
	GD(Q2)	0.5285	1.9994
	GD(Q3)	0.3075	0.8976
Q1 – Solutions obtained using EHNN Q2 – Solutions obtained using EWHNN Q3 – Solutions obtained using PEWN			

Testing the Diversity Among Optimal Solutions

The following metrics measures the diversification between the optimal solutions in the set.

Spacing (S)

The relative distance between consecutive solutions in the portfolio set is calculated using spacing. If d_i is the minimum value obtained for the sum of the absolute difference in objective function values between the ith solution and any other solution in the obtained portfolio set Q and $\overline{d}$ is the mean value, then

$$S = \sqrt{\frac{1}{|Q|}\sum_{i=1}^{|Q|}(d_i - \bar{d})^2} \qquad (26)$$

Spread (Δ)

The extent of diversification in the solution set is measured using spread. If d_i is the distance between neighboring solutions, d_m^e is the distance between the extreme solutions of P* and Q for m[th] objective function and $\bar{d}$ is the mean distance measure, then

$$\Delta = \frac{\sum_{m=1}^{M} d_m^e + \sum_{i=1}^{|Q|}\left|d_i - \bar{d}\right|}{\sum_{m=1}^{M} d_m^e + |Q|\bar{d}} \qquad (27)$$

Maximum Spread (D)

The length of the diagonal of a hyper box obtained by the extreme function values in the portfolio set is measured using maximum spread (D). For a bi-objective problem, this metric refers to the Euclidean distance between the two extreme solutions in the objective space.

$$D = \sqrt{\sum_{m=1}^{M}\left(\max_{i=1}^{|Q|} f_m^i - \min_{i=1}^{|Q|} f_m^i\right)^2} \qquad (28)$$

Table 2. Comparison of spacing, spread and maximum spread for the solutions obtained using three algorithms (EHNN, EWHNN and PEWN) with respect to BSE200 and Nikkei225 data sets

Metrics		BSE200 index	Nikkei225 index
Spacing (S)	S(Q1)	1.3842	1.5302
	S(Q2)	3.2481	2.1154
	S(Q3)	1.1144	1.2999
Spread (Δ)	Δ(Q1)	0.9972	1.0200
	Δ(Q2)	0.9503	1.2451
	Δ(Q3)	0.8554	0.7823
Maximum spread (D)	D(Q1)	0.9467	0.7837
	D(Q2)	0.9328	0.7937
	D(Q3)	0.9654	0.8083
Q1 – Solutions obtained using EHNN Q2 – Solutions obtained using EWHNN Q3 – Solutions obtained using PEWN			

Table 2 gives a comparison of spacing, spread and maximum spread for the solutions obtained using the three strategies (EHNN, EWHNN and PEWN) with respect to BSE200 and

Nikkei225 data sets. The smaller values of spacing and spread indicate better diverse set of solution. Alternatively, the larger value of maximum spread represents its better distribution.

EXPERIMENTAL ANALYSIS OF SOLUTION STRATEGIES

The weighted single objective optimization problem model with all the listed constraints is solved using EHNN and EWHNN solution strategies for an investable universe obtained using k-means cluster analysis. Similarly, the multi objective optimization problem model with same constraint and same investable universe is solved using PEWN strategy.

The efficient frontier of the constrained portfolio optimization problem obtained using EHNN, EWHNN and PEWN solution strategies for the BSE200 and Nikkei225 data set are illustrated in Figures 7(a) and 7(b). In BSE200 data set, the investable universe containing maximum of 20 assets (K=20) is chosen representing small portfolio sets. The limits for short sales constraint is chosen as -0.1 and 1.1 for R- and R+ respectively as a typical instance. The upper and lower bounds for BSE SENSEX 30 listed (blue chip) stocks in the investable universe is uniformly fixed as 0.01 and 1.1. However, for the rest of the assets L_i=-0.1 and U_i=1.1. Considering that the investor prefers to invest non-zero weights in three sectors (banks, technology and oil and gas), the proportion range is decided as 0.01 to 0.15, 0.01 to 0.2 and 0.01 to 0.15 in banks, technology and oil and gas sectors respectively.

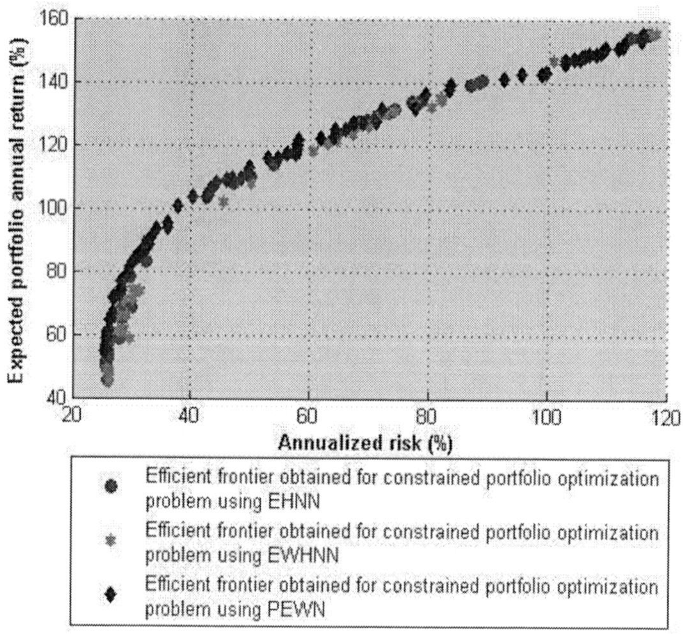

Figure 7(a). Efficient frontier traced by EHNN, EWHNN and PEWN on k-means clustered portfolio sets of BSE200 (July 2001-July 2006) data set for K=20 satisfying all the listed constraints.

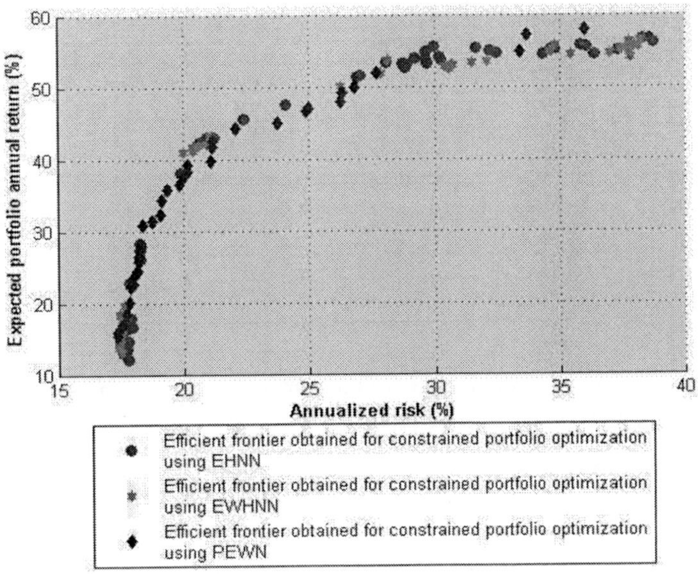

Figure 7(b). Efficient frontier traced by EHNN, EWHNN and PEWN on k-means clustered portfolio sets of Nikkei225 data set for K=50 satisfying all the listed constraints.

In Nikkei225 data set, K is chosen as 50 assets representing large portfolio set. The same short sales limits are employed. Bounding limits are implemented on industrial goods and services, construction, material and basic resources sectors. The same sectors are chosen for class constraint. The proportion range is decided as 0.01 to 0.15, 0.01 to 0.2 and 0.01 to 0.15 in industrial goods and services, construction and material and basic resources respectively.

DATA ENVELOPMENT ANALYSIS (DEA)

Although a visual inspection of the efficient frontiers traced by the three algorithms in Figures 7(a) and 7(b) reveals close proximity between the frontiers, there is a need to study the quality of the efficient frontiers. Data envelopment analysis is one such methodology based on a mathematical programming model applied to experimental data for determining the efficient frontier by evaluating the performance of set of peer entities called decision making units (DMUs) which convert inputs to outputs. Let $(X,Y)=\{(x_1,y_1),(x_2,y_2),\ldots,(x_n,y_n)\}$ are the input and output of n DMUs, θ is the efficiency score, $\lambda=(\lambda_1,\lambda_2,\ldots,\lambda_n)$ are the n dual variables. Then the mathematical model which can be solved using linear programming is given as follows.

Minimize θ

Subject to $-y_i + Y\lambda \geq 0$

$$\theta x_i - X\lambda \geq 0 \tag{29}$$

$\sum \lambda = 1$, $\lambda \geq 0$, θ - unrestricted in sign

The key objectives of DEA are (a) to study the efficiency of portfolio sets yielded by EHNN, EWHNN and PEWN (b) to analyse the significant difference in the efficiency scores of the portfolios obtained with respect to mean and variance obtained using the three strategies and (c) to investigate the robust portfolios, by computing their efficiency scores. This analysis is undertaken on the optimal portfolios obtained for the investable universe of the BSE200 and Nikkei225 data sets, which includes all the listed constraints. The portfolios on the efficient frontiers illustrated in Figures 7(a) and (b) are the inputs to DEA. The number of DMUs considered is 50 points and 38 points for BSE200 and Nikkei225 data sets respectively. Efficiency scores obtained using the linear programming system given in (29).

Table 3 illustrates the summary of DEA for the three strategies (EHNN, EWHNN and PEWN) with respect to BSE200 and Nikkei225 data sets. The results show that all the three strategies are accepted with respect to mean and standard deviation for both BSE200 and Nikkei225 data sets. But, the coefficient of variation on efficiency scores of PEWN is less when compared with that of EWHNN and EHNN for both BSE200 and Nikkei225 data sets. This concludes that PEWN is more robust with regard to their efficiencies when compared with that of EWHNN and EHNN solution strategies.

Table 3. Summary of the DEA analysis of three strategies (EHNN, EWHNN and PEWN) for constrained optimal portfolios of the BSE200 and Nikkei225 data sets

Data set	Method	Number of DMUs (n)	Average efficiency scores, (μ)	Standard deviation of the efficiency scores, (σ)	Coefficient of Variation (%) $\left(\frac{\sigma}{\mu}\times 100\right)$
BSE200 index	EHNN	50	0.9794	0.0269	2.7432
	EWHNN	50	0.9825	0.0205	2.0868
	PEWN	50	0.9804	0.0186	1.8961
Nikkei225 index	EHNN	38	0.9725	0.0272	2.7935
	EWHNN	38	0.9765	0.0268	2.7404
	PEWN	38	0.9858	0.0178	1.8017

CONCLUSION

In this chapter, wavelet concepts are employed for financial constrained portfolio optimization problem. However, not much work is reported in the literature on the application of wavelets for denoising covariance matrix and solution of portfolio optimization problem using wavelet networks. Hence, knowing the efficiency and richness of mathematical concepts in wavelets, several experiments are undertaken on the suitability of wavelets to the portfolio optimization problem. Finally, wavelet shrinkage denoising technique using 'symlet 3' wavelet function is adopted to denoise the financial covariance matrix, which is a key input to the portfolio optimization problem. Also, two different wavelet network based solution strategies EWHNN and PEWN are developed for the solution of constrained weighted single objective and multi objective portfolio optimization problem respectively. The main advantages of wavelet based heuristic solution strategies are a) produces more robust portfolio sets b) computationally less expensive in constraint handling c) irrespective of diversification in small and large portfolios and with the inclusion of constraints, yields portfolios in realistic time d) wavelet concept enhances faster convergence of the solutions towards the optimal solutions e) flexible to include any number of constraints.

REFERENCES

[1] H. Markowitz, *J. Financ.*, 7, 77 (1952).

[2] L. Laloux, P. Cizeau, J.P. Bouchaud and M. Potters, *Phys. Rev. Lett.* 83, 1467 (1999).

[3] L. Laloux, P. Cizeau, M. Potters and J.P. Bouchaud, *Int. J. Theoret. Appl. Finance,* 3, 391 (2000).

[4] S. Malgorzata and J. Krzych, *Acta Phys. Pol. B*, 37,3145 (2006).

[5] G. Papp, S.Z. Pafka, M.A. Nowak and I. Kandor, *Acta Phys. Pol. B*, 36, 2767 (2005).

[6] R.N. Mantegna and H.E. Stanley, *An Introduction to Econophysics: Correlations and Complexity in Finance*, Cambridge University Press, Cambridge (1999).

[7] L. Giada and M. Marsili, *Phys. Rev. E,* 63, 061101 (2001).

[8] S. Maslov, *Physica A*, 301, 397 (2001).

[9] M. Bernaschi, L. Grilli and D. Vergni, *Physica A* 308, 381 (2002).

[10] L. Kullmann, J. Kertesz and K. Kaski, *Phys. Rev. E,* 66, 026125 (2002).

[11] N. Basalto, R. Bellotti, F. De Carlo, P. Facchi and S. Pascazio, *Physica A* 345, 196 (2005).

[12] T. Vincenzo, L. Fabrizio, M. Gallegati and R.N. Mantegna, *J. Econ. Dyn. Control*, 32, 235 (2008).

[13] G.A. Vijayalakshmi Pai and T. Michel, *IEEE Trans. Evolut. Comp.* 13, 1030 (2009).

[14] N.C. Suganya and G.A. Vijayalakshmi Pai, *Digit. Tech. J.* 2, 30 (2009).

[15] N.C. Suganya and G.A. Vijayalakshmi Pai, "Wavelet based Noise dressing for reliable cardinality constrained financial portfolio sets", In *Proceedings of the International Workshop on Machine Intelligence Research* (2009).

[16] R. Armananzas and J.A. Lozano, "A multi objective approach to the portfolio optimization problem", IEEE congress on Evolutionary Computation, 2, (2005), pp. 1388-1395.

[17] T.J. Chang, N. Meade, J. Beasley and Y. Sharaiha, *Comput. Oper. Res.* 27, 1271 (2000).

[18] A. Fernandez and S. Gomez, *Comput. Oper. Res.* 34, 1177 (2007).

[19] F. Streichert, H. Ulmer and A. Zell, "Evolutionary Algorithms and the cardinality constrained portfolio optimization problem", In *Proceedings of the International Conference on Operations*, (2003).

[20] G. Di. Tollo and A. Roli, *Int. J. Operat. Res.* 5, 13 (2008).

[21] N.C. Suganya and G.A. Vijayalakshmi Pai, *Dyn. Contin. Discret. I* 17, 175 (2010).

[22] N.C. Suganya and G.A. Vijayalakshmi Pai, *Intell. Syst. Accounting, Finance and Management,* 17, 59 (2010).

[23] N.C. Suganya and G.A. Vijayalakshmi Pai, "Evolution based Portfolio selection using Hopfield Neural Network", *In Proceedings of Mathematical and computational models*, (2007), pp. 262-274.

[24] N.C. Suganya, G.A. Vijayalakshmi Pai and T. Michel, "Evolution based Hopfield Neural Network for constrained optimization of financial portfolios", In *Proceedings of IEEE Conference on AI Tools in Engineering* (2008).
[25] V. Chopra and W. T. Ziemba, *J. Portfolio Manag.* 19, 6 (1993).
[26] P. Frost and J. Savarino, *J. Financ. Quant. Anal.* 21, 293 (1986).
[27] P. Jorion, *J. Financ. Quant. Anal.* 21, 279 (1986).
[28] O. Ledoit and M. Wolf, *J. Empirical Finance*, 10, 603 (2003).
[29] J.L. Farrell, Jr., *Portfolio Management Theory and Applications*, McGraw Hill, New York (1997).
[30] K. Deb, *Multi objective optimization using Evolutionary Algorithms*, John Wiley and Sons, Chichester (2003).

In: Wavelets: Classification, Theory and Applications
Editors: Manel del Valle et al.
ISBN: 978-1-62100-252-9

Chapter 14

THREE- DIMENSIONAL ESTIMATION IN 1-D WAVELET PROFILOMETRY BY ESTABLISHING A COMPARISON BETWEEN PHASE UNWRAPPING ALGORITHMS

Jesús Carlos Pedraza-Ortega*, Efren Gorrostieta-Hurtado, Marco A. Aceves Fernandez, Sandra L. Canchola-Magdaleno, Juan Manuel Ramos-Arreguin, Saul Tovar-Arrigaga, Emilio Vargas-Soto and Artemio Sotomayor-Olmedo
Facultad de Informática, Universidad Autónoma de Querétaro, México

ABSTRACT

Three-dimensional (3-D) surface shape measurement and sensing by using a projected fringe pattern has been widely used considering the Transform Based profilometry. Several articles show that these methods require a single image with a sinusoidal fringe pattern projected on it. Usually, the projected pattern has a known spatial frequency and this information is used to avoid discontinuities in the fringes with high frequency. Among single projected fringe pattern, most of the methods use Fourier or Wavelet transforms to extract the phase information. However, they focused only on the transform method and not on the phase unwrapping algorithms. It is well known that wavelet transforms are used to analyze signals whose frequency contents varies over the duration of the signal (wrapped phase), also, wavelet techniques provide a flexible, elegant, and a computational efficient solution to these type of problems. In this work, a 1-D wavelet profilometry method is presented considering two different wavelet transforms. Later, different phase unwrapping algorithms are used to extract the depth information considering local and global analysis. Several computer simulations and experiments are carried out to validate the proposed method. The merits and limitations of each of these variations on the method are indicated and the error is estimated.

* E-mail address: caryoko@yahoo.com

1. INTRODUCTION

Since a few years ago, the 3D information of an object is required and the main idea is to extract the depth information from an image or set of images in an efficient and automatic way. As a result of the process, the useful depth information can be used in several tasks such as synthetic aperture radar (SAR), automatic inspection, reverse engineering, 3D robot navigation magnetic resonance imaging (MRI), optical metrology, interferometry and so on [1]. Among all the contact measurement techniques to carry out this process, the most widely used are a vision-based system together with a tool which is in contact with the object, like laser, fringe projection and so on. Moreover, the most widely used is the fringe projection. The fringe processing methods are widely used because of its characteristics like high accuracy, noise-immunity and fast processing speed. Inside the fringe projection methods, the mostly used and known is the Fourier Transform Profilometry (FTP) method [2], Phase-shifting or phase stepping [3, 4], digital phase locked loop [5], direct phase detection [6], and a variation of the FTP method, the one called Wavelet Transform Profilometry [12, 16]. One of the main challenges of the previously mentioned methods is the problem statement of the wrapped phase information caused by the fact that the phase of a periodically varying intensity pattern is encoded or wrapped and it contains the depth information of the object. The phase unwrapping problem has been tackled by several researchers in the past; dozens of algorithms are fully dedicated to solve this problem in many ways. One of the first algorithms to deal with this problem was proposed by Takeda and Mutoh in 1982 [2], which used a single phase unwrapping algorithm. Later Berryman [7] and Pedraza [8, 17, 18] proposed a modified Fourier Transform Profilometry by carrying out global and local analyses as a solution of the phase wrapped information. Then, unwrapping algorithms (temporal and spatial) were introduced and modified [7-10]. Phase unwrapping techniques use exhaustive data computations and approximations; however, these approaches have a slight contribution to understand the cause of failure in the phase unwrapping process. This research presents an implementation of phase unwrapping algorithms, considering the problem of residues.

Generally, most of the cited methods used a Modified Fourier Transform Profilometry, and another suitable solution is to use the wavelet transform to extract the phase information.

Wavelet transform also offers multi-resolution in time and space frequency, and it is an alternative that offers some advantages over the Fourier transform [9-10]. Using a wavelet transform can carry out the fringe pattern analysis. The main idea of this analysis consists in demodulating the deformed fringe patterns and extracting the phase information encoded into it, therefore estimating the height profile of the object quite similarly to Fourier transform.

Different wavelet algorithms are used in the demodulation process to extract the phase of the deformed fringe patterns. Those algorithms can be classified into two categories: phase estimation and frequency estimation techniques. The phase estimation algorithm uses a complex mother wavelet; here, the extracted phase is wrapped into 2π discontinuities consequently a phase unwrapping algorithm is required to remove these 2π jumps. Zhong et al. [9] have applied Gabor wavelets to extract the phase distribution where a phase unwrapping algorithm is required. The frequency estimation technique estimates the instantaneous frequencies in a fringe pattern, which are then integrated to estimate the phase. The phase extracted using this technique is continuous; consequently, phase unwrapping algorithms are not required for 2D Wavelet Profilometry. Complex or real mother wavelet can be used to estimate the instantaneous frequencies in the fringe pattern. Dursun et al. [14] and Afifi et al. [15] have used Morlet or Paul wavelets, separately, to obtain the phase distribution of projected fringes. Also, Gdeisat et al. [16] have proposed a 1D continuous wavelet transform approach to retrieve phase information in temporally and spatially fringe patterns.

Most of the previous research are focused on using the Fourier and wavelet transforms to obtain the 3D information from an object; pre-filtering the images, extracting the phase information of fringe patterns, using phase unwrapping algorithms, and so on.

In the present research, a comparison between two phase unwrapping algorithms is presented in 1D Wavelet Profilometry, implemented in order to obtain the 3D information from an object. First, the spatial frequency of the projected fringe pattern is obtained; later the mathematical model is obtained and used together with the spatial frequency in order to establish the problem. Then, a 1D Wavelet Profilometry is applied considering the most suitable wavelets for the fringe analysis. Later, two phase unwrapping algorithms are used to obtain the desired 3D information. One contribution in this research is the proposed methodology, because in previous works there are no comparisons among different phase unwrapping (PU) methods in 1D Wavelet based profilometry. The results show that it is suitable to compare the present work with other similar researches. To test the method, some virtual objects were created for use in computer simulations and also some experiments were carried out.

2. Profilometry Basics

As were described early, there are several fringe projection techniques that are used to extract the three-dimensional information from the objects. The most widely used techniques are Fourier Transform Profilometry and Wavelet Profilometry, which are presented in this work.

2.1. Fourier Transform Profilometry (FTP)

The projected fringe pattern and an object with projected fringes on it, as shown on Figure 3, can be represented as:

$$g(x,y) = a(x,y) + b(x,y) * \cos[2 * \pi f_0 x + \varphi(x,y)] \tag{1}$$

$$g_0(x,y) = a(x,y) + b(x,y) * \cos[2 * \pi f_0 x + \varphi_0(x,y)] \tag{2}$$

where$g(x,y)$ and $g_0(x,y)$ are the intensities of the images at the point (x,y), $a(x,y)$ represents the background illumination, $b(x,y)$ is the contrast between the light and dark fringes, f_0 is the spatial-carrier frequency and $\varphi(x,y)$ and $\varphi_0(x,y)$ are the corresponding phase to the fringe and distorted fringe pattern.

The phase $\varphi(x,y)$ contains the desired information. This angle $\varphi(x,y)$ is the phase shift caused by the object surface end the angle of projection, and its expressed as:

$$\varphi(x,y)=\varphi_0(x,y)+\varphi_z(x,y) \tag{3}$$

where$\varphi_0(x,y)$ is the reference plane projected phase angle, and $\varphi_z(x,y)$ is the object's height distribution phase.

In Pedraza et al work [17, 18], the Equation 3 can be rewritten as:

$$\varphi_z(x,y)=\frac{h(x,y)2\pi f_0 d_0}{h(x,y)-l_0}\;;h(x,y)=\frac{l_0\phi_z(x,y)}{\phi_z(x,y)-2\pi f_0 d_0} \tag{4}$$

where the value of $h(x,y)$ is measured and considered as positive to the left side of the reference plane. Also, the Equation 4 expresses the height distribution as a function of the phase distribution.

The Equation 1 can be rewritten as:

$$g(x,y)=\sum_{n=-\infty}^{\infty}A_n r(x,y)\exp(in\varphi(x,y))*\exp(i2\pi n f_0 x) \tag{5}$$

where $r(x,y)$ is the reflectivity distribution on the diffuse object [3,4]. Then, a *FFT* (Fast Fourier Transform) is applied to the signal in the x direction only. Thus, the following equation is obtained:

$$G(f,y)=\sum_{-\infty}^{\infty}Q_n(f-nf_0,y) \tag{6}$$

where Q_n is the 1D Fourier Transform of $A_n\, exp[in\varphi(x,y)]$.

Here $\varphi(x,y)$ and $r(x,y)$ vary very slowly in comparison with the fringe spacing, then the Q peaks in the spectrum are separated from each other. It is also necessary to consider that if a high spatial fringe pattern is chosen, the *FFT* will have a wider spacing among the frequencies; this behavior helps to identify the fundamental peak f_0. In *FTP*, next step is to remove all signals but positive fundamental peak f_0. Then, the result is shifted and centered. Later, the *IFFT* (Inverse Fast Fourier Transform) is applied in the x direction only. Here, is necessary to separate the phase part of the result from the rest because it contains the depth information:

$$\begin{aligned}\varphi_z(x,y)&=\varphi(x,y)+\varphi_0(x,y)\\&=\operatorname{Im}\{\log(\hat{g}(x,y)\hat{g}_0^*(x,y))\}\end{aligned} \tag{7}$$

The whole phase map is obtained by applying the same procedure for each x line. The result is that the values of the phase map are wrapped at some specific values whose range lie between π and $-\pi$. Then, to recover the true phase it is necessary to restore the measured wrapped phase by an unknown multiple of $2\pi f_0$ [17]. However, to analyze and describe signals, it requires information from both domains time and frequency, therefore Fourier is not a suitable solution to express those signals and another way is proposed, wavelet transform.

2.2. Wavelet Transform Profilometry

The wavelet transform (WT) is considered an appropriate tool to analyze non-stationary signals. This technique has been developed as an alternative approach to the most common transforms, such as Fourier transform, to analyze fringe patterns. Furthermore, WT has a multi-resolution property in both time and frequency domains which solves a commonly know problem in other transforms like the resolution.

A wavelet is a small wave of limited duration (this can be real or complex). For this, two conditions must be satisfied: firstly, it must have a finite energy. Secondly, the wavelet must have an average value of zero (admissibility condition). It is worth noting that many different types of mother wavelets are available for phase evaluation applications. The most suitable mother wavelet is probably the complex Morlet one[2]. The Morlet wavelet is a plane wave modulated by a Gaussian function, and is defined as:

$$\psi(x) = \pi^{1/4} \exp(icx)\exp(-x^2/2) \tag{8}$$

where c is a fixed spatial frequency, and chosen to be about 5 or 6 to satisfy an admissibility condition [11]. Figure 1 shows the real part (dashed line) and the imaginary part (solid line) of the Morlet wavelet.

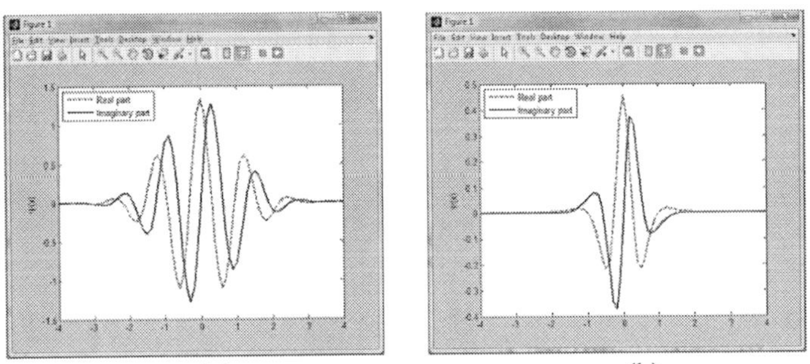

(a) (b)

Figure 1. Mother wavelets: (a) Complex Morlet and (b) Paul.

Also, Paul Wavelet is considered as one choice to perform the phase evaluation and is defined as:

$$\psi(x) = \frac{2^n n!(1-ix)^{-(n+1)}}{2\pi\sqrt{\frac{(2n)!}{2}}} \tag{9}$$

where n is the order of the Paul mother wavelet and chosen to have the value of 5.

The one-dimensional continuous wavelet transform (1D-CWT) of a row $f(x)$ of a fringe pattern is obtained by translation on the x axis by b (with y fixed) and dilation by s of the mother wavelet ψ(x) as given by:

$$W(s,b)=\frac{1}{\sqrt{s}}\int_{-\infty}^{\infty} f(x)\psi^{*}\left(\frac{x-b}{s}\right)dx \tag{10}$$

here, * denotes complex conjugation and *W(s,b)* is the calculated CWT coefficients which refers to the closeness of the signal to the wavelet at a particular scale.

In this research, the phase estimation and frequency estimation methods are used to extract the phase distribution from two dimensional fringe patterns. In the phase estimation method, a complex Morlet and Paul wavelets will be applied to a row of the fringe pattern. The resultant wavelet transform is a two dimensional complex array, where the phase arrays can be calculated as follows:

$$abs(s,b)=|W(s,b)| \tag{11}$$

and

$$\varphi(s,b)=\tan^{-1}\left(\frac{\Im\{W(s,b)\}}{\Re\{W(s,b)\}}\right) \tag{12}$$

To compute the phase of the row, the maximum value of each column of the modulus array is determined and then its corresponding phase value is found from the phase array. By repeating this process on all rows of the fringe pattern, a wrapped phase map results and an unwrapping algorithm is then needed to unwrap it.

In the frequency estimation method, a complex Morlet wavelet and Paul wavelet are applied to a row of the fringe pattern. The resultant wavelet transform is a two dimensional complex array. The modulus array can be found using Equation (13) and hence the maximum value for each column and its corresponding scale value can be determined. Considering that we are interested in the 1D signal:

$$f(x)=a(x)+b(x)\cos(2\pi f_0 x+\varphi(x)) \tag{13}$$

Considering the Euler identity for *cos(x)*, then the Equation 13 could be re-formulated as:

$$f(x)=a(x)+b(x)\cos(\varphi(x))=a(x)+b(x)\frac{e^{i\varphi(x)}}{2}+b(x)\frac{e^{-i\varphi(x)}}{2} \tag{14}$$

The analytic function *f* in an open interval *A*, where $z_0 \in A$ can be decomposed into Taylor series:

$$f(z)=\sum_{k=0}^{\infty}\frac{f^{(k)}(z_0)}{k!}(z-z_0)^k \tag{15}$$

Therefore:

$$\varphi(x)=\varphi(x)+\varphi'(x)(b)(x-b)+\frac{\varphi''(x)}{2!}(x-b)^2+\frac{\varphi'''(x)}{3!}(x-b)^3+\ldots \tag{16}$$

Considering:

$$\varphi(x)=\varphi(b)+\varphi''(b)\approx 0,\ \varphi'''(b)\approx 0,\ \ \varphi'^{v}(b)\approx 0,\cdots,\varphi^{k}(b)\approx 0 \tag{17}$$

Then, the function can be reduced as:

$$\varphi(x)=\varphi(b)+\varphi'(b)(x-b) \tag{18}$$

Moreover, the Morlet Wavelet is defined in the Equation (8) this wavelet will be applied to the mother wavelet, as shown in Equation (10):

If s − 1, then:

$$W(s,b)=\frac{1}{s}\int_{-\infty}^{\infty}f(x)\psi^*\left(\frac{x-b}{s}\right)dx$$
$$=\int_{-\infty}^{\infty}\left[a(x)+b(x)\frac{e^{i\varphi(x)}}{2}+b(x)\frac{e^{-i\varphi(x)}}{2}\right]\psi^*\left(\frac{x-b}{s}\right)dx$$
$$=a\int_{-\infty}^{\infty}\psi^*\left(\frac{x-b}{s}\right)dx+\frac{b}{2}\int_{-\infty}^{\infty}\frac{e^{i[\varphi(b)+\varphi'(b)(x-b)]}}{2}\psi^*\left(\frac{x-b}{s}\right)dx+\frac{b}{2}\int_{-\infty}^{\infty}\frac{e^{-i[\varphi(b)+\varphi'(b)(x-b)]}}{2}\psi^*\left(\frac{x-b}{s}\right)dx \tag{19}$$

By solving Equation 19, the following equation is obtained:

$$a\sqrt{2\pi e}^{-\frac{1}{2}\varpi_0^2}+\frac{b}{2}\sqrt{2\pi e}^{-\frac{1}{2}(\varpi_0+s\varpi_s)^2}e^{-i\varpi_s b}+\frac{b}{2}\sqrt{2\pi e}^{-\frac{1}{2}(\varpi_0-s\varpi_s)^2}e^{i\varpi_s b} \tag{20}$$

Then the instantaneous frequencies are computed using the next Equation [11]:

$$\hat{f}(b)=\frac{c+\sqrt{c^2+2}}{2s_{\max}(b)}-2\pi f_o \tag{21}$$

where f_0 is the spatial frequency. At the end, the phase ditribution can be extracted by integrating the estimated frequencies.

The same procedure can be developed to get the instantaneous frequencies, which lead us to have the wrapped phase map and therefore it is necessary to apply a phase unwrapping algorithm.

3. Phase Unwrapping

Since two decades ago, phase unwrapping has been a research area and many papers have been published, presenting some ideas that solves the problem. Several phase unwrapping algorithms have been proposed, implemented and tested.

The phase unwrapping process is not a trivial problem due to the presence of phase singularities (points in 2D, and lines in 3D) generated by local or global undersampling. The correct 2D branch cut lines and 3D branch cut surfaces should be placed where the gradient of the original phase distribution exceeded π rad value. However, this important information is lost due to undersampling and cannot be recovered from the sampled wrapped phase distribution alone. Also, is important to notice that finding a proper surface, or obtaining a minimal area or using a gradient on a wrapped phase will not work and one could not find the correct branch in cut surfaces.

The phase unwrapping has many applications in applied optics that require an unwrapping process, and hence many phase unwrapping algorithms has been developed specifically for data with a particular application. Moreover, there is no universal phase unwrapping algorithm that can solve wrapped phase data from any application. Therefore, phase unwrapping algorithms are considered as a trade-off problem between accuracy of solution and computational requirements. However, even the most robust and complete phase unwrapping algorithm cannot guarantee in giving successful or acceptable unwrapped results without a good set of initial parameters. Unfortunately, there is no standard or technique to define the parameters that guarantee a good performance on phase unwrapping.

To face the phase unwrapping problems, algorithms can be divided in two main categories: local and global phase unwrapping. Local phase unwrapping algorithms find the unwrapped phase values by integrating the phase along a certain path[6].

Global phase unwrapping algorithms locate the unwrapped phase by minimizing a global error function and are also called local phase unwrapping algorithm and a global phase unwrapping algorithm, by following the methodology proposed by Pedraza in [1]. The unwrapped phase values and the wrapped phase can be related with each other as:

$$\Psi(n) = \varphi(\pi) + 2\pi\, k(n)$$

With constraints (22)

$$-\pi < \Psi(n) \leq \pi$$

and:

$$\varphi(n) = \Psi(n) + 2\pi\, v(n)$$

with constraints (23)

$-\infty < \varphi(n) \leq \infty$

Here Ψ(n) holds the wrapped phase values and $\varphi(n)$ holds the unwrapped phase values, k(n) is the function containing the integers that must be added to the wrapped phase φ to be unwrapped, n is an integer and v(n) is the function containing a set of integers that must be added to the wrapped phase Ψ.

The wrapping operation ω, which converts the unwrapped phase, is defined by:

$$\omega\{\phi(n)\} = \arctan\left[\frac{\sin(\phi(n))}{\cos(\phi(n))}\right] \tag{24}$$

3.1. Local Phase Unwrapping

Local phase unwrapping algorithms finds the unwrapped phase values by integrating the phase along certain path that covers the whole wrapped phase map. The local phase unwrapping defines the quality of each pixel in the phase map to unwrap the highest quality pixels first and the lowest quality pixels last (quality-guided phase unwrapping). For this purpose, the methods known as residue-balancing are proposed, which attempts to prevent error propagation by identifying residues (the source of noise in the wrapped phase). The residues must be balanced and isolated by using barriers (branch-cuts), therefore, it aims to produce a path-independent wrapped phase map. Path-dependency occurs to the existence of residues.

Residue-balancing algorithms search for residues in a wrapped-phase map and attempt to balance positive and negative residues by placing cut lines between them to prevent the unwrapping path breaking the mesh created. The residue is identified for each pixel in the phase map by estimating the wrapped gradients in a 2 × 2 closed loop, as shown in Figure 2.

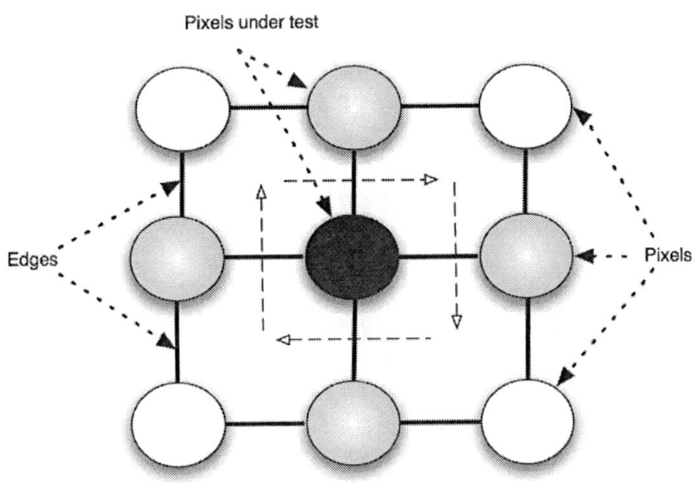

Figure 2. Identifying residues in a 2 × 2 closed path.

This is carried out using the following equation:

$$r = \Re\left[\frac{\Psi_{i,j} - \Psi_{i+1,j}}{2\pi}\right] + \Re\left[\frac{\Psi_{i+1,j} - \Psi_{i+1,j+1}}{2\pi}\right] + \Re\left[\frac{\Psi_{i,j+1} - \Psi_{i,j}}{2\pi}\right] + \Re\left[\frac{\Psi_{i,j+1} - \Psi_{i,j+1}}{2\pi}\right] \quad (25)$$

where $\Re[]$ rounds its argument to the nearest integer, $\Psi_{x,y}$ is the wrapped pixel. The equation 25 can only take three possible results: 0, +1, and -1. A pixel under test is considered to be a positive residue if the value of r is +1, and it is considered to be a negative residue if the value is -1. Conversely, the pixel is not a residue if the value of r is zero. After identifying all residues in the wrapped phase map, these residues have to be balanced by means of branch cuts. Branch-cuts act as barriers to prevent the unwrapping path going thorough them. If these branch cuts are avoided during the unwrapping process, no errors propagate and the unwrapping path is considered to be path independent. On the other hand, if these branch cuts are penetrated during the unwrapping, errors propagate throughout the whole phase map, and in this case the unwrapping path is considered to be path dependent.

3.2. Global Phase Unwrapping

In the previous section, it was stated that local phase unwrapping algorithms follow a certain unwrapping path in order to unwrap the phase. They begin at a grid point and integrate the wrapped phase differences over that path, which ultimately covers the entire phase map. Local phase unwrapping algorithms (residue-balancing algorithms) generate branch cuts and define the unwrapping path around these cuts in order to minimize error propagation.

In contrast, global phase unwrapping algorithms formulate the phase unwrapping problem in a generalized minimum-norm sense [6]. Global phase unwrapping algorithms attempt to find the unwrapped phase by minimizing the global error function as shown in Eq 26:

$$\varepsilon^2 = ||\text{solution} - \text{problem}||^2 \quad (26)$$

Global phase unwrapping algorithms seek the unwrapped phase whose local gradients in the *x* and *y* direction match, as closely as possible and shown in Figure 3.

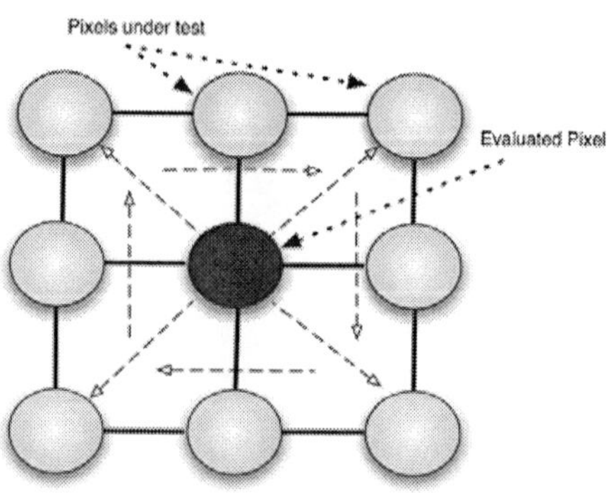

Figure 3. Matching pixel gradients in *x* and *y* direction.

$$\varepsilon^2 = \sum_{i=0}^{M-2}\sum_{j=0}^{N-1}\left|\Delta^x\varphi(i,j)-\hat{\Delta}^x\psi(i,j)\right|^p + \sum_{i=0}^{M-1}\sum_{j=0}^{N-2}\left|\Delta^y\varphi(i,j)-\hat{\Delta}^y\psi(i,j)\right|^p \quad (27)$$

where $\Delta^x\phi(i,j)$ and $\Delta^x\varphi(i,j)$ are unwrapped phase gradients in the x and y directions respectively, which are given by:

$$\Delta^x\phi(i,j) = \phi(i+1,j) - \phi(i,j) \quad (28)$$

$$\Delta^x\phi(i,j) = \phi(i,j+1) - \phi(i,j) \quad (29)$$

$\hat{\Delta}^x\psi(i,j)$ and $\hat{\Delta}^y\psi(i,j)$ are the wrapped values of the phase gradients in the x and y directions respectively, and they are given by:

$$\hat{\Delta}^x\psi(i,j) = \omega\{\psi(i+1,j) - \psi(i,j)\} \quad (30)$$

$$\hat{\Delta}^y\psi(i,j) = \omega\{\psi(i,j+1) - \psi(i,j)\} \quad (31)$$

Finally, the wrapping operator defined in Equation (24) is applied.

4. Setup and Proposed Methodology

Considering Figure 4, we have a fringe pattern, which is projected from the projector; the fringe reaches the object at point H and will cross the reference plane at the point C. By observation, the triangles *DpHDc* and *CHF* are similar and since:

$$\frac{CD}{-h} = \frac{d_0}{l_0} \quad (32)$$

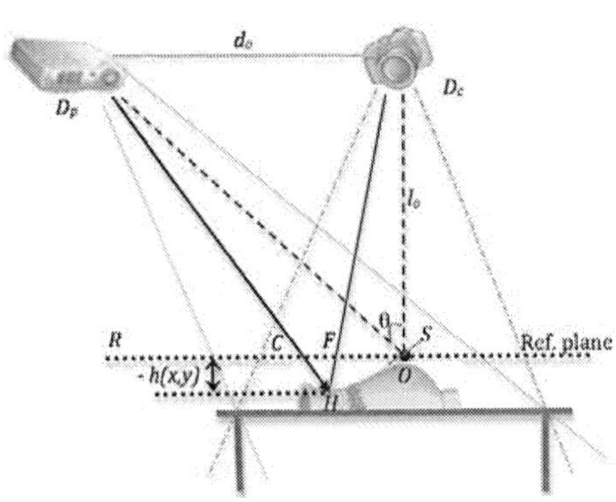

Figure 4. Experimental setup.

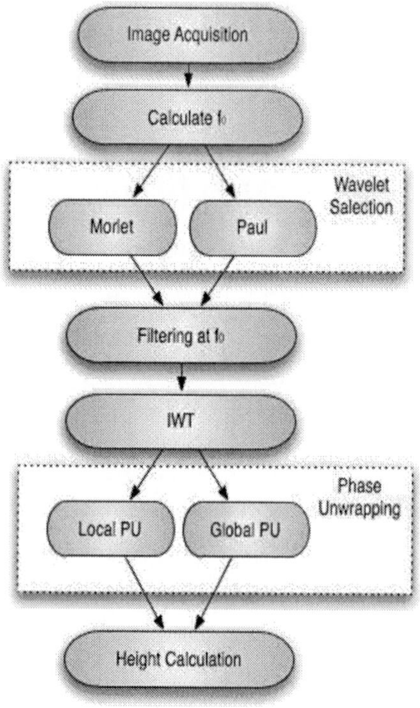

Figure 5. Proposed methodology.

The experimental setup shown in Figure 4 is proposed and during the experiments the methodology in Figure 5 is applied. The first step is to acquire the image. Due to the nature of the image, sometimes a filtering to eliminate the noise is necessary, and a filter is used. Next, the fundamental frequency *fo* is estimated. Later, the mother is selected (Morlet or Paul) and applied.

The filter at *fo* is carried out and the Inverse Wavelet transformation is done. At this stage, the information of the height is phase wrapped and two phase unwrapping algorithms are proposed: Local and Global Analysis Algorithm and Graph Cuts Algorithm. The final step is to obtain the object reconstruction and in some cases to determine the error (in case of virtual created objects). The experimental setup uses a high-resolution digital CCD camera and a high resolution digital projector.

The object of interest can be any three-dimensional object and for this work, three objects are considered, which are shown on Figure 6.

It is also important to develop software able to produce several different fringe patterns. To create several patterns, it is necessary to modify the spatial frequency (number of fringes per unit area), and resolution (number of levels to create the sinusoidal pattern) of the fringe pattern.

It may also be necessary to include into the software development a routine capable of performing phase shifting as well as to include the horizontal or vertical orientation projection of the fringe pattern.

(a) (b) (c)

Figure 6. Virtual objects used in the test: (a) Hand, (b) Fly, and (c) Airplane.

5. RESULTS

To test the methodology, first an object with Hand shape is used. Then, a sinusoidal fringe pattern of known spatial frequency is created with 128 fringes and added to the shape of the created object. The resulting image is shown in Figure 7(c). It is worth noting the distortions of the fringe pattern due to the object's shape.

The wrapped phase and its mesh are shown in Figure 8. The reconstructed Hand using the Morlet Wavelet Transform and applying the Local PU Analysis and the Global PU Analysis can be seen in Figure 9. Notice that, by applying this method, the shape of the Hand looks almost equal, but it has an error magnitude around 3.2 and 2.1% respectively.

The results on Figure 9 show that the whole volume presents an acceptable error and the shape is well defined. The mother wavelet used was the Morlet but the same experiment was conducted for the Paul wavelet and the results obtained were similar to those presented on Figure 9. The computer simulation allowed us to test the proposed methodology. On Figures 7, 8 and 9 the results of the methodology are presented step by step.

To validate the whole methodology, more experiments were conducted considering the objects observed on Figure 6(b) and (c). Those objects have different shapes (computer created), where the height is known in every point in the object.

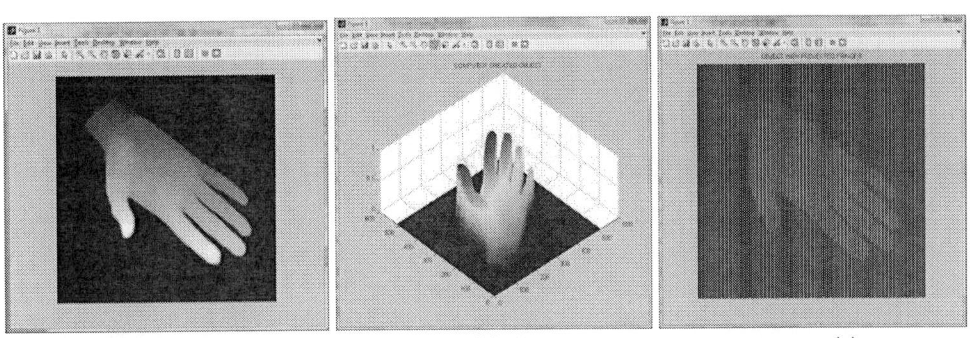

(a) (b) (c)

Figure 7. Computer created Hand: (a) Object image, (b) Object mesh, and (c) fringes projected on it.

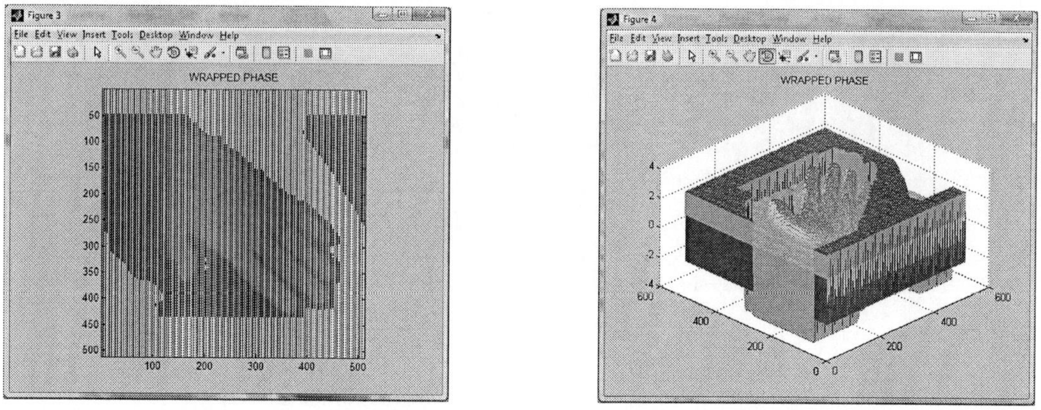

Figure 8. Wrapped phase (image and mesh).

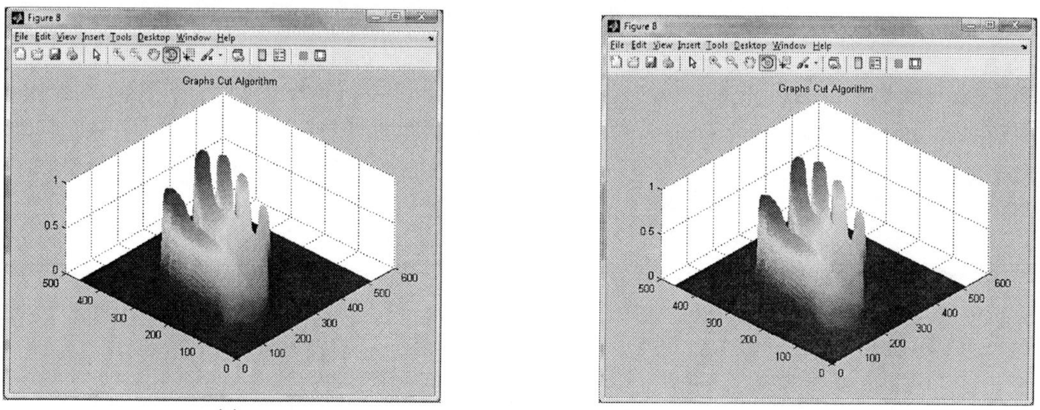

(a) (b)

Figure 9. Reconstructed objects by using Morlet Wavelet and (a) Local Phase Unwrapping Algorithm, and (b) Global Phase Unwrapping Algorithm.

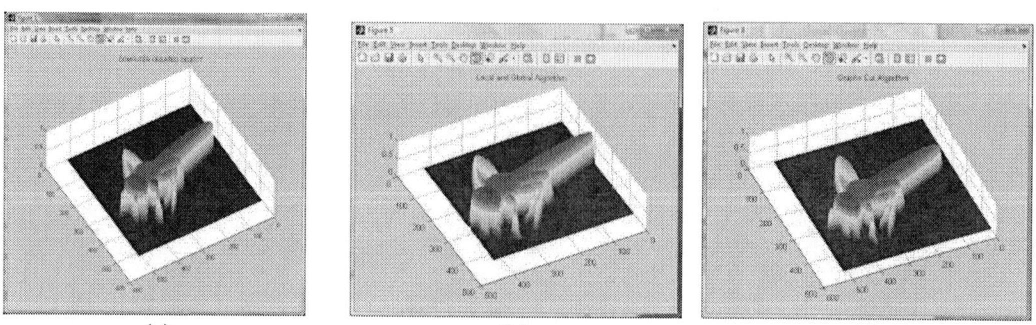

(a) (b) (c)

Figure 9. Fly object and its reconstruction using Paul Wavelet: (a) Object, (b) Local PU Algorithm, and (c) Global PU Algorithm.

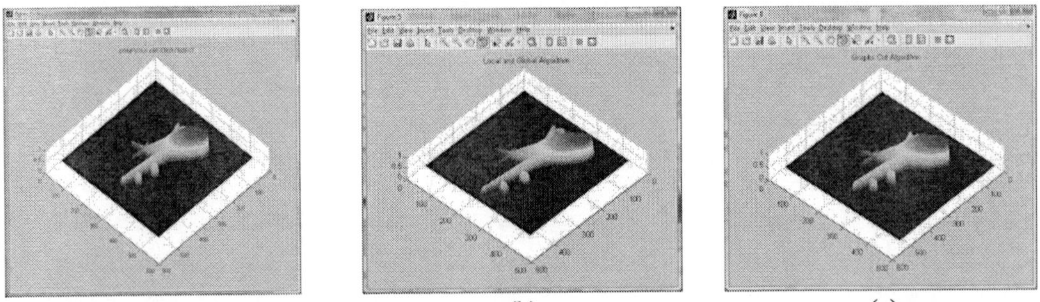

(a) (b) (c)

Figure 10. Airplane object and reconstruction using Morlet Wavelet: (a) Object, (b) Local PU Algorithm, and (c) Global PU Algorithm.

Then, the Morlet and Paul mother Wavelets are considered as well as the two different phase unwrapping algorithms. As a second experiment, Paul Wavelet is used and height of the virtual object was compared with each one of the analysis and the results are presented in Figure 9.

Table 1. Error table using Morlet Wavelet

Object	Local PU	Global PU
Hand	3.26	2.11
Fly	3.47	2.21
Airplane	3.51	2.18

Table 2. Error table using Paul Wavelet

Object	Local PU	Graph PU
Hand	4.37	3.43
Fly	4.65	3.76
Airplane	4.71	3.55

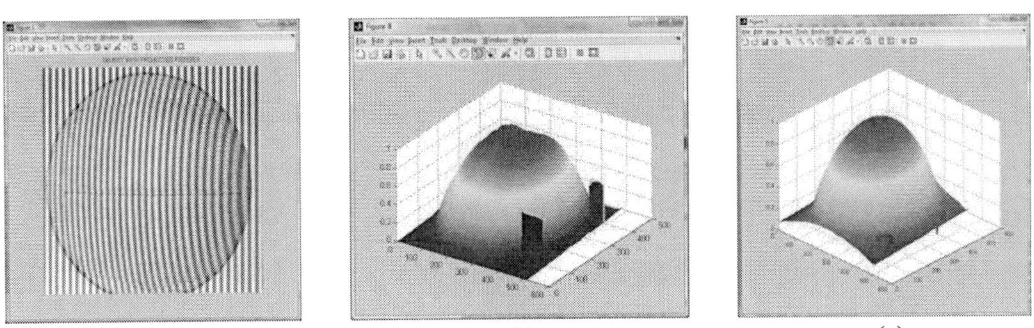

(a) (b) (c)

Figure 11. Real object and reconstruction: (a) Object, (b) Using Morlet Wavelet, and (c) Paul Wavelet with Graph Cuts Algorithm.

Table 3. Error table analysis of real object

Object	Wavelet	Local PU	Global PU
Volleyball's ball	Morlet	4.51	3.24
Volleyball's ball	Paul	4.49	3.13

Third experiment was conducted with Morlet Wavelet and the object used was the airplane, and the respective results of the reconstruction process can be seen in Figure 10.Later, all the experiments are joined and the error magnitude is enclosed on Tables 1 and 2. The results show that the better performance was obtained by using the Morlet wavelet together with the Global Phase Unwrapping Algorithms in final step to perform the 3D reconstruction process.

Finally, the performance of the proposed methodology was tested in a real object (a volleyball's ball) and both Morlet and Paul wavelets were used considering the Global Phase Unwrapping Algorithm for the phase unwrapping and the results are observed in Figure 11.

Conclusion

In this work, an enhanced wavelet based profilometry was presented and tested. Both Morlet and Paul mother wavelets were used in conjunction with local and global methods, in the phase unwrapping process. Three different objects generated by computer were utilized (hand, fly and airplane). The object's projected fringe pattern has a known spatial frequency. Additionally, a real object was taken in order to implement the proposed methodology and accurate reconstruction of the object. Among the proposed wavelets, the one who shown better performance was Morlet wavelet over Paul wavelet, since Morlet was the one that produced a minimal error. As a conclusion, we can say that the proposed methodology could be used to digitize diverse objects with good results. As a future work, the software performance can be improved in order to implement it inside an embedded system.

References

[1] Gokstorp, M. Depth Computation in Robot Vision, Ph.D. Thesis, *Department of Electrical Engineering*, Linkoping University, Linkoping, Sweden, (1995).

[2] M. Takeda, H. Ina and S. Kobayashi, *J. Opt. Soc. Am.* 72, 156 (1982).

[3] J.E. Grevenkamp and J.H. Bruning, Phase-shifting interferometry. In D. Malacara (Ed.), *Optical Shop Testing*, Wiley, New York (1992).

[4] K. Creath, in: W.R. Robinson and G.T. Reid (Eds.), *Interferogram Analysis: Digital Fringe Pattern Measurement Techniques*, Institute of Physics, Philadelphia, PA (1993).

[5] M. Gdeisat, D. Burton, M. Lalor, *Appl. Opt.* 39, 5326 (2000).

[6] Y. Ichioka, M. Inuiya, *Appl. Opt.* 11, 1507 (1972).

[7] F.Berryman, P. Pynsent, J. Cubillo, *Opt. Lasers Eng.* 39, 35 (2003).

[8] J.C. Pedraza, Image Processing for 3D Reconstruction using a Modified Fourier Transform Profilometry Method, *Lecture Notes on Artificial Intelligence*, Springer-Verlag, Berlin, (2007) pp. 705.
[9] P.K. Rastogi, *Digital Speckle Pattern Interferometry and Related Techniques*, Wiley, New York (2001).
[10] K. Itoh, *Appl. Opt.* 21, 2470 (1982).
[11] L.S. Wu, *J. Zhejiang Univ. Sci. A,* 7, 1026 (2006).
[12] J.G. Zhong, J.W. Wang, *Appl. Opt.* 43, 4993 (2004).
[13] Q. Zang, W.J.Chen, Y. Tang, *Opt. Commun.* 282, 778 (2009).
[14] A. Dursun, S. Ozder, N. Ecevit, *Meas. Sci. Tech.* 15, 1768 (2004).
[15] M. Afifi, A. Fassi-Fihri, M. Marjane, K. Nassim, M. Sidki and S. Rachafi, *Opt. Commun.* 211, 47 (2002).
[16] M.A. Gdeisat, D.R. Burton and M.J. Lalor, *Appl. Opt.* 45, 8722 (2006).
[17] J.C. Pedraza-Ortega, et al., Three-Dimensional Reconstruction System based on a Segmentation Algorithms and a Modified Fourier Transform Profilometry. *In Proceedings of the IEEE Electronics, Robotics and Automotive Mechanics Conference,* Cuernavaca, Morelos, September (2009).
[18] J.C. Pedraza-Ortega, et al., A Profilometric Approach for 3D Reconstruction Using Fourier and Wavelet Transforms. *Lecture Notes on Artificial Intelligence*, Springer-Verlag, Berlin, pp. 313-323 (2009).
[19] M.A. Gdeisat, D.R. Burton and M.J. Lalor, *Opt. Lasers Eng.* 47, 1348 (2009).

In: Wavelets: Classification, Theory and Applications
Editors: Manel del Valle et al.
ISBN: 978-1-62100-252-9

Chapter 15

DIFFUSION WAVELET FOR EFFICIENT REAL-TIME MONITORING OF ACOUSTIC EMISSION SIGNAL IN NANOMANUFACTURING

Rajesh Ganesan*

Department of Systems Engineering and Operations Research,
George Mason University, Fairfax, Virginia, US

ABSTRACT

Acoustic emission (AE) signal has been used in several process monitoring applications in manufacturing. AE is generated when atoms are torn from the surface during the material removal process in manufacturing. Of particular interest is the AE signal captured from nanomanufacturing processes, which have been shown to contain critical information related to the health of the process. Understanding the properties of AE signal could yield a myriad of process information, however, the task to process AE signal is challenging due to its non-stationary, high frequency, and heavy noise characteristics. This chapter presents a novel diffusion wavelet based AE signal processing method that is capable of efficiently extracting the underlying nano-scale process characteristics better than the well-known classical wavelets such as Haar and Daubechies wavelets. Results from applying diffusion wavelet in analyzing AE signal from a Chemical Mechanical Planarization (CMP) process indicate that the wavelet and scaling signals generated by the diffusion wavelet method has higher magnitude and sharp features to clearly detect nanomanufacturing process anomalies, and therefore provides greater insight into the AE signal characteristics. In particular, the classical wavelets fail to detect the end-point of the CMP process using AE signal. However, the end-point is clearly detected by the diffusion wavelet methodology. The diffusion wavelet method is made suitable for online application by developing a moving block data processing strategy, which matches the rate of data acquisition.

* E-mail address: rganesan@gmu.edu

1. INTRODUCTION

Some of the critical requirements facing semiconductor device manufacturing are continual feature size reduction, introduction of new materials for higher processing speeds and improved reliability, multilevel metallization (MLM) or interconnections, and increased productivity through larger wafer sizes. Wafer polishing using Chemical Mechanical Planarization (CMP) is a key nanoscale manufacturing step that can significantly impact on how the above requirements are met by the industry. However, the increased sophistication of CMP process has brought difficult manufacturing challenges, which include defects (delamination, dishing, under/over polishing) identification, process monitoring, and process control. In recent years, acoustic emission (AE) technology has emerged as a viable means to examine and characterize CMP processes [1]. The AE signal arising out of the CMP process, where material removal takes place at the nanoscale, has a very high signal to noise ratio (sensitivity). Also, the frequency of the AE signal is very high compared to those of machine vibrations and other environmental noises, and thus provides a good representation of the material removal process of planarization. The above properties of AE signal make it very suitable in extracting critical information about the state of the CMP process, and in building detection/control strategies.

The AE signal is non-stationary and its strength (measured in Volts) is high at the beginning of the CMP process, which subsides as polishing progresses. The signal spans over a broad range of high frequencies, and contains much time based characteristics, and noise. Consequently, visual inspection of time plots of AE signals fails to provide any meaningful information, thus requiring a systematic time-frequency analysis. Wavelet based approaches have the potential to analyze non-stationary AE signals [2]-[9], and provide useful insight into the CMP process. Some of the more recent applications of AE signal monitoring have been in laser shock cleaning of wafer surface to remove nano-scale residue in CMP [10]. AE has been used in the characterization of AFM nanomanufacturing [11].

Classical wavelet analysis of AE signal for delamination defect detection in the CMP process has been presented in [12], and for end-point detection in CMP using the coefficient of friction (CoF) signal has been presented in [13]. However, as presented in [13], the wavelet analysis of AE signal was inefficient in end-point detection. Some of the limitations in the above applications of classical wavelet to study AE signal properties in nanomanufacturing include the following. 1) The wavelet and scaling coefficients that arise from the time-frequency decomposition of the AE signal have very small magnitude. Therefore, detecting shifts in the mean and variance requires a highly sensitive process control chart. 2) Due to the small magnitude of the coefficients (typically in the range of -0.2 to 0.2), there is a possibility that the AE signal features could get thresholded during the denoising process. 3) Finally, classical wavelet methodology requires several levels of decomposition to obtain the significant wavelet coefficients. This results in a coarse and weak reconstructed signal that may not be very effective for process monitoring.

The primary objective of this chapter is to overcome the above limitations of classical wavelet based AE signal processing methods. The chapter presents a novel diffusion wavelet based methodology, which has superior AE signal feature extraction capabilities when compared with the classical wavelets. The strengths of the diffusion wavelet procedure is its ability to generate best scaling and wavelet basis functions as explained in Section 2.. The advantage of having the best basis functions results in superior function approximation characteristics, which can then assist in accurate feature extraction. The best basis functions are level dependent, which is unlike the classical wavelets where the same wavelet is translated (in time) and dilated (in frequency) at different levels of decomposition. The wavelet and scaling signals generated by the diffusion wavelet process is several times higher in magnitude as compared to those generated by classical wavelets. Therefore, highly sensitive control charts such as cumulative sum (CUSUM), exponentially weighted moving average (EWMA), and sequential probability ratio tests (SPRT) are not required. The mean and variance shifts in the wavelet and scaling signals generated by the diffusion wavelet process is very dramatic, which is a very crucial element in designing monitoring schemes for nano-scale machining processes. The mathematical construction of the diffusion wavelet procedure inherently removes noise from the AE signal, which is unlike the classical wavelets where the levels 1 and 2 of the wavelet signal usually contains noise that must be further thresholded [14].

The chapter is organized as follows. Section 2 discusses the classical and diffusion wavelet theory briefly. Experimental set-up details of CMP are provided in Section 3. Section 4 provides details of applying diffusion wavelet to AE signal for delamination defect detection and end-point detection in a nanomanufacturing CMP process. Conclusions are provided in Section 5.

2. Overview of Multiresolution Approach

The AE data used for the analysis is sampled at a fairly high frequency (1kHz) and is corrupted with noise (See top of Figure 3). Hence, it is must be 1) denoised, 2) broken up into many frequency bands, and 3) analyzed using time domain methods at each band. Thus, a direct statistical analysis of time domain AE data would yield poor results unless the noise component is removed and the significant features are extracted. The conventional time domain analysis methods, which are sensitive to impulsive oscillations, have limited utility in extracting hidden patterns and frequency related information in these signals ([15], [16]). This problem is partially overcome by spectral (frequency) analysis such as Fourier transform, the power spectral density, and the coherence function analysis. However, many spectral methods rely on the implicit fundamental assumption of signals being periodic and stationary, and are also inefficient in extracting time related features. Moreover, Fourier transform of non-stationary signals results in averaging of the frequency components over the entire duration of the signal. This problem has been addressed to a large extent through the use of time-frequency based short time Fourier transform (STFT) methods. However, this method uses a fixed tiling scheme, i.e., it maintains a constant aspect ratio (the width of the time window to the width of the frequency band) throughout the analysis. As a result, one must choose multiple window widths to analyze different data features localized in time and frequency domains in order to determine the suitable width of the time window. STFT is also inefficient in resolving short time phenomena associated with high frequencies since it has a limited choice of wave forms [3]. In recent years, another time-frequency (or timescale) method known as wavelet based multiresolution analysis have gained popularity in the analysis of both stationary and non-stationary signals. These methods provide excellent time-frequency localized information, which is achieved by varying the aspect ratio. This means that multiple frequency bands can be analyzed simultaneously in the form of details and approximations plotted over time, as described later in this section. Hence, different time and frequency localized features are revealed simultaneously with high resolution. This scheme is more adaptable (compared to STFT) to signals with short time features occurring at higher frequencies.

2.1. Classical Wavelet

The basic idea behind signal processing with wavelets is that the signal can be decomposed into its constituent elements through the use of basis functions. These basis functions can be obtained from the scaled (dilated) and shifted (translated) versions of the mother wavelet (w). The wavelet analysis uses linear combinations of basis functions (wavelets), localized both in time and frequency, to represent any function in the $L^2(\mathfrak{R})$ space (Euclidean-norm). For example,

$$f(t) = \sum_{j=-\infty}^{\infty} \sum_{k=-\infty}^{\infty} b_{j,k} w_{j,k}(t), \qquad j,k \in Z, \tag{1}$$

where j and k are dilation (or scale) and translation indices respectively, $w_{j,k}$ denotes a collection of basis functions, and $b_{j,k}$ are the coefficients of these functions. The wavelet basis functions can also be derived from the dilation and translation of scaling functions (ϕ) that span the $L^2(\mathfrak{R})$ subspace. By combining the scaling and the wavelet functions, we can represent any class of signals in $L^2(\mathfrak{R})$ as

$$f(t) = \sum_{j=-\infty}^{\infty} c_{j_0,k} \phi(t-k) \sum_{k=-\infty}^{\infty} \sum_{j=j_0}^{\infty} d_{j,k} w(2^j t - k), \tag{2}$$

where $c_{j_0,k}$ and $d_{j,k}$ are coefficients for the scaling and wavelet functions respectively. They are also called the discrete wavelet transform (DWT) of the function $f(t)$, and its customary to start with $j_0 = 0$. If the wavelet system is orthogonal then the coefficients can be calculated by

$$c_{j_0,k} = \langle f(t), \phi_{j_0,k}(t) \rangle = \int f(t) \phi_{j_0,k}(t)\, dt \tag{3}$$

$$d_{j,k} = \langle f(t), w_{j,k}(t) \rangle = \int f(t) w_{j,k}(t)\, dt \tag{4}$$

If the signal is smooth, the coefficients have small magnitude with uniform variance over time; however, if there is a jump in the signal, the magnitude of the coefficients will temporarily increase. Such an abrupt change in a process can be detected using the *extrema* of the wavelet coefficients. A more detailed theory on multiresolution analysis can be found in [17].

2.2. Diffusion Wavelets

Diffusion Wavelet is a compact multi-level representation of a Markov diffusion process on graphs and is used in analyzing n-D data. Classical wavelet is a special case of diffusion wavelet in which the basis functions ϕ and w are derived from the mother wavelet, which must be specified. In contrast to classical wavelet, the diffusion wavelet method builds best-basis functions to approximate the original raw data that is obtained from a process. In the classical wavelet analysis, which is performed on 1-D Euclidean spaces, dilations by powers of two and translations by integers are applied to a mother wavelet to obtain the wavelet bases. However, for diffusion wavelet, the diffusion operators T acting on functions of the space formed by the raw data are used to build the wavelet basis functions.

The best basis functions will be obtained as follows. Using a sample set (of dyadic length 2^n, where n is the level of decomposition) of the raw AE data, a Gaussian kernel is used to build a graph (G,E,W^g) using spectral graph theory. The graph can be imagined as a cloud of interconnected points. The graph (G,E,W^g) represents the structure of space spanned by the raw data, where E is the edge and W^g is the weight on the edge. The graph is obtained as follows

$$W^g_{u\sim v} = e^{-(\frac{\| u-v \|}{\delta})} \tag{5}$$

$$D_{u\sim u} = \sum_{v\in G} W^g_{u\sim v} \tag{6}$$

$$D_{u\sim v} = 0, \quad \forall u \neq v \tag{7}$$

where u and v are indices that refer to position numbers of any two data points in f(t), in which the data is arranged in rows and the columns represent the dimension of the data. Indices u and v also indicate the corresponding row and column position respectively in matrices Wg and D, where D is a diagonal matrix, and _ is a normalizing parameter.

The Laplacian operator L of the graph is obtained using spectral graph theory, and $I - L = T^1$ is obtained, where I is an identity matrix and T^1 is the diffusion operator at level $j = 1$. This is given as follows

$$I - L = D^{-\frac{1}{2}} W D^{-\frac{1}{2}} = T^1 \tag{8}$$

Sparse factorization (QR factorization) of T^1 yields the scaling basis functions ϕ, which are obtained from Q. The wavelet basis functions w are then obtained from the scaling basis functions ϕ by sparse factorization of $(I-\phi*\phi)$, where I is an identity matrix. The concept is that the basis functions in T^1 intrinsically represent the structure of the space spanned by the raw data and its geometric constraints. To obtain basis functions at the next level, the diffusion operator T^2 is first obtained from R as follows: $T^2 = R \times R^*$ where R^* is the complex conjugate of R. Sparse factorization of T^2 yields the scaling basis functions at the next level, and the procedure is continued until no further decomposition is possible. The number of elements in the basis function is automatically selected by the size of the basis function, and the diffusion wavelet theory guarantees compact representation. The details of diffusion wavelets are in [18].

3. Experimental Set-Up

In this section, the details of the experimental set-ups and process conditions under which test data were collected are provided.

The CMP task has been made more challenging in recent years due to the complex wafer topographies, and the introduction of copper (instead of aluminum) and low-k dielectrics. The multilevel metallization process consists of etching, deposition, and planarization using a copper damascene process. Figure 1 shows the schematic of a CMP setup, which synergistically combines both tribological (abrasion) and chemical (etching) effects to achieve planarization. The material removal rate (MRR) in CMP is usually in the range of 100-800 nm/min in thickness.

The AE data used in [12] and [13] were used in order to compare the results from diffusion and classical wavelets. The data comes from the planarization of several wafers under different combinations of rotational speed (rpm) and downward polishing pad pressure (psi) while maintaining same slurry composition, wafer type, and pad materials. A first set of wafers were polished using rotational speeds of 100, 150, 200, and 250 rpm under constant downward pressure of 3 psi. Subsequently, more wafers were polished using downward pressures of 2, 4, and 6 psi while maintaining a constant rotational speed. In each wafer polishing trial, acoustic emission (AE) data were collected. The types of wafers used were oxide, and blanket and patterned copper wafers backed with low-k dielectric material, and the polishing pad used was of type IC 1000/SUBA IV. AE signals were recorded from a good planarization process (in-control data), and from a wafer delamination process (out of-control data).

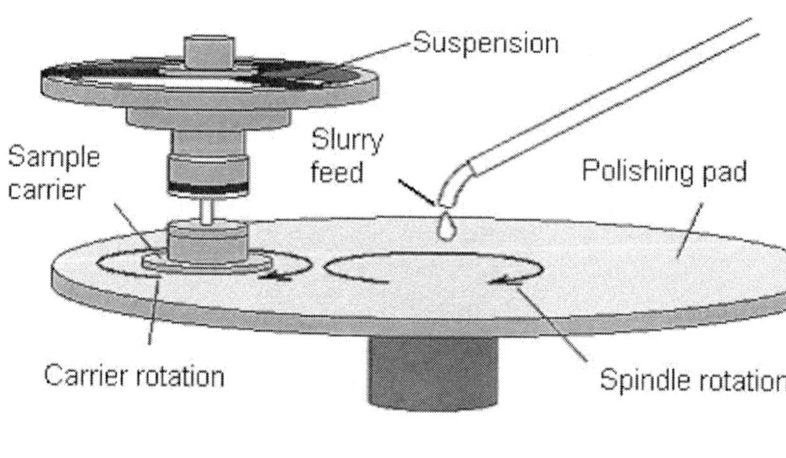

Figure 1. Schematic diagram of the CMP process.

4. RESULTS

Diffusion wavelets were applied to acoustic emission signals captured during CMP operations. Data is obtained at 1kHz and is processed using MATLAB software as fast as they are collected, which makes the monitoring suitable for real-time in-situ implementation. The collected data is processed in a moving window of dyadic length (2n), where n specifies the maximum level of decomposition using both diffusion wavelet and classical wavelets such as Haar or Daubechies wavelets. In this study an arbitrary moving window length of 128 (27) was chosen, which allows for 7 levels of decomposition within the window. After processing the data in the current window, the window is moved to the next set of 128 points.

4.1. Delamination Defect Detection

In this study the diffusion wavelet is compared to classical Haar and Daubechies wavelets in detecting delamination of silicon wafers. Delamination refers to the peeling of wafer surface during polishing, which results in uneven wafer surface. This often results in thinning or a complete loss of circuitry below the peeled surface. Acoustic emission can be used to detect the occurrence of delamination in CMP in real-time using classical wavelets [12]. In the above study, thresholded wavelet signals at decomposition levels 3, 4 and 5 were found to detect the delamination defect. However, it should be noted that the threshold values are dependent on the variance of the wavelet coefficients as provided by the Donoho's level dependent thresholding scheme [14]. A few large wavelet coefficients can increase the threshold values to a point where the wavelet coefficients could be completely contained within the threshold, which could make the delamination defect undetectable. As observed in Figure 2, the wavelet coefficients barely exceed the threshold for the out-of-control (delaminated) wafer during the first few thousand data points in which delamination occurs. An alternative approach to thresholding would be to use a very sensitive control chart such as a sequential probability ratio test (SPRT) for variance chart to detect variance changes that indicate the presence of delamination defects.

Processing acoustic emission signals using diffusion wavelets yields a far more sensitive wavelet signal as shown in Figure 3 (a) and (b), where the amplitude of the wavelet signal (DW level 1) is much higher than the reconstructed wavelet signal from unthresholded wavelet coefficients that are obtained using classical wavelets. For the purpose of comparison, the raw acoustic emission data, wavelet signal from level 1 of the diffusion wavelet process (DWlevel 1), and the reconstructed wavelet signal from levels 3, 4 and 5 of wavelet decomposition by the classical Haar wavelet are presented for both good (in-control) and delaminated (out-of-control) CMP wafer polishing process (Figure 3). For in-control data the wavelet signal from diffusion wavelet has a uniform variance (no-defects), whereas for out-of-control the wavelet signal has very high amplitude in the beginning, which indicates delamination defect. The amplitude of the wavelet signal in levels 3, 4, and 5 is much smaller in comparison to diffusion wavelet (level 1) for both in-control and out-of-control AE data that was analyzed using Haar wavelet (Note scale of -0.1 to 0.1 for in-control and -10 to 10 for out-of-control AE data).

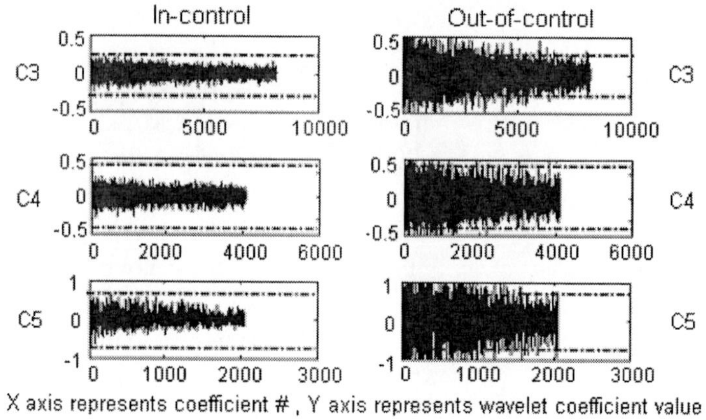

Figure 2. Haar wavelet decomposition of acoustic emission signal from in-control and out of-control (delaminated) CMP process.

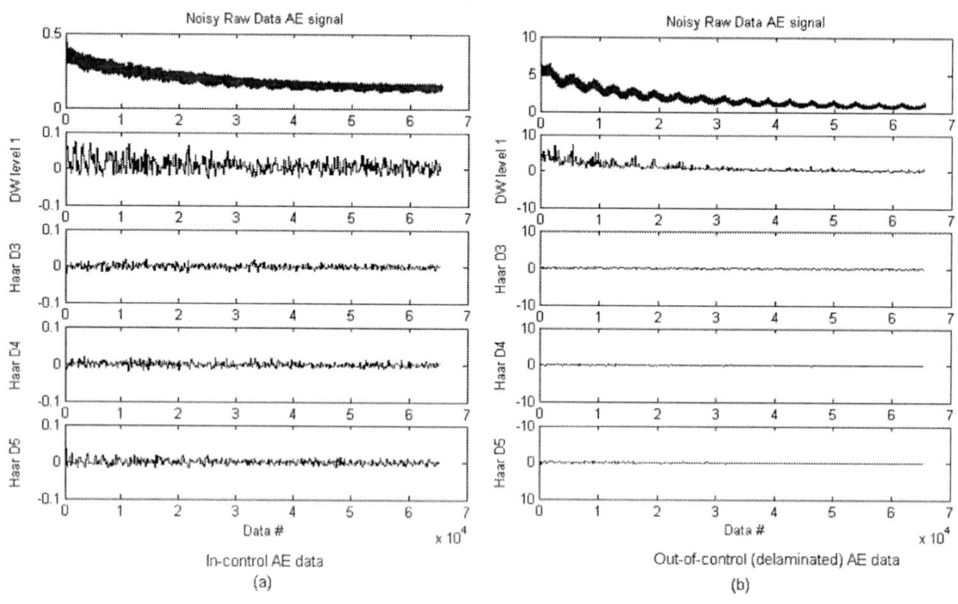

Figure 3. Comparison between Haar wavelet and diffusion wavelet decomposition of AE signal from in-control and out-of-control (delaminated) CMP process.

A similar comparison between Daubechies wavelet (3 levels) and diffusion wavelet (level 1) decomposition of AE signal from in-control and out-of-control (delaminated) CMP process is shown in Figure 4. The above procedure was repeated with different settings of pad pressure and rpms during both oxide and copper metal CMP, and similar results were noted. The diffusion wavelet was far more sensitive in capturing the delamination defect using AE signal than the classical wavelet analysis. This is because of the excellent function representation (feature extraction) capability of diffusion wavelet due to it best orthogonal basis functions.

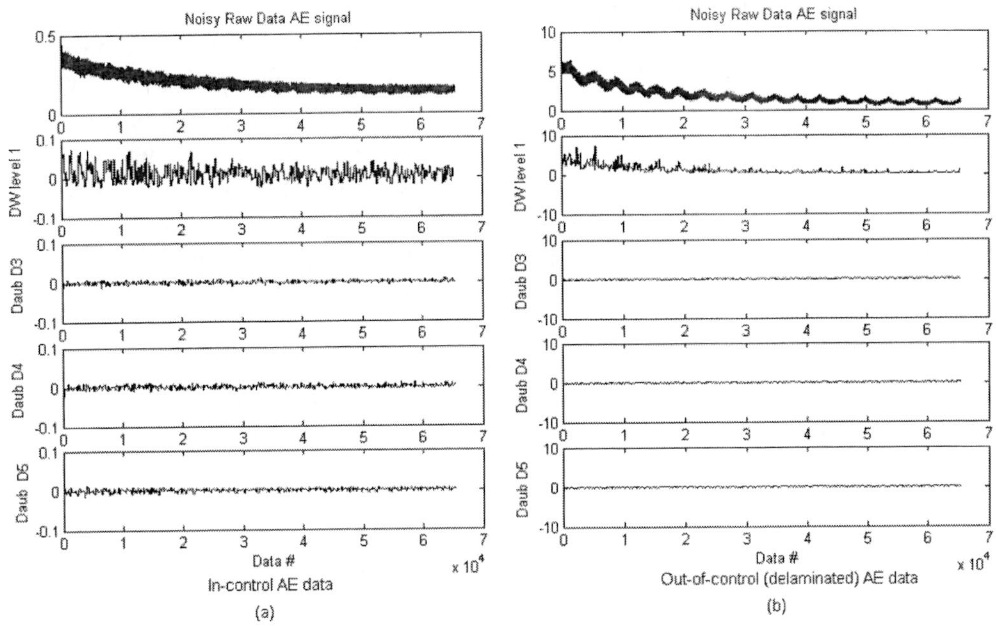

Figure 4. Comparison between Daubechies wavelet and diffusion wavelet decomposition of AE signal from in-control and out-of-control (delaminated) CMP process.

4.2. End Point Detection

End point detection is a very critical aspect of CMP process, which is required to precisely stop the CMP process when polishing is completed. This prevents under- and over-polishing of the wafer. As shown in an earlier research [13], end point was better detected by using coefficient-of-friction (CoF) signal than by acoustic emission signal when the raw CoF data was decomposed to the 9th level of decomposition using Haar classical wavelet. This is because at end point, the material properties change between the top layer and the next layer beneath it, which produces a different CoF against the polishing pad. Despite the change in material properties, which also produces a different AE signal, the change in AE signal was not well detected by classical wavelets. The 9th level was determined using energy analysis of each level of decomposition. It should be noted that the wavelet coefficients are fewer in number at that level and the signal is very coarse, which could result in inaccurate end point detection. Moreover, the wavelet coefficients are very small is magnitude, and it requires a highly sensitive SPRT for variance chart to detect the exact start and finish time of the end point event [13].

In contrast to the above patented procedure [13], the diffusion wavelets are capable of detecting end point of CMP using the AE signal. Several experiments were conducted using different pad pressure (in psi) and plate rotational speed (in rpms) on oxide, blanket copper-metal, and patterned (with circuits) copper-metal CMP processes to detect the end point event. Figure 5 shows end point detection in a blanket copper-metal CMP process (2 psi and 100 rpm) by the scaling signal (DS level 1) obtained from the diffusion wavelet procedure in comparison with scaling signal at levels 4 (Haar A4) and 5 (Haar A5) from Haar classical wavelet. Clearly on DS level 1, the start and end of the mean shift (start and end of end point event) can be observed, which can be monitored using standard control charts for mean shifts [19]. The scaling signals at level 4 and 5 does not indicate any change. It should be also noted that the magnitude of the scaling signal (DS level 1) is much higher than the others, which aids in efficient and accurate detection of the end point event. The wavelet signals from both classical (Haar D4 and Haar D5) and diffusion wavelets (DW level1) do not provide much information about the end point event. It must be noted that AE signal analysis using classical wavelets have been shown to be ineffective in end point detection at all levels of decomposition [13].

Similar results as above can be observed in Figure 6, which shows end point detection in an oxide CMP process. It should be noted that the end point event is longer for copper metal CMP than oxide CMP because of the presence of a barrier layer between the metal and the substrate below, and polishing is done through the barrier layer as well.

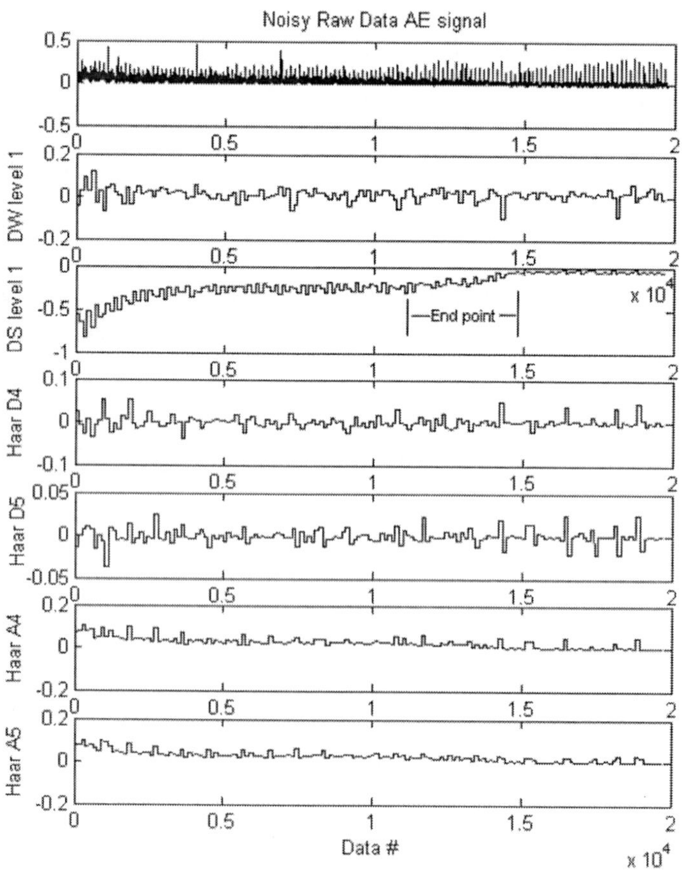

Figure 5. Comparison between Haar wavelet and diffusion wavelet decomposition of AE signal to detect end-point in a Cu-metal CMP process.

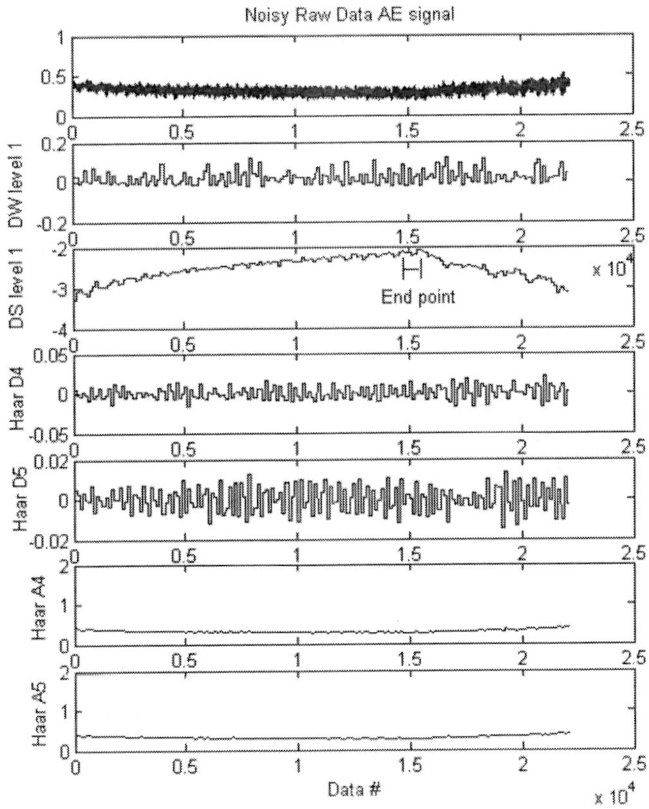

Figure 6. Comparison between Haar wavelet and diffusion wavelet decomposition of AE signal to detect end-point in an oxide CMP process.

CONCLUSION

This chapter presented the application of diffusion wavelet to analyze acoustic emission signal properties in a CMP process. Results from analyzing AE signals collected from a delamination process, and to detect end point of a CMP process were shown. In general, the AE signals have several applications in CMP such as monitoring polishing pad wear, laser shock cleaning of wafer surface to remove nano-sized residue particles, and in nanoscratch monitoring in AFM nanomachining. Therefore, it is important to develop highly effective methods that can assist in understanding the AE signal properties of a nanomanufacturing process to make accurate process related decisions. In this chapter, the diffusion wavelet has been shown to be more versatile and powerful than the classical wavelets such as Haar and Daubechies wavelet in AE signal processing. The primary advantages of using a diffusion wavelet based methodology is the greater insight into AE signal properties from superior feature representation that is displayed through the high amplitude and strength of the wavelet and scaling signals, and its sharp features such as vivid mean shifts, which are extremely useful and desirable in critical nano-scale process monitoring and control applications. The diffusion wavelet can be used as a general purpose tool for processing and analyzing the complex high-frequency AE signals for efficient monitoring of several nanomanufacturing process characteristics in different applications.

References

[1] D.A. Dornfeld. Process monitoring and control for precision manufacturing. Technical report, Laboratory for Manufacturing Automation, University of California, Berkeley, CA, USA, 1999.

[2] G. Qi. *NDTE International*, 33, 133 (2000).

[3] X. Li, *Int. J. Machine Tools Manufact.,* 42, 157 (2002).

[4] X. Li, *Advanced Manufact. Technol.,* 14, 539, (1998).

[5] X. Li, Y. Yao, and Z. Yuan. *J. Intelligent Manufact.*, 8, 271 (1997).

[6] X. Li, S. Dong, and Z. Yuan. *Int. J. Machine Tools Manufact.*, 39, 1935 (1999).

[7] X. Chen, J. Tang, and D. Dornfeld. Monitoring and analysis of ultra-precision metal cutting with acoustic emission. In *Proceedings of the ASME Dynamics Systems and Control Division*, volume DSC-58, New York, (1996) pp. 387–393.

[8] W.J. Staszewski and K.M. Holford, *Key Eng. Mater.,* 204/205, 351 (2001).

[9] Q.Wang and F. Chu *J. Sound Vibration*, 248(1), 91 (2001).

[10] T. Kim, J.M. Lee, S.H. Cho and T.H. Kim, *Opt. Laser Eng.*, 43(9), 1010 (2005).

[11] B.W. Ahn and S.H. Lee. *J. Micromech. Microeng.*, 19, 045028 (2009).

[12] R. Ganesan, T.K. Das, A.K. Sikder and A. Kumar, *IEEE Trans. Semiconduct. Manufact.,* 16(4), 677 (2003).

[13] T.K. Das, R. Ganesan, A.K. Sikder and A. Kumar, *IEEE Trans. Semiconduct. Manufact.*, 18(3), 440 (2005).

[14] D.L. Donoho. *IEEE Trans. Inform. Theory*, 41, 613 (1995).

[15] A.Terchi and Y.H.J. Au. Meas. Control, 34, 240 (2001).

[16] I.N. Tansel, C. Mekdesi, O. Rodriguez, and B. Uragun. Monitoring microdrilling operations with wavelets. *Quality Assurance Through Integration of Manufacturing Processes and Systems - ASME*, p. 151 (1992).

[17] G. Strang and T. Nguyen. *Wavelets and Filter Banks*. Wellesley Cambridge Press, Wellesley MA (1996).

[18] R.R. Coifman and M. Maggioni. Diffusion wavelets, (2004). http://www.math.duke.edu/mauro/diffusionwavelets.html.

[19] D.C. Montgomery. *Introduction to Statistical Quality Control*, Wiley, New York, (2001).

INDEX

A

B

C

D

K

L

M

S

T

U

V

W

Y